Optimal Covariate Designs

Premadhis Das · Ganesh Dutta
Nripes Kumar Mandal · Bikas Kumar Sinha

Optimal Covariate Designs

Theory and Applications

Springer

Premadhis Das
Department of Statistics
University of Kalyani
Kalyani
India

Nripes Kumar Mandal
Department of Statistics
University of Calcutta
Kolkata
India

Ganesh Dutta
Department of Statistics
Basanti Devi College (Affiliated to
   University of Calcutta)
Kolkata
India

Bikas Kumar Sinha
Indian Statistical Institute
Kolkata
India

ISBN 978-81-322-2991-9   ISBN 978-81-322-2461-7   (eBook)
DOI 10.1007/978-81-322-2461-7

Springer New Delhi Heidelberg New York Dordrecht London

Printed on acid-free paper

Springer (India) Pvt. Ltd. is part of Springer Science+Business Media (www.springer.com)

*The greatest gifts we ever had were the gifts
from god we call them parents*

*In memory of My Father Late Jadunandan
Das and My Mother Late Sankari Das*

Premadhis Das

*Dedicating to My Father Shri Amarendra
Prasad Dutta and to My Mother Shrimati
Sankari Dutta for their Love, Affection &
Blessings*

Ganesh Dutta

*In Living Memory of My Parents Late
Jatindranath Mandal and Late Bidyutlata
Mandal*

Nripes Kumar Mandal

*Remembering My Parents Late Birendra
Nath Sinha and Late Jogmaya Sinha for their
Love, Affection & Blessings*

Bikas Kumar Sinha

Foreword

It is a standard classroom exercise to assert that in a simple linear regression model involving only one regressor [or, covariate] x, viz., $y = \alpha + \beta x + \text{error}$, the covariate-values (x), assumed to be continuous and to lie in a finite nondegenerate interval $a \le x \le b$, should allow for maximum dispersion in order that the regression parameters can be estimated with the highest efficiency. This suggests a 50–50 split of the total number of observations, i.e., the set of observations are to be generated by setting the covariate (x) at the two extreme values, viz., $x = a$ and $x = b$, equally often.

Going beyond this, there are basic results, when more than one covariate like this are involved. On the other hand, in the absence of any such covariates, we have available standard ANOVA models involving 'design parameters'.

The ANCOVA models introduced in the textbooks and in the literature are based on the study of models in situations wherein regression parameters and design parameters are both present.

Naturally the question of the most efficient estimation of the regression parameter(s) in the presence of design parameters needs to be studied in very general terms, and also under very specialized experimental settings.

Lopes Troya initiated this study and BKS (Bikas Kumar Sinha) followed it up with his research collaborators [Kalyan Das (KD), Nripes Kumar Mandal (NKM), Premadhis Das (PD), Ganesh Dutta (GD), S.B. Rao, P.S.S.N.V.P. Rao, G.M. Saha (GMS)]. It is amazing to note that so much was hidden in this topic of research, and that their successful collaboration over these years had culminated in a Research Monograph.

I had an opportunity to collaborate with BKS several years back and I am thankful to the authors for approaching me to write this Foreword.

This monograph aims at providing an up-to-date account of the research findings in various experimental settings. As the authors describe and admit, mostly they confine to 'idealistic scenarios' in order to develop and apply tools and techniques for the study of optimal estimation of covariates' parameters. In the introductory chapter as also in Chapter 9, they discuss about 'real life' examples and provide a detailed study of optimality.

The authors have taken up a thorough study of the problems associated with this area of research. I personally thank them for their tremendous efforts and congratulate them for this remarkable achievement.

June 2015   Gour Mohan Saha
                                                          Retired Professor of Statistics
                                                            Indian Statistical Institute
                                                                            Kolkata
                                                                              India

Preface

Three of us are 'Senior Citizens' in the context of 'Statistics Learning' and we are ever-grateful to our revered postgraduate teachers for highlighting the fundamental and basic contributions of R.A. Fisher and Frank Yates in such areas as Design of Experiments [DoE]. We had the opportunity to read their books, so much so that we went through Fisher's original book published in the 1930s. These are indeed 'Treasured Collections'! Our fascination for DoE started from that point of time and it has continued to be intriguing for more than 40 years! We thoroughly enjoy reading, learning and discussing all aspects of DoE—theory and applications.

There are two incidences to be told in real-time experience underlying this project.

First, around 2002 one of the co-authors was trying to make a 'dent' into a paper on Optimal Covariates Designs [OCDs] with a colleague of him with very little success primarily because the notations were difficult to follow. Fortunately for the rest of us and for the optimal design community at large, they did not give up altogether. Instead, at the earliest opportunity they approached one of the other co-authors for looking into this paper. That was one positive development indeed and together, they could digest the paper and go forward as a 'high speed jet'! On another occasion around 2003, again one of the co-authors was struggling with a constructional problem involving OCDs and this time he was accompanied by one enthusiastic graph-theorist and one matrix-specialist. While they were in 'seemingly deep' trouble and in a 'confused state of mind', one of their colleagues—a design specialist—suddenly 'peeped in' and made a very casual observation, 'it seems … you are discussing some aspects of Mixed Orthogonal Arrays' and that was it to give again another big push to this work.

In a nutshell, these two incidences gave a boost to our group and we did not have to look back any more! We have enjoyed working on this project. We have derived much pleasure working in a group discussing, arguing and counter-arguing, till the time that we thought we came to understand enough of this fascinating topic of research to prepare a Research Monograph.

We must hasten to add that the youngest member of our group [GD] kept the others in toe with his frequent 'claims' and 'counter-claims' and 'proofs' and 'counter-examples'! Working with him was a matter of great pleasure for us. His enthusiastic and provocative statements/claims frequently served as 'make-belief' prophecies which were to be verified by the other three; it was not easy all the time anyway.

Finally, we are here with a comprehensive account of what we believe to be a Treatise on OCDs, more from the viewpoint of 'Idealistic Scenarios' in different experimental situations. The emphasis all through is about 'optimal' choice of what are called 'controllable covariates' in continuous domain(s). Only in the last chapter, we dwell on 'realistic experimental situations' and provide solutions to some well-posed problems.

Confusion continued to follow us and it gave us a scope for generating arguments and counter-arguments till we reached *Clarity* with our own understanding of the findings.

Kolkata, West Bengal, India   Premadhis Das
June 2015   Ganesh Dutta
                                                          Nripes Kumar Mandal
                                                            Bikas Kumar Sinha

Acknowledgment

During the course of our work on this fascinating topic, we had the opportunity to interact with three enthusiastic researchers: Prof. Kalyan Das [Calcutta University], Profs. S.B. Rao and Prasad Rao [both from Indian Statistical Institute, Kolkata]. Their collaboration enriched our thoughts and we are grateful to them for their enthusiasm and encouragement.

During this period, we had the opportunity to be invited to some conferences and make technical presentations of our papers before learned gatherings. We were benefited by the remarks/comments/criticisms received therefrom.

We are thankful to the authorities at Indian Statistical Institute, Kolkata [BKS], Department of Statistics, Calcutta University [NKM], Department of Statistics, Kalyani University [PD] and Basanti Devi College, Kolkata [GD] for providing excellent research opportunities for our team.

We are thankful to Prof. Gour Mohan Saha of the Indian Statistical Institute, Kolkata for his insightful comments during a critical phase of this study and for complying with our request for writing the Foreword of this monograph.

We have freely consulted available published literature and books and journals and benefited immensely from the collective wisdom of researchers worldwide on topics such as Hadamard Matrices, Mutually Orthogonal Latin Squares [MOLS], Orthogonal Arrays [OAs], Mixed Orthogonal Arrays [MOAs], Linear Models, ANOVA Models, Regression designs and ANCOVA models.

We fondly hope this monograph will be received enthusiastically by the statistical design theorists in general and combinatorial design theorists in particular and they will identify open problems in the broad area of Optimal Covariates' Designs.

Kolkata
June 2015

Premadhis Das
Ganesh Dutta
Nripes Kumar Mandal
Bikas Kumar Sinha

Contents

About the Authors

Prof. Premadhis Das is Senior Professor in the Department of Statistics, University of Kalyani, India. He has been working in the area of Design of Experiments for more than 30 years, and has published research articles in many national and international journals of repute. Professor Das has co-authored a Springer-Verlag Lecture Notes Series in Statistics: Monograph on Optimal mixture experiments, Vol. 1028, 2014.

Dr. Ganesh Dutta is Assistant Professor, Department of Statistics, Basanti Devi College affiliated to the University of Calcutta, India. He completed his Ph.D. degree in the area of Design of Experiments from the University of Calcutta in 2009 and has published research articles in the area of Optimum Covariate Designs in reputed peer-reviewed journals.

Prof. Nripes Kumar Mandal is a senior faculty member of the Department of Statistics, University of Calcutta, India. He has visited many countries for collaborative research, and has published about 70 research articles in peer-reviewed journals. He has been associated with many international statistical journals of repute as a reviewer. Professor Mandal has co-authored two Springer-Verlag Lecture Notes Series in Statistics: Monograph on Optimal Designs, Vol. 163, 2002 and Monograph on Optimal mixture experiments, Vol. 1028, 2014.

Prof. Bikas Kumar Sinha was attached to the Indian Statistical Institute [ISI], Kolkata, India for more than 30 years until his retirement on March 31, 2011. He has traveled extensively within USA and Europe for collaborative research and with teaching assignments. He has more than 120 research articles published in peer-reviewed journals and has acted as a referee for many international journals. Professor Sinha has served on the Editorial Board of statistical journals including *Sankhya, Journal of Statistical Planning and Inference* and *Calcutta Statistical Association Bulletin*. He has co-authored three Springer-Verlag Lecture Notes Series in Statistics: Monographs on Optimal Designs [Vol. 54, 1989, Vol. 163, 2002 and Vol. 1028, 2014].

Chapter 1
Optimal Covariate Designs (OCDs): Scope of the Monograph

1.1 Preamble: A Reflection on the Choice of Covariates

Most standard textbooks in the area of linear models and design of experiments provide discussions on what are known as analysis of covariance models applied to completely randomized designs (CRD), randomized block designs (RBD) and latin square designs (LSD). It is a well-accepted practice in experimental design contexts to use one or more available and meaningful covariates together with *local control* to reduce the experimental error. Such a model comprises three components: local control parameter(s) (if any), 'treatment' parameters, and the covariate parameter(s), apart from the error. This generates a family of 'covariate models'—serving as a 'blend' of 'regression models' (in the absence of treatment parameters) and 'varietal design models' (in the absence of covariates). These are the so-called analysis of covariance (ANCOVA) Models. Generally, for such models, emphasis is given on analysis of the data. Inference-related procedures are fairly routine exercises and are well discussed in the texts.

At times there lies a (possibly huge) potential for improving the experimental results by suitably classifying/reclassifying the existing experimental units through a study of the associated covariate values or by first suitably choosing the covariate values from a larger lot and then, hopefully, identifying the associated experimental units from a larger pool.

Here we cite a motivating example from Snedecor and Cochran (1989, p. 377), suitably presented to explain our point. There are 30 patients for a study of leprosy and there are three drugs (two antibiotics A and D, and one control F) to be compared—each to be applied to 10 patients. For each patient we have available a pretreatment score (count of bacilli) which may be used as a covariate. Table 1.1 shows the allocation of the three treatments covering all the 30 patients as against their covariate values. There is nothing wrong with this and the data analysis is fairly routine using an ANCOVA Model, once the CRD is implemented.

We now ask an intriguing question: How was the allocation of treatments (A, D and F) across the pool of 30 patients decided? Was it purely 'ad hoc'? Could

© Springer India 2015
P. Das et al., *Optimal Covariate Designs*, DOI 10.1007/978-81-322-2461-7_1

Table 1.1 Original allocation of patients based on covariate values (patient serial number, covariate value)

1	Treatment A	(P1, 3), (P2, 5), (P3, 6), (P4, 6), (P5, 8), (P6, 10), (P7, 11), (P8, 11), (P9, 14), (P10, 19)
2	Treatment D	(P11, 5), (P12, 6), (P13, 6), (P14, 7), (P15, 8), (P16, 8), (P17, 8), (P18, 15), (P19, 18), (P20, 19)
3	Control F	(P21, 7), (P22, 9), (P23, 11), (P24, 12), (P25, 12), (P26, 12), (P27, 13), (P28, 16), (P29, 16), (P30, 21)

Table 1.2 'Improved' allocation of patients based on covariate values

1	Treatment A	(P1, 3), (P3, 6), (P4, 6), (P22, 9), (P6, 10), (P7, 11), (P24, 12), (P9, 14), (P19, 18), (P10, 19)
2	Treatment D	(P2, 5), (P12, 6), (P14, 7), (P5, 8), (P15, 8), (P8, 11), (P25, 12), (P18, 15), (P28,16), (P20, 19)
3	Control F	(P11, 5), (P13, 6), (P21, 7), (P16, 8), (P17, 8), (P23, 11), (P26, 12), (P27, 13), (P29, 16), (P30, 21)

we do anything 'better'? It would be an interesting exercise to compare different conceivable allocations for say, 'most efficient estimation' of the covariate parameter in the ANCOVA model underlying the CRD. A trial and error solution is given in Table 1.2 and it turns out that we can achieve 12.28 % gain in efficiency by following this plan.

There is much more to it. If we had a larger pool of patients to choose from, what would have been our strategy for most efficient estimation of the covariate parameter? It turns out that our optimal choice would necessarily accumulate all those patients having equal split between the smallest and the largest pretreatment scores and that would be needed for each treatment! If it turns out that in the pool, 3 and 21 are the lowest and highest pretreatment counts of bacilli, then we would recruit five patients with the lowest and five patients with the highest count for each of the three treatments A, D and F. By doing so, we would have gained access over a group of 15 patients—each with the lowest count and another group of 15 patients—each with the highest count! Then it would be a matter of dividing each group of 15 equally into three so that there are five patients for each treatment from each group. This would have resulted in the best possible allocation design with 309.78 % efficiency as against the original allocation in Table 1.1 above and 264.97 % efficiency as against the allocation indicated in Table 1.2 above. We revisit this example in Chap. 9.

We take up a second example now. The data in Table 1.3 are from an experimental piggery arranged for individual feeding of six pigs in each of five pens. From each of five litters, six young pigs, three males (M) and three females (F), were selected and allotted to one of the pens. Three feeding treatments denoted by A, B, C, containing increasing proportions ($p_A < p_B < p_C$) of protein, were used and each was given to one male and one female in each pen. The pigs were individually weighed each

Table 1.3 Data for analysis

Pen	Treatment	Sex	Initial weight	Growth rate in pounds per week
I	A	F	48	9.94
	B	F	48	10.00
	C	F	48	9.75
	C	M	48	9.11
	B	M	39	8.51
	A	M	38	9.52
II	B	F	32	9.24
	C	F	28	8.66
	A	F	32	9.48
	C	M	37	8.50
	A	M	35	8.21
	B	M	38	9.95
III	C	F	33	7.63
	A	F	35	9.32
	B	F	41	9.34
	B	M	46	8.43
	C	M	42	8.90
	A	M	41	9.32
IV	C	F	50	10.37
	A	M	48	10.56
	B	F	46	9.68
	A	F	46	10.90
	B	M	40	8.86
	C	M	42	9.51
V	B	F	37	9.67
	A	F	32	8.82
	C	F	30	8.57
	B	M	40	9.20
	C	M	40	8.76
	A	M	43	10.42

Data Source: Rao (1973), p. 291 and Scheffé (1999), p. 217

week for 16 weeks. For each pig the growth rate in pounds per week was calculated. The weight at the beginning of the experiment is also given in Table 1.3.

There are 15 female pigs and 15 male pigs available for this study and we arrange their initial weights separately into two 5 × 3 arrays. The arrangements are shown in Table 1.4a, b respectively.

Now one may consider the standard covariate model (ANCOVA) for two-way RBD Pen×Treatment layout with a single covariate (here covariate is initial weights of pigs) separately for females and males. The notations are standard and we use

Table 1.4 Initial weight distribution as per allocation of pigs

Pen	Treatment			Totals
	A	B	C	
(a) Female				
1	48	48	48	144
2	32	32	28	92
3	35	41	33	109
4	46	46	50	142
5	32	37	30	99
Totals	193	204	189	586
(b) Male				
1	38	39	48	125
2	35	38	37	110
3	41	46	42	129
4	48	40	42	130
5	43	40	40	123
Totals	205	203	209	617

γ_F and γ_M to, respectively, denote the covariate effect for female and male pigs. These are routine computations and for the given allocation design in Table 1.4, to be denoted by d_0, $I_{d_0}(\gamma_F) = 57.8667$ and $I_{d_0}(\gamma_M) = 116.2667$.

Again, we ask an intriguing question: For the given collection of 15 female/male pigs, is it possible to identify an improved reallocation plan across the two-way Pen × Treatment table in the sense of increased precision in the estimation of the covariate parameters? Another related question also makes some sense: If the experimenter is given a 'free choice' of the 15 pigs (both female and male) from a larger pool, what would have been an 'optimal choice', given that initial weight distribution is perfectly known for the pool of pigs? Note that this question has embedded in it (i) selection of pigs with suitable initial weights and (ii) their distribution across the two-way table.

Below we provide answers to the two questions raised above. In Table 1.5a, b we provide improved allocation designs (based on the given collection of pigs), separately for female and male pigs with respective percent gain in efficiency given by 1375.345 and 76.4908 %. In Table 1.6a, b, we provide optimal allocation designs based on free choice of the experimenter, assuming that the initial weight distribution for female pigs lies in the closed interval [28 lbs, 50 lbs] and for males it is in the closed interval [35 lbs, 48 lbs].

We skip the details and will take up this example again in Chap. 9.

Remark 1.1.1 Our purpose in this monograph is to give the readers a taste of such comparative results in diverse experimental contexts and with one or more covariates being encountered simultaneously.

Table 1.5 Improved allocation designs

Pen	Treatment			Totals
	A	B	C	
(a) Improved allocation for female pigs				
	46	28	48	122
	30	37	50	117
	48	35	32	115
	41	46	32	119
	32	48	33	113
Totals	197	194	195	586
(b) Improved allocation for male pigs				
	A	B	C	Totals
	37	38	48	123
	38	46	40	124
	40	42	41	124
	43	39	42	123
	48	40	35	123
Totals	206	205	206	617

Table 1.6 Optimal initial weights for female and male pigs

	Treatment			Totals
	A	B	C	
(a) Female				
	28	50	50	128
	50	28	50	128
	50	50	28	128
	28	50	50	128
	50	28	28	106
Totals	206	206	206	$618 = G$
(b) Male				
	35	48	48	131
	48	35	48	131
	48	48	35	131
	35	48	48	131
	48	35	35	118
Totals	214	214	214	$642 = G$

Much of the theory of OCDs has grown out of the 'convenient proposition/
supposition' that the experimenters have a 'free' choice in the selection of the experimental units with any preassigned covariate values whatsoever! Notwithstanding
the fact that such a situation rarely arises in practice, the published literature is vast

and varied in respect of all kinds of experimental design settings, with the proviso that the optimal design theorists/statisticians are the masterminds in the whole business and they have the 'ultimate say' in the choice of the experimental units from a conceivably larger 'pool' with designated covariate values.

We will dwell on the developments toward characterization and construction of the OCDs as we have witnessed in the published literature, in the contexts of what are identified as 'ideal' scenarios. This study will be taken up systematically in Chaps. 2, 3, 4, 5, 6, 7 and 8. In Chap. 9, we will consider application areas and discuss some examples. Our understanding of the OCDs in the so-called ideal scenarios will guide us towards identification of optimal/nearly optimal covariate designs in real-life applications and some applications are discussed in Chap. 9.

We briefly trace the history of development of OCDs below.

The choice of experimental units possessing suitably defined/chosen values of the covariates for a given experimental set-up so as to attain minimum variance/maximum precision for estimation of the regression parameters has attracted the attention of researchers only in recent times. In the context of ANCOVA models where both qualitative and quantitative factors are present, the problem of inference on varietal contrasts corresponding to qualitative factors was studied by Harville (1974, 1975), Haggstrom (1975) and Wu (1981). The problem of determining optimum designs for the estimation of regression parameters corresponding to controllable covariates was first considered by [1] Troya Lopes (1982a, b). She restricted investigations in the set-up of completely randomized design (CRD). Das et al. (2003) extended it to the block design set-up, viz. randomized block design (RBD) and some series of balanced incomplete block designs (BIBDs) and constructed OCDs for the estimation of covariate parameters. Rao et al. (2003) revisited the problem in CRD and RBD set-ups and identified the solutions as mixed orthogonal arrays (MOAs), thereby providing further insights and some new solutions. Dutta (2004, 2009) and Dutta et al. (2007, 2009b, 2010a, c) considered optimal estimation of the regression coefficients under different experimental set-ups where the analysis of variance (ANOVA) effects are non-orthogonally estimable. Dutta et al. (2009a) also considered optimal estimation of the regression coefficients in the set-ups of split-plot and strip-plot designs where the ANOVA effects are orthogonally estimable. These were subsequently generalized in Dutta and Das (2013a) to multi-factor set-up. For one-way set-up, D-optimal designs were proposed by Dey and Mukerjee (2006) and, these were further studied in Dutta et al. (2014). Dutta et al. (2010b) also considered D-optimal covariate designs for estimation of regression coefficients in incomplete block design set-up when global optimal designs do not exist. The other related

---

[1]Late Professor Jack Kiefer pioneered the study of optimal experimental designs in standard ANOVA models as well as in regression designs. He guided Lopes Troya for her Doctoral Dissertation in a topic which was to bridge ANOVA and regression designs into what are known as ANCOVA models. The unfortunate premature death of Professor Kiefer was a blow to the design theorists in general. His expertise and insightful contributions could have gone a long way in this direction.

references are Wierich (1984), Kurotschka and Wierich (1984), Chadjiconstantini-dis and Moyssiadis (1991), Chadjiconstantinidis and Chadjipadelis (1996), Liski et al. (2002), Dutta (2009), Sinha (2009) and Das (2011).

1.2 Basic Set-Up and Optimality Conditions

Let the following covariate model be considered:

$$\mathbf{Y} = \mathbf{X}\theta + \mathbf{Z}\gamma + \mathbf{e} \tag{1.2.1}$$

where $\mathbf{Y}^{n \times 1}$ denotes the observation vector, $\mathbf{X}^{n \times p}$ denotes the coefficient matrix for the ANOVA effects parameters $\theta' = (\theta_1, \theta_2, \ldots, \theta_p)$ and $\mathbf{Z}^{n \times c}$ denotes the matrix of the values given to c covariates, viz. $\mathbf{Z} = (\mathbf{z}^1, \mathbf{z}^2, \ldots, \mathbf{z}^c)$. In the above, $\mathbf{Z}$ is also called the covariate design matrix of the vector of covariate effects $\gamma = (\gamma_1, \gamma_2, \ldots, \gamma_c)'$. As usual, $\mathbf{e}$ is the random error component with $E(\mathbf{e}) = \mathbf{0}$, $Disp(\mathbf{e}) = \sigma^2 \mathbf{I}_n$, where $\mathbf{I}_n$ is the identity matrix of order n. We represent the above set-up by the triplet:

$$\left(\mathbf{Y}, \ \mathbf{X}\theta + \mathbf{Z}\gamma, \ \sigma^2 \mathbf{I}_n \right). \tag{1.2.2}$$

Here the observations are uncorrelated and variances of each of the observations are equal to σ^2. In addition to the comparison of ANOVA effects and in particular, of the underlying treatment effects, interest lies in accommodating as many covariates as possible, subject to these being optimally estimated. Situations where the covariates are *not* under the control of the experimenter, were discussed by Harville (1974, 1975), Haggstrom (1975) and Wu (1981) in the context of comparison of treatment effects. These are also briefly discussed in Shah and Sinha (1989). Traditionally, in a study of linear regression design involving non-stochastic regressors, we tacitly call for homogeneous experimental units so that the assumed model for the $n \times 1$ observation vector $\mathbf{Y}$ is of the form

$$\left(\mathbf{Y}, \ \mu \mathbf{1}_n + \mathbf{Z}\gamma, \ \sigma^2 \mathbf{I}_n \right) \tag{1.2.3}$$

where μ represents the intercept term, γ is the vector of covariate effects, $\mathbf{Z}$ is, as before, the design matrix of covariate values and $\mathbf{1}_n$ is a vector of order n with all elements unity. Understandably, the homogeneous nature of the experimental units safeguards the same intercept term as indicated in the model (1.2.3) for every expectation.

Here Z's are assumed to be controllable/given non-stochastic covariates. The n values $z_{i1}, z_{i2}, \ldots, z_{in}$, assumed by the ith covariate Z_i are such that they belong to a finite interval $[a_i, b_i]$ for each i and j, i.e.

$$a_i \leq z_{ij} \leq b_i$$
$$\text{i.e. } z_{ij} = \frac{a_i + b_i}{2} + \frac{b_i - a_i}{2} z_{ij}^* \qquad\qquad (1.2.4)$$

so that z_{ij}^* lies in $[-1, 1]$ for each i, j. Then replacing z_{ij} by z_{ij}^*'s we get the same covariate model in a reparameterized scenario, i.e. the regression coefficients can be suitably adjusted and the constant part will be adjusted with μ. Thus this transformation does not hamper our optimality study. So, without loss of generality, we can assume in (1.2.3), the covariate values z_{ij}'s to vary within $[-1, 1]$. It is well known that the experimental domain of the regressors being a c-dimensional cube of the form: $[-1, 1]^c$, the most efficient design for estimation of the regression coefficients (i.e. the γ-parameters) is derived from a Hadamard matrix (defined in Chap. 2), whenever the latter exists. When $n > c$ and $n \equiv 0 \pmod 4$, it is enough to start with a Hadamard matrix $\mathbf{H}_n$ of order n (in its standard form) and select any c of its columns for the $\mathbf{Z}$-matrix, leaving the first column which contains 1's only. This yields an optimum design for the (joint) estimation of μ and γ on the basis of n observations. *Optimality here, refers to attaining the least possible value $\frac{\sigma^2}{n}$ of the individual variances simultaneously for all the covariate parameter estimates.* It is known that the maximum number of covariates (i.e. $c_{\max}$) cannot exceed the error degrees of freedom (d.f.) of a given set-up. Therefore $c_{\max} = (n - 1)$ under the model in (1.2.3); $c_{\max} = n - v$ for a CRD set-up and for a block design set-up $c_{\max} = n - b - \text{Rank}(\mathbf{C})$, where $\mathbf{C}$ is the characteristic matrix of a block design.

In general, the experimental set-ups are much more complicated and so are the models much different from (1.2.3). Use of Hadamard matrices and other tools and techniques has to be introduced in a systematic manner. The points to be noted are:

(i) We want optimal estimation of the covariates parameters.
(ii) We want to know how many covariates can be optimally accommodated.

We mostly confine to the 'idealistic' situations wherein there exist conceivably larger pools of experimental units with experimenter's choice of the covariates' values. This should serve as a basis and a guideline for actual experimental situations.

1.3 Chapter-Wise Summary

In Chap. 2, we study the choice of optimum covariate design in CRD set-up. Troya Lopes (1982a, b) first studied the problem of choice of the $\mathbf{Z}$-matrix in a CRD model when the treatment allocation matrix $\mathbf{X}$ corresponds to an equal allocation number, i.e. when n is a multiple of v. We will write as $n = vb$ so that b is the common allocation number of the v treatments under investigation. Here we discussed some results from Troya Lopes (1982a) with reference to the $\mathbf{W}$-matrices. If n is not an integral multiple of v, this allows us to study situations where no equireplicate design exists. In this situation, it is not possible to find designs attaining minimum variance for the estimated covariate parameters. This problem has been considered by Dey and Mukerjee (2006) and Dutta et al. (2014). They provided optimum designs with respect to ANOVA effect and covariate effects using D-optimality criterion. We also deal with this issue in this chapter.

In Chap. 3, we discuss optimum covariate design in RBD set-up. For an RBD set-up, (Das et al. 2003) studied for the first time, the problem of OCDs. They exploited mutually orthogonal latin squares (MOLS) and Hadamard matrices to construct such designs which attain the upper bound for the number of covariates which can be incorporated in the covariate model for RBD. Rao et al. (2003) re-visited the problem in CRD and RBD set-ups and identified the solutions as mixed orthogonal arrays (MOAs) (defined in Chap. 2), thereby providing further insights.

For BIBD set-up, Das et al. (2003) also initiated the construction of optimal designs for covariates in some series of symmetric balanced incomplete block designs (SBIBD) constructed through Bose's difference technique and some BIBDs with repeated blocks. Dutta (2004) dealt with the problem of constructing OCDs in some other classes of BIBDs which may or may not have cyclic structure. However, he dealt with the problem with the restriction $n \equiv 0$ (mod 4). But such designs cannot always be obtained because of the restriction $n \equiv 0$ (mod 4). Dutta et al. (2010b) found optimum designs with respect to covariate effects using D-optimality criterion retaining orthogonality with the treatment and block effect contrasts, where $n \equiv 2$ (mod 4). Results given by Das et al. (2003), Dutta (2004) and Dutta et al. (2010b) are included in Chap. 4.

In a BIBD set-up, we have noticed that the scope of construction of OCDs becomes limited as the parametric relations do not always permit the existence of Hadamard matrices. Also, the stringency of equal occurrence of each pair of treatments limits the scope of OCDs. For this, Dutta et al. (2009b) extended their research to the partially balanced incomplete block design (PBIBD) set-up. Moreover, PBIBDs are popular among practitioners and OCDs in this set-up will be of help to them. However, in Chap. 5, we restrict to an important subclass of PBIBDs viz., the group divisible designs (GDDs) and discuss about existence and constructional aspects of OCDs. We have given a catalogue at the end of Chap. 5 which shows that the method covers a large number of GDDs obtained from Clatworthy (1973).

Binary proper equireplicate block designs (BPEBDs) form a rich class of block designs and this class encompasses designs beyond those considered in the previous chapters. In Chap. 6, we venture into the constructional aspects of OCDs for such designs. General cyclic and non-cyclic BPEBDs as also t-fold BPEBDs having OCD structure have been studied and a catalogue has also been provided at the end.

In Chap. 7, following Dutta and Das (2013b), we discuss the OCD problem in balanced treatment incomplete block (BTIB) design set-up using Hadamard matrices and other techniques described in previous chapters.

In Chap. 8, we start with a discussion of the OCD problems in crossover design set-up and multi-factor set-up. For these designs, key references are Dutta and SahaRay (2013) and Dutta and Das (2013a). Mixed orthogonal arrays and their generalized version are very useful to construct OCDs in multi-factor set-ups.

In all the above cases so far discussed in various chapters, the observations are naturally uncorrelated. But if they are not, then difficulty arises for the choice of OCDs.

It becomes even more difficult for arbitrary variance-covariance matrix. However, if the variance-covariance matrix has a nice structure, it is possible to construct OCDs. In particular, Dutta et al. (2009a) considered the set-ups of the split-plot and strip-plot designs where the correlations among the observations follow a definite pattern. Further, they have seen that a generalized version of the mixed orthogonal array has a close relationship with the OCDs for such set-ups. They have exploited it to construct OCDs for such experimental contexts. In this Chap. 8, we also discuss this aspect at length.

In the concluding chapter (Chap. 9), we turn back to the questions raised in Chap. 1 and deal with a number of application areas wherein optimality study in the context of uses of covariates has a natural scope for enhancing the experimental results. We rework on the two motivating examples and provide details of the computations. We also take up four other examples arising in experiments involving covariates in natural sciences.

References

Chadjiconstantinidis S, Chadjipadelis T (1996) D-optimal cyclic complex linear designs and supplementary difference sets with association vector. J Stat Plan Inference 53:93–115

Chadjiconstantinidis S, Moyssiadis C (1991) Some D-optimal odd equi-eplicated designs for a covariate model. J Stat Plan Inference 28:83–93

Clatworthy WH (1973) Tables of two-associate class partially balanced designs. US Department of Commerce, National Bureau of Standards

Das K, Mandal NK, Sinha BK (2003) Optimal experimental designs with covariates. J Stat Plan Inference 115:273–285

Das P (2011) A review on optimum covariate designs. Calcutta Stat Assoc Bull, 63, 249–252. (Proceedings of the seventh international triennial calcutta symposium on probability statistics, December 28–31, 2009)

Dey A, Mukerjee R (2006) D-optimal designs for covariate models. Statistics 40:297–305

Dutta G (2004) Optimum choice of covariates in BIBD set-up. Calcutta Stat Assoc Bull 55:39–55

Dutta G (2009) Optimum designs for covariates models. (Unpublished Ph.D. Thesis)

Dutta G, Das P, Mandal NK (2007) Optimum choice of covariates for a series of SBIBDs obtained through projective geometry. J Mod Appl Stat Methods 6:649–656

Dutta G, Das P, Mandal NK (2009a) Optimum covariate designs in split-plot and strip-plot design set-ups. J Appl Stat 36:893–906

Dutta G, Das P, Mandal NK (2009b) Optimum covariate designs in partially balanced incomplete block (PBIB) design set-ups. J Stat Plan Inference 139:2823–2835

Dutta G, Das P, Mandal NK (2010a) Optimum covariate designs in binary proper equi-replicate block design set-up. Discret Math 310:1037–1049

Dutta G, Das P, Mandal NK (2010b) D-optimal Designs for covariate parameters in block design set-up. Commun Stat Theory Methods 39:3434–3443

Dutta G, Das P, Mandal NK (2010c) Tables for optimum covariate designs in PBIBD set-ups. J Indian Soc Agric Stat 64:375–389

Dutta G, Das P (2013a) Optimum design for estimation of regression parameters in multi-factor set-up. Commun Stat Theory Methods 42:4431–4443

Dutta G, Das P (2013b) Optimum designs for estimation of regression parameters in a balanced treatment incomplete block design set-up. J Stat Plan Inference 143:1203–1214

Dutta G, Das P, Mandal NK (2014) D-optimal designs for covariate models. Commun Stat Theory Methods 43:165–174

Dutta G, SahaRay R (2013) Optimal choice of covariates in the set-up of crossover designs. Stat Appl 11(1–2):93–109 (Special Issue in Memory of Professor M.N. Das)

Haggstrom GW (1975) Pitfalls Manpow Exp. RAND Corporation, Santa Monica

Harville DA (1974) Nearly optimal allocation of experimental units using observed covariate values. Technometrics 16:589–599

Harville, DA (1975) Computing optimum designs for covariate models. In: Srivastava JN (ed) A survey of statistical design and linear models. 209–228, Amsterdam, North Holland

Kurotschka V, Wierich W (1984) Optimale planung eines kovarianzanalyse und eines (Intraclass regressions experiments). Metrika 31:361–378

Liski EP, Mandal NK, Shah KR, Sinha BK (2002). Topics in optimal design. Lecture notes in statistics. Series 163, Springer, New York

Rao CR (1973) Linear statistical inference and its applications. 2nd edn, Wiley, USA

Rao PSSNVP, Rao SB, Saha GM, Sinha BK (2003) Optimal designs for covariates' models and mixed orthogonal arrays. Electron Notes Discret Math 15:157–160

Scheffé H (1999) The analysis of variance. A Wiley-Interscience Publication, John Wiley & Sons, Inc., New York

Shah KR, Sinha BK (1989) Theory of optimal designs. Lecture notes in statistics. Series 54, Springer, New York

Sinha BK (2009) A reflection on the choice of covariates in the planning of experimental designs. J Indian Soc Agric Stat 63:219–225

Snedecor GW, Cochran WG (1989) Stat Methods, 8th edn. Iowa State University Press, Ames, Iowa

Troya LJ (1982a) Optimal designs for covariate models. J Stat Plan Inference 6:373–419

Troya LJ (1982b) Cyclic designs for a covariate model. J Stat Plan Inference 7:49–75

Wierich W (1984) Konkrete optimale Versuchsplne fr ein lineares Modell mit einem qualitativen und zwei quantitativen Einflussfaktoren. Metrika 31:285–301

Wu CFJ (1981) Iterative construction of nearly balanced assignments I: Categorical covariates. Technometrics 23:37–44

Chapter 2
OCDs in Completely Randomized Design Set-Up

2.1 Introduction

We consider in this chapter the one-way linear model with v treatments, c covariates and a total of n experimental units. We work under the linear model

$$y_{ij} = \tau_i + \sum_{t=1}^{c} \gamma_t z_{ij}^{(t)} + e_{ij}, \quad 1 \le j \le n_i, \ 1 \le i \le v. \tag{2.1.1}$$

where $n_i (> 1)$ is the number of times the ith treatment is replicated; clearly

$$\sum_{i=1}^{v} n_i = n. \tag{2.1.2}$$

For $1 \le j \le n_i, 1 \le i \le v$, here y_{ij} is the observation arising from the jth replication of the ith treatment, τ_i effect due to the ith treatment.

In matrix notation the above model can be represented as

$$\left(\mathbf{Y}, \ \mathbf{X}\tau + \mathbf{Z}\gamma, \ \sigma^2 \mathbf{I}_n \right), \tag{2.1.3}$$

where, $\mathbf{Y}$ is an observation vector and $\mathbf{X}$ is the design matrix corresponding to vector of treatment effects $\tau^{v \times 1}$ and $\mathbf{Z} = ((z_{ij}^{(t)}))$ is the design matrix corresponding to vector of covariate effects $\gamma^{c \times 1} = (\gamma_1, \gamma_2, \ldots, \gamma_c)'$. This is referred to as *one-way model with covariates* (*without the general mean*).

Troya Lopes (1982a, b) studied the nature of optimal allocation of treatments and covariates in the above set-up for simultaneous estimation of the (fixed) treatment effects (in the absence of the general effect) and the covariate effects with maximum efficiency in the sense of minimum generalized variance. This is to note that the information matrix with respect to model (2.1.3) is given by $\sigma^{-2} \mathbf{I}(\eta)$, where

© Springer India 2015

P. Das et al., *Optimal Covariate Designs*, DOI 10.1007/978-81-322-2461-7_2

$$\mathbf{I}(\eta) = \begin{pmatrix} \mathbf{X'X} & \mathbf{X'Z} \\ \mathbf{Z'X} & \mathbf{Z'Z} \end{pmatrix} \tag{2.1.4}$$

and $\eta' = (\tau', \gamma')$.

The problem is to suggest an optimal allocation scheme (for given design parameters n, v, c) for efficient estimation of the treatment effects as well as the covariate effects by ascertaining the values of the covariates for each one of them, assuming that each one is controllable and quantitative within a stipulated finite closed interval.

The information matrix of γ is given by

$$\sigma^{-2} I(\gamma) = \mathbf{Z'Z} - \mathbf{Z'X(X'X)^- X'Z} \tag{2.1.5}$$

where $(\mathbf{X'X})^-$ is a generalised inverse of $\mathbf{X'X}$ satisfying
$$\mathbf{X'X\,(X'X)^-\,X'X = X'X}$$

(cf. Rao 1973, p. 24). It is evident that $\mathbf{Z'X(X'X)^- X'Z}$ is a positive semi-definite matrix. So from (2.1.5), it follows that

$$\sigma^{-2} I(\gamma) \leq \mathbf{Z'Z} \tag{2.1.6}$$

in the Loewner order sense (vide Pukelsheim 1993) where for two non-negative definite matrices $\mathbf{A}$ and $\mathbf{B}$, $\mathbf{A}$ is said to dominate $\mathbf{B}$ in the Lowener order sense if $\mathbf{A} - \mathbf{B}$ is a non-negative definite matrix.

Equality in (2.1.6) is attained whenever

$$\mathbf{X'Z = 0}. \tag{2.1.7}$$

If $\mathbf{Z}$ satisfies (2.1.7), then treatment effects and covariate effects are orthogonally estimated. Again under condition (2.1.7), the information matrix $\mathbf{I}(\gamma)$ reduces to $\mathbf{I}(\gamma) = \mathbf{Z'Z}$. The z-values are so chosen that $\mathbf{Z'Z}$ is positive definite so that from (2.1.6)

$$Var(\widehat{\gamma}_t) \geq \frac{\sigma^2}{\sum\limits_{i=1}^{v} \sum\limits_{j=1}^{n_i} z_{ij}^{(t)2}} \geq \frac{\sigma^2}{n} \tag{2.1.8}$$

as $z_{ij}^{(t)} \in [-1, 1]; \ \forall\, i, j, t$.

Now equality in (2.1.8) holds for all i if and only if the $\mathbf{Z}$-matrix is such that

$$\mathbf{z}^{(s)'} \mathbf{z}^{(t)} = 0 \ \forall\, s \neq t. \tag{2.1.9}$$

and

$$z_{ij}^{(t)} = \pm 1 \tag{2.1.10}$$

Condition (2.1.7) implies that the estimators of ANOVA effects parameters or parametric contrasts do not interfere with those of the covariate effects and conditions (2.1.9) and (2.1.10) imply that the estimators of each of the covariate effects are such that these are pairwise uncorrelated, attaining the minimum possible variance.

Thus the covariate effects are estimated with the maximum efficiency if and only if

$$\mathbf{Z}'\mathbf{Z} = n\mathbf{I}_c \qquad (2.1.11)$$

along with (2.1.7). The designs allowing the estimators with the minimum variance are called *globally optimal designs* (cf. Shah and Sinha 1989, p. 143). Henceforth, we shall only be concerned with such optimal estimation of regression parameters and by optimal covariate design, *to be abbreviated as OCD hereafter*, we shall only mean *globally optimal design*, unless otherwise mentioned.

It is clear that conditions (2.1.7) and (2.1.11) hold simultaneously if and only if z_{ij}'s are necessarily $+1$ or -1 and that condition (2.1.7) is satisfied.

It is difficult to visualize the $\mathbf{Z}$-matrix satisfying conditions (2.1.7) and (2.1.11). In the set-up of the model (2.1.3), it transpires from Troya Lopes (1982a) that optimal estimation of the treatment effects and the covariates effects is possible when the treatment replications are all necessarily equal, assuming that n is a multiple of v, the number of treatments. We set $n = bv$, where b is the common replication of treatments, henceforth. Das et al. (2003) had represented each column of the $\mathbf{Z}$-matrix by a $v \times b$ matrix $\mathbf{W}$ with elements of ± 1, where the rows of $\mathbf{W}$ correspond to the v treatments and the columns of $\mathbf{W}$ correspond to different replication numbers. Condition (2.1.7) implies that the sum of each row of $\mathbf{W}$ should vanish. Again, condition (2.1.11) implies that the sum of products of the corresponding elements, i.e. the Hadamard product of $\mathbf{W}^{(s)}$ and $\mathbf{W}^{(t)}$, defined in (2.1.13) should also vanish, $1 \le s < t \le c$. The above two facts can be represented in the following schematic forms through the row totals and Hadamard product.

Row Totals:

$$
\mathbf{W}^{(s)} =
\begin{array}{c}
\text{Tr.} \quad \text{Repl. no.} \rightarrow \text{Row} \\
\downarrow \quad 1 \ 2 \ \ldots \ b \ \text{Totals} \\
\begin{array}{c}
1 \\
2 \\
\vdots \\
v
\end{array}
\left|
\begin{array}{c}
\\
(\pm 1) \\
\\
\end{array}
\right|
\begin{array}{c}
0 \\
0 \\
\vdots \\
0
\end{array}
\end{array}
\qquad (2.1.12)
$$

Hadamard product of $\mathbf{W}^{(s)}$ and $\mathbf{W}^{(t)}$ (cf. Rao 1973, p. 30):

$$
\mathbf{W}^{(s)} * \mathbf{W}^{(t)} =
\begin{array}{c}
\text{Tr.} \quad \text{Repl. no.} \rightarrow \\
\downarrow \quad 1 \ 2 \ \ldots \ b \\
\begin{array}{c}
1 \\
2 \\
\vdots \\
v
\end{array}
\left|
\begin{array}{c}
\\
(w_{ij}^{(s)} w_{ij}^{(t)}) \\
\\
\end{array}
\right|
\end{array}
\qquad (2.1.13)
$$

where '*' denotes Hadamard product. For orthogonality of sth and tth columns of $\mathbf{Z}$, it is required that $\sum\limits_{i=1}^{v} \sum\limits_{j=1}^{b} w_{ij}^{(s)} w_{ij}^{(t)} = 0$.

The schematic representation (2.1.12), (2.1.13) of Das et al. (2003) is a break-through in the sense that handling of $\mathbf{Z}$-matrix has been made much easier and it has been followed throughout the monograph.

Troya Lopes (1982a) first studied the nature of optimal allocation of treatments and covariates in the above set-up when $\frac{n}{v}$ is an integer. It may be noted that whenever condition (2.1.7) is ensured, presence of the covariates in model (2.1.3) does not pose any threat to the usual "optimal treatment allocation" problem. In Sect. 2.2, following Troya Lopes, we intend to discuss about the availability of $\mathbf{Z}$-matrices satisfying (2.1.7) and (2.1.11) when the treatment allocation matrix $\mathbf{X}$ corresponds to equal allocation number, i.e. in situations where n is a multiple of v. We will write $n = vb$ so that b is the common allocation number of the v treatments under investigation. The situations where (2.1.7), (2.1.11) and $b = \frac{n}{v}$ = integer are satisfied, are identified as *regular* cases. Otherwise it is called a *non-regular* case. If the situation is non-regular, then it is not possible to allocate simultaneously the treatments and covariates optimally. For non-regular situation, efficient allocation of treatments and covariates simultaneously can be done by using other specific optimality criteria. Dey and Mukerjee (2006) and Dutta et al. (2014) considered this problem in non-regular situations and found D-optimal designs in this context. Details are presented in Sect. 2.3.

It has been seen that Hadamard matrix plays a key role for constructing OCDs. Definition of Hadamard matrix (cf. Hedayat et al. 1999, p. 145) is given below:

Definition 2.1.1 A Hadamard matrix $\mathbf{H}_t$ of order t is a $t \times t$ matrix with elements ± 1 satisfying

$$\mathbf{H}_t \mathbf{H}_t' = t\mathbf{I}_t.$$

2.2 Covariate Designs Under Regular Cases

Consider the case when n is a multiple of v, that is $n = vb$ where b is such that $\mathbf{H}_b$, Hadamard matrix of order b, exists. We shall also consider some cases where b is even. Then ANOVA parameters as well as the covariate effect-parameters can be estimated orthogonally and/or most efficiently. This holds simultaneously for c covariates and one can deduce maximum possible value of c for this to happen. As already mentioned, the most efficient estimation of γ-components is possible when (2.1.7) and (2.1.11) are simultaneously satisfied and these conditions reduce, in terms of $\mathbf{W}$-matrices defined in above, to C_1, C_2 where

C_1. Each of the c **W**-matrices has all row-sums equal to zero;
C_2. The grand total of all the entries in the Hadamard product of any two distinct **W**-matrices reduces to zero. $\qquad$ (2.2.1)

Now we define optimum **W**-matrices for covariate designs in CRD set-up.

Definition 2.2.1 With respect to model (2.1.3), the c **W**-matrices corresponding to the c covariates are said to be optimum if they satisfy conditions C_1 and C_2 of (2.2.1).

In this context, the following results were deduced in Troya Lopes (1982a).

Theorem 2.2.1 *Let c^* be the maximum number of covariates that can be optimally accommodated. Then a lower bound to c^* is given by*

(a) *$b-1$ when $v = odd$, $\mathbf{H}_b$ exists;*
(b) *$2(b-1)$ when $v \equiv 2 \ (mod \ 4)$, $\mathbf{H}_b$ exists;*
(c) *$4(b-1)$ when $v \equiv 0 \ (mod \ 4)$, $\mathbf{H}_b$ exists;*
(d) *$3v$ when $b \equiv 0 \ (mod \ 4)$, $\mathbf{H}_v$ exists;*
(e) *v when $b \equiv 2 \ (mod \ 4)$, $\mathbf{H}_v$ exists.*

Proof Hadamard matrix $\mathbf{H}_b$ is given to exist and we write it as

$$\mathbf{H}_b = (\mathbf{h}_1, \ \mathbf{h}_2, \ldots, \mathbf{h}_{b-1}, \mathbf{1}).\qquad(2.2.2)$$

The choice of optimum **W**-matrices is indicated below one by one. The verification of (2.2.1) is immediate and we leave it to the reader. The Kronecker product of two matrices is formally defined in Chap. 5 (Definition 5.1.1) and it is used in the constructions below.

(a)
$$\mathbf{W}^{(j) \, v \times b} = \mathbf{1}_v \otimes \mathbf{h}'_j, \ \ 1 \le j \le b-1; \qquad(2.2.3)$$

(b)
$$\left. \begin{aligned} \mathbf{W}^{(j) \, v \times b} &= (1, \ 1)' \otimes \mathbf{1}_{\frac{v}{2}} \otimes \mathbf{h}'_j, \ \ 1 \le j \le b-1; \\ \mathbf{W}^{(b-1+j) \, v \times b} &= (1, \ -1)' \otimes \mathbf{1}_{\frac{v}{2}} \otimes \mathbf{h}'_j, \ \ 1 \le j \le b-1. \end{aligned} \right\} \qquad(2.2.4)$$

(c)
$$\left. \begin{aligned} \mathbf{W}^{(j) \, v \times b} &= (1, \ 1, \ 1, \ 1)' \otimes \mathbf{1}_{\frac{v}{4}} \otimes \mathbf{h}'_j, \ \ 1 \le j \le b-1; \\ \mathbf{W}^{(b-1+j) \, v \times b} &= (1, \ -1, \ 1, \ -1)' \otimes \mathbf{1}_{\frac{v}{4}} \otimes \mathbf{h}'_j, \ \ 1 \le j \le b-1; \\ \mathbf{W}^{(2(b-1)+j) \, v \times b} &= (1, \ -1, \ -1, \ 1)' \otimes \mathbf{1}_{\frac{v}{4}} \otimes \mathbf{h}'_j, \ \ 1 \le j \le b-1; \\ \mathbf{W}^{(3(b-1)+j) \, v \times b} &= (1, \ 1, \ -1, \ -1)' \otimes \mathbf{1}_{\frac{v}{4}} \otimes \mathbf{h}'_j, \ \ 1 \le j \le b-1. \end{aligned} \right\} \qquad(2.2.5)$$

(d) Let us represent a Hadamard matrix $\mathbf{H}_v$ of order v as

$$\mathbf{H}_v = \left(\mathbf{h}_1^*, \ \mathbf{h}_2^*, \ldots, \mathbf{h}_v^*\right).\qquad(2.2.6)$$

$$\left.\begin{array}{l} \mathbf{W}^{(j)\ v\times b} = (1,\ -1,\ 1,\ -1)\otimes\mathbf{1}'_{\frac{b}{4}}\otimes\mathbf{h}^*_j,\ 1\le j\le v; \\ \mathbf{W}^{(v+j)\ v\times b} = (1,\ -1,\ -1,\ 1)\otimes\mathbf{1}'_{\frac{b}{4}}\otimes\mathbf{h}^*_j,\ 1\le j\le v; \\ \mathbf{W}^{(2v+j)\ v\times b} = (1,\ 1,\ -1,\ -1)\otimes\mathbf{1}'_{\frac{b}{4}}\otimes\mathbf{h}^*_j,\ 1\le j\le v. \end{array}\right\} \qquad (2.2.7)$$

(e)

$$\mathbf{W}^{(j)\ v\times b} = (1,\ -1)\otimes\mathbf{1}'_{\frac{b}{2}}\otimes\mathbf{h}^*_j,\ 1\le j\le v. \qquad (2.2.8)$$

$\square$

Remark 2.2.1 In case (c), we can assume existence of $\mathbf{H}_v$ for all practical purposes as $v\equiv 0\ (\mathrm{mod}\ 4)$. So in this case, an optimal design for maximum possible $v(b-1)$ optimum $\mathbf{W}$-matrices can easily be constructed as

$$\mathbf{W}^{((b-1)(i-1)+j)} = \mathbf{h}^*_i\otimes\mathbf{h}'_j,\ i=1,2,\ldots,v,\ j=1,2,\ldots,b-1. \qquad (2.2.9)$$

This was obtained in (Rao et al. 2003) where it was observed that OCDs in CRD and RBD have one to one correspondences with mixed orthogonal array (MOA) (definition given in Chap. 3). This fact will be discussed in Sect. 3.3 of Chap. 3 in some further details.

2.3 Covariate Designs Under Non-regular Cases

Now we examine the situations where at least any one of the conditions (2.1.7), (2.1.11) and $b=\frac{n}{v}=$ integer is violated. In that case, it is not possible to estimate simultaneously ANOVA parameters and γ-parameters orthogonally and/or most efficiently. Thus we consider D-optimality criterion to give an efficient allocation of treatments and covariates in Set-up (2.1.1). Dey and Mukerjee (2006) and Dutta et al. (2014) have considered this situation and found D-optimal design. Here we discuss their contributions in this direction in details.

The vector of parameters θ, where

$$\theta = (\mu_1,\ \mu_2,\ldots,\mu_v,\gamma_1,\ldots,\gamma_c)' \qquad (2.3.1)$$

is assumed to be estimable.

The information matrix for θ is given by $\sigma^{-2}\mathbf{I}(\theta)$, where

$$\mathbf{I}(\theta) = \begin{pmatrix} \mathbf{N} & \mathbf{T} \\ \mathbf{T}' & \mathbf{Z}'\mathbf{Z} \end{pmatrix}, \qquad (2.3.2)$$

$$\mathbf{N} = \mathrm{Diag}(n_1, n_2,\ldots, n_v), \qquad (2.3.3)$$

$$\mathbf{T} = (\mathbf{T}'_1, \mathbf{T}'_2,\ldots, \mathbf{T}'_v)',\ \mathbf{T}_i = \mathbf{1}'_{n_i}\mathbf{Z}_i, \qquad (2.3.4)$$

$$\mathbf{Z}^{n \times c} = (\mathbf{Z}_1', \mathbf{Z}_2', \ldots, \mathbf{Z}_v')' \tag{2.3.5}$$

and

$$\mathbf{Z}_i^{n_i \times c} = \begin{pmatrix} z_{i1}^{(1)} & z_{i1}^{(2)} & \cdots & z_{i1}^{(c)} \\ z_{i2}^{(1)} & z_{i2}^{(2)} & \cdots & z_{i2}^{(c)} \\ \vdots & \vdots & \ddots & \vdots \\ z_{in_i}^{(1)} & z_{in_i}^{(2)} & \cdots & z_{in_i}^{(c)} \end{pmatrix}. \tag{2.3.6}$$

For D-optimality, we have to maximize the determinant of $\mathbf{I}(\theta)$, denoted as $\det(\mathbf{I}(\theta))$, with respect to the design variables $\{z_{ij}^{(t)}\}$ satisfying $z_{ij}^{(t)} \in [-1, 1]$, $1 \le j \le n_i$, $1 \le i \le v$ and n_i's satisfying (2.1.2).

From (2.3.2) it is easy to see that

$$
\begin{aligned}
\det(\mathbf{I}(\theta)) &= \left(\prod_{i=1}^{v} n_i \right) \det(\mathbf{Z}'\mathbf{Z} - \mathbf{T}'\mathbf{N}^{-1}\mathbf{T}) \\
&= \left(\prod_{i=1}^{v} n_i \right) \det\left(\mathbf{Z}'\mathbf{Z} - \sum_i n_i^{-1}\mathbf{T}_i'\mathbf{T}_i\right) \\
&= \det(\mathbf{N})\det(\mathbf{C}),
\end{aligned}
\tag{2.3.7}
$$

where

$$\mathbf{C} = \mathbf{Z}'\mathbf{Z} - \sum_i n_i^{-1}\mathbf{T}_i'\mathbf{T}_i. \tag{2.3.8}$$

Note that $\mathbf{C}$ is the information matrix for the regression coefficients $\gamma_1, \gamma_2, \ldots, \gamma_c$. The maximization of $\det(\mathbf{I}(\theta))$ is done in two stages. In the first stage, the maximization is done for varying z-values for fixed n_i's. This leads to an upper bound for $\det(\mathbf{I}(\theta))$ obtained through completely symmetric $\mathbf{C}$-matrices. At the second stage, maximization is done for varying n_i's subject to $\sum_i n_i = n$, and this leads to a sufficiently small class $\mathcal{N}$ of contending $\mathbf{n} = (n_1, n_2, \ldots, n_v)$'s wherein the overall upper bound to $\det(\mathbf{I}(\theta))$ belongs.

2.3.1 First Stage of Maximization

Maximisation of $\mathbf{I}(\theta)$ with respect to $z_{ij}^{(t)} \in [-1, 1]$ is based on the following lemma.

Lemma 2.3.1 *A necessary condition for maximization of* $det(\mathbf{C})$ *of (2.3.8) with respect to* $z_{ij}^{(t)} \in [-1, 1]$, *for fixed* n_i's *is that* $z_{ij}^{(t)} = \pm 1$ $\forall i, j$ *and* t.

Proof From (2.3.8), **C** can be expressed as

$$\mathbf{C} = \mathbf{Z'MZ} = \mathbf{Z^{*'}Z^*} \tag{2.3.9}$$

where

$$\mathbf{Z^*} = \mathbf{MZ}, \quad \mathbf{M} = \text{diag}(\mathbf{M}_1, \mathbf{M}_2, \ldots, \mathbf{M}_v), \quad \mathbf{M}_i = (\mathbf{I}_{n_i} - n_i^{-1}\mathbf{1}_{n_i}\mathbf{1}'_{n_i}). \tag{2.3.10}$$

It is known that (cf. Galil and Kiefer 1980; Wojtas 1964), $\det(\mathbf{Z^{*'}Z^*})$ is maximum at the extreme entries of $\mathbf{Z^*}$. Again, as $z_{ij}^{(t)*}$'s are linear in $z_{ij}^{(t)}$'s, the determinant is maximum at the extreme values of $z_{ij}^{(t)}$'s for all i, j and t. Hence the lemma follows. $\square$

Theorem 2.3.1 *For fixed $\{n_i\}$'s satisfying* (2.1.2),

$$det(\mathbf{I}(\theta)) \leq \left(\prod_{i=1}^{v} n_i\right)\{a + (c-1)b\}(a-b)^{c-1} \tag{2.3.11}$$

where

$$a = n - \delta, \quad b = |\xi - \delta| \tag{2.3.12}$$

$$\delta = \sum_{i=1}^{v} n_i^{-1}\delta_i, \quad \delta_i = 1(0) \text{ if } n_i = \text{odd(even)} \tag{2.3.13}$$

$$\xi = \xi(n, \delta) = \begin{cases} \lfloor\delta\rfloor & \text{if both of } n, \ \lfloor\delta\rfloor \text{ are odd or even} \\ \\ \lfloor\delta\rfloor + 1 & \text{if } n = \text{odd}, \ \lfloor\delta\rfloor = \text{even} \text{ or n=even}, \ \lfloor\delta\rfloor = \text{odd} \end{cases} \tag{2.3.14}$$

$\lfloor\delta\rfloor = $ *greatest integer less than equal to δ.*

Proof Because of Lemma 2.3.1, we restrict $z_{ij}^{(t)}$ to the class $\chi = \{z_{ij}^{(t)} : z_{ij}^{(t)} = \pm 1\}$. From the Eq. (2.3.8), we note that, $c_{t,t'}$, the $(t, t')^{\text{th}}$ element of the **C**-matrix is given by

$$c_{t,t'} = \sum_i \left\{ \sum_j z_{ij}^{(t)}z_{ij}^{(t')} - \frac{\left(\sum_j z_{ij}^{(t)}\right)\left(\sum_j z_{ij}^{(t')}\right)}{n_i} \right\}, \quad 1 \leq t, t' \leq c. \tag{2.3.15}$$

It follows from Wojtas (1964) that $\det(\mathbf{C})$ is maximum when **C** is completely symmetric with all the diagonal elements equal to a and all off-diagonal elements equal to b where a and b are given by $\max_{1\leq t\leq c} c_{tt}$ and $\min_{1\leq t\neq t'\leq c} |c_{tt'}|$ respectively. Again as

$z_{ij}^{(t)} = \pm 1 \; \forall i, j$ and t, for fixed n_i's, it can be deduced that

$$\max_{1 \le t \le c} c_{tt} = n - \delta = a, \quad \min_{1 \le t \neq t' \le c} |c_{tt'}| = |\xi - \delta| = b \qquad (2.3.16)$$

where δ and ξ are given in (2.3.13) and (2.3.14) respectively. Therefore the theorem follows. □

2.3.2 Second Stage of Maximization

In view of Theorem 2.3.1, we now consider the problem of maximizing

$$g(\mathbf{n}) = g(n_1, n_2, \dots, n_v) = \left(\prod_{i=1}^{v} n_i \right) \{a + (c-1)b\}(a-b)^{c-1} \qquad (2.3.17)$$

with respect to n_i's subject to $\sum_{i=1}^{v} n_i = n$, where a and b are given by (2.3.12)–(2.3.14), so as to find the overall upper bound of $\det(\mathbf{I}(\theta))$. The following lemma helps to reduce the class $\mathcal{N}$ of $\mathbf{n}$'s where $\mathbf{n} = (n_1, n_2, \dots, n_v)$, satisfying (2.1.2), to a subclass in which maximum of $g(\mathbf{n})$ lies.

Lemma 2.3.2 *Let* $\mathbf{n}^* = (n_1^*, n_2^*, \dots, n_v^*)$ *be a maximizer of* $g(\mathbf{n})$ *of* (2.3.17) *subject to the condition* (2.1.2). *Then* $\mathbf{n}^*$ *cannot have*

(i) *two unequal odd elements;*
(ii) *two even elements that differ by more than 2;*
(iii) *an even and an odd element that differ by more than 1.* □ ·

Proof (i) Without loss of generality it is assumed that n_1^* and n_2^* be odd and $n_1^* \le n_2^* - 2$. Define $\tilde{\mathbf{n}} = (\tilde{n}_1, \tilde{n}_2, \dots, \tilde{n}_v)$, where $\tilde{n}_1 = n_1^* + 1$, $\tilde{n}_2 = n_2^* - 1$ and $\tilde{n}_i = n_i^* \; \forall i \neq 1, 2$. Note that $\tilde{n}_i$'s satisfy condition (2.1.2). Then by Eq. (2.3.17),

$$\frac{g(\tilde{\mathbf{n}})}{g(\mathbf{n}^*)} = \left(\prod_{i=1}^{v} \tilde{n}_i \right) \bigg/ \left(\prod_{i=1}^{v} n_i^* \right) \left(\frac{\{\tilde{a}+(c-1)\tilde{b}\}(\tilde{a}-\tilde{b})^{c-1}}{\{a^*+(c-1)b^*\}(a^*-b^*)^{c-1}} \right)$$

$$= \frac{(n_1^*+1)(n_2^*-1)}{n_1^* n_2^*} \frac{\{\tilde{a}+(c-1)\tilde{b}\}(\tilde{a}-\tilde{b})^{c-1}}{\{a^*+(c-1)b^*\}(a^*-b^*)^{c-1}}, \qquad (2.3.18)$$

where,

$$\tilde{a} = n - \sum_{i=1}^{v} \tilde{n}_i^{-1} \tilde{\delta}_i = n - \sum_{i=1}^{v} n_i^{*-1} \delta_i^* + \frac{1}{n_1^*} + \frac{1}{n_2^*} = a^* + \frac{1}{n_1^*} + \frac{1}{n_2^*} \qquad (2.3.19)$$

Again,

$$\widetilde{b} = \left|\widetilde{\xi} - \sum_{i=1}^{v} \widetilde{n}_i^{-1} \widetilde{\delta}_i\right| \le \left|\xi^* - \sum_{i=1}^{v} n_i^{*-1} \delta_i^* + \left(\frac{1}{n_1^*} + \frac{1}{n_2^*}\right)\right| \le b^* + \left(\frac{1}{n_1^*} + \frac{1}{n_2^*}\right). \quad (2.3.20)$$

We consider the two cases $\widetilde{b} \le b^*$ and $\widetilde{b} > b^*$ separately.

(a) Let $\widetilde{b} \le b^*$. Then, as by (2.3.19), $\widetilde{a} > a^*$, it follows that $g(\widetilde{\mathbf{n}}) > g(\mathbf{n}^*)$, which is impossible.

(b) Let $\widetilde{b} > b^*$ and let $\widetilde{b}$ assume the highest possible value given in (2.3.20). Then from (2.3.18)–(2.3.20), it is seen that

$$\frac{g(\widetilde{\mathbf{n}})}{g(\mathbf{n}^*)} > \frac{(n_1^* + 1)(n_2^* - 1)}{n_1^* n_2^*} > 1 \quad (2.3.21)$$

which is again a contradiction. As the inequality (2.3.21) is true for the highest value of $\widetilde{b}$, it will be true for all values of $\widetilde{b}$ in $[b^*, b^* + \frac{1}{n_1^*} + \frac{1}{n_2^*}]$ as $\widetilde{a} > a^*$.

(ii) If possible, let $\mathbf{n}^*$ have two even elements, say $n_1^* < n_2^*$ which differ by more than 2. Then as in (i) above, we reach at a contradiction by increasing n_1^* by two and decreasing n_2^* by two.

(iii) If possible, let $\mathbf{n}^*$ have an even element n_1^* and an odd element n_2^* which differ by more than 1.

Case A: Let $n_1^* > n_2^*$. Satisfying (2.1.2), define $\widetilde{\mathbf{n}} = (\widetilde{n}_1, \widetilde{n}_2, \ldots, \widetilde{n}_v)$, where $\widetilde{n}_1 = n_1^* - 2$, $\widetilde{n}_2 = n_2^* + 2$ and $\widetilde{n}_i = n_i^* \ \forall i \neq 1, 2$. Then by Eq. (2.3.17), we have

$$\frac{g(\widetilde{\mathbf{n}})}{g(\mathbf{n}^*)} = \frac{(n_1^* - 2)(n_2^* + 2)\{\widetilde{a} + (c-1)\widetilde{b}\}(\widetilde{a} - \widetilde{b})^{c-1}}{(n_1^* n_2^*)\{a^* + (c-1)b^*\}(a^* - b^*)^{c-1}} \quad (2.3.22)$$

where,

$$\widetilde{a} = n - \sum_i \widetilde{n}_i^{-1} \widetilde{\delta}_i = \left(n - \sum_i n_i^{*-1} \delta_i^*\right) + \left(\frac{1}{n_2^*} - \frac{1}{n_2^* + 2}\right) = a^* + \left(\frac{1}{n_2^*} - \frac{1}{n_2^* + 2}\right) \quad (2.3.23)$$

$$\widetilde{b} = |\widetilde{\xi} - \widetilde{\delta}| \le |(\xi^* - \delta^*) + \left(\frac{1}{n_2^*} - \frac{1}{n_2^* - 2}\right)| \le b^* + \left(\frac{1}{n_2^*} - \frac{1}{n_2^* - 2}\right). \quad (2.3.24)$$

We consider two cases when $\widetilde{b} \le b^*$ and $\widetilde{b} > b^*$. For $\widetilde{b} \le b^*$, it follows, from (2.3.22) that $g(\widetilde{\mathbf{n}}) > g(\mathbf{n}^*)$ as $\widetilde{a} > a^*$. Again, for $\widetilde{b} > b^*$, we assume its highest value viz. $b^* + (\frac{1}{n_2^*} - \frac{1}{n_2^* - 2})$ from (2.3.24) and use it in (2.3.22). It is seen that $g(\widetilde{\mathbf{n}}) > g(\mathbf{n}^*)$, which obviously holds for all other values of $\widetilde{b} > b^*$ as $\widetilde{a} > a^*$.

So we reach at a contradiction that $\mathbf{n}^*$ is a maximizer of $g(\mathbf{n})$.

Case B: Let $n_1^* < n_2^*$ (i.e. $n_1^* \leq n_2^* - 3$), then we have the following two cases:

(a) n_2^* is not the only odd element of $\mathbf{n}^*$.

(b) n_2^* is the only odd element of $\mathbf{n}^*$.

For (a), let $\mathbf{n}^*$ have another odd element n_3^*. Then by part (i) of this lemma, $n_2^* = n_3^*$. Define $\widetilde{\mathbf{n}} = (\widetilde{n}_1, \widetilde{n}_2, \ldots, \widetilde{n}_v)$, where $\widetilde{n}_1 = n_1^* + 2$, $\widetilde{n}_2 = \widetilde{n}_3 = n_2^* - 1$ and $\widetilde{n}_i = n_i^* \ \forall \ i \neq 1, 2, 3$. Then by (2.3.17)

$$\frac{g(\widetilde{\mathbf{n}})}{g(\mathbf{n}^*)} = \frac{(n_1^* + 2)(n_2^* - 1)^2}{(n_1^* n_2^* n_3^*)} \frac{\{\widetilde{a} + (c-1)\widetilde{b}\}(\widetilde{a} - \widetilde{b})^{c-1}}{\{a^* + (c-1)b^*\}(a^* - b^*)^{c-1}}. \tag{2.3.25}$$

where,

$$\widetilde{a} = n - \sum_i \widetilde{n}_i^{-1} \delta_i = \left(n - \sum_i n_i^{*-1}\delta_i\right) + \frac{2}{n_2^*} = a^* + \frac{2}{n_2^*} \tag{2.3.26}$$

$$\widetilde{b} = |\widetilde{\xi} - \widetilde{\delta}| \leq |(\xi^* - \delta^*) + \frac{2}{n_2^*}| \leq b^* + \frac{2}{n_2^*}. \tag{2.3.27}$$

If $\widetilde{b} \leq b^*$, then from (2.3.25) and (2.3.26) $g(\widetilde{\mathbf{n}}) > g(\mathbf{n}^*)$ which is a contradiction.

If $\widetilde{b} > b^*$, the above contradiction also holds by the same reasons as given in Case A.

For (b), let us define $\widetilde{\mathbf{n}} = (\widetilde{n}_1, \widetilde{n}_2, \ldots, \widetilde{n}_v)$ satisfying (1.2) where $\widetilde{n}_1 = n_1^* + 2$, $\widetilde{n}_2 = n_2^* - 2$ and $\widetilde{n}_i = n_i^* \ \forall \ i \neq 1, 2$. Proceeding as before, it can be proved that

$$\widetilde{a} = a^* + \left(\frac{1}{n_2^*} - \frac{1}{n_2^* - 2}\right), \quad \widetilde{b} = \left(1 - \frac{1}{n_2^* - 2}\right). \tag{2.3.28}$$

Using (2.3.28) in (2.3.17), it is seen that

$$\frac{g(\widetilde{\mathbf{n}})}{g(\mathbf{n}^*)} = \frac{(n_1^* + 2)(n_2^* - 2)}{n_1^* n_2^*} \frac{\left(n + c - 1 - \frac{c}{n_2^* - 2}\right)}{\left(n + c - 1 - \frac{c}{n_2^*}\right)} > 1 \quad \text{as } (n_2^* - n_1^*) \geq 3.$$

This is again a contradiction. Therefore the lemma follows. $\qquad\square$

From Lemma 2.3.2, we get the following theorem whose proof is immediate.

Theorem 2.3.2 *Let $\bar{o}$ be an odd integer, where $\bar{o} = \lfloor \frac{n}{v} \rfloor$ or $\lfloor \frac{n}{v} \rfloor + 1$ according as $\lfloor \frac{n}{v} \rfloor$ is odd or even and $\mathbf{n}^* = (n_1^*, n_2^*, \ldots, n_v^*)$ be a maximizer of $g(\mathbf{n})$ of (2.3.17) subject to $\sum_i n_i = n$. Then $n_i^* \in \{\bar{o} - 1, \bar{o}, \bar{o} + 1\}$.*

Lemma 2.3.3 *If f, f^- and f^+ be the frequencies of $\bar{o}$, $\bar{o}-1$ and $\bar{o}+1$ respectively, then the following relations*

$$f + f^- + f^+ = v; \quad \bar{o}f + (\bar{o}-1)f^- + (\bar{o}+1)f^+ = n, \tag{2.3.29}$$

minimize considerably the search for optimum **n**, *for which* $g(\mathbf{n})$ *is a maximum.*

Let $\mathcal{N}^*$ ($\subset \mathcal{N}$) denote the class of **n**'s satisfying Theorem 2.3.2 and Lemma 2.3.2.

Remark 2.3.1 For given n, v and c, let $g(\mathbf{n}^*)$ be the maximum of $g(\mathbf{n})$ of (2.3.17) over $\mathbf{n} = (n_1, n_2, \ldots, n_v)$ subject to $\sum_i n_i = n$. Then by Theorem 2.3.1

$$\det(\mathbf{I}(\theta) \leq g(\mathbf{n}^*). \tag{2.3.30}$$

If a choice of $\{z_{ij}^{(t)}\}$ exists corresponding to $\mathbf{n}^*$, such that equality in (2.3.30) holds, then $\mathbf{n}^*$ together with $\{z_{ij}^{(t)}\}$ gives a D-optimal design.

Remark 2.3.2 If all n_i's are even, so that all the $\mathbf{T}_i$'s of (2.3.4) may be made equal to zero, then it is possible to estimate the regression parameters γ's orthogonally to the μ_i's. In that case, γ's are estimated most efficiently with the minimum possible variance when $\mathbf{Z}'\mathbf{Z} = n\mathbf{I}_c$.

Remark 2.3.3 If $n_i = \frac{n}{v}$ = an even integer for all i, the situation reduces to regular case and then Remark 2.3.1 is in full agreement with Troya Lopes (1982a) and in that case γ's can be estimated most efficiently so that each estimator has minimum possible variance when $\mathbf{Z}'\mathbf{Z} = n\mathbf{I}_c$.

Remark 2.3.4 If the v levels of the single factor set-up are assumed to be the v level combinations of m factors $F_1, \ldots, F_m$ having $s_1, \ldots, s_m$ levels, respectively ($v = \prod s_i$), then the optimum design for the single factor set-up is also optimum for the estimation of γ and all effects up to m-factor interactions which can be obtained through an orthogonal transformation of γ and the mean vector μ corresponding to the v level combinations.

2.3.3 Examples

Now we consider following examples to illustrate the above method.

Example 2.3.1 Let us consider the one-way set-up with $n = 12$, $v = 4$. It follows that $\mathcal{N}^* = \{(3, 3, 3, 3), (2, 3, 3, 4), (2, 2, 4, 4)\} \equiv \{(3^4), (2, 3^2, 4), (2^2, 4^2)\}$.

(a) For $c = 1$, $\mathbf{n}^* = (3^4)$ is the unique maximizer of $g(\mathbf{n})$ and this $\mathbf{n}^*$ together with $\mathbf{Z}_1' = (1, 1, -1)$, $\mathbf{Z}_2' = (1, 1, -1)$, $\mathbf{Z}_3' = (1, 1, -1)$, $\mathbf{Z}_4' = (1, 1, -1)$ gives a D-optimal design.

(b) For $c = 2$ both $\mathbf{n}^* = (2, 3^2, 4)$ and $(2^2, 4^2)$ are maximizers of $g(\mathbf{n})$.

(i) $\mathbf{n}^* = (2^2, 4^2)$ and

$$\mathbf{Z}_1 = \begin{pmatrix} 1 & 1 \\ -1 & -1 \end{pmatrix}, \mathbf{Z}_2 = \begin{pmatrix} -1 & 1 \\ 1 & -1 \end{pmatrix}, \mathbf{Z}_3 = \begin{pmatrix} 1 & 1 \\ 1 & -1 \\ -1 & 1 \\ -1 & -1 \end{pmatrix}, \mathbf{Z}_4 = \begin{pmatrix} -1 & 1 \\ -1 & -1 \\ 1 & 1 \\ 1 & -1 \end{pmatrix},$$

give a D-optimal design.

(ii) $\mathbf{n}^* = (2, 3^2, 4)$ and

$$\mathbf{Z}_1 = \begin{pmatrix} 1 & 1 \\ -1 & -1 \end{pmatrix}, \mathbf{Z}_2 = \begin{pmatrix} -1 & 1 \\ 1 & -1 \\ 1 & 1 \end{pmatrix}, \mathbf{Z}_3 = \begin{pmatrix} 1 & -1 \\ -1 & 1 \\ -1 & -1 \end{pmatrix}, \mathbf{Z}_4 = \begin{pmatrix} -1 & 1 \\ -1 & -1 \\ 1 & 1 \\ 1 & -1 \end{pmatrix}, \text{ also}$$

give a D-optimal design.

(c) For $c = 3, 4$, $\mathbf{n}^* = (2^2, 4^2)$ is the unique maximizer of $g(\mathbf{n})$.

Example 2.3.2 In one-way set-up with $n = 9$, $v = 3$, $c = 3$, D-optimal design should be searched within the set $\{(2, 3, 4), (3^3)\}$ of $\mathbf{n}$. It is seen that for $\mathbf{n} = (3^3)$ and

$$D_1: \quad \mathbf{Z}^{(1)} = \begin{pmatrix} -1 & -1 & -1 \\ 1 & 1 & 1 \\ 1 & 1 & 1 \end{pmatrix}, \mathbf{Z}^{(2)} = \begin{pmatrix} -1 & + & 1 \\ 1 & -1 & 1 \\ 1 & 1 & -1 \end{pmatrix} \text{ and } \mathbf{Z}^{(3)} = \begin{pmatrix} -1 & 1 & 1 \\ 1 & -1 & 1 \\ 1 & 1 & -1 \end{pmatrix},$$

$\mathbf{C} = \text{diag}(8, 8, 8)$ and $g(\mathbf{n}) = 3^3.8^3$. But for $\mathbf{n} = (2, 3, 4)$, and

$$D_2: \quad \mathbf{Z}^{(1)} = \begin{pmatrix} 1 & 1 & 1 \\ -1 & -1 & -1 \end{pmatrix}, \mathbf{Z}^{(2)} = \begin{pmatrix} 1 & -1 & 1 \\ -1 & 1 & -1 \\ 1 & 1 & 1 \end{pmatrix} \text{ and } \mathbf{Z}^{(3)} = \begin{pmatrix} 1 & 1 & -1 \\ -1 & -1 & 1 \\ 1 & -1 & -1 \\ -1 & 1 & 1 \end{pmatrix}$$

It can be seen that $\mathbf{C} = 8\mathbf{I}_3 + \frac{2}{3}\mathbf{J}_3$, where $\mathbf{J}_3$ is a 3×3 matrix containing elements one only. Also $g(2, 3, 4)$ which is equal to 15360, attains the upper bound in (2.3.14) and $g(2, 3, 4) > g(3, 3, 3)$ implying that D_2 is D-optimal.

Again for $n = 9$, $v = 3$, $c = 4$, it is noted that $\mathbf{n}^* = (2, 3, 4)$ together with

$$D_3: \quad \mathbf{Z}^{(1)} = \begin{pmatrix} 1 & 1 & 1 & 1 \\ -1 & -1 & -1 & -1 \end{pmatrix}, \mathbf{Z}^{(2)} = \begin{pmatrix} 1 & -1 & 1 & -1 \\ -1 & 1 & -1 & 1 \\ 1 & 1 & 1 & 1 \end{pmatrix}$$

and

$$\mathbf{Z}^{(3)} = \begin{pmatrix} 1 & 1 & -1 & -1 \\ -1 & -1 & 1 & 1 \\ 1 & -1 & -1 & 1 \\ -1 & 1 & 1 & -1 \end{pmatrix}$$

maximizes $g(\mathbf{n})$ of (2.3.17) and hence gives a D-optimal design.

Remark 2.3.5 It is seen from the examples that the choice of optimum **n** depends on the number of the covariates used apart from the number of cells v in the set-up. Again it is noted from (2.3.7) that $\det(\mathbf{I}(\theta))$ depends on two factors viz. $\det(\mathbf{N})(= \prod_i n_i)$ and $\det(\mathbf{C})$. Determinant of **N** increases as the homogeneity between the n_i's increases subject to $\sum_i n_i = n$. On the other hand $\det(\mathbf{C})$ increases, apart from c, with the largeness of a and the smallness of b, which again are achieved by inclusion of maximum number of even n_i's closed to $\lfloor \frac{n}{v} \rfloor$. The number of odd n_i's subject to $\sum_i n_i = n$, in between the even ones with proper homogeneity, actually strikes a balance between $\det(\mathbf{N})$ and $\det(\mathbf{C})$. It is also seen that, when c is small, $\det(\mathbf{N})$ is the dominant factor, while, if c is large $\det(\mathbf{C})$ becomes the dominant factor.

Incidentally, the above analysis is based on the work in Dutta et al. (2014) and it improves over what was achieved in Dey and Mukerjee (2006).

References

Das K, Mandal NK, Sinha BK (2003) Optimal experimental designs with covariates. J Stat Plan Inference 115:273–285

Dey A, Mukerjee R (2006) D-optimal designs for covariate models. Statistics 40:297–305

Dutta G, Das P, Mandal NK (2014) D-Optimal designs for covariate models. Commun Stat Theory Methods 43:165–174

Galil Z, Kiefer J (1980) D-optimum weighing designs. Ann Stat 8:1293–1306

Hedayat AS, Sloane NJA, Stufken J (1999) Orthogonal arrays: theory and applications. Springer, New York

Pukelsheim F (1993) Optimal design of experiments. Wiley, New York

Rao CR (1973) Linear statistical inference and its applications. Wiley, New York

Rao PSSNVP, Rao SB, Saha GM, Sinha BK (2003) Optimal designs for covariates' models and mixed orthogonal arrays. Electron Notes Discret Math 15:157–160

Shah KR, Sinha BK (1989) Theory of optimal designs. lecture notes in statistics, Series 54. Springer, New York

Troya Lopes J (1982a) Optimal designs for covariate models. J Stat Plan Inference 6:373–419

Troya Lopes J (1982b) Cyclic designs for a covariate model. J Stat Plan Inference 7:49–75

Wojtas M (1964) On Hadamard's inequality for the determinants of order non-divisible by 4. Colloq Math 12:73–83

Chapter 3
OCDs in Randomized Block Design Set-Up

3.1 Introduction

For two-way layout, the set-up can be written as

$$\left(\mathbf{Y}, \ \mu\mathbf{1} + \mathbf{X}_1\tau + \mathbf{X}_2\beta + \mathbf{Z}\gamma, \ \sigma^2\mathbf{I}_n \right) \tag{3.1.1}$$

where μ, as usual, stands for the general effect, $\tau^{v\times1}$, $\beta^{b\times1}$ represent vectors of treatment and block effects, respectively, and $\mathbf{X}_1^{n\times b}$ and $\mathbf{X}_2^{n\times v}$ are, respectively, the corresponding incidence matrices. $\mathbf{Y}$, $\mathbf{Z}$ as usual, represent an observation vector of order $n \times 1$ and the design matrix of order $n \times c$ corresponding to vector of covariate effects $\gamma^{c\times1}$ respectively. It should be noted that each column of $\mathbf{Z}$-matrix has a natural interpretation in terms of the correspondence of the covariate values with the experimental units in the RBD set-up we start with.

We straightway compute the form of the information matrix for the whole set of parameters $\eta = \left(\mu, \ \beta', \ \tau', \ \gamma' \right)'$ underlying a design d with $\mathbf{X}_{1d}$, $\mathbf{X}_{2d}$ and $\mathbf{Z}_d$ as the versions of $\mathbf{X}_1$, $\mathbf{X}_2$ and $\mathbf{Z}$ in (3.1.1):

$$\mathbf{I}_d(\eta) = \begin{pmatrix} n & \mathbf{1}'\mathbf{X}_{1d} & \mathbf{1}'\mathbf{X}_{2d} & \mathbf{1}'\mathbf{Z}_d \\ & \mathbf{X}_{1d}'\mathbf{X}_{1d} & \mathbf{X}_{1d}'\mathbf{X}_{2d} & \mathbf{X}_{1d}'\mathbf{Z}_d \\ & & \mathbf{X}_{2d}'\mathbf{X}_{2d} & \mathbf{X}_{2d}'\mathbf{Z}_d \\ & & & \mathbf{Z}_d'\mathbf{Z}_d \end{pmatrix}. \tag{3.1.2}$$

For the covariates, as before, we assume, without loss of generality, the (location-scale)-transformed version: $|z_{ij}^{(t)}| \leq 1; i, j, t$.

It is evident from (3.1.2) that orthogonal estimation of treatment and block effect contrasts on one hand and covariate effects on the other is possible when the conditions

$$\mathbf{X}_{1d}'\mathbf{Z}_d = \mathbf{0}, \quad \mathbf{X}_{2d}'\mathbf{Z}_d = \mathbf{0} \tag{3.1.3}$$

© Springer India 2015
P. Das et al., *Optimal Covariate Designs*, DOI 10.1007/978-81-322-2461-7_3

are satisfied. It is to be noted that under (3.1.3), $\mathbf{1}'\mathbf{Z}_d = \mathbf{0}'$ also holds. Further, as before, most efficient estimation of γ-components is possible whenever, in addition to (3.1.3), we can also ascertain

$$\mathbf{Z}_d'\mathbf{Z}_d = n\mathbf{I}_n. \tag{3.1.4}$$

It is also true that, whenever (3.1.3) is ensured, presence of the covariates in (3.1.1) does not pose any threat to the usual optimal design problem in a block design set-up as the covariate parameters and the block design parameters are orthogonally estimable.

As before for an RBD set-up, following Das et al. (2003), we recast each column of the $\mathbf{Z}^{n \times c} = (\pm 1)$ matrix by a $\mathbf{W}$-matrix of order $v \times b$. Corresponding to the treatment $\times$ block classifications, conditions (3.1.3) and (3.1.4) reduce, in terms of $\mathbf{W}$-matrices, to $C_1 - C_3$ where

C_1. Each $\mathbf{W}$-matrix has all column-sums equal to zero;
C_2. Each $\mathbf{W}$-matrix has all row-sums equal to zero;
C_3. The grand total of all the entries in the Hadamard product   (3.1.5)
    of any two distinct $\mathbf{W}$-matrices reduces to zero.

Now we define optimum $\mathbf{W}$-matrix for covariate design, in an RBD set-up.

Definition 3.1.1 With respect to model (3.1.1), the c $\mathbf{W}$-matrices corresponding to the c covariates are said to be optimum if they satisfy the conditions $C_1 - C_3$ of (3.1.5).

We arrange the remaining sections of this chapter as follows. In Sect. 3.2, we consider the constructional methods of optimum $\mathbf{W}$-matrices given by Das et al. (2003) and in Sects. 3.3 and 3.4, we discuss the relationships between OCDs and MOAs and construction of optimum $\mathbf{Z}$s given in Rao et al. (2003).

3.2 Construction of Optimum W-Matrices

Here we consider the following method for constructing optimum $\mathbf{W}$-matrices given in Das et al. (2003). They used mutually orthogonal latin squares (MOLS) for construction of optimum $\mathbf{W}$s. The method is given in the following theorem.

Theorem 3.2.1 *Suppose $\mathbf{H}_v$ and m MOLS of order v exist. Then $m(v-1)$ optimum $\mathbf{W}$-matrices can be constructed for an RBD with $b = v$ blocks and v treatments.*

Proof For the construction of the optimum $\mathbf{W}$-matrices, we will proceed as follows:

Step 1 We set the Hadamard Matrix $\mathbf{H}_v$ in the following form:

$$\mathbf{H}_v = (\mathbf{h}_1, \mathbf{h}_2, \ldots, \mathbf{h}_{v-1}, \mathbf{1}) \tag{3.2.1}$$

where $\mathbf{h}_j$ denotes the jth column of $\mathbf{H}_v$.

Step 2 We can construct the ith member L_i of the set of m MOLS of order v by using the symbols

$$a_{i1}, a_{i2}, \ldots, a_{iv}; \quad 1 \le i \le m. \tag{3.2.2}$$

Step 3 Take L_i and replace the symbols $a_{i1}, a_{i2}, \ldots, a_{iv}$ by the elements of $\mathbf{h}_j$ successively and we get a **W**-matrix. By varying i, j we get $m(v-1)$ **W**-matrices. We can easily check from the properties of MOLS and Hadamard matrices that these are optimum **W**s. □

Remark 3.2.1 When $b = v = 2^p$, $p = $ integer, we have a complete set of MOLS of order v. Then we can construct $(b-1)(v-1)$ optimum **W**-matrices. In this situation, it exhausts the error degrees of freedom in RBD model.

Example 3.2.1 We illustrate the above method of construction by citing an example. Take $b = v = 2^2$ and replacing a_{ij} by other suitable symbols, we write down the MOLS of order 4 as follows:

$$L_1 = \begin{pmatrix} a\,b\,c\,d \\ b\,a\,d\,c \\ c\,d\,a\,b \\ d\,c\,b\,a \end{pmatrix}, \quad L_2 = \begin{pmatrix} \alpha\,\delta\,\beta\,\gamma \\ \beta\,\gamma\,\alpha\,\delta \\ \gamma\,\beta\,\delta\,\alpha \\ \delta\,\alpha\,\gamma\,\beta \end{pmatrix}, \quad L_3 = \begin{pmatrix} p\,s\,r\,q \\ q\,r\,s\,p \\ s\,p\,q\,r \\ r\,q\,p\,s \end{pmatrix}.$$

We write $\mathbf{H}_4$ as

$$\mathbf{H}_4 = \begin{pmatrix} 1 & 1 & 1 & 1 \\ -1 & -1 & 1 & 1 \\ 1 & -1 & -1 & 1 \\ -1 & 1 & -1 & 1 \end{pmatrix} = (\mathbf{h}_1, \mathbf{h}_2, \mathbf{h}_3, \mathbf{1}). \tag{3.2.3}$$

Using $\mathbf{h}_1$, $\mathbf{h}_2$ and $\mathbf{h}_3$ in L_1, we get the following three optimum **W**-matrices:

$$\mathbf{W}^{(1)} = \begin{pmatrix} 1 & -1 & 1 & -1 \\ -1 & 1 & -1 & 1 \\ 1 & -1 & 1 & -1 \\ -1 & 1 & -1 & 1 \end{pmatrix}, \quad \mathbf{W}^{(2)} = \begin{pmatrix} 1 & -1 & -1 & 1 \\ -1 & 1 & 1 & -1 \\ -1 & 1 & 1 & -1 \\ 1 & -1 & -1 & 1 \end{pmatrix}, \quad \mathbf{W}^{(3)} = \begin{pmatrix} 1 & 1 & -1 & -1 \\ 1 & 1 & -1 & -1 \\ -1 & -1 & 1 & 1 \\ -1 & -1 & 1 & 1 \end{pmatrix}.$$

Similarly, using $\mathbf{h}_1$, $\mathbf{h}_2$ and $\mathbf{h}_3$ in L_2 and L_3 respectively, we get six more optimum **W**-matrices as

$$\mathbf{W}^{(4)} = \begin{pmatrix} 1 & -1 & 1 & -1 \\ 1 & -1 & 1 & -1 \\ -1 & 1 & -1 & 1 \\ -1 & 1 & -1 & 1 \end{pmatrix}, \quad \mathbf{W}^{(5)} = \begin{pmatrix} 1 & -1 & -1 & 1 \\ -1 & 1 & 1 & -1 \\ 1 & -1 & -1 & 1 \\ -1 & 1 & 1 & -1 \end{pmatrix}, \quad \mathbf{W}^{(6)} = \begin{pmatrix} 1 & 1 & -1 & -1 \\ -1 & -1 & 1 & 1 \\ -1 & -1 & 1 & 1 \\ 1 & 1 & -1 & -1 \end{pmatrix},$$

$$\mathbf{W}^{(7)} = \begin{pmatrix} 1 & -1 & 1 & -1 \\ -1 & 1 & -1 & 1 \\ -1 & 1 & -1 & 1 \\ 1 & -1 & 1 & -1 \end{pmatrix}, \quad \mathbf{W}^{(8)} = \begin{pmatrix} 1 & -1 & -1 & 1 \\ 1 & -1 & -1 & 1 \\ -1 & 1 & 1 & -1 \\ -1 & 1 & 1 & -1 \end{pmatrix}, \quad \mathbf{W}^{(9)} = \begin{pmatrix} 1 & 1 & -1 & -1 \\ -1 & -1 & 1 & 1 \\ 1 & 1 & -1 & -1 \\ -1 & -1 & 1 & 1 \end{pmatrix}.$$

Remark 3.2.2 When $b = pv$, $v = 0$ (mod 4), $p \geq 1$, $\mathbf{H}_v$ and m MOLS of order v exist, then by writing the $\mathbf{W}$-matrices of order $v \times b$ side by side p times, we can get $m(v - 1)$ optimum $\mathbf{W}$-matrices. If in addition $\mathbf{H}_p$ exists, then we can construct $pm(v - 1)$ optimum $\mathbf{W}$-matrices. Below in Theorem 3.3.1 we provide non-trivial generalization of these results using mixed orthogonal arrays.

3.3 Relationship Between OCDs and MOAs

Orthogonal arrays (OA) introduced by Rao (1947) were generalized by Rao (1973) to Mixed orthogonal arrays (MOA) which have wide applications specially in the construction of designs. There are various results on constructions of OAs and MOAs. We refer to the books of Hedayat et al. (1999) and Dey and Mukerjee (1999) for details. Also a website of Sloane is available for ready reference and we also have a catalogue of potential sources on OAs and MOAs (cf. http://neilsloane.com/oadir/index.html). Definition of MOA (cf. Hedayat et al. (1999), p. 200) is given below:

Definition 3.3.1 An MOA($N, s_1^{k_1} s_2^{k_2} \ldots s_v^{k_v}, t$) is an array of size $k \times N$, where $k = \sum_{i=1}^{v} k_i$ is the total number of factors, in which the first k_1 rows have symbols from $\{0, 1, \ldots, s_1 - 1\}$, the next k_2 rows have symbols from $\{0, 1, \ldots, s_2 - 1\}$, and so on, with the property that in any $t \times N$ sub-array every t-tuple occurs an equal number of times as a column.

Rao et al. (2003) identifies the construction of OCDs to that of MOAs. In this chapter, we will discuss the relationship between the OCDs in the set-ups of CRD, RBD and MOAs. This was established in Rao et al. (2003).

We consider the following theorem given in Rao et al. (2003).

Theorem 3.3.1 *A set of c optimum $\mathbf{W}$-matrices of order $v \times b$ under the RBD set-up co-exist with an MOA($vb, v \times b \times 2^c, 2$).*

Proof For $i = 1, 2, \ldots, c$, let $\mathbf{W}^{(i)}$ matrix be written as

$$\mathbf{W}^{(i)\, v \times b} = \left(\mathbf{w}_1^{(i)}, \mathbf{w}_2^{(i)}, \ldots, \mathbf{w}_b^{(i)} \right)$$

where $\mathbf{w}_j^{(i)}$ is the jth column of $\mathbf{W}^{(i)}$. Now we consider an array $\mathbf{A}$ with $2 + c$ rows and vb columns where the first two rows of $\mathbf{A}$ form the following $2 \times vb$ sub-array

$$\begin{matrix} 1\,2 \ldots v\, 1\,2 \ldots v \ldots 1\,2 \ldots v \\ 1\,1 \ldots 1\,2\,2 \ldots 2 \ldots b\,b \ldots b \end{matrix} \qquad (3.3.1)$$

corresponding to the vb level combinations of the treatment and block factors and the $(2+i)$th row of $\mathbf{A}$ is given by $\left(\mathbf{w}_1^{(i)\prime}, \mathbf{w}_2^{(i)\prime}, \ldots, \mathbf{w}_b^{(i)\prime}\right), i = 1, 2, \ldots, c$. Note that first row and second row of $\mathbf{A}$ have v and b symbols respectively and the remaining rows have two symbols $+1$ and -1. From the properties of optimum $\mathbf{W}$-matrices it can be easily proved that $\mathbf{A}$ is an MOA(vb, $v \times b \times 2^c$, 2).

Conversely, given any MOA (vb, $v \times b \times 2^c$, 2), we can take, without loss of generality, the first two rows in the form (3.3.1).

We construct $\mathbf{W}^{(i)}$-matrix by using the elements of $(i+2)$th row of $\mathbf{A}$, where $w_{m,m'}^{(i)}$, the (m, m')th element of $\mathbf{W}^{(i)}$ = the element in the $(i+2)$th row of $\mathbf{A}$ corresponding to the ordered pair (m, m') in the first and second rows of $\mathbf{A}$, $m \neq m' = 1, 2, \ldots, c$. $\qquad \square$

Corollary 3.3.1 *A set of c optimum $\mathbf{W}$-matrices of order $v \times b$ under the CRD set-up co-exist with an MOA(vb, $v \times 2^c$, 2).*

Proof Given a set of c optimum $\mathbf{W}$-matrices $\mathbf{W}^{(1)}, \mathbf{W}^{(2)}, \ldots, \mathbf{W}^{(c)}$ of order $v \times b$ under the CRD set-up, observe that in this situation, the column sums, corresponding to the blocks, of the $\mathbf{W}$-matrices need not be zero. Hence the array $\mathbf{A}$ in the above result without the second row can be shown to be a mixed orthogonal array MOA(vb, $v \times 2^c$, 2). $\qquad \square$

Remark 3.3.1 Theorem 3.3.1 and Corollary 3.3.1 help us to construct OCDs for the set-ups of CRDs and RBDs from the list of suitable orthogonal arrays.

3.4 Some Further Constructions of Optimum W-Matrices

In this subsection we exploit the properties of Hadamard matrices and conference matrices to construct OCDs in CRD and RBD set-ups.

Theorem 3.4.1 *If there exist $\mathbf{H}_b$ and $\mathbf{H}_v$, then $(b-1)(v-1)$ optimum $\mathbf{W}$-matrices can be constructed for an RBD with b blocks and v treatments.*

Proof Write

$$\mathbf{H}_v = \left(\mathbf{h}_1, \mathbf{h}_2, \ldots, \mathbf{h}_{v-1}, \mathbf{1}\right) \tag{3.4.1}$$

and

$$\mathbf{H}_b = \left(\mathbf{h}_1^*, \mathbf{h}_2^*, \ldots, \mathbf{h}_{b-1}^*, \mathbf{1}\right). \tag{3.4.2}$$

Let us write

$$\mathbf{W}^{((b-1)(i-1)+j)} = \mathbf{h}_i \otimes \mathbf{h}_j^{*\prime}, \quad i = 1, 2, \ldots, v-1, \quad j = 1, 2, \ldots, b-1. \tag{3.4.3}$$

We can easily check that these $\mathbf{W}$-matrices satisfy conditions $C_1 - C_3$ of (3.1.5) giving $c = (b-1)(v-1)$ OCDs. These $\mathbf{W}$s exhaust the error degrees of freedom of the RBD.   □

Remark 3.4.1 We note that, by Theorems 3.3.1 and 3.4.1 we can construct an MOA $(vb, \ v \times b \times 2^{(v-1)(b-1)}, \ 2)$ from these $(v-1)(b-1)$ optimum $\mathbf{W}$-matrices and conversely for given this MOA$(vb, \ v \times b \times 2^{(v-1)(b-1)}, \ 2)$, we can also construct $(b-1)(v-1)$ optimum $\mathbf{W}$-matrices, $\mathbf{H}_v$ and $\mathbf{H}_b$.

Corollary 3.4.1 *If there exist* $\mathbf{H}_b$ *and* $\mathbf{H}_v$, *then* $v(b-1)$ *optimum* $\mathbf{W}$-*matrices can be constructed for an CRD with* v *treatments, each being replicated* b *times.*

Proof The matrices defined in Eq. (3.4.3) can also be treated as optimum $\mathbf{W}$-matrices of CRD set-up considered in Corollary 3.4.1. In this situation, we can construct an additional number of $(b-1)$ optimum $\mathbf{W}$-matrices for this CRD set-up given by

$$\mathbf{W}^{((b-1)(v-1)+j)} = \mathbf{1}_v \otimes \mathbf{h}_j^{*\prime}, \quad j = 1, 2, \ldots, b-1. \tag{3.4.4}$$

Thus in total, we get $v(b-1)$ optimum $\mathbf{W}$-matrices for this CRD set-up. These exhaust the error degrees of freedom of the CRD. As stated Corollary 3.3.1 an MOA$(vb, \ v \times 2^c, \ 2)$ can be constructed from the above $\mathbf{W}$-matrices in usual way.   □

Example 3.4.1 Let $b = v = 4$. Consider $\mathbf{H}_4$ of (3.2.3). From (3.4.3), we can construct optimum $\mathbf{W}$-matrices as follows:

$$\mathbf{W}^{(1)} = \mathbf{h}_1 \otimes \mathbf{h}_1' = \begin{pmatrix} 1 & -1 & 1 & -1 \\ -1 & 1 & -1 & 1 \\ 1 & -1 & 1 & -1 \\ -1 & 1 & -1 & 1 \end{pmatrix}; \quad \mathbf{W}^{(2)} = \mathbf{h}_1 \otimes \mathbf{h}_2' = \begin{pmatrix} 1 & -1 & -1 & 1 \\ -1 & 1 & 1 & -1 \\ 1 & -1 & -1 & 1 \\ -1 & 1 & 1 & -1 \end{pmatrix};$$

$$\mathbf{W}^{(3)} = \mathbf{h}_1 \otimes \mathbf{h}_3' = \begin{pmatrix} 1 & 1 & -1 & -1 \\ -1 & -1 & 1 & 1 \\ 1 & 1 & -1 & -1 \\ -1 & -1 & 1 & 1 \end{pmatrix}; \quad \mathbf{W}^{(4)} = \mathbf{h}_2 \otimes \mathbf{h}_1' = \begin{pmatrix} 1 & -1 & 1 & -1 \\ -1 & 1 & -1 & 1 \\ -1 & 1 & -1 & 1 \\ 1 & -1 & 1 & -1 \end{pmatrix};$$

$$\mathbf{W}^{(5)} = \mathbf{h}_2 \otimes \mathbf{h}_2' = \begin{pmatrix} 1 & -1 & -1 & 1 \\ -1 & 1 & 1 & -1 \\ -1 & 1 & 1 & -1 \\ 1 & -1 & -1 & 1 \end{pmatrix}; \quad \mathbf{W}^{(6)} = \mathbf{h}_2 \otimes \mathbf{h}_3' = \begin{pmatrix} 1 & 1 & -1 & -1 \\ -1 & -1 & 1 & 1 \\ -1 & -1 & 1 & 1 \\ 1 & 1 & -1 & -1 \end{pmatrix};$$

$$\mathbf{W}^{(7)} = \mathbf{h}_3 \otimes \mathbf{h}_1' = \begin{pmatrix} 1 & -1 & 1 & -1 \\ 1 & -1 & 1 & -1 \\ -1 & 1 & -1 & 1 \\ -1 & 1 & -1 & 1 \end{pmatrix}; \quad \mathbf{W}^{(8)} = \mathbf{h}_3 \otimes \mathbf{h}_2' = \begin{pmatrix} 1 & -1 & -1 & 1 \\ 1 & -1 & -1 & 1 \\ -1 & 1 & 1 & -1 \\ -1 & 1 & 1 & -1 \end{pmatrix};$$

$$\mathbf{W}^{(9)} = \mathbf{h}_3 \otimes \mathbf{h}_3' = \begin{pmatrix} 1 & 1 & -1 & -1 \\ 1 & 1 & -1 & -1 \\ -1 & -1 & 1 & 1 \\ -1 & -1 & 1 & 1 \end{pmatrix}.$$

Note that here $h_j^* = h_j$ for all j=1, 2, 3. Therefore, MOA(16, $4 \times 4 \times 2^9$, 2) can be constructed in the lines of Theorem 3.4.1:

$$\mathbf{A} = \begin{pmatrix} 1 & 2 & 3 & 4 & 1 & 2 & 3 & 4 & 1 & 2 & 3 & 4 & 1 & 2 & 3 & 4 \\ 1 & 1 & 1 & 1 & 2 & 2 & 2 & 2 & 3 & 3 & 3 & 3 & 4 & 4 & 4 & 4 \\ 1 & -1 & 1 & -1 & -1 & 1 & -1 & 1 & 1 & -1 & 1 & -1 & -1 & 1 & -1 & 1 \\ 1 & -1 & 1 & -1 & -1 & 1 & -1 & 1 & -1 & 1 & -1 & 1 & 1 & -1 & 1 & -1 \\ 1 & -1 & 1 & -1 & 1 & -1 & 1 & -1 & -1 & 1 & -1 & 1 & -1 & 1 & -1 & 1 \\ 1 & -1 & -1 & 1 & -1 & 1 & 1 & -1 & 1 & -1 & -1 & 1 & -1 & 1 & 1 & -1 \\ 1 & -1 & -1 & 1 & -1 & 1 & 1 & -1 & -1 & 1 & 1 & -1 & 1 & -1 & -1 & 1 \\ 1 & -1 & -1 & 1 & 1 & -1 & -1 & 1 & 1 & -1 & -1 & 1 & 1 & -1 & -1 & 1 \\ 1 & 1 & -1 & -1 & -1 & -1 & 1 & 1 & 1 & 1 & -1 & -1 & -1 & -1 & 1 & 1 \\ 1 & 1 & -1 & -1 & -1 & -1 & 1 & 1 & -1 & -1 & 1 & 1 & 1 & 1 & -1 & -1 \\ 1 & 1 & -1 & -1 & 1 & 1 & -1 & -1 & -1 & -1 & 1 & 1 & -1 & -1 & 1 & 1 \end{pmatrix}.$$

The above **W**-matrices are also optimum in CRD set-up with 4 treatments each being replicated 4 times. However as mentioned in Corollary 3.4.1 three additional **W**-matrices can be constructed and these are given below:

$$\mathbf{W}^{(10)} = \mathbf{1}_3 \otimes \mathbf{h}_1' = \begin{pmatrix} 1 & -1 & 1 & -1 \\ 1 & -1 & 1 & -1 \\ 1 & -1 & 1 & -1 \\ 1 & -1 & 1 & -1 \end{pmatrix}; \quad \mathbf{W}^{(11)} = \mathbf{1}_3 \otimes \mathbf{h}_2' = \begin{pmatrix} 1 & -1 & -1 & 1 \\ 1 & -1 & -1 & 1 \\ 1 & -1 & -1 & 1 \\ 1 & -1 & -1 & 1 \end{pmatrix};$$

$$\mathbf{W}^{(12)} = \mathbf{1}_3 \otimes \mathbf{h}_2' = \begin{pmatrix} 1 & 1 & -1 & -1 \\ 1 & 1 & -1 & -1 \\ 1 & 1 & -1 & -1 \\ 1 & 1 & -1 & -1 \end{pmatrix}.$$

The corresponding MOA(16, 4×2^{12}, 2) for the CRD set-up is given below:

$$\begin{pmatrix} 1 & 2 & 3 & 4 & 1 & 2 & 3 & 4 & 1 & 2 & 3 & 4 & 1 & 2 & 3 & 4 \\ 1 & -1 & 1 & -1 & -1 & 1 & -1 & 1 & 1 & -1 & 1 & -1 & -1 & 1 & -1 & 1 \\ 1 & -1 & 1 & -1 & -1 & 1 & -1 & 1 & -1 & 1 & -1 & 1 & 1 & -1 & 1 & -1 \\ 1 & -1 & 1 & -1 & 1 & -1 & 1 & -1 & -1 & 1 & -1 & 1 & -1 & 1 & -1 & 1 \\ 1 & -1 & -1 & 1 & -1 & 1 & 1 & -1 & 1 & -1 & -1 & 1 & -1 & 1 & 1 & -1 \\ 1 & -1 & -1 & 1 & -1 & 1 & 1 & -1 & -1 & 1 & 1 & -1 & 1 & -1 & -1 & 1 \\ 1 & -1 & -1 & 1 & 1 & -1 & -1 & 1 & -1 & 1 & 1 & -1 & -1 & 1 & 1 & -1 \\ 1 & 1 & -1 & -1 & -1 & -1 & 1 & 1 & 1 & 1 & -1 & -1 & -1 & -1 & 1 & 1 \\ 1 & 1 & -1 & -1 & -1 & -1 & 1 & 1 & -1 & -1 & 1 & 1 & 1 & 1 & -1 & -1 \\ 1 & 1 & -1 & -1 & 1 & 1 & -1 & -1 & -1 & -1 & 1 & 1 & -1 & -1 & 1 & 1 \\ 1 & 1 & 1 & 1 & -1 & -1 & -1 & -1 & 1 & 1 & 1 & 1 & -1 & -1 & -1 & -1 \\ 1 & 1 & 1 & 1 & -1 & -1 & -1 & -1 & -1 & -1 & -1 & -1 & 1 & 1 & 1 & 1 \\ 1 & 1 & 1 & 1 & 1 & 1 & 1 & 1 & -1 & -1 & -1 & -1 & -1 & -1 & -1 & -1 \end{pmatrix}.$$

Theorem 3.4.2

(a) *If* $\mathbf{H}_{2b}$ *and* $\mathbf{H}_{\frac{v}{2}}$ *both exist, then* $v(b-1)$ *optimum* **W***-matrices can be constructed for a CRD with* v *treatments and* b *replications.*

(b) *If* $\mathbf{H}_{2b}$ *and* $\mathbf{H}_{\frac{v}{2}}$ *both exist, then* $(b-1)(v-1) - (b-2)$ *optimum* **W***-matrices can be constructed for an RBD with* b *blocks and* v *treatments.*

Proof of (a) Write $\mathbf{H}_{2b}$ as a $2b \times 2b$ matrix with the last column as $(1, 1, \ldots, 1)'$ and the last but one column as $(\mathbf{1}'_b, -\mathbf{1}'_b)'$. Further write $\mathbf{H}_{\frac{v}{2}}$ as a matrix with the last column as $\mathbf{1}'_{\frac{v}{2}}$. Let $\mathbf{H}^{**}_{2b}$ be a matrix of order $2b \times 2(b-1)$ obtained from $\mathbf{H}_{2b}$ by deleting the last two columns. It follows that in each column of $\mathbf{H}^{**}_{2b}$ both the top b elements and the bottom b elements have equal number of 1's and -1's. Now we construct a matrix $\mathbf{A}_1$ of order $v(b-1) \times vb$ as:

$$\mathbf{A}_1 = \mathbf{H}^{**'}_{2b} \otimes \mathbf{H}_{\frac{v}{2}}.$$

We convert $\mathbf{A}_1$ into an MOA(vb, $v \times 2^{v(b-1)}$, 2) by appending the row: $(1, 2, \ldots, b, 1, 2, \ldots, b, \ldots, 1, 2, \ldots, b)$ of length vb. This establishes the result via Corollary 3.3.1.   □

Proof of (b) Let $\mathbf{H}^{*}_{\frac{v}{2}}$ be a matrix of order $\frac{v}{2} \times (\frac{v}{2} - 1)$ obtained from $\mathbf{H}_{\frac{v}{2}}$ ignoring the last column consisting of all 1s. Now we construct $\mathbf{A}_2$ of order $(v-2)(b-1) \times vb$ as follows:

$$\mathbf{A}_2 = \mathbf{H}^{**'}_{2b} \otimes \mathbf{H}^{*'}_{\frac{v}{2}}.$$

From $\mathbf{A}_2$, we can construct an MOA(vb, $v \times b \times 2^{(v-2)(b-1)+1}$, 2) by adjoining three more rows; the first two rows are used for coordinatisation and third row is $\mathbf{1}'_{\frac{v}{2}} \otimes (1, -1) \otimes (-\mathbf{1}'_{\frac{b}{2}}, \mathbf{1}'_{\frac{b}{2}})$. The proof follows from the method of construction of Theorem 3.3.1.   □

Remark 3.4.2 Theorem 3.4.2(b) strengthens and generalises Theorem 4.3.4, p. 54 in Dey and Mukerjee (1999).

Example 3.4.2 Let $b = 6$, $v = 4$. We take $\mathbf{H}_{12}$ in accordance with the proof of Theorem 3.4.2:

$$\mathbf{H}_{12} = \begin{pmatrix} 1 & -1 & 1 & -1 & -1 & -1 & 1 & 1 & 1 & -1 & 1 & 1 \\ -1 & 1 & 1 & -1 & 1 & -1 & -1 & -1 & 1 & 1 & 1 & 1 \\ 1 & -1 & 1 & 1 & -1 & 1 & -1 & -1 & -1 & 1 & 1 & 1 \\ 1 & 1 & -1 & 1 & 1 & -1 & 1 & -1 & -1 & -1 & 1 & 1 \\ -1 & -1 & -1 & 1 & 1 & 1 & -1 & 1 & 1 & -1 & 1 & 1 \\ -1 & 1 & -1 & -1 & -1 & 1 & 1 & 1 & -1 & 1 & 1 & 1 \\ -1 & -1 & -1 & -1 & -1 & -1 & -1 & -1 & -1 & -1 & -1 & 1 \\ 1 & 1 & -1 & 1 & -1 & -1 & -1 & 1 & 1 & 1 & -1 & 1 \\ 1 & 1 & 1 & -1 & 1 & 1 & -1 & 1 & -1 & -1 & -1 & 1 \\ -1 & 1 & 1 & 1 & -1 & 1 & 1 & -1 & 1 & -1 & -1 & 1 \\ -1 & -1 & 1 & 1 & 1 & -1 & 1 & 1 & -1 & 1 & -1 & 1 \\ 1 & -1 & -1 & -1 & 1 & 1 & 1 & -1 & 1 & 1 & -1 & 1 \end{pmatrix} = (\mathbf{H}^{**}_{12}, \mathbf{h}_{11}, \mathbf{1}),$$

and

$$H_2 = \begin{pmatrix} 1 & 1 \\ -1 & 1 \end{pmatrix} = (H_2^*, 1).$$

Therefore using $A_1 = H_{12}^{**'} \otimes H_2$, we can construct MOA(24, 4×2^{20}, 2) as described in Corollary 3.3.1. Using $A_2 = H_{12}^{**'} \otimes H_2^{*'}$ and the row $(1, 1) \otimes (1, -1) \otimes (-1, -1, -1, 1, 1, 1) = (-1, -1, -1, 1, 1, 1, 1, 1, 1, -1, -1, -1, -1, -1, -1, 1, 1, 1, 1, 1, -1, -1, -1)$, we can construct MOA(24, $4 \times 4 \times 2^{11}$, 2) as described in Theorem 3.3.1. Now it is routine task to construct optimum W-matrices for CRD and RBD from these MOAs.

For further construction of MOAs we need the concept of Conference Matrices which is introduced below (cf. Hedayat et al. (1999), p. 152).

Definition 3.4.1 A symmetric matrix S of order n with elements $+1, -1$ and 0 is said to be a conference matrix (CM) if it can expressed in the form

$$S = \begin{pmatrix} 0 & 1'_{n-1} \\ 1_{n-1} & A \end{pmatrix} \tag{3.4.5}$$

satisfying $SS' = (n-1)I_n$.

In such a representation of S the matrix A in (3.4.5) is called the core matrix of the CM. It can be easily checked that this A satisfies the conditions

$$AA' = (n-1)I_{n-1} - 1_{n-1}1'_{n-1}, \quad A = A' \text{ and } A1_{n-1} = 0.$$

Note that CMs are known to exist for the following values of n (cf. Wallis et al. (1972)):

(1) $n = p^s + 1$ where p is a prime and s is a positive integer such that $p^s \equiv 1 \pmod 4$.
(2) $n = (h-1)^2 + 1$ where h is the order of a Skew-Hadamard matrix.
(3) $n = (h-1)^\mu + 1$ where h is the order of a CM and $\mu > 0$ is an odd integer.

Set $n - 1 = p$ and let A be as in (3.4.5) and let A^* be the matrix of order $p^2 \times p^2$ obtained by taking the Kronecker product of A with itself. Define a matrix X as

$$X = A^* + \begin{pmatrix} J_p - I_p & -I_p & \cdots & -I_p \\ -I_p & J_p - I_p & \cdots & -I_p \\ \vdots & \vdots & \vdots & \vdots \\ -I_p & -I_p & \cdots & J_p - I_p \end{pmatrix}$$

where each block is of order $p \times p$ and $J_p = 1_p 1'_p$.

Theorem 3.4.3 X is a core matrix of a CM of order p^2.

The proof is given in the Appendix. In the following theorem we give a method of constructing OCDs from CMs.

Theorem 3.4.4

(a) *If $b \equiv 2 \pmod 4$, $(b - 1)$ is a prime or a prime power and $\mathbf{H}_v$ exists, then $c = v(b - 1)$ optimum $\mathbf{W}$-matrices can be constructed for a CRD with v treatments and b replications.*

(b) *If $b \equiv 2 \pmod 4$, $(b - 1)$ is a prime or a prime power and $\mathbf{H}_v$ exists, then $c = (b - 1)(v - 1) - (b - 2)$ optimum $\mathbf{W}$-matrices can be constructed for an RBD with b blocks and v treatments.*

Proof We will construct OCDs through the following steps.

Step I: We start with $\mathbf{S}$, a CM of order $(p + 1) \times (p + 1)$ where $\mathbf{A} = (a_{ij})$ be the core matrix of $\mathbf{S}$ of order $p \times p$.

Step II: Define a matrix $\mathbf{B}$ of order $(p + 1) \times p$ such that the (i, j)th element b_{ij} is given by

$$
\begin{aligned}
b_{ii} &= -\beta \quad \text{for } i = 1, 2, \ldots, p, \\
b_{ij} &= a_{ij}\alpha \text{ for } i, j = 1, 2, \ldots, p, \text{ and } i \neq j \\
b_{(p+1),j} &= \beta \quad \text{for } j = 1, 2, \ldots, p,
\end{aligned}
$$

where α and β are elements satisfying $1.\alpha = \alpha$; $-1.\alpha = -\alpha$; $1.\beta = \beta$; $-1.\beta = -\beta$; $\alpha.\alpha = (-\alpha).(-\alpha) = \beta.\beta = (-\beta).(-\beta) = c$ a constant; $\alpha.\beta = \beta.\alpha = -(\alpha).\beta = (-\beta).\alpha = \alpha.(-\beta) = \beta.(-\alpha) = 0$, $\alpha.(-\alpha) = (-\alpha).\alpha = \beta.(-\beta) = (-\beta).\beta = -c$.

Define another a matrix $\mathbf{C}$ of order $(p + 1) \times p$ such that the (i, j)th element c_{ij} is given by

$$
\begin{aligned}
c_{ii} &= \alpha \quad \text{for } i = 1, 2, \ldots, p, \\
c_{ij} &= a_{ij}\beta \text{ for } i, j = 1, 2, \ldots, p, \text{ and } i \neq j \\
c_{(p+1),j} &= -\alpha \quad \text{for } j = 1, 2, \ldots, p.
\end{aligned}
$$

Now we construct a matrix $\mathbf{D}$ of order $(p + 1) \times 2p$ as

$$\mathbf{D} = (\mathbf{B} : \mathbf{C}).$$

It is observed that the columns of the matrix $\mathbf{D}$ of order $(p + 1) \times 2p$ are orthogonal (cf. Theorem 3.4.5 in the Appendix).

Example 3.4.3 For $n = 6$, let the core matrix $\mathbf{A}$ be

$$
\begin{pmatrix}
0 & 1 & -1 & -1 & 1 \\
1 & 0 & 1 & -1 & -1 \\
-1 & 1 & 0 & 1 & -1 \\
-1 & -1 & 1 & 0 & 1 \\
1 & -1 & -1 & 1 & 0
\end{pmatrix}.
$$

From the definitions of **B**, **C** and **D**, we have

$$\mathbf{D} = (\mathbf{B} : \mathbf{C}) = \begin{pmatrix} -\beta & \alpha & -\alpha & -\alpha & \alpha & \alpha & \beta & -\beta & -\beta & \beta \\ \alpha & -\beta & \alpha & -\alpha & -\alpha & \beta & \alpha & \beta & -\beta & -\beta \\ -\alpha & \alpha & -\beta & \alpha & -\alpha & -\beta & \beta & \alpha & \beta & -\beta \\ -\alpha & -\alpha & \alpha & -\beta & \alpha & -\beta & -\beta & \beta & \alpha & \beta \\ \alpha & -\alpha & -\alpha & \alpha & -\beta & \beta & -\beta & -\beta & \beta & \alpha \\ \beta & \beta & \beta & \beta & \beta & -\alpha & -\alpha & -\alpha & -\alpha & -\alpha \end{pmatrix}. \qquad (3.4.6)$$

Step III: By assumption, $\mathbf{H}_v$ exists and we write it as

$$\mathbf{H}_v = (\mathbf{h}_1,\ \mathbf{h}_2, \ldots, \mathbf{h}_{v-1},\ \mathbf{h}_v = \mathbf{1}_v). \qquad (3.4.7)$$

Take one pair, say $(\mathbf{h}_i, \mathbf{h}_j)$ and replace the two symbols α, β by $\mathbf{h}'_i$, $\mathbf{h}'_j$ respectively in the matrix $\mathbf{D}$ of order $b \times 2(b-1)$. Then each column of $\mathbf{D}$ will give a matrix of order $b \times v$ and so we can get $2(b-1)$ matrices using all the columns of $\mathbf{D}$ for the fixed pair $(\mathbf{h}_i, \mathbf{h}_j)$. Now using ith column of $\mathbf{D}$ and jth pair $(\mathbf{h}_{2j-1}, \mathbf{h}_{2j})$ of columns of $\mathbf{H}_v$, we get a matrix of order $b \times v$ which is denoted by $\mathbf{U}^{((i-1)v/2+j)}$. Now varying i over $1, 2, \ldots, 2(b-1)$ and j over $1, 2, \ldots, \frac{v}{2}$, we get $v(b-1)$ matrices $\mathbf{U}^{(1)}, \mathbf{U}^{(2)}, \ldots, \mathbf{U}^{(v(b-1))}$. We can easily check that $\mathbf{W}^{(1)} = \mathbf{U}^{(1)\,\prime}$, $\mathbf{W}^{(2)} = \mathbf{U}^{(2)\,\prime}, \ldots,$ $\mathbf{W}^{(v(b-1))} = \mathbf{U}^{(v(b-1))\,\prime}$ are $v(b-1)$ optimum **W**-matrices for CRD set-up.

However in RBD set-up, we cannot use the last column $\mathbf{h}_v$ as the sum of elements of the last column is not zero. So leaving it out we have only $\frac{v-2}{2}$ distinct pairs of columns $(\mathbf{h}_1, \mathbf{h}_2, \ldots, \mathbf{h}_{v-2})$ and an extra column, $\mathbf{h}_{v-1}$. By using these distinct pairs of columns we can construct $(b-1)(v-2)$ **W**-matrices from $\mathbf{D}$ in the same manner as described in above. Here we can also construct one more optimum **W**-matrix using the residual column $\mathbf{h}_{v-1}$ as

$$\mathbf{W}^{((b-1)(v-2)+1)\,\prime} = \begin{pmatrix} \mathbf{1}_{\frac{b}{2}} \\ -\mathbf{1}_{\frac{b}{2}} \end{pmatrix} \otimes \mathbf{h}'_{v-1}.$$

Therefore for RBD set-up, we can construct $(b-1)(v-2)+1$ optimum **W**-matrices in all. $\qquad\square$

Now we illustrate the above method by considering the following example.

Example 3.4.4 Let $b = 6$ and $v = 4$. Then $\mathbf{H}_4$ is

$$\mathbf{H}_4 = \begin{pmatrix} 1 & 1 & 1 & 1 \\ -1 & -1 & 1 & 1 \\ 1 & -1 & -1 & 1 \\ -1 & 1 & -1 & 1 \end{pmatrix} = (\mathbf{h}_1,\ \mathbf{h}_2,\ \mathbf{h}_3,\ \mathbf{h}_4 = \mathbf{1}_4).$$

Then take $\alpha = \mathbf{h}'_1$ and $\beta = \mathbf{h}'_2$ and using the first column of $\mathbf{D}$ of (3.4.6), we get the following **W**-matrix:

$$\begin{pmatrix} -1 & 1 & 1 & -1 \\ 1 & -1 & 1 & -1 \\ -1 & 1 & -1 & 1 \\ -1 & 1 & -1 & 1 \\ 1 & -1 & 1 & -1 \\ 1 & -1 & -1 & 1 \end{pmatrix}.$$

Similarly, using the above methods we get other **W**-matrices for CRD and RBD.

Appendix

Proof of Theorem 3.4.3 It is observed that **X** is a symmetric matrix. Now (i, i)th block matrix of **XX**$'$ is

$$(\mathbf{XX}')_{ii} = \sum_{k=1,\ k\neq i}^{p} (a_{ik}\mathbf{A} - \mathbf{I}_p)(a_{ki}\mathbf{A} - \mathbf{I}_p) + (a_{ii}\mathbf{A} + \mathbf{J}_p - \mathbf{I}_p)(a_{ii}\mathbf{A} + \mathbf{J}_p - \mathbf{I}_p)$$

$$= \sum_{k=1\ k\neq i}^{p} (a_{ik}^2\mathbf{A}^2 - 2a_{ik}\mathbf{A} + \mathbf{I}) + (\mathbf{J}_p - \mathbf{I}_p)(\mathbf{J}_p - \mathbf{I}_p) \text{ since } a_{ij} = a_{ji} \text{ and } a_{ii} = 0 \ \forall i$$

$$= (p\mathbf{I}_p - \mathbf{J}_p)\sum_{k=1}^{p} a_{ik}^2 - 2\mathbf{A}\sum_{k=1}^{p} a_{ik} + (p-1)\mathbf{I}_p + p\mathbf{J}_p - 2\mathbf{J}_p + \mathbf{I}_p$$

$$= (p\mathbf{I}_p - \mathbf{J}_p)(p-1) - 2\mathbf{A}.0 + (p-1)\mathbf{I}_p + (p-2)\mathbf{J}_p + \mathbf{I}_p$$

$$= (p(p-1) + p)\mathbf{I}_p + (p-2)\mathbf{J}_p - (p-1)\mathbf{J}_p$$

$$= p^2\mathbf{I}_p - \mathbf{J}_p$$

(i, j)th block matrix of **XX**$'$ is

$$(\mathbf{XX}')_{ij} = \sum_{k=1,\ k\neq i,j}^{p} (a_{ik}\mathbf{A} - \mathbf{I}_p)(a_{kj}\mathbf{A} - \mathbf{I}_p) + (a_{ii}\mathbf{A} + \mathbf{J}_p - \mathbf{I}_p)(a_{ji}\mathbf{A} - \mathbf{I}_p)$$

$$+ (a_{ij}\mathbf{A} - \mathbf{I}_p)(a_{jj}\mathbf{A} + \mathbf{J}_p - \mathbf{I}_p)$$

$$= \sum_{k=1,\ k\neq i,j}^{p} (a_{ik}a_{jk}\mathbf{A}^2 - a_{ik}\mathbf{A} - a_{jk}\mathbf{A} + \mathbf{I}_p) + (\mathbf{J}_p - \mathbf{I}_p)(a_{ji}\mathbf{A} - \mathbf{I}_p)$$

$$+ (a_{ij}\mathbf{A} - \mathbf{I}_p)(\mathbf{J}_p - \mathbf{I}_p)$$

$$= -(p\mathbf{I}_p - \mathbf{J}_p) + a_{ij}\mathbf{A} + a_{ij}\mathbf{A} + (p-2)\mathbf{I}_p - \mathbf{J}_p - a_{ij}\mathbf{A} - \mathbf{I}_p - \mathbf{J}_p - a_{ij}\mathbf{A} + \mathbf{I}_p$$

$$= -\mathbf{J}_p$$

Thus **XX**$' = p^2\mathbf{I}_{p^2} - \mathbf{J}_{p^2}$. $\square$

Theorem 3.4.5 *The columns of the matrix* **D** *are orthogonal.*

Proof The cross product of ith and jth elements of **B**,

$$\sum_{k=1}^{p} b_{ki}b_{kj} = \sum_{k=1,\ k\neq i,j}^{p} b_{ki}b_{kj} + b_{ii}b_{ij} + b_{ji}b_{jj} + b_{(p+1),i}b_{(p+1),j}$$

$$= \sum_{k=1,\ k\neq i,j}^{p} (a_{ki}\alpha)(a_{kj}\alpha) + (-\beta)(a_{ij}\alpha) + (a_{ji}\alpha)(-\beta) + \beta.\beta$$

$$= \sum_{k=1,\ k\neq i,j}^{p} a_{ki}a_{kj}(\alpha.\alpha) + a_{ij}((-\beta).\alpha) + a_{ji}(\alpha.(-\beta)) + c$$

$$= c\sum_{k=1}^{p} a_{ki}a_{kj} + 0 + 0 + c$$

$$= -c + c = 0.$$

Similarly, it can be shown that the columns of $\mathbf{C}$ are also orthogonal. Now we want to show that any column of $\mathbf{B}$ is orthogonal to any column of $\mathbf{C}$. For this, we consider the cross product of ith column of $\mathbf{B}$ and jth column of $\mathbf{C}$:

$$\sum_{k=1}^{p} b_{ki}c_{kj} = \sum_{k=1,\ k\neq i,j}^{p} b_{ki}c_{kj} + b_{ii}c_{ij} + b_{ji}c_{jj} + b_{(p+1),i}c_{(p+1),j}$$

$$= \sum_{k=1,\ k\neq i,j}^{p} (a_{ki}\alpha)(a_{kj}\beta) + (-\beta)(a_{ij}\beta) + (a_{ji}\alpha)(\alpha) + \beta.(-\alpha)$$

$$= \sum_{k=1,\ k\neq i,j}^{p} a_{ki}b_{kj}(\alpha.\beta) + a_{ij}((-\beta).\beta) + a_{ji}(\alpha.(-\alpha)) + 0 = 0 - c + c + 0 = 0$$

$\square$

References

Das K, Mandal NK, Sinha BK (2003) Optimal experimental designs with covariates. J Stat Plan Inference 115:273–285

Dey A, Mukerjee R (1999) Fractional factorial plans. Wiley, Hoboken

Hedayat AS, Sloane NJA, Stufken J (1999) Orthogonal arrays: theory and applications. Springer, New York

Rao CR (1947) Factorial experiments derivable from combinatorial arrangements of arrays. J R Stat Soc (Suppl) 9:128–139

Rao CR (1973) Some combinatorial problems of arrays and applications to design of experiments. In: Srivastava JN (ed) A survey of combinatorial theory. University of North Carolina Press, Chapel Hill, pp 349–359

Rao PSSNVP, Rao SB, Saha GM, Sinha BK (2003) Optimal designs for covariates' models and mixed orthogonal arrays. Electron Notes Discret Math 15:157–160

Troya Lopes J (1982a) Optimal designs for covariate models. J Stat Plan Inference 6:373–419

Troya Lopes J (1982b) Cyclic designs for a covariate model. J Stat Plan Inference 7:49–75

Wallis WD, Street AP, Wallis JS (1972) Combinatorics: room squares, sum-free sets, Hadamard matrices. Lecture notes in mathematics, vol 292. Springer, New York

Chapter 4
OCDs in Balanced Incomplete Block Design Set-Up

4.1 Introduction

A balanced incomplete block design (BIBD) as an arrangement of v treatments into b blocks each of k ($< v$) treatments, satisfying the conditions:

1. Every symbol occurs at most once in each block.
2. Every treatment occurs in exactly r blocks.
3. Every pair of symbols occurs together in exactly λ blocks.

Let us consider a BIBD (b, v, r, k, lambda) satisfying (3.1.3) and (3.1.4) where $\mathbf{X}_{2d}$ has a similar structure as in RBD. But now the structure of $\mathbf{X}_{1d}$ is somewhat different. In the $\mathbf{W}$-matrix corresponding to the incidence matrix of the said design the non-zero elements (± 1) appear only in the r positions in every row and the k positions in every column. So the situation is more complex than before in the sense that in the case of an RBD, we were to place ± 1's in all the vb cells of the $\mathbf{W}$-matrices while here, we have to place ± 1 in the non-zero cells of the incidence matrix $\mathbf{N}^{v \times b}$. Thus, the construction of optimum $\mathbf{W}$-matrix or equivalently the $\mathbf{Z}$-matrix depends on the method of construction of the corresponding BIBD.

The elements of optimum $\mathbf{W}$-matrices for a BIBD set-up should satisfy following conditions

$$
\left.
\begin{array}{ll}
\displaystyle\sum_{i=1}^{v} w_{ij}^{(s)} = 0 \; \forall j; & \displaystyle\sum_{j=1}^{b} w_{ij}^{(s)} = 0 \; \forall i \\
\text{and} & \\
\displaystyle\sum_{i=1}^{v}\sum_{j=1}^{b} w_{ij}^{(s)} w_{ij}^{(s')} = 0 \; \forall s \neq s'.
\end{array}
\right\}
\tag{4.1.1}
$$

© Springer India 2015
P. Das et al., *Optimal Covariate Designs*, DOI 10.1007/978-81-322-2461-7_4

Condition (4.1.1) can be presented schematically as

$$
\mathbf{W}^{(s)} =
\begin{array}{c}
\text{Tr.} \quad \text{bl. no.} \rightarrow \text{Row} \\
\downarrow \quad 1\,2\,\ldots\,b \quad \text{Totals} \\
\begin{array}{c|c|c}
1 & & 0 \\
2 & & 0 \\
\vdots & (\pm n_{ij}) & \vdots \\
v & & 0 \\
\end{array}
\end{array}
\quad 1 \le s \le c \qquad (4.1.2)
$$

Column
Total   0 0 ... 0

and

$$
\mathbf{W}^{(s)v \times b} * \mathbf{W}^{(t)v \times b} =
\begin{array}{c}
\text{Tr.} \quad \text{bl. no.} \rightarrow \\
\downarrow \quad 1\,2\,\ldots\,b \\
\begin{array}{c|c}
1 & \\
2 & \\
\vdots & (\pm n_{ij}) \\
v & \\
\end{array}
\end{array}
\qquad (4.1.3)
$$

$$
0 = \sum_{i=1}^{b} \sum_{j=1}^{v} w_{ij}^{(s)} w_{ij}^{(s')}
$$

In the set-up of a binary design, we use the notations n_{ij}s to indicate the incidence pattern of the treatments across different blocks. Naturally, $n_{ij} = 1$ or 0 according as the $(i, \ j)$ combination is present or absent. When $n_{ij} = 1$, we need to ascertain the value of w_{ij} (+1 or −1). It is observed that the condition (4.1.1) does not require any other property of BIBD other than that it is proper, binary and equireplicate. So it follows that the above principle of constructing optimum $\mathbf{W}$-matrices equally applies for any binary proper equireplicate block design (BPEBD) not necessarily for the BIBD set-up only. So in this context we shall consider also BPEBD set-up whenever necessary.

In a BIBD set-up, $\mathbf{W}$-matrices of order $v \times b$ can be constructed from the incidence matrix of the BIBD by placing ± 1's in the non-zero r positions in every row and in the non-zero k positions in every column such that $\mathbf{W}$-matrices satisfy condition (4.1.1). For the BIBD set-up, Das et al. (2003) initiated the construction of OCDs in the following experimental designs:

(i) some series of SBIBDs constructed through Bose's difference technique (cf. Bose 1939),
(ii) some BIBDs with repeated blocks.

Later Dutta (2004) dealt with a number of classical series of BIBDs having incidence matrices derived essentially through Bose's difference technique which was not covered in Das et al. (2003). With this, he covered a large class of existing

BIBDs. Again Dutta et al. (2007) considered the problem of OCDs for a series of complements of SBIBDs obtained through projective geometry.

It may be mentioned that in the series considered in Sect. 4.2, the layouts have cyclical pattern which simplified the choice of $\mathbf{W}$-matrices. But the series of SBIBDs considered in Sect. 4.3 does not possess the above cyclical property.

When $n \neq 0 \pmod 4$, it is impossible to find designs attaining minimum variance for estimated covariate parameters. Dutta et al. (2010) considered this problem and instead of using the criterion of attaining the lower bound (viz. $\frac{\sigma^2}{n}$) to the variance of each of the estimated covariate parameters γ, they found optimum designs with respect to covariate effects using D-optimality criterion retaining orthogonality with respect to treatment and block effect contrasts, where $n \equiv 2 \pmod 4$. We consider their work in Sect. 4.4.

4.2 BIBDs Through Bose's Difference Technique

In this section, we consider some series of BIBDs constructed by applying Bose's difference technique (Bose 1939) and present construction procedures given by Das et al. (2003) and Dutta (2004) for $\mathbf{W}$-matrices satisfying (4.1.2) and (4.1.3).

Theorem 4.2.1 *Suppose a SBIBD* $(v = b, \ r = k, \ \lambda)$ *is obtained by applying Bose's difference technique and a Hadamard matrix* $\mathbf{H}_k$ *of order k exists. Then* $(k - 1)$ *optimum* $\mathbf{W}$*-matrices can be constructed.*

Proof $\mathbf{H}_k$ exists by assumption and it can be represented as

$$\mathbf{H}_k = (\mathbf{h}_1, \ \mathbf{h}_2, \ldots, \mathbf{h}_{k-1}, \ \mathbf{1}). \tag{4.2.1}$$

Without loss of generality take the initial block of SBIBD as the first block and transform it into the form of the first column vector of the incidence matrix. Then we replace the non-zero positions of this column vector successively by the elements of $\mathbf{h}_t$. This gives the first column vector of $\mathbf{W}^{(t)}$. Now we develop this column into the full form of $\mathbf{W}^{(t)}$ cyclically. If the above method is carried out for each of the vectors $\mathbf{h}_1, \ \mathbf{h}_2, \ldots, \mathbf{h}_{k-1}$, then we get $(k - 1)$ $\mathbf{W}$-matrices. We can easily check that these $\mathbf{W}$-matrices satisfy condition (4.1.1) and are optimum. $\qquad\square$

Example 4.2.1 Consider SBIBD $(7, \ 4, \ 2)$ obtained by cyclical development of the initial block $(0, 3, 5, 6)$ mod 7. Note that the first column of the incidence matrix is given by $\begin{pmatrix} 1 & 0 & 0 & 1 & 0 & 1 & 1 \end{pmatrix}'$ and others are obtained by cyclic permutations of this column. As block size is 4 we consider the 3 columns of $\mathbf{H}_4$ viz. $\mathbf{h}_1' = \begin{pmatrix} 1 & -1 & 1 & -1 \end{pmatrix}$, $\mathbf{h}_2' = \begin{pmatrix} 1 & 1 & -1 & -1 \end{pmatrix}$ and $\mathbf{h}_3' = \begin{pmatrix} 1 & -1 & -1 & 1 \end{pmatrix}$ excluding $\begin{pmatrix} 1 & 1 & 1 & 1 \end{pmatrix}'$. Let us consider $\mathbf{h}_1$ and construct $\mathbf{W}^{(1)}$ by replacing the non-zero elements of the first column of $\mathbf{N}$ by the elements of $\mathbf{h}_1$ in that order and permute cyclically. $\mathbf{W}^{(1)}$ is given by

$$\mathbf{W}^{(1)} = \begin{pmatrix} 1 & -1 & 1 & 0 & -1 & 0 & 0 \\ 0 & 1 & -1 & 1 & 0 & -1 & 0 \\ 0 & 0 & 1 & -1 & 1 & 0 & -1 \\ -1 & 0 & 0 & 1 & -1 & 1 & 0 \\ 0 & -1 & 0 & 0 & 1 & -1 & 1 \\ 1 & 0 & -1 & 0 & 0 & 1 & -1 \\ -1 & 1 & 0 & -1 & 0 & 0 & 1 \end{pmatrix}^{7 \times 7}$$

and the corresponding column of $\mathbf{Z}$ is $(1, -1, 1, -1, -1, 1, -1, 1, 1, -1, 1, -1, 1,$ $-1, 1, -1, -1, 1, -1, 1, -1, 1, -1, 1, -1, 1, -1, 1)$.

Similarly, construct $\mathbf{W}^{(2)}$ and $\mathbf{W}^{(3)}$ by using $\mathbf{h}_2$ and $\mathbf{h}_3$ respectively and the corresponding columns of $\mathbf{Z}$ accordingly. It can be seen that all the conditions in (4.1.1) are satisfied by $\mathbf{W}$s. $\mathbf{Z}$ gives the OCD in the design format. Thus an OCD for three covariates is obtained.

If the blocks of such BIBD is repeated m times each where $\mathbf{H}_m$ exists then we can increase the number of covariates in the new BIBD with repeated blocks and the result is represented in following corollary.

Corollary 4.2.1 *Suppose an SBIBD (b, r, λ) is available as per the description in Theorem 4.2.1. Suppose further that $\mathbf{H}_m$ exists for some m. Then for the BIBD $(v, B = mb, R = mr, k, \Lambda = m\lambda)$ obtained by repeating the blocks of the SBIBD, we can construct $c^* = m(k - 1)$ optimum $\mathbf{W}$-matrices.*

Proof Let us write $\mathbf{H}_m$ as

$$\mathbf{H}_m = (\mathbf{h}_1^*, \mathbf{h}_2^*, \dots, \mathbf{h}_m^*) = (h_{rt}^*).$$

Denote the $\mathbf{W}^{(t)}$-matrices of Theorem 4.2.1 by $\mathbf{W}_{v \times b}^{(t)}$ and the required $\mathbf{W}$-matrices by $\mathbf{G}_{v \times B}$-matrices as follows:

$$\mathbf{G}_{v \times B}^{(t,r)} = \left(h_{1t}^* \mathbf{W}_{v \times b}^{(t)}, \; h_{2t}^* \mathbf{W}_{v \times b}^{(t)}, \dots, h_{mt}^* \mathbf{W}_{v \times b}^{(t)} \right) = \mathbf{h}_r^{*\prime} \otimes \mathbf{W}_{v \times b}^{(t)}. \quad (4.2.2)$$

It is now a routine task to verify the claim of the corollary. □

Example 4.2.2 Consider BIBD $(7, 28, 16, 4, 8)$ obtained by repeating 4 times each of 7 blocks of SBIBD $(7, 4, 2)$ of Example 4.2.1. $\mathbf{H}_4$ can be written as

$$\mathbf{H}_4 = \begin{pmatrix} 1 & 1 & 1 & 1 \\ -1 & -1 & 1 & 1 \\ 1 & -1 & -1 & 1 \\ -1 & 1 & -1 & 1 \end{pmatrix} = (\mathbf{h}_1^*, \; \mathbf{h}_2^*, \; \mathbf{h}_3^*, \; \mathbf{h}_4^*) = (h_{rt}^*).$$

Take $\mathbf{W}^{(1)}$ of Example 4.2.1 and the corresponding $\mathbf{G}$-matrices are as follows:

$$G_{v \times B}^{(1,1)} = \left(\mathbf{W}_{7 \times 7}^{(1)}, -\mathbf{W}_{7 \times 7}^{(1)}, \mathbf{W}_{7 \times 7}^{(1)}, -\mathbf{W}_{7 \times 7}^{(1)} \right);$$

$$G_{v \times B}^{(1,2)} = \left(\mathbf{W}_{7 \times 7}^{(1)}, -\mathbf{W}_{7 \times 7}^{(1)}, -\mathbf{W}_{7 \times 7}^{(1)}, \mathbf{W}_{7 \times 7}^{(1)} \right);$$

$$G_{v \times B}^{(1,3)} = \left(\mathbf{W}_{7 \times 7}^{(1)}, \mathbf{W}_{7 \times 7}^{(1)}, -\mathbf{W}_{7 \times 7}^{(1)}, -\mathbf{W}_{7 \times 7}^{(1)} \right);$$

$$G_{v \times B}^{(1,4)} = \left(\mathbf{W}_{7 \times 7}^{(1)}, \mathbf{W}_{7 \times 7}^{(1)}, \mathbf{W}_{7 \times 7}^{(1)}, \mathbf{W}_{7 \times 7}^{(1)} \right).$$

Similarly, we construct other **G**-matrices using other **W**-matrices of Example 4.2.1 and the columns of **H**$_4$.

Remark 4.2.1 If a BIBD (v, mv, mk, k, λ) is formed by developing m initial blocks each of size k, then $m(k-1)$ optimum **W**-matrices can be constructed whenever $\mathbf{H}_m$ and $\mathbf{H}_k$ exist. The result follows by noting that the above principle may be applied when the blocks are not repeated but are obtained by developing m initial blocks.

Remark 4.2.2 Let for a BIBD (v, b, r, k, λ) t optimum **W**-matrices be available. Then for the BIBD $(V = v, B = mb, R = mr, K = k, \Lambda = m\lambda)$ obtained by repeating each block m times, mt optimum **W**-matrices can be constructed whenever $\mathbf{H}_m$ exists. A similar but a more general result is discussed in Chap. 6.

When a BIBD is not necessarily cyclic, we can always accommodate $c^* = k - 1$ covariates optimally if each block of the design is repeated twice and $\mathbf{H}_k$ exists.

Theorem 4.2.2 *Suppose a BIBD (v, b, r, k, λ) exists which is not necessarily cyclic. Then if $\mathbf{H}_k$ exists, we can construct $c^* = k - 1$ optimum **W**-matrices for the BIBD $(V = v, B = 2b, R = 2r, K = k, \Lambda = 2\lambda)$.*

Proof Let $\mathbf{N}^{v \times b}$ denote the incidence matrix of the former BIBD. Let $\mathbf{H}_k = (\mathbf{h}_1, \mathbf{h}_2, \dots, \mathbf{h}_{k-1}, \mathbf{1})$. In order to construct $\mathbf{W}_{V \times B}^{(t)}$-matrix, we fill up the non-empty positions in $\mathbf{N}^{v \times b}$, the incidence matrix, by placing the elements of $\mathbf{h}_t$ successively in each column and in the order the positions appear. We denote the resultant matrix as $\mathbf{W}_t^{v \times b}$. Then

$$\mathbf{W}_{V \times B}^{(t)} = (\mathbf{W}_t^{v \times b}, -\mathbf{W}_t^{v \times b}).$$

It is now easy to assert the claim.   □

Now we consider some other series of BIBDs which are not necessarily symmetric but are constructed by Bose's difference technique and give the constructional method of OCDs as given in Dutta (2004). At first we consider the complementary designs of the Steiner's triple system (cf. Bose 1939) obtained by difference technique

$$\text{BIBD}(v = 3(2t + 1), b = (3t + 1)(2t + 1), r = 3t + 1, k = 3, \lambda = 1). \quad (4.2.3)$$

Theorem 4.2.3 *Let t be an even positive integer such that $2t + 1$ be a prime number or a prime power and further let $\mathbf{H}_{2t}$ and $\mathbf{H}_{6t}$ exist. Then we can construct $(2t - 1)$ optimum **W**-matrices for the following complementary design of (4.2.3)*

$$\text{BIBD}(v' = 3(2t + 1), \ b' = (3t + 1)(2t + 1), \ r' = 2t(3t + 1), \ k' = 6t,$$
$$\lambda' = (3t + 1)(2t - 1) + 1). \tag{4.2.4}$$

Proof Let $0, 1, \ldots, 2t$ be the elements of GF $(2t + 1)$. To each element a of GF $(2t + 1)$, we associate three symbols $1, 2, 3$ to have three treatments a_1, a_2, a_3. It is well known that the initial blocks for the series (4.2.3) are given by (cf. Bose 1939, p. 373)

$$S'_1 = \{(1_1, (2t)_1, 0_2), \ (2_1, (2t - 1)_1, 0_2), \ldots, (t_1, (t + 1)_1, 0_2)\};$$
$$S'_2 = \{(1_2, (2t)_2, 0_3), \ (2_2, (2t - 1)_2, 0_3), \ldots, (t_2, (t + 1)_2, 0_3)\};$$
$$S'_3 = \{(1_3, (2t)_3, 0_1), \ (2_3, (2t - 1)_3, 0_1), \ldots, (t_3, (t + 1)_3, 0_1)\};$$
$$S'_4 = (0_1, 0_2, 0_3).$$

We divide the initial blocks of the design (4.2.4) which is the complementary design of (4.2.3) into the following four sets:

$S_1 = \{(0_1, 2_1, 3_1, \ldots, (2t - 1)_1, 1_2, 2_2, \ldots, (2t)_2, 0_3, 1_3, \ldots, (2t)_3), \ (0_1, 1_1, 3_1, \ldots, (2t - 2)_1, (2t)_1, 1_2, 2_2,$
$\ldots, (2t)_2, 0_3, 1_3, \ldots, (2t)_3), \ldots, (0_1, 1_1, \ldots, (t - 1)_1, (t + 2)_1, \ldots, (2t)_1, 1_2, 2_2, \ldots, (2t)_2, 0_3, 1_3, \ldots, (2t)_3)\};$

$S_2 = \{(0_2, 2_2, 3_2, \ldots, (2t - 1)_2, 1_3, 2_3, \ldots, (2t)_3, 0_1, 1_1, \ldots, (2t)_1), \ (0_2, 1_2, 3_2, \ldots, (2t - 2)_2, (2t)_2, 1_3, 2_3,$
$\ldots, (2t)_3, 0_1, 1_1, \ldots, (2t)_1), \ldots, (0_2, 1_2, \ldots, (t - 1)_2, (t + 2)_2, \ldots, (2t)_2, 1_3, 2_3, \ldots, (2t)_3, 0_1, 1_1, \ldots, (2t)_1)\};$

$S_3 = \{(0_3, 2_3, 3_3, \ldots, (2t - 1)_3, 1_1, 2_1, \ldots, (2t)_1, 0_2, 1_2, \ldots, (2t)_2), \ (0_3, 1_3, 3_3, \ldots, (2t - 2)_3, (2t)_3, 1_1, 2_1,$
$\ldots, (2t)_1, 0_2, 1_2, \ldots, (2t)_2), \ldots, (0_3, 1_3, \ldots, (t - 1)_3, (t + 2)_3, \ldots, (2t)_3, 1_2, 2_2, \ldots, (2t)_1, 0_2, 1_2, \ldots, (2t)_2)\};$

$S_4 = \{(1_1, 2_1, \ldots, (2t)_1, 1_2, 2_2, \ldots, (2t)_2, 1_3, 2_3, \ldots, (2t)_3)\}.$

Let us assume the existence of $\mathbf{H}_{k'}$, where $k' = 6t$ and write it as

$$\mathbf{H}_{k'} = (\mathbf{h}_1, \mathbf{h}_2, \ldots, \mathbf{h}_{k'-1}, \mathbf{1}). \tag{4.2.5}$$

Consider the first $\frac{t}{2}$ initial blocks of the set $S_i (i = 1, 2, 3)$ and display them in the form of column vectors of the incidence matrix. Let us replace the non-zero elements of the jth column by the elements of $\mathbf{h}_s$, where $\mathbf{h}_s$ is any one of the first $(2t - 1)$ columns of $\mathbf{H}_{k'}$. We develop this initial block by cyclically permuting the elements to form a matrix $\mathbf{U}_{is}^j$ of order $v' \times (2t+1)$, $j = 1, 2, \ldots, \frac{t}{2}$. Using the same procedure we transform $(\frac{t}{2} + j)$th block of S_i by $-\mathbf{h}_s$ and develop in the same manner. We denote this matrix of order $v' \times (2t + 1)$ by $\mathbf{U}_{is}^{\frac{t}{2}+j}$ $(i = 1, 2, 3, \ j = 1, 2, \ldots, \frac{t}{2})$. In this way we can construct $\mathbf{U}_{is}^j$ and $\mathbf{U}_{is}^{\frac{t}{2}+j}$ $(j = 1, 2, \ldots, \frac{t}{2})$ for different s $(s = 1, 2, \ldots, (2t-1))$. Again, for an even integer t, we assume that $\mathbf{H}_{2t}$ exists and write it as

$$\mathbf{H}_{2t} = (\mathbf{h}_1^*, \mathbf{h}_2^*, \ldots, \mathbf{h}_{2t-1}^*, \mathbf{1}). \tag{4.2.6}$$

Consider the single initial block S_4. Note that the $2t$ elements except zero of S_4 correspond to each of the symbols $1, 2$ and 3. Transform the elements of this block into the form of a column vector of the incidence matrix. Then we replace the non-zero elements of each class of this column by the elements of $\mathbf{h}_s^*$ $(s = 1, 2, \ldots, (2t - 1))$ and develop this column into full form of $\mathbf{U}_s^{(4)}$ as

$$U_s^{(4)\prime} = (V_{1s}^{(4)\prime}, \; V_{2s}^{(4)\prime}, \; V_{3s}^{(4)\prime}),$$

(4.2.7)

where, for $i = 1, 2, 3$, $V_{is}^{(4)}$ ($i = 1, 2, 3$) is a matrix of order $(2t + 1) \times (2t + 1)$ obtained by cyclical permutation of the elements of the column vector after replacing the non-zero elements of ith class of the initial block of S_4 by h_s^*.

Schematically, the form of the $W^{(s)}$-matrix, $s = 1, 2, \ldots, 2t - 1$, can be written as

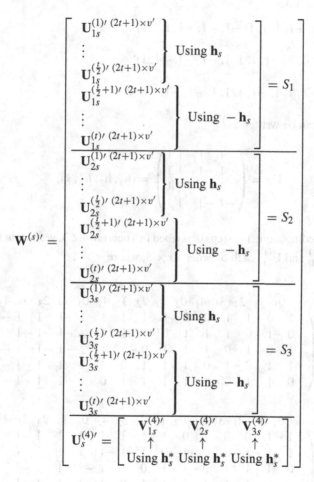

So varying $s = 1, 2, \ldots, 2t - 1$, we can construct $(2t - 1)$ optimum W-matrices. This establishes the claim. $\square$

We shall illustrate the construction through the following example.

Example 4.2.3 For $t = 2$, BIBD (15, 35, 28, 12, 22) is the complementary design of BIBD (15, 35, 7, 3, 1). Thus we have three optimum W-matrices, each of order 15×35. We exhibit the construction in detail.

Note that the four sets are:

$S_1 = \{(0_1, 2_1, 3_1, 1_2, 2_2, 3_2, 4_2, 0_3, 1_3, 2_3, 3_3, 4_3),\ (0_1, 1_1, 4_1, 1_2, 2_2, 3_2, 4_2, 0_3, 1_3, 2_3, 3_3, 4_3)\};$

$S_2 = \{(0_2, 2_2, 3_2, 1_3, 2_3, 3_3, 4_3, 0_1, 1_1, 2_1, 3_1, 4_1),\ (0_2, 1_2, 4_2, 1_3, 2_3, 3_3, 4_3, 0_1, 1_1, 2_1, 3_1, 4_1)\};$

$S_3 = \{(0_3, 2_3, 3_3, 1_1, 2_1, 3_1, 4_1, 0_2, 1_2, 2_2, 3_2, 4_2),\ (0_3, 1_3, 4_3, 1_1, 2_1, 3_1, 4_1, 0_2, 1_2, 2_2, 3_2, 4_2)\};$

$S_4 = \{(1_1, 2_1, 3_1, 4_1, 1_2, 2_2, 3_2, 4_2, 1_3, 2_3, 3_3, 4_3)\};$

$\mathbf{H}_{12}$ is available in standard literature (cf. Hedayat et al. 1999, p. 151). Without loss of generality, we take

$$\mathbf{h}_1 = (-1, 1, 1, -1, 1, 1, 1, -1, -1, -1, 1, -1)';$$

$$\mathbf{h}_2 = (-1, -1, 1, 1, -1, 1, 1, 1, -1, -1, -1, 1)';$$

$$\mathbf{h}_3 = (-1, 1, -1, 1, 1, -1, 1, 1, 1, -1, -1, -1)'.$$

$\mathbf{H}_{2t} = \mathbf{H}_4$ can be written as

$$\mathbf{H}_4 = \begin{pmatrix} 1 & 1 & 1 & 1 \\ 1 & -1 & -1 & 1 \\ -1 & 1 & -1 & 1 \\ -1 & -1 & 1 & 1 \end{pmatrix} = (\mathbf{h}_1^*, \mathbf{h}_2^*, \mathbf{h}_3^*, \mathbf{1}).$$

Then proceeding along the steps described in Theorem 4.2.3, we obtain $\mathbf{U}_{11}^{(1)}, \mathbf{U}_{11}^{(2)}$, $\mathbf{U}_{21}^{(1)}, \mathbf{U}_{21}^{(2)}, \mathbf{U}_{31}^{(1)}$ and $\mathbf{U}_{31}^{(1)}$ each of order 15×5, where

$$\mathbf{U}_{11}^{(1)'} = \begin{array}{c} \begin{array}{ccccccccccccccc} 0_1 & 1_1 & 2_1 & 3_1 & 4_1 & 0_2 & 1_2 & 2_2 & 3_2 & 4_2 & 0_3 & 1_3 & 2_3 & 3_3 & 4_3 \end{array} \\ \begin{pmatrix} -1 & 0 & 1 & 1 & 0 & 0 & -1 & 1 & 1 & 1 & -1 & -1 & -1 & 1 & -1 \\ 0 & -1 & 0 & 1 & 1 & 1 & 0 & -1 & 1 & 1 & -1 & -1 & -1 & -1 & 1 \\ 1 & 0 & -1 & 0 & 1 & 1 & 1 & 0 & -1 & 1 & 1 & -1 & -1 & -1 & -1 \\ 1 & 1 & 0 & -1 & 0 & 1 & 1 & 1 & 0 & -1 & -1 & 1 & -1 & -1 & -1 \\ 0 & 1 & 1 & 0 & -1 & -1 & 1 & 1 & 1 & 0 & -1 & -1 & 1 & -1 & -1 \end{pmatrix} \end{array},$$

$$\mathbf{U}_{11}^{(2)'} = \begin{array}{c} \begin{array}{ccccccccccccccc} 0_1 & 1_1 & 2_1 & 3_1 & 4_1 & 0_2 & 1_2 & 2_2 & 3_2 & 4_2 & 0_3 & 1_3 & 2_3 & 3_3 & 4_3 \end{array} \\ \begin{pmatrix} 1 & -1 & 0 & 0 & -1 & 0 & 1 & -1 & -1 & -1 & 1 & 1 & 1 & -1 & 1 \\ -1 & 1 & -1 & 0 & 0 & -1 & 0 & 1 & -1 & -1 & 1 & 1 & 1 & 1 & -1 \\ 0 & -1 & 1 & -1 & 0 & -1 & -1 & 0 & 1 & -1 & -1 & 1 & 1 & 1 & 1 \\ 0 & 0 & -1 & 1 & -1 & -1 & -1 & -1 & 0 & 1 & 1 & -1 & 1 & 1 & 1 \\ -1 & 0 & 0 & -1 & 1 & 1 & -1 & -1 & -1 & 0 & 1 & 1 & -1 & 1 & 1 \end{pmatrix} \end{array},$$

$$\mathbf{U}_{21}^{(1)'} = \begin{pmatrix} -1 & 1 & 1 & -1 & 1 & 1 & 0 & 1 & -1 & 0 & 0 & -1 & -1 & 1 & -1 \\ 1 & -1 & 1 & 1 & -1 & 0 & 1 & 0 & 1 & -1 & -1 & 0 & -1 & -1 & 1 \\ -1 & 1 & -1 & 1 & 1 & -1 & 0 & 1 & 0 & 1 & 1 & -1 & 0 & -1 & -1 \\ 1 & -1 & 1 & -1 & 1 & 1 & -1 & 0 & 1 & 0 & -1 & 1 & -1 & 0 & -1 \\ 1 & 1 & -1 & 1 & -1 & 0 & 1 & -1 & 0 & 1 & -1 & -1 & 1 & -1 & 0 \end{pmatrix},$$

$$
\mathbf{U}_{21}^{(2)\prime} = \begin{pmatrix}
1 & -1 & -1 & 1 & -1 & -1 & -1 & 0 & 0 & 1 & 0 & 1 & 1 & -1 & 1 \\
-1 & 1 & -1 & -1 & 1 & 1 & -1 & -1 & 0 & 0 & 1 & 0 & 1 & 1 & -1 \\
1 & -1 & 1 & -1 & -1 & 0 & 1 & -1 & -1 & 0 & -1 & 1 & 0 & 1 & 1 \\
-1 & 1 & -1 & 1 & -1 & 0 & 0 & 1 & -1 & -1 & 1 & -1 & 1 & 0 & 1 \\
-1 & -1 & 1 & -1 & 1 & -1 & 0 & 0 & 1 & -1 & 1 & 1 & -1 & 1 & 0
\end{pmatrix},
$$

$$
\mathbf{U}_{31}^{(1)\prime} = \begin{pmatrix}
0 & -1 & 1 & 1 & -1 & 1 & 1 & 1 & -1 & -1 & -1 & 0 & 1 & -1 & 0 \\
-1 & 0 & -1 & 1 & 1 & -1 & 1 & 1 & 1 & -1 & 0 & -1 & 0 & 1 & -1 \\
1 & -1 & 0 & -1 & 1 & -1 & -1 & 1 & 1 & 1 & -1 & 0 & -1 & 0 & 1 \\
1 & 1 & -1 & 0 & -1 & 1 & -1 & -1 & 1 & 1 & 1 & -1 & 0 & -1 & 0 \\
-1 & 1 & 1 & -1 & 0 & 1 & 1 & -1 & -1 & 1 & 0 & 1 & -1 & 0 & -1
\end{pmatrix}
$$

and

$$
\begin{array}{c}
\phantom{\mathbf{U}_{31}^{(2)\prime} =}\ \ 0_1\ \ 1_1\ \ 2_1\ \ 3_1\ \ 4_1\ \ 0_2\ \ 1_2\ \ 2_2\ \ 3_2\ \ 4_2\ \ 0_3\ \ 1_3\ \ 2_3\ \ 3_3\ \ 4_3 \\[4pt]
\mathbf{U}_{31}^{(2)\prime} = \begin{pmatrix}
0 & 1 & -1 & -1 & 1 & -1 & -1 & -1 & 1 & 1 & 1 & -1 & 0 & 0 & 1 \\
1 & 0 & 1 & -1 & -1 & 1 & -1 & -1 & -1 & 1 & 1 & 1 & -1 & 0 & 0 \\
-1 & 1 & 0 & 1 & -1 & 1 & 1 & -1 & -1 & -1 & 0 & 1 & 1 & -1 & 0 \\
-1 & -1 & 1 & 0 & 1 & -1 & 1 & 1 & -1 & -1 & 0 & 0 & 1 & 1 & -1 \\
1 & -1 & -1 & 1 & 0 & -1 & -1 & 1 & 1 & -1 & -1 & 0 & 0 & 1 & 1
\end{pmatrix}
\end{array}
$$

Using $\mathbf{h}_1^*$, the matrices $\mathbf{V}_{11}^{(4)}, \mathbf{V}_{21}^{(4)}$ and $\mathbf{V}_{31}^{(4)}$ each of order 5×5 are obtained as

$$
\begin{array}{cc}
\begin{array}{c}
0_1\ \ 1_1\ \ 2_1\ \ 3_1\ \ 4_1 \\[4pt]
\mathbf{V}_{11}^{(4)\prime} = \begin{pmatrix}
0 & 1 & 1 & -1 & -1 \\
-1 & 0 & 1 & 1 & -1 \\
-1 & -1 & 0 & 1 & 1 \\
1 & -1 & -1 & 0 & 1 \\
1 & 1 & -1 & -1 & 0
\end{pmatrix},
\end{array}
&
\begin{array}{c}
0_2\ \ 1_2\ \ 2_2\ \ 3_2\ \ 4_2 \\[4pt]
\mathbf{V}_{21}^{(4)\prime} = \begin{pmatrix}
0 & 1 & 1 & -1 & -1 \\
-1 & 0 & 1 & 1 & -1 \\
-1 & -1 & 0 & 1 & 1 \\
1 & -1 & -1 & 0 & 1 \\
1 & 1 & -1 & -1 & 0
\end{pmatrix},
\end{array}
\end{array}
$$

$$
\begin{array}{c}
0_3\ \ 1_3\ \ 2_3\ \ 3_3\ \ 4_3 \\[4pt]
\mathbf{V}_{31}^{(4)\prime} = \begin{pmatrix}
0 & 1 & 1 & -1 & -1 \\
-1 & 0 & 1 & 1 & -1 \\
-1 & -1 & 0 & 1 & 1 \\
1 & -1 & -1 & 0 & 1 \\
1 & 1 & -1 & -1 & 0
\end{pmatrix}.
\end{array}
$$

Thus $\mathbf{W}^{(1)}$ is obtained by suitably arranging the $\mathbf{U}$ and $\mathbf{V}$-matrices as

$$
\mathbf{W}^{(1)\prime\ 35 \times 15} = (\mathbf{U}_{11}^{(1)\prime}, \mathbf{U}_{11}^{(2)\prime}, \mathbf{U}_{21}^{(1)\prime}, \mathbf{U}_{21}^{(2)\prime}, \mathbf{U}_{31}^{(1)\prime}, \mathbf{U}_{31}^{(2)\prime}, \mathbf{U}_{1}^{(4)\prime}),
$$

where

$$
\mathbf{U}_{1}^{(4)\prime} = (\mathbf{V}_{11}^{(4)\prime}, \mathbf{V}_{21}^{(4)\prime}, \mathbf{V}_{31}^{(4)\prime}).
$$

Similarly, $\mathbf{W}^{(2)}$ and $\mathbf{W}^{(3)}$ can be constructed by using $(\mathbf{h}_2, \mathbf{h}_2^*)$ and $(\mathbf{h}_3, \mathbf{h}_3^*)$ respectively.

Dutta (2004) also constructed OCDs for the following series of BIBD. For detailed discussion, readers are referred to the original paper.

$$v' = 5(4t + 1), \; b' = (5t + 1)(4t + 1), \; r' = 4t(5t + 1), \; k' = 20t, \; \lambda' = (5t + 1)(4t - 1) + 1,$$
$$(4.2.8)$$

and

$$v' = 4(3t + 1), \; b' = (4t + 1)(3t + 1), \; r' = 3t(4t + 1), \; k' = 12t, \; \lambda' = (4t + 1)(3t - 1) + 1.$$
$$(4.2.9)$$

4.3 BIBDs Through Projective Geometry

As mentioned earlier in the series considered in Sect. 2.2, the layouts had cyclical patterns which simplified the choice of optimum $\mathbf{W}$-matrices. Now we consider complementary designs of the SBIBDs obtained through projective geometry. However, by suitable partition of the blocks into different sets, and by judicious choice of the covariate values, it is possible to construct OCDs for the series with parameters $v' = b' = s^N + s^{N-1} + \cdots + s + 1, r' = k' = s^N, \lambda' = s^N - s^{N-1}$.

4.3.1 Partitioning of the Blocks

With the help of the Galois field GF (s), we can construct the finite projective geometry of N dimensions, to be written as PG (N, s), where, $s = p^n$, p is a prime number and n is any positive integer. Any ordered set of $(N + 1)$ elements $(x_0, x_1, \ldots, x_N)$ where the x_i's belong to GF (s) and are not simultaneously zero, is called a point of the projective geometry PG (N, s). $(x_0, x_1, \ldots, x_N) = \mathbf{x}'$ and $\rho \mathbf{x}'$ represent the same point, where $\rho(\neq 0) \in$GF (s). It is known that the number of points in PG (N, s) is equal to $\phi(N, m, s)$, where $\phi(N, m, s) = \frac{(s^{N+1}-1)(s^N-1)\ldots(s^{N-m+1}-1)}{(s^{m+1}-1)(s^m-1)\ldots(s-1)}$. For more detailed discussions in this respect one is referred to Bose (1939).

By making a correspondence between the points and the m-flats of PG (N, s) with the varieties and the blocks respectively, we get a BIBD with parameters (cf. Bose 1939, p. 362): $v = \phi(N, 0, s), \; b = \phi(N, m, s), \; r = \phi(N - 1, m - 1, s), \; k = \phi(m, 0, s), \lambda = \phi(N - 2, m - 2, s)$. For $m = N - 1$, the following SBIBD is obtained:

$$v = b = s^N + s^{N-1} + \cdots + s + 1, \quad r = k = s^{N-1} + s^{N-2} + \cdots + s + 1$$
$$\lambda = s^{N-1} + s^{N-2} + \cdots + s + 1$$
$$(4.3.1)$$

We consider the complementary design given in (4.3.1) which is also an SBIBD with the following parameters:

$$v' = b' = s^N + s^{N-1} + \cdots + s + 1, \ r' = k' = s^N, \ \lambda = s^N - s^{N-1}. \quad (4.3.2)$$

It was mentioned earlier that the choice of the levels of the covariates in a BIBD set-up depends on the method of construction of the BIBD and the maximum number of covariates satisfying condition (3.1.5) varies from series to series. The blocks of the SBIBD with parameters given in (4.3.2) are partitioned into $(s^{N-1} + s^{N-3} + \cdots + s^2 + 1)(=t$, say) disjoint sets; each set contains $(s + 1)$ blocks such that the portion of the incidence matrix of the complementary design corresponding to these $(s + 1)$ sets conforms to that of the incidence matrix of an RBD with suitable parameters. This fact has been used in the choice of the $\mathbf{Z}$-matrix.

We note that the number of $(N-1)$-flats passing through a particular $(N-2)$-flat is given by $\phi(1, 0, s) = s+1$. Such $(s+1)$, $(N-1)$-flats passing through a particular $(N-2)$-flat can be obtained as follows:

Consider an $(N-2)$-flat, given by

$$\mathbf{a}'\mathbf{x} = 0, \ \ \mathbf{b}'\mathbf{x} = 0 \quad (4.3.3)$$

where, $\mathbf{a}$ and $\mathbf{b}$ are two column vectors with elements from GF (s) such that rank $(\mathbf{A}') = \text{rank}(\mathbf{a}, \mathbf{b}) = 2$.

The $(s + 1)$, $(N-1)$-flats containing the $(N-2)$-flat given in (4.3.3) are given by $(\lambda_1\mathbf{a}' + \lambda_2\mathbf{b}')\mathbf{x} = 0$; $(\lambda_1, \lambda_2) \neq (0, 0)$ and $(\lambda_1, \lambda_2) \equiv \rho(\lambda_1, \lambda_2)$ where, ρ is a non-zero element of GF (s). If N is odd, then the full set of $\phi(N, N-1, s)$, $(N\text{-}1)$-flats can be partitioned into $\frac{s^{N+1}-1}{(s-1)(s+1)} = \frac{\phi(N,N-1,s)}{s+1} = (s^{N-1} + s^{N-3} + \cdots + s^2 + 1)$ sets each containing $(s + 1)$, $(N-1)$-flats having a common $(N-2)$-flat, are disjoint. As the blocks correspond to $(N-1)$-flats, so through one to one correspondence, we can partition the blocks into $(s^{N-1} + s^{N-3} + \cdots + s^2 + 1)$ disjoint sets each containing $(s + 1)$ blocks. It will be clear from the following two examples from Dutta et al. (2007) covering both the situations where s is a prime number and a prime power.

Example 4.3.1 $N = 3$, $m = 2$, $s = 2$. There are 15 blocks which can be partitioned into five sets each of size 3:

$$
\begin{array}{lll}
\quad x_0 = 0 & \quad x_1 = 0 & \quad x_2 = 0 \\
S_1 : x_1 + x_2 = 0 & S_2 : x_0 + x_3 = 0 & S_3 : x_1 + x_3 = 0 \\
\quad x_0 + x_1 + x_2 = 0 & \quad x_0 + x_1 + x_3 = 0 & \quad x_1 + x_2 + x_3 = 0
\end{array}
$$

$$
\begin{array}{ll}
\quad x_3 = 0 & \quad x_0 + x_1 = 0 \\
S_4 : x_0 + x_2 = 0 & S_5 : x_2 + x_3 = 0 \\
\quad x_0 + x_2 + x_3 = 0 & \quad x_0 + x_1 + x_2 + x_3 = 0.
\end{array}
$$

It is to be noted that only two equations in each set S_i are independent and these can conveniently be represented as $\mathbf{Ax} = \mathbf{0}$. It is clear that the choice of $\mathbf{A}$-matrix in S_1 is given by:

$$\begin{pmatrix} 1 & 0 & 0 & 0 \\ 0 & 1 & 1 & 0 \end{pmatrix}.$$

The choice of $\mathbf{A}$-matrices for other S's are obvious.

Example 4.3.2 $N = 3$, $m = 2$ and $s = 2^2$. There are 85 blocks which can be partitioned into 17 sets each of size 5. Let the elements of GF (2^2) be $\alpha_0 = 0$, $\alpha_1 = 1$, $\alpha_2 = x$, $\alpha_3 = 1 + x$, where x is a primitive root of GF (2^2). Then the 17 sets are:

$$
\begin{array}{lll}
x_0 = 0 & x_2 = 0 & x_0 + x_2 = 0 \\
x_1 = 0 & x_3 = 0 & x_1 + x_3 = 0 \\
S_1 : x_0 + x_1 = 0 & S_2 : x_2 + x_3 = 0 & S_3 : x_0 + x_1 + x_2 + x_3 = 0 \\
x_0 + \alpha_2 x_1 = 0 & x_2 + \alpha_2 x_3 = 0 & x_0 + \alpha_2 x_1 + x_2 + \alpha_2 x_3 = 0 \\
x_0 + \alpha_3 x_1 = 0 & x_2 + \alpha_3 x_3 = 0 & x_0 + \alpha_3 x_1 + x_2 + \alpha_3 x_3 = 0
\end{array}
$$

$$
\begin{array}{ll}
x_0 + \alpha_2 x_2 = 0 & x_0 + \alpha_3 x_2 = 0 \\
x_1 + \alpha_3 x_3 = 0 & x_1 + \alpha_2 x_3 = 0 \\
S_4 : x_0 + x_1 + \alpha_2 x_2 + \alpha_3 x_3 = 0 & S_5 : x_0 + x_1 + \alpha_3 x_2 + \alpha_2 x_3 = 0 \\
x_0 + \alpha_2 x_1 + \alpha_2 x_2 + x_3 = 0 & x_0 + \alpha_2 x_1 + \alpha_3 x_2 + \alpha_3 x_3 = 0 \\
x_0 + \alpha_3 x_1 + \alpha_2 x_2 + \alpha_2 x_3 = 0 & x_0 + \alpha_3 x_1 + \alpha_3 x_2 + x_3 = 0
\end{array}
$$

$$
\begin{array}{ll}
x_0 + x_3 = 0 & x_0 + \alpha_2 x_3 = 0 \\
x_1 + x_2 + x_3 = 0 & x_1 + \alpha_3 x_2 + \alpha_2 x_3 = 0 \\
S_6 : x_0 + x_1 + x_2 = 0 & S_7 : x_0 + x_1 + \alpha_3 x_2 = 0 \\
x_0 + \alpha_2 x_1 + \alpha_2 x_2 + \alpha_3 x_3 = 0 & x_0 + \alpha_2 x_1 + x_2 + x_3 = 0 \\
x_0 + \alpha_3 x_1 + \alpha_3 x_2 + \alpha_2 x_3 = 0 & x_0 + \alpha_3 x_1 + \alpha_2 x_2 + \alpha_3 x_3 = 0
\end{array}
$$

$$
\begin{array}{ll}
x_0 + \alpha_3 x_3 = 0 & x_0 + x_2 + x_3 = 0 \\
x_1 + \alpha_2 x_2 + \alpha_3 x_3 = 0 & x_1 + x_2 = 0 \\
S_8 : x_0 + x_1 + \alpha_2 x_2 = 0 & S_9 : x_0 + x_1 + x_3 = 0 \\
x_0 + \alpha_2 x_1 + \alpha_3 x_2 + \alpha_2 x_3 = 0 & x_0 + \alpha_2 x_1 + \alpha_3 x_2 + x_3 = 0 \\
x_0 + \alpha_3 x_1 + x_2 + x_3 = 0 & x_0 + \alpha_3 x_1 + \alpha_2 x_2 + x_3 = 0
\end{array}
$$

$$
\begin{array}{ll}
x_0 + \alpha_2 x_2 + \alpha_3 x_3 = 0 & x_0 + \alpha_3 x_2 + \alpha_2 x_3 = 0 \\
x_1 + \alpha_2 x_2 = 0 & x_1 + \alpha_3 x_2 = 0 \\
S_{10} : x_0 + x_1 + \alpha_3 x_3 = 0 & S_{11} : x_0 + x_1 + \alpha_2 x_3 = 0 \\
x_0 + \alpha_2 x_1 + x_2 + \alpha_3 x_3 = 0 & x_0 + \alpha_2 x_1 + \alpha_2 x_2 + \alpha_2 x_3 = 0 \\
x_0 + \alpha_3 x_1 + \alpha_3 x_2 + \alpha_3 x_3 = 0 & x_0 + \alpha_3 x_1 + x_2 + \alpha_2 x_3 = 0
\end{array}
$$

$$x_0 + \alpha_3 x_2 + \alpha_3 x_3 = 0 \qquad\qquad x_0 + \alpha_2 x_2 + \alpha_2 x_3 = 0$$
$$x_1 + \alpha_2 x_2 + x_3 = 0 \qquad\qquad x_1 + \alpha_3 x_2 + x_3 = 0$$
$$S_{12} : x_0 + x_1 + x_2 + \alpha_2 x_3 = 0 \quad S_{13} : x_0 + x_1 + x_2 + \alpha_3 x_3 = 0$$
$$x_0 + \alpha_2 x_1 + x_3 = 0 \qquad\qquad x_0 + \alpha_2 x_1 + \alpha_3 x_2 = 0$$
$$x_0 + \alpha_3 x_1 + \alpha_2 x_2 = 0 \qquad\qquad x_0 + \alpha_3 x_1 + x_3 = 0$$

$$x_0 + x_2 + \alpha_3 x_3 = 0 \qquad\qquad x_0 + x_2 + \alpha_2 x_3 = 0$$
$$x_1 + \alpha_2 x_2 + \alpha_2 x_3 = 0 \qquad\qquad x_1 + \alpha_3 x_2 + \alpha_3 x_3 = 0$$
$$S_{14} : x_0 + x_1 + \alpha_3 x_2 + x_3 = 0 \quad S_{15} : x_0 + x_1 + \alpha_2 x_2 + x_3 = 0$$
$$x_0 + \alpha_2 x_1 + \alpha_2 x_2 = 0 \qquad\qquad x_0 + \alpha_2 x_1 + \alpha_3 x_3 = 0$$
$$x_0 + \alpha_3 x_1 + \alpha_2 x_3 = 0 \qquad\qquad x_0 + \alpha_3 x_1 + \alpha_3 x_2 = 0$$

$$x_0 + \alpha_2 x_2 + x_3 = 0 \qquad\qquad x_0 + \alpha_3 x_2 + x_3 = 0$$
$$x_1 + x_2 + \alpha_2 x_3 = 0 \qquad\qquad x_1 + x_2 + \alpha_3 x_3 = 0$$
$$S_{16} : x_0 + x_1 + \alpha_3 x_2 + \alpha_3 x_3 = 0 \quad S_{17} : x_0 + x_1 + \alpha_2 x_2 + \alpha_2 x_3 = 0$$
$$x_0 + \alpha_2 x_1 + \alpha_2 x_3 = 0 \qquad\qquad x_0 + \alpha_2 x_1 + x_2 = 0$$
$$x_0 + \alpha_3 x_1 + x_2 = 0 \qquad\qquad x_0 + \alpha_3 x_1 + \alpha_3 x_3 = 0$$

where, (x_0, x_1, x_2, x_3) is a point of PG $(3, 2^2)$.

As an illustration, the choice of **A**-matrix corresponding to S_1 and S_4 are given, respectively, by

$$\begin{pmatrix} 1 & 0 & 0 & 0 \\ 0 & 1 & 0 & 0 \end{pmatrix}, \quad \begin{pmatrix} 1 & 0 & \alpha_2 & 0 \\ 0 & 1 & 0 & \alpha_3 \end{pmatrix}.$$

Similarly, **A**-matrices for other S_i's can be written.

4.3.2 Optimum Covariate Designs

From (4.3.1), we see that any block of the design contains $k = (s^{N-1} + \lambda)$ treatments and any two blocks have exactly λ treatments in common. As any two blocks of the set S_i $(i = 1, 2, \ldots, t; \ t = (s^{N-1} + s^{N-3} + \cdots + s^2 + 1))$, have the same λ treatments in common, without loss of any generality, we can write the portion $\mathbf{N}_i$, the incidence matrix corresponding to the blocks in S_i $(i = 1, 2, \ldots, t)$ in the following form (with some rearrangement of blocks if necessary):

$$\mathbf{N}_i' = \begin{pmatrix} \mathbf{1}'_{s^{N-1}} & \mathbf{0}' & \cdots & \mathbf{0}' & \mathbf{1}'_{\lambda} \\ \mathbf{0}' & \mathbf{1}'_{s^{N-1}} & \cdots & \mathbf{0}' & \mathbf{1}'_{\lambda} \\ \vdots & \vdots & \vdots & \vdots & \vdots \\ \mathbf{0}' & \mathbf{0}' & \cdots & \mathbf{1}'_{s^{N-1}} & \mathbf{1}'_{\lambda} \end{pmatrix} \begin{matrix} (s+1) \times v \end{matrix}. \qquad (4.3.4)$$

The part of the incidence matrix of the design with parameters in (4.3.2) corresponding to the part $\mathbf{N}_i$ of the design with parameters in (4.3.1) is obtained by replacing ones by zeros and zeros by ones in (4.3.4) and is given by:

$$
\mathbf{N}_i^{c\,\prime} =
\begin{pmatrix}
\mathbf{0}' & \mathbf{1}'_{s^{N-1}} & \cdots & \mathbf{1}'_{s^{N-1}} & \mathbf{0}'_\lambda \\
\mathbf{1}'_{s^{N-1}} & \mathbf{1}'_{s^{N-1}} & \cdots & \mathbf{1}'_{s^{N-1}} & \mathbf{0}'_\lambda \\
\vdots & \vdots & \vdots & \vdots & \vdots \\
\mathbf{1}'_{s^{N-1}} & \mathbf{1}'_{s^{N-1}} & \cdots & \mathbf{0}' & \mathbf{0}'_\lambda
\end{pmatrix}^{(s+1)\times v}
.
\tag{4.3.5}
$$

Using the structure (4.3.5) above, we develop a method for choosing covariates optimally for the series of complementary designs of (4.3.1). The precise statement follows.

Theorem 4.3.1 *If $s = 2^p$ where p is any positive integer, then $(s^{N-1} - 1)(s - 1) + (s - 1)$ optimum $\mathbf{W}$-matrices can be constructed for the design with parameters in (4.3.2), where N is an odd integer.*

Proof Since s is a power of 2, $\mathbf{H}_{s^{N-1}}$ and $\mathbf{H}_s$ exist and we write them as

$$
\mathbf{H}_{s^{N-1}} = \left(\mathbf{h}_1, \ldots, \mathbf{h}_{s^{N-1}-1}, \mathbf{1}\right)
$$

$$
\mathbf{H}_s = \left(\mathbf{h}_1^*, \ldots, \mathbf{h}_{s-1}^*, \mathbf{1}\right).
$$

Again, the matrix (4.3.5) can be written as

$$
\mathbf{N}_i^{c\,\prime} = \left(\mathbf{A}_{1i}, \mathbf{A}_{2i}, \ldots, \mathbf{A}_{ji}, \ldots, \mathbf{A}_{(s+1)i}, \mathbf{0}_i\right)
$$

where $\mathbf{A}_{ji}$ is the matrix in the jth column block of $\mathbf{N}_i^{c\,\prime}$, $j = 1, 2, \ldots, (s + 1)$. We replace kth non-null row of $\mathbf{A}_{ji}$ by the kth row of $\mathbf{h}_m^* \mathbf{h}_n'$; $k = 1, 2, \ldots, s$, $m = 1, 2, \ldots, (s - 1)$ and $n = 1, 2, \ldots, (s^{N-1} - 1)$ and denote the resultant matrix by $\mathbf{A}_{ji}^*$. We repeat the procedure for each $\mathbf{A}_{ji}$ with the same m, n. This leads to a matrix $\mathbf{W}_{i;m,n}^*$ with elements ± 1 satisfying the properties C_1 and C_2 of condition (3.1.5). Using the same $\mathbf{h}_m$ and $\mathbf{h}_n^*$ we get different $\mathbf{W}_{i;m,n}^*$'s corresponding to different $\mathbf{N}_i^{c\prime}$'s. Therefore, for fixed m, n

$$
\mathbf{W}_{m,n}^* = \left(\mathbf{W}_{1;m,n}^{*\prime}, \mathbf{W}_{2;m,n}^{*\prime}, \ldots, \mathbf{W}_{t;m,n}^{*\prime}\right)
$$

satisfies the properties C_1 and C_2 of condition (3.1.5). For different choices of $\mathbf{h}_m$ and $\mathbf{h}_n^*$ we get $(s^{N-1} - 1)(s - 1)$, $\mathbf{W}_{m,n}^*$-matrices which satisfy condition (3.1.5). The transformation required to be applied on (4.3.5) to get back the corresponding portion of the incidence matrix of the design may also be applied on the elements of the above $\mathbf{W}^*$-matrices to get the original $\mathbf{W}$-matrices.

Again, note that the number of unit vectors in the rows of $\mathbf{N}_i^{c\,\prime}$ is s which is the same as that of the elements of $\mathbf{h}_m^*$. We replace the qth vector $\mathbf{1}'_{s^{N-1}}$ in the first column block matrix of $\mathbf{N}_i^{c\,\prime}$ by $+\mathbf{1}'_{s^{N-1}}$ or by $-\mathbf{1}'_{s^{N-1}}$ according as the qth element of $\mathbf{h}_m^*$ is $+1$ or -1, respectively, to get $\mathbf{A}_1^{**}$. Now we permute $+\mathbf{1}'_{s^{N-1}}, -\mathbf{1}'_{s^{N-1}}$ and $\mathbf{0}'_{s^{N-1}}$

in the rows of $\mathbf{A}_1^{**}$ cyclically to get $\mathbf{A}_2^{**}$, $\mathbf{A}_3^{**}$, ..., $\mathbf{A}_{s+1}^{**}$ and hence can construct a new $\mathbf{W}$-matrix viz. $\mathbf{W}_m^{**}$. By taking different $\mathbf{h}_m^*$, we can construct $(s-1)$, $\mathbf{W}_m^{**}$-matrices. It is easy to show that these $\mathbf{W}_m^{**}$-matrices together with the $\mathbf{W}_{m,n}^*$-matrices, $m = 1, 2, \ldots, (s-1)$, $n = 1, 2, \ldots, (s^{N-1} - 1)$ satisfy condition (3.1.5). Thus in all, we get $(s^{N-1} - 1)(s - 1) + (s - 1)$ optimum $\mathbf{W}$-matrices. $\square$

Example 4.3.3 We consider the SBIBD whose blocks are the 2-flats of PG (3,2), so that the parameters of the SBIBD are $v = b = 15, r = k = 7, \lambda = 3$. Now for the complementary design, the parameters are: $v' = b' = 15$, $r' = k' = 8, \lambda' = 4$.

According to Example 4.3.1, the sets of blocks of the complementary design, where the treatment corresponding to the point (x_0, x_1, x_2, x_3) is indexed by $2^3 x_0 + 2^2 x_1 + 2x_2 + x_3$, are:

$S_1 = [(8, 9, 10, 11, 12, 13, 14, 15), (2, 3, 4, 5, 10, 11, 12, 13), (2, 3, 4, 5, 8, 9, 14, 15)]$

$S_2 = [(4, 5, 6, 7, 12, 13, 14, 15), (1, 3, 5, 7, 8, 10, 12, 14), (1, 3, 4, 6, 8, 10, 13, 15)]$

$S_3 = [(2, 3, 6, 7, 10, 11, 14, 15), (1, 3, 4, 6, 9, 11, 12, 14), (1, 2, 4, 7, 9, 10, 12, 15)]$

$S_4 = [(1, 3, 5, 7, 9, 11, 13, 15), (2, 3, 6, 7, 8, 9, 12, 13), (1, 2, 5, 6, 8, 11, 12, 15)]$

$S_5 = [(4, 5, 6, 7, 8, 9, 10, 11), (1, 2, 5, 6, 9, 10, 13, 14), (1, 2, 4, 7, 8, 11, 13, 14)]$

We write $\mathbf{H}_2$ and $\mathbf{H}_4$ as

$$\mathbf{H}_2 = \begin{pmatrix} 1 & 1 \\ -1 & 1 \end{pmatrix} = (\mathbf{h}_1, \ 1) \quad \text{and} \quad \mathbf{H}_4 = \begin{pmatrix} 1 & 1 & 1 & 1 \\ -1 & -1 & 1 & 1 \\ 1 & -1 & -1 & 1 \\ -1 & 1 & -1 & 1 \end{pmatrix} = (\mathbf{h}_1^*, \mathbf{h}_2^*, \mathbf{h}_3^*, \mathbf{1}).$$

Using $\mathbf{h}_1$ and $\mathbf{h}_i^*$ $(i = 1, 2, 3)$ and proceeding according to Theorem 4.3.1 we can construct three optimum $\mathbf{W}$-matrices. Below we give $\mathbf{W}_{1,1}^*$-matrix which is constructed by using $\mathbf{h}_1$ and $\mathbf{h}_1^*$.

$$\begin{pmatrix}
0 & 0 & 0 & 0 & 0 & 0 & 0 & 1 & -1 & 1 & -1 & 1 & -1 & 1 & -1 \\
0 & 1 & -1 & 1 & -1 & 0 & 0 & 0 & 0 & -1 & 1 & -1 & 1 & 0 & 0 \\
0 & -1 & 1 & -1 & 1 & 0 & 0 & -1 & 1 & 0 & 0 & 0 & 0 & -1 & 1 \\
0 & 0 & 0 & 1 & 1 & -1 & -1 & 0 & 0 & 0 & 0 & 1 & 1 & -1 & -1 \\
1 & 0 & -1 & 0 & -1 & 0 & 1 & 1 & 0 & -1 & 0 & -1 & 0 & 1 & 0 \\
-1 & 0 & 1 & -1 & 0 & 1 & 0 & -1 & 0 & 1 & 0 & 0 & -1 & 0 & 1 \\
0 & 1 & 1 & 0 & 0 & -1 & -1 & 0 & 0 & 1 & 1 & 0 & 0 & -1 & -1 \\
1 & 0 & -1 & -1 & 0 & 1 & 0 & 0 & 1 & 0 & -1 & -1 & 0 & 1 & 0 \\
-1 & -1 & 0 & 1 & 0 & 0 & 1 & 0 & -1 & -1 & 0 & 1 & 0 & 0 & 1 \\
1 & 0 & 1 & 0 & -1 & 0 & -1 & 0 & 1 & 0 & 1 & 0 & -1 & 0 & -1 \\
0 & 1 & -1 & 0 & 0 & -1 & 1 & 1 & -1 & 0 & 0 & -1 & 1 & 0 & 0 \\
-1 & -1 & 0 & 0 & 1 & 1 & 0 & -1 & 0 & 0 & -1 & 1 & 0 & 0 & 1 \\
0 & 0 & 0 & 1 & 1 & -1 & -1 & 1 & 1 & -1 & -1 & 0 & 0 & 0 & 0 \\
1 & -1 & 0 & 0 & -1 & 1 & 0 & 0 & -1 & 1 & 0 & 0 & 1 & -1 & 0 \\
-1 & 1 & 0 & -1 & 0 & 0 & 1 & -1 & 0 & 0 & 1 & 0 & -1 & 1 & 0
\end{pmatrix}'.$$

Similarly, by taking the combinations $(\mathbf{h}_1, \mathbf{h}_2^*)$ and $(\mathbf{h}_1, \mathbf{h}_3^*)$ we can construct $\mathbf{W}_{1,2}^*$ and $\mathbf{W}_{1,3}^*$ respectively. Using $\mathbf{h}_1$, we can get another matrix $\mathbf{W}_1^{**}$ which is given below:

$$
\begin{pmatrix}
0 & 0 & 0 & 0 & 0 & 0 & 0 & -1 & -1 & 1 & 1 & 1 & 1 & -1 & -1 \\
0 & 1 & 1 & 1 & 1 & 0 & 0 & 0 & 0 & -1 & -1 & -1 & -1 & 0 & 0 \\
0 & -1 & -1 & -1 & -1 & 0 & 0 & 1 & 1 & 0 & 0 & 0 & 0 & 1 & 1 \\
0 & 0 & 0 & -1 & 1 & -1 & 1 & 0 & 0 & 0 & 0 & 1 & -1 & 1 & -1 \\
1 & 0 & 1 & 0 & -1 & 0 & -1 & 1 & 0 & 1 & 0 & -1 & 0 & -1 & 0 \\
-1 & 0 & -1 & 1 & 0 & 1 & 0 & -1 & 0 & -1 & 0 & 0 & 1 & 0 & 1 \\
0 & -1 & 1 & 0 & 0 & 1 & -1 & 0 & 0 & -1 & 1 & 0 & 0 & 1 & -1 \\
1 & 0 & -1 & 1 & 0 & -1 & 0 & 0 & 1 & 0 & -1 & 1 & 0 & -1 & 0 \\
-1 & 1 & 0 & -1 & 0 & 0 & 1 & 0 & -1 & 1 & 0 & -1 & 0 & 0 & 1 \\
-1 & 0 & 1 & 0 & -1 & 0 & 1 & 0 & 1 & 0 & -1 & 0 & 1 & 0 & -1 \\
0 & 1 & -1 & 0 & 0 & 1 & -1 & 1 & -1 & 0 & 0 & 1 & -1 & 0 & 0 \\
1 & -1 & 0 & 0 & 1 & -1 & 0 & -1 & 0 & 0 & 1 & -1 & 0 & 0 & 1 \\
0 & 0 & 0 & -1 & 1 & 1 & -1 & -1 & 1 & 1 & -1 & 0 & 0 & 0 & 0 \\
1 & 1 & 0 & 0 & -1 & -1 & 0 & 0 & -1 & -1 & 0 & 0 & 1 & 1 & 0 \\
-1 & -1 & 0 & 1 & 0 & 0 & 1 & 1 & 0 & 0 & 1 & 0 & -1 & -1 & 0
\end{pmatrix}'
.
$$

Thus four optimum $\mathbf{W}$-matrices are constructed.

Remark 4.3.1 In Sects. 4.2 and 4.3, OCDs have been constructed for BIBD set-ups. The series of BIBDs considered here are either constructed through Bose's method of difference (cf. Bose 1939) or through projective geometry. As mentioned earlier, it is very difficult to find OCDs for arbitrary BIBDs. But for the particular case when $b = mv$, where m is any positive integer, OCDs can be constructed for arbitrary BIBDs. More generally, in such situation, OCDs can be constructed for any BPEBD which will be considered in Chap. 6. The class of BPEBDs contains cyclic designs which also contain a number of BIBDs. Though the method described in Chap. 6 covers a large class of BIBDs, but the methods applied in these sections are illustrative and important in their own merit.

4.4 D-Optimal Covariate Designs in Block Design Set-Up

The optimal designs considered in previous sections of this chapter are necessarily D-optimal. But such designs cannot always be obtained because of the restriction $n \equiv 0 \pmod 4$. When $n \neq 0 \pmod 4$, finding optimal design is very difficult. Dutta et al. (2010) consider D-optimal design in this set-up when $n \equiv 2 \pmod 4$. In this case of a block design for given b and v, the reduced normal equation for estimation of γ is given by

$$(\mathbf{Z}'\mathbf{Q}\mathbf{Z})\gamma = \mathbf{Z}'\mathbf{Q}\mathbf{y}$$

which yields

$$\hat{\gamma} = (\mathbf{Z}'\mathbf{Q}\mathbf{Z})^{-1}\mathbf{Z}'\mathbf{Q}\mathbf{y}$$

where

$$\mathbf{Q} = (\mathbf{I} - \mathbf{X}(\mathbf{X}'\mathbf{X})^{-}\mathbf{X}'), \quad \mathbf{X} = (\mathbf{X}_1, \mathbf{X}_2).$$

Hence, the information matrix for γ is given by $\mathbf{I}(\gamma) = \mathbf{Z}'\mathbf{Q}\mathbf{Z}$. Since $\mathbf{Q}$ is non-negative definite, it follows that

$$\mathbf{Z}'\mathbf{Q}\mathbf{Z} \leq \mathbf{Z}'\mathbf{Z} \text{ (in Lowener order sense; Pukelsheim 1993)}$$

'=' if and only if $\mathbf{Z}'\mathbf{X} = \mathbf{0}$, i.e., if and only if

$$\mathbf{Z}'\mathbf{X}_1 = \mathbf{0}, \quad \mathbf{Z}'\mathbf{X}_2 = \mathbf{0}. \tag{4.4.1}$$

Thus the problem is that of selecting $\mathbf{Z}$-matrix with $|z_{ij}^{(t)}| \leq 1$ satisfying (4.4.1) such that the covariate design is D-optimal, i.e., $det(\mathbf{Z}'\mathbf{Z})$ is maximum when $\mathbf{Z} \in \mathcal{Z}$, $\mathcal{Z} = \{\mathbf{Z} : z_{ij}^{(t)} \in [-1, 1] \, \forall \, i, j\}$.

4.4.1 Conditions for D-Optimality

We have already observed that when $n \equiv 2 \pmod 4$, it is impossible to estimate γ-components most efficiently in the sense of attaining the lower bound $\frac{\sigma^2}{n}$ to the variance of the estimated covariate parameters. Thus, in the case $n \equiv 2 \pmod 4$, the problem is that of choosing a matrix $\mathbf{Z}^{n \times c} = (z_{ij}^{(t)})$ with $z_{ij}^{(t)} \in [-1, 1] \, \forall \, i, j$ such that $det(\mathbf{Z}'\mathbf{Z})$ is a maximum subject to the orthogonality condition (4.4.1). Towards this, we state the following lemma giving a necessary condition for maximization of $det(\mathbf{Z}'\mathbf{Z})$, $\mathbf{Z} \in \mathcal{Z}$ (cf. Galil and Kiefer 1980; Wojtas 1964).

Lemma 4.4.1 *A necessary condition for maximization of $det(\mathbf{Z}'\mathbf{Z})$ where $\mathbf{Z} \in \mathcal{Z}$, is that $z_{ij}^{(t)} = \pm 1 \, \forall \, i, j, t$.*

From the above lemma, it is clear that we can restrict to the class $\mathcal{Z}^* = \{\mathbf{Z} : z_{ij}^{(t)} = \pm 1 \, \forall \, i, j, t\}$ for finding the D-optimum design. In this direction, we have the following theorem.

Theorem 4.4.1 *A covariate design $\mathbf{Z}^* \in \mathcal{Z}^*$ is D-optimal in the sense of maximizing $det(\mathbf{Z}'\mathbf{Z})$ subject to the condition (4.4.1), if it satisfies*

$$\mathbf{Z}^{*'}\mathbf{Z}^* = (n - 2)\mathbf{I}_c + 2\mathbf{J}_c \tag{4.4.2}$$

where $\mathbf{I}_c$ is the identity matrix of order c and $\mathbf{J}_c$ is the matrix of order c with all elements equal to unity.

Proof Because of Lemma 4.4.1, we can restrict to the class $\mathcal{Z}^*$ for maximization of $det(\mathbf{Z}'\mathbf{Z})$. For any $\mathbf{Z} \in \mathcal{Z}^*$, we can write

$$det(\mathbf{Z}'\mathbf{Z}) = det \begin{pmatrix} n & s_{12} & \dots & s_{1c} \\ s_{12} & n & \dots & s_{2c} \\ \vdots & \vdots & \vdots & \vdots \\ s_{1c} & s_{2c} & \dots & n \end{pmatrix}, \qquad (4.4.3)$$

where $s_{tt'} = \sum_i \sum_j z_{ij}^{(t)} z_{ij}^{(t')}, t \neq t' = 1, 2, \dots, c$. Because of (4.4.1), each column of $\mathbf{Z}$ is orthogonal to $\mathbf{1}_n$, and hence orthogonality of any pair of columns of $\mathbf{Z}$ implies that $n \equiv 0 \pmod 4$ which violates our assumption that $n \equiv 2 \pmod 4$. So, no off-diagonal element of $\mathbf{Z}'\mathbf{Z}$ can be zero. From Wojtas (1964) the determinant in (4.4.2) is maximum if all s_{ij}'s are equal to s, where

$$0 \leq s \leq \min_{i \neq j} |s_{ij}|. \qquad (4.4.4)$$

As $z_{ij}^{(t)} = \pm 1$ and $n \equiv 2 \pmod 4$, $|s_{ij}|$ can not be equal to 0 or 1 $\forall i \neq j$. Therefore, the minimum value of $|s_{ij}|$ is 2. So the theorem is proved. $\square$

Now we can represent any column of $\mathbf{Z}^*$ (which is a column vector of order $n \times 1$) in the form of a matrix $\mathbf{U}^{v \times b}$ corresponding to the $v \times b$ incidence matrix of the block design.

With the conditions (4.4.1) and (4.4.2) in terms of $\mathbf{U}$-matrix, the conditions reduce to:

C_1. Each $\mathbf{U}$-matrix has all column-sums equal to zero;
C_2. Each $\mathbf{U}$-matrix has all row-sums equal to zero;
C_3. The grand total of all the entries in the Hadamard product of any two distinct $\mathbf{U}$-matrices reduces to 2.   (4.4.5)

4.4.2 Construction of the D-Optimal Covariate Design in a SBIBD Set-Up

In Sect. 4.4.1, we have established that a $\mathbf{Z}$-matrix is D-optimal subject to condition (4.4.1) if it satisfies (4.4.2). Now in a BIBD set-up, the $\mathbf{U}$-matrices defined in Sect. 4.4.1 can be constructed by suitably replacing the non-zero elements of the incidence matrix of BIBD by ± 1 such that the conditions in (4.4.5) are satisfied. Here, we consider the series of irreducible SBIBD (cf. Raghavarao 1971) with parameters

$v = b$, $r = k = v - 1$, $\lambda = v - 2$, where $k \equiv 2 \pmod 4$, b is an odd integer. To start with, we consider the following lemma which gives a method of construction for particular value of the parameters viz. $v = b = 7$, $r = k = 6$ and $\lambda = 5$. This will help understand the method for the general case.

Lemma 4.4.2 *Three **U**-matrices can be constructed for the irreducible SBIBD with parameters $v = b = 7$, $r = k = 6$, $\lambda = 5$.*

Proof Without loss of generality the incidence matrix $\mathbf{N}^{7 \times 7}$ can be written in the following partitioned form:

$$\mathbf{N} = \left(\begin{array}{ccccc|cc} 0 & 1 & 1 & 1 & 1 & 1 & 1 \\ 1 & 0 & 1 & 1 & 1 & 1 & 1 \\ 1 & 1 & 0 & 1 & 1 & 1 & 1 \\ 1 & 1 & 1 & 0 & 1 & 1 & 1 \\ 1 & 1 & 1 & 1 & 0 & 1 & 1 \\ \hline 1 & 1 & 1 & 1 & 1 & 0 & 1 \\ 1 & 1 & 1 & 1 & 1 & 1 & 0 \end{array}\right). \tag{4.4.6}$$

Let us denote the 5×5 top left-hand matrix by $\mathbf{N}_{11}$; the 4×2 top right-hand matrix by $\mathbf{N}_{12}$; the 2×4 bottom left-hand matrix by $\mathbf{N}_{21}$; the 3×3 bottom right-hand matrix by $\mathbf{N}_{22}$. Then, we can write

$$\mathbf{N}_{11} = \begin{pmatrix} 0 & 1 & 1 & 1 & 1 \\ 1 & 0 & 1 & 1 & 1 \\ 1 & 1 & 0 & 1 & 1 \\ 1 & 1 & 1 & 0 & 1 \\ 1 & 1 & 1 & 1 & 0 \end{pmatrix}, \quad \mathbf{N}_{21} = \mathbf{N}'_{12} = \begin{pmatrix} 1 & 1 & 1 & 1 \\ 1 & 1 & 1 & 1 \end{pmatrix}, \quad \mathbf{N}_{22} = \begin{pmatrix} 0 & 1 & 1 \\ 1 & 0 & 1 \\ 1 & 1 & 0 \end{pmatrix}. \tag{4.4.7}$$

We see that '0' the element of 5th row and 5th column is common to both $\mathbf{N}_{11}$ and $\mathbf{N}_{22}$. We shall see later on that this particular element always remains static in this position during the process of construction. Such bordering of an element which is common both in $\mathbf{N}_{11}$ and $\mathbf{N}_{22}$ does not, in any way, hamper the construction of optimum $\mathbf{Z}$-matrix. Consider a Hadamard matrix $\mathbf{H}_4$ of order 4, where the first two columns are $\mathbf{h}_1 = (1, -1, 1, -1)'$ and $\mathbf{h}_2 = (1, 1, -1, -1)'$. Now we replace the non-zero elements of the first column of $\mathbf{N}_{11}$ by the elements of $\mathbf{h}_1$ and through cyclical development of this column we generate $\mathbf{U}_{11}^{(1)}$ of order 5×5 as

$$\mathbf{U}_{11}^{(1)} = \begin{pmatrix} 0 & -1 & 1 & -1 & 1 \\ 1 & 0 & -1 & 1 & -1 \\ -1 & 1 & 0 & -1 & 1 \\ 1 & -1 & 1 & 0 & -1 \\ -1 & 1 & -1 & 1 & 0 \end{pmatrix}. \tag{4.4.8}$$

Again with $\mathbf{h}_2$, we generate another matrix $\mathbf{U}_{11}^{(2)}$ in the same way. It can be checked that the row sums and column sums of each of $\mathbf{U}_{11}^{(1)}$ and $\mathbf{U}_{11}^{(2)}$ are equal to zero and the sum of all elements of the Hadamard product of these two matrices also vanishes. Next, by replacing the non-zero elements in the first column of $\mathbf{N}_{22}$ in (4.4.7) by $(1, -1)$, we get a column vector $(0, 1, -1)'$. By cyclically permutation of this column, we generate a 3×3 matrix $\mathbf{U}_{22}$ where

$$\mathbf{U}_{22} = \begin{pmatrix} 0 & -1 & 1 \\ 1 & 0 & -1 \\ -1 & 1 & 0 \end{pmatrix}. \tag{4.4.9}$$

Finally, we construct three 7×7 matrices $\mathbf{U}_1$, $\mathbf{U}_2$ and $\mathbf{U}_3$ corresponding to the incidence matrix $\mathbf{N}$, by replacing the matrices $\mathbf{N}_{11}$, $\mathbf{N}_{12}$, $\mathbf{N}_{21}$ and $\mathbf{N}_{22}$ in (4.4.7), respectively, by:

(a) $\mathbf{U}_{11}^{(1)}, \mathbf{U}_{12}^{(1)}, \mathbf{U}_{12}^{(1)\prime}, \mathbf{U}_{22}$;
(b) $\mathbf{U}_{11}^{(2)}, \mathbf{U}_{12}^{(2)}, \mathbf{U}_{12}^{(1)\prime}, -\mathbf{U}_{22}$; and
(c) $-\mathbf{U}_{11}^{(2)}, \mathbf{U}_{12}^{(2)}, \mathbf{U}_{12}^{(1)\prime}, -\mathbf{U}_{22}$.

Thus, finally, corresponding to (a)–(c) above, we have the following three U-matrices:

$$\mathbf{U}_1 = \begin{pmatrix} 0 & -1 & 1 & -1 & 1 & 1 & -1 \\ 1 & 0 & -1 & 1 & -1 & -1 & 1 \\ -1 & 1 & 0 & -1 & 1 & 1 & -1 \\ 1 & -1 & 1 & 0 & -1 & -1 & 1 \\ -1 & 1 & -1 & 1 & 0 & -1 & 1 \\ 1 & -1 & 1 & -1 & 1 & 0 & -1 \\ -1 & 1 & -1 & 1 & -1 & 1 & 0 \end{pmatrix}, \tag{4.4.10}$$

$$\mathbf{U}_2 = \begin{pmatrix} 0 & -1 & -1 & 1 & 1 & 1 & -1 \\ 1 & 0 & -1 & -1 & 1 & -1 & 1 \\ 1 & 1 & 0 & -1 & -1 & 1 & -1 \\ -1 & 1 & 1 & 0 & -1 & -1 & 1 \\ -1 & -1 & 1 & 1 & 0 & 1 & -1 \\ 1 & 1 & -1 & -1 & -1 & 0 & 1 \\ -1 & -1 & 1 & 1 & 1 & -1 & 0 \end{pmatrix}, \tag{4.4.11}$$

$$\mathbf{U}_3 = \begin{pmatrix} 0 & 1 & 1 & -1 & -1 & 1 & -1 \\ -1 & 0 & 1 & 1 & -1 & -1 & 1 \\ -1 & -1 & 0 & 1 & 1 & 1 & -1 \\ 1 & -1 & -1 & 0 & 1 & -1 & 1 \\ 1 & 1 & -1 & -1 & 0 & 1 & -1 \\ 1 & 1 & -1 & -1 & -1 & 0 & 1 \\ -1 & -1 & 1 & 1 & 1 & -1 & 0 \end{pmatrix}, \tag{4.4.12}$$

It can be easily checked that $\mathbf{U}_1$, $\mathbf{U}_2$ and $\mathbf{U}_3$ satisfy all of condition (4.4.5) and these constitute the required D-optimal covariate design.   □

Theorem 4.4.2 If a Hadamard matrix of order $(v - 7)$ exists, then we can construct three U-matrices for an irreducible SBIBD $(v = b,\ r = k = v - 1,\ \lambda = v - 2)$ where k is 2 (mod 4), $k > 6$.

Proof As in (4.4.6), we partition the incidence matrix $\mathbf{N}$ as

$$
\mathbf{N} = \left(
\begin{array}{ccccc|ccccccc}
0 & 1 & \cdots & 1 & 1 & 1 & 1 & 1 & 1 & 1 & 1 \\
1 & 0 & \cdots & 1 & 1 & 1 & 1 & 1 & 1 & 1 & 1 \\
\vdots & \vdots & \vdots & \vdots & \vdots & \vdots & \vdots & \vdots & \vdots & \vdots & \vdots \\
1 & 1 & \cdots & 0 & 1 & 1 & 1 & 1 & 1 & 1 & 1 \\ \hline
1 & 1 & \cdots & 1 & 0 & 1 & 1 & 1 & 1 & 1 & 1 \\
1 & 1 & \vdots & 1 & 1 & 0 & 1 & 1 & 1 & 1 & 1 \\
1 & 1 & \vdots & 1 & 1 & 1 & 0 & 1 & 1 & 1 & 1 \\
1 & 1 & \vdots & 1 & 1 & 1 & 1 & 0 & 1 & 1 & 1 \\
1 & 1 & \vdots & 1 & 1 & 1 & 1 & 1 & 0 & 1 & 1 \\
1 & 1 & \vdots & 1 & 1 & 1 & 1 & 1 & 1 & 0 & 1 \\
1 & 1 & \vdots & 1 & 1 & 1 & 1 & 1 & 1 & 1 & 0 \\
\end{array}
\right). \tag{4.4.13}
$$

As in Lemma 4.4.2, we denote the $(v - 6) \times (v - 6)$ top left-hand matrix by $\mathbf{N}_{11}^*$; the $(v - 7) \times 6$ top right-hand matrix by $\mathbf{N}_{12}^*$; the $6 \times (v - 7)$ bottom left-hand matrix by $\mathbf{N}_{21}^*$; the 7×7 bottom right-hand matrix by $\mathbf{N}_{22}^*$. Then we can write

$$\mathbf{N}_{11}^* = \mathbf{J}_{v-6} - \mathbf{I}_{v-6}, \quad \mathbf{N}_{12}^* = \mathbf{J}_{(v-7)\times 6} = \mathbf{N}_{21}^{*\prime}, \quad \mathbf{N}_{22}^* = \mathbf{J}_7 - \mathbf{I}_7, \tag{4.4.14}$$

where $\mathbf{I}_*$ is the identity matrix of order $(*)$, $\mathbf{J}_*$ is the matrix of order $(*)$ with all elements equal to unity.

Let the first three columns of a Hadamard matrix of order $(v - 7)$ be $\mathbf{h}_1^*$, $\mathbf{h}_2^*$ and $\mathbf{h}_3^*$. Following the same steps as in Lemma 4.4.1, we construct three matrices $\mathbf{U}_{11}^{(1)*}$, $\mathbf{U}_{11}^{(2)*}$ and $\mathbf{U}_{11}^{(3)*}$ each of order $(v - 6) \times (v - 6)$ corresponding to the matrix $\mathbf{N}_{11}^*$ of (4.4.14) with the help of $\mathbf{h}_1^*$, $\mathbf{h}_2^*$ and $\mathbf{h}_3^*$, respectively. Again for $\mathbf{N}_{12}^*$, we construct three matrices $\mathbf{V}_1$, $\mathbf{V}_2$ and $\mathbf{V}_3$ each of order $(v - 7) \times 6$ as

$$\mathbf{V}_1 = \mathbf{h}_1^* \otimes \mathbf{a}', \ \mathbf{V}_2 = \mathbf{h}_2^* \otimes \mathbf{a}' \text{ and } \mathbf{V}_3 = \mathbf{h}_3^* \otimes \mathbf{a}', \text{ where } \mathbf{a}' = (1, -1, 1, -1, 1, -1).$$

Now using $\mathbf{U}_i$ from (4.4.10) to (4.4.12) for $\mathbf{N}_{22}^*$ and $\mathbf{U}_{11}^{(i)*}$, $\mathbf{V}_i$, $\mathbf{V}_i'$ for $\mathbf{N}_{11}^*$, $\mathbf{N}_{12}^*$ and $\mathbf{N}_{21}^*$ of (4.4.14), respectively, $i = 1, 2, 3$, we get the D-optimal design.   □

Example 4.4.1 Let us consider a SBIBD with parameters $v = b = 11$, $r = k = 10$, $\lambda = 9$ where the initial block is $(1, 2, 3, 4, 5, 6, 7, 8, 9, 10)$ mod 11. The incidence matrix can be displayed as

$$
\mathbf{N} = \left(\begin{array}{ccccc|cccccc}
0 & 1 & 1 & 1 & 1 & 1 & 1 & 1 & 1 & 1 & 1 \\
1 & 0 & 1 & 1 & 1 & 1 & 1 & 1 & 1 & 1 & 1 \\
1 & 1 & 0 & 1 & 1 & 1 & 1 & 1 & 1 & 1 & 1 \\
1 & 1 & 1 & 0 & 1 & 1 & 1 & 1 & 1 & 1 & 1 \\
1 & 1 & 1 & 1 & 0 & 1 & 1 & 1 & 1 & 1 & 1 \\ \hline
1 & 1 & 1 & 1 & 1 & 0 & 1 & 1 & 1 & 1 & 1 \\
1 & 1 & 1 & 1 & 1 & 1 & 0 & 1 & 1 & 1 & 1 \\
1 & 1 & 1 & 1 & 1 & 1 & 1 & 0 & 1 & 1 & 1 \\
1 & 1 & 1 & 1 & 1 & 1 & 1 & 1 & 0 & 1 & 1 \\
1 & 1 & 1 & 1 & 1 & 1 & 1 & 1 & 1 & 0 & 1 \\
1 & 1 & 1 & 1 & 1 & 1 & 1 & 1 & 1 & 1 & 0
\end{array}\right).
$$

The first $\mathbf{U}$-matrix is given by:

$$
\mathbf{U}_1 = \left(\begin{array}{ccccc|cccccc}
0 & -1 & 1 & -1 & 1 & 1 & -1 & 1 & -1 & 1 & -1 \\
1 & 0 & -1 & 1 & -1 & -1 & 1 & -1 & 1 & -1 & 1 \\
-1 & 1 & 0 & -1 & 1 & 1 & -1 & 1 & -1 & 1 & -1 \\
1 & -1 & 1 & 0 & -1 & -1 & 1 & -1 & 1 & -1 & 1 \\
-1 & 1 & -1 & 1 & 0 & -1 & 1 & -1 & 1 & 1 & -1 \\ \hline
1 & -1 & 1 & -1 & 1 & 0 & -1 & 1 & -1 & -1 & 1 \\
-1 & 1 & -1 & 1 & -1 & 1 & 0 & -1 & 1 & 1 & -1 \\
1 & -1 & 1 & -1 & 1 & -1 & 1 & 0 & -1 & -1 & 1 \\
-1 & 1 & -1 & 1 & -1 & 1 & -1 & 1 & 0 & -1 & 1 \\
1 & -1 & 1 & -1 & 1 & -1 & 1 & -1 & 1 & 0 & -1 \\
-1 & 1 & -1 & 1 & -1 & 1 & -1 & 1 & -1 & 1 & 0
\end{array}\right).
$$

Similarly, we can construct the other two.

Remark 4.4.1 The proposed design is also optimal with respect to any Type I criteria in the class of $\mathcal{Z}^* = \{\mathbf{Z}^{*n \times c} : z_{ij}^{(t)} = \pm 1 \,\forall\, i, j, t\}$, rank$(\mathbf{Z}^*) = c$ (cf. Cheng 1980).

References

Bose RC (1939) On the construction of balanced incomplete block designs. Ann Eugen 9:353–399
Cheng CS (1980) Optimality of some weighing and 2^n fractional factorial designs. Ann Stat 8:436–446
Das K, Mandal NK, Sinha BK (2003) Optimal experimental designs with covariates. J Stat Plan Inference 115:273–285
Dutta G (2004) Optimum choice of covariates in BIBD set-up. Calcutta Stat Assoc Bull 55:39–55
Dutta G, Das P, Mandal NK (2007) Optimum choice of covariates for a series of SBIBDs obtained through projective geometry. J Mod Appl Stat Methods 6:649–656

Dutta G, Das P, Mandal NK (2010) D-optimal designs for covariate parameters in block design set-up. Commun Stat Theory Methods 39:3434–3443

Galil Z, Kiefer J (1980) D-optimum weighing designs. Ann Stat 8:1293–1306

Hedayat AS, Sloane NJA, Stufken J (1999) Orthogonal arrays: theory and applications. Springer, New York

Pukelsheim F (1993) Optimal design of experiments. Wiley, New York

Raghavarao D (1971) Construction and combinatorial problems in design of experiments. Wiley, New York

Wojtas M (1964) On Hadamard's inequality for the determinants of order non-divisible by 4. Colloq Math 12:73–83

Chapter 5
OCDs in Group Divisible Design Set-Up

5.1 Introduction

It is observed that the BIBDs are restrictive as every pair of treatments should occur equal number of times. As a result the availability of OCDs in this set-up becomes limited. In this context, it is observed that PBIBDs are less restrictive and at the same time are popular among practitioners. So it is desirable to have OCDs involving these set-ups. Dutta et al. (2009) have considered the problem of construction of OCDs in the series of PBIBDs which are obtained not only through the method of differences but also are obtained by other methods as described by Bose et al. (1953), Zelen (1954) and Vartak (1954). In this chapter, we will only confine to GDDs and discuss methods of construction of the OCDs based on GDDs.

To construct OCDs we have often applied two matrix-products, viz. Kronecker product and Khatri-Rao product. The definitions of the matrix-products can be found in Rao (1973), p. 29–30, where the Khatri-Rao product has been termed as 'New Product'. For completeness we reproduce the two definitions below:

Definition 5.1.1 (*Kronecker-Product*) Let $\mathbf{A} = (a_{ij})$ and $\mathbf{B} = (b_{ij})$ be two matrices of orders $m \times n$ and $p \times q$ respectively. Then the Kronecker product of $\mathbf{A}$ and $\mathbf{B}$, denoted by $\mathbf{A} \otimes \mathbf{B}$, is defined to be an $mp \times nq$ matrix expressible as a partitioned matrix with $a_{ij}\mathbf{B}$ as the (i, j)th partition, $i = 1, 2, \ldots, m$ and $j = 1, 2, \ldots, n$, i.e.

$$\mathbf{A} \otimes \mathbf{B} = (a_{ij}\mathbf{B}). \tag{5.1.1}$$

Definition 5.1.2 (*Khatri-Rao Product*) Let $\mathbf{A} = (\mathbf{A}_1, \ldots, \mathbf{A}_k)$ and $\mathbf{B} = (\mathbf{B}_1, \ldots, \mathbf{B}_k)$ be two partitioned matrices with the same number of partitions. Then the Khatri-Rao product of $\mathbf{A}$ and $\mathbf{B}$, denoted as $\mathbf{A} \odot \mathbf{B}$, is defined by

$$\mathbf{A} \odot \mathbf{B} = (\mathbf{A}_1 \otimes \mathbf{B}_1, \ldots, \mathbf{A}_k \otimes \mathbf{B}_k). \tag{5.1.2}$$

© Springer India 2015
P. Das et al., *Optimal Covariate Designs*, DOI 10.1007/978-81-322-2461-7_5

5.2 Optimum Covariate Designs

In this section, we mainly confine to the work in Dutta et al. (2009) and describe the construction of OCDs described therein for different series of GDDs.

These designs are based on the concept of association scheme with respect to PBIBDs, which is defined below for the sake of completeness.

Definition 5.2.1 Given v symbols $1, 2, \ldots, v$, a relation satisfying the following conditions is said to be an association scheme with m classes:

1. Any two treatments are either 1st, 2nd, ..., or mth associates, the relation of association being symmetrical; that is, if the symbol α is the ith associate of the symbol β, then β is the ith associate of α.
2. Each treatment α has n_i ith associates, the number n_i being independent of α.
3. If any two treatments α and β are ith associates, then the number of symbols that are jth associates of α, and kth associates of β, is p^i_{jk} and is independent of the pair of ith associates α and β.

The numbers v, n_i $(i = 1, 2, \ldots, m)$ and p^i_{jk} $(i, j, k = 1, 2, \ldots, m)$ are called the parameters of the association scheme.

Given an association scheme for the v treatments, we define a PBIBD as follows:

Definition 5.2.2 Given an association scheme with m classes and given parameters as above, we get a PBIBD with m associate classes if the v symbols are arranged into b blocks of size k $(< v)$ such that

1. Every symbol occurs at most once in a block.
2. Every symbol occurs in exactly r blocks.
3. If two symbols α and β are ith associates, then they occur together in λ_i blocks, the number λ_i being independent of the particular pair of ith associates α and β.

The numbers v, b, r, k, λ_i $(i = 1, 2, \ldots, m)$ are called the parameters of the design. Two-associate class PBIBDs were classified by Bose and Shimamoto (1952) in the following types depending on the association schemes:

1. Group divisible (GD)
2. Simple (SI)
3. Triangular (T)
4. Latin-square type (L_i)
5. Cyclic (C).

In the context of cyclic design, more refined definition has been suggested by Nandi and Adhikari (1966). However, our consideration of OCDs will be based only on the GDDs. For the other types, we refer to Dutta et al. (2009).

Definition 5.2.3 (*GD association scheme and design*) For integers $m \geq 2$ and $n \geq 2$, consider $v = mn$ treatments, which are divided in an m groups is containing n treatments. Any two treatments of the same group are called first associate and any two treatments for different groups are called 2nd associate. The parameters of the GD association scheme are as follows:

$$v = mn, \; n_1 = n - 1, n_2 = n(m - 1),$$

$$\mathbf{P}_1 = (p_{ij}^1) = \begin{pmatrix} n - 2 & 0 \\ 0 & n(m - 1) \end{pmatrix}, \; \mathbf{P}_2 = (p_{ij}^2) = \begin{pmatrix} 0 & n - 1 \\ n - 1 & n(m - 1) \end{pmatrix}. \quad (5.2.1)$$

A PBIBD is said to be group-divisible if it is based on the GD association scheme.

If $\mathbf{N}$ be the incidence matrix of GD design then the characteristic roots θ_i of the $\mathbf{NN'}$ matrix and the respective multiplicity α_i, $i = 0, 1, 2$ are given by

$$\begin{aligned} \theta_0 &= rk, \quad \alpha_0 = 1 \\ \theta_1 &= r - \lambda_1, \quad \alpha_1 = m(n - 1), \\ \theta_2 &= rk - v\lambda_2, \quad \alpha_2 = m - 1. \end{aligned} \quad (5.2.2)$$

A GD design is called

(a) singular if $r = \lambda_1$;
(b) semi-regular, if $r > \lambda_1$ and $rk = v\lambda_2$;
(c) regular, if $r > \lambda_1$ and $rk > v\lambda_2$.

Note 5.2.1 In what follows, the incidence matrices of the relevant designs are represented in terms of their transposes, keeping the same style as in the case of BIBDs followed in earlier chapters.

5.2.1 Singular Group Divisible Design (SGDD) Set-Up

It had been shown in Bose et al. (1953) that if in a BIBD with parameters v^*, b^*, r^*, k^* and λ^* each treatment is replaced by a group of n treatments, an SGDD can be obtained with parameters

$$v = nv^*, \; b = b^*, \; r = r^*, \; k = nk^*, \lambda_1 = r^*, \; \lambda_2 = \lambda^*, \; m = v^*, \; n = n. \quad (5.2.3)$$

Here m stands for the number of groups in the corresponding association scheme. It will be seen that $\mathbf{W}$-matrices for such an SGDD with parameters in (5.2.3) can be constructed and the construction of $\mathbf{W}$ in this case does not depend on the method of construction of the corresponding BIBD.

Theorem 5.2.1 *A set of t optimum $\mathbf{W}$-matrices can be constructed for the SGDD with parameters in (5.2.3), where*

(i) *$t = c$, if c optimum $\mathbf{W}$-matrices exist for an RBD with n treatments and r blocks;*
(ii) *$t = v^*(n - 1)(r - 1)$, if $\mathbf{H}_{v^*}$, $\mathbf{H}_n$ and $\mathbf{H}_r$ exist;*

(iii) $t = v^*((n-1)(r-1) - (n-2))$,

   (a) *if $n \equiv 2 \ (mod \ 4)$, $(n-1)$ is a prime or a prime power and $\mathbf{H}_{v^*}$ and $\mathbf{H}_r$ exist;*
   or
   (b) *if $\mathbf{H}_{v^*}$, $\mathbf{H}_{2n}$ and $\mathbf{H}_{\frac{r}{2}}$ exist;*

(iv) $t = v^*$ *if $n = even$, $r = even$ and $\mathbf{H}_{v^*}$ exists.*

Proof Consider the SGDD with parameters in (5.2.3) obtained by replacing each treatment of the BIBD(v, b, r, k, λ) by a group of n treatments. Let the n treatments of the SGDD corresponding to the treatment θ_i $(i = 1, 2, \ldots, v^*)$ of the BIBD be denoted by $(\theta_{1i}, \theta_{2i}, \ldots, \theta_{ni})$ and the transpose of the partitioned incidence matrix of the SGDD be denoted by

$$\mathbf{N}' = \left(\mathbf{N}'_1, \ \mathbf{N}'_2, \ldots, \mathbf{N}'_i, \ldots, \mathbf{N}'_{v^*} \right) \tag{5.2.4}$$

where $\mathbf{N}_i$ is the incidence matrix corresponding to $(\theta_{1i}, \theta_{2i}, \ldots, \theta_{ni})$; $i = 1, 2, \ldots, v^*$. If the rows of $\mathbf{N}_i$ containing the null elements only are omitted, then the reduced matrix corresponds to the incidence matrix of an RBD with n treatments arranged in r blocks. We denote an RBD with r blocks and n treatments by RBD(n, r). This is true for all i. For the time being, let it be assumed that c optimum W-matrices for an RBD(n, r) exist and let them be denoted by $\mathbf{W}_1, \mathbf{W}_2, \ldots, \mathbf{W}_c$. Putting the elements of $\mathbf{W}_j$ of RBD(n, r) in the corresponding non-zero positions of each $\mathbf{N}_i$, a matrix $\mathbf{W}_j^*$ is obtained and let its transpose be written as

$$\mathbf{W}_j^{*'} = (\mathbf{W}_{1j}^{*'}, \ \mathbf{W}_{2j}^{*'}, \ldots, \mathbf{W}_{v^*j}^{*'}). \tag{5.2.5}$$

It is easy to verify that each of $\mathbf{W}_1^*, \mathbf{W}_2^*, \ldots, \mathbf{W}_c^*$ give optimum W-matrices for the SGDD (5.2.3) and thus (i) of the theorem follows.

   Again if $\mathbf{H}_{v^*}$ exists then the number of optimum W-matrices can be increased by application of Khatri-Rao product. Let $\mathbf{H}_{v^*}$ be written as

$$\mathbf{H}_{v^*} = (h_{lm}), \ \text{where } h_{lm} \text{ is the } (l, m)\text{th element of } \mathbf{H}_{v^*}.$$

   For $l = 1, 2, \ldots, v^*$, a matrix $\mathbf{W}_{lj}^{**}$ is constructed by Khatri-Rao product where the transpose of $\mathbf{W}_{lj}^{**}$ is

$$\mathbf{W}_{lj}^{**'} = \mathbf{h}_l \odot \mathbf{W}_j^{*'} = (h_{l1}\mathbf{W}_{1j}^{*'}, h_{l2}\mathbf{W}_{2j}^{*'}, \ldots, h_{li}\mathbf{W}_{ij}^{*'}, \ldots, h_{lv^*}\mathbf{W}_{v^*j}^{*'}), \tag{5.2.6}$$

where $\mathbf{h}_l$ is the lth row of $\mathbf{H}_{v^*}$. Now varying l and j, v^*c optimum $\mathbf{W}_{lj}^{**}$-matrices can be constructed and it can be easily checked that these matrices satisfy the condition (3.1.5).

   It is proved in Chap. 3 that the values of c are (a_1) $(n-1)(r-1)$, if $\mathbf{H}_n$ and $\mathbf{H}_r$ exist; (a_2) $(n-1)(r-1) - (n-2)$, if $n \equiv 2 \ (mod \ 4)$, $(n-1)$ is a prime or a prime power and $\mathbf{H}_r$ exists and (a_3) $(n-1)(r-1) - (n-2)$, if $\mathbf{H}_{2n}$ and $\mathbf{H}_{\frac{r}{2}}$ exist. These values imply, respectively, (ii), (iii) of the theorem when $\mathbf{H}_{v^*}$ exists.

Again if n and r are even, we can write a $n \times r$ matrix $\mathbf{W}_1$ as

$$\mathbf{W}_1 = \begin{pmatrix} \mathbf{J} & -\mathbf{J} \\ -\mathbf{J} & \mathbf{J} \end{pmatrix}$$

where $\mathbf{J}$ is a $\frac{n}{2} \times \frac{r}{2}$ matrix with all elements unity. It is easy to see that $\mathbf{W}_1$ gives an optimum $\mathbf{W}$-matrix for an RBD(n, r). Thus (iv) of the theorem follows. □

Remark 5.2.1 Exchanging the roles of r and n in (iii) of Theorem 5.2.1, we may get $t = v^*((n-1)(r-1) - (r-2))$ optimum $\mathbf{W}$-matrices for the SGDD with parameters in (5.2.3) if $r \equiv 2 \pmod 4$, $(r-1)$ is a prime or a prime power, $\mathbf{H}_{v^*}$ and $\mathbf{H}_n$ exist or $\mathbf{H}_{v^*}$, $\mathbf{H}_{2r}$ and $\mathbf{H}_{\frac{n}{2}}$ exist.

Remark 5.2.2 If v^* is an even integer, then a set of t optimum $\mathbf{W}$-matrices can be constructed for the SGDD with parameters (5.2.3) by using $\mathbf{1}'_{v^*}$ and $(\mathbf{1}'_{\frac{v^*}{2}}, -\mathbf{1}'_{\frac{v^*}{2}})$ respectively in place of the rows of $\mathbf{H}_{v^*}$ in (5.2.6). Again from (ii)–(iii) of Theorem 5.2.1 it follows that

(i) $t = 2(n-1)(r-1)$ if $\mathbf{H}_n$ and $\mathbf{H}_r$ exist;
(ii) $t = 2((n-1)(r-1) - (n-2))$ If $n \equiv 2 \pmod 4$, $(n-1)$ is a prime or a prime power and $\mathbf{H}_r$ exists or if $\mathbf{H}_{2n}$ and $\mathbf{H}_{\frac{r}{2}}$ exist;

respectively.

Remark 5.2.3 It is easily seen that for the construction of optimum $\mathbf{W}$-matrices for RBD(n, r), it is necessary that r and n must be even. If r, n and v^* are even but none of them are multiple of 4, then 2 optimum $\mathbf{W}$-matrices can always be constructed for the SGDD with parameters (5.2.3) by using two orthogonal rows as in Remark 5.2.2.

Remark 5.2.4 Suppose t_1 optimum $\mathbf{W}$-matrices exist for the BIBD$(v^*, b^*, r^*, k^*, \lambda^*)$; then additional t_1 optimum $\mathbf{W}$-matrices, orthogonal to the previous ones, can be constructed for a SGDD with parameters given in (5.2.3).

We give some examples illustrating (i), (ii) and (iv) of Theorem 5.2.1 and Remark 5.2.4.

Example 5.2.1 Consider a BIBD with parameters $v^* = b^* = 3$, $r^* = k^* = 2$, $\lambda^* = 1$ with the incidence matrix

$$\mathbf{N}^* = \begin{pmatrix} 1 & 1 & 0 \\ 0 & 1 & 1 \\ 1 & 0 & 1 \end{pmatrix}.$$

Now for $n = 2$, the SGDD with parameters $v = 6$, $b = 3$, $r = 2$, $k = 4$, $\lambda_1 = 2$, $\lambda_2 = 1$, $m = 3$, $n = 2$ has the transpose of the incidence matrix,

$$\mathbf{N}' = \begin{pmatrix} 1 & 1 & 1 & 1 & 0 & 0 \\ 0 & 0 & 1 & 1 & 1 & 1 \\ 1 & 1 & 0 & 0 & 1 & 1 \end{pmatrix} = \begin{pmatrix} \mathbf{1}'_2 & \mathbf{1}'_2 & \mathbf{0}' \\ \mathbf{0}' & \mathbf{1}'_2 & \mathbf{1}'_2 \\ \mathbf{1}'_2 & \mathbf{0}' & \mathbf{1}'_2 \end{pmatrix}.$$

H_2 is written as

$$H_2 = \begin{pmatrix} 1 & 1 \\ 1 & -1 \end{pmatrix} = (h_1^*, h_2^*).$$

Applying the method described in Theorem 3.4.1 and using h_2^*, only one W-matrix for RBD(2,2) can be constructed and it is given by

$$W_1 = \begin{pmatrix} 1 & -1 \\ -1 & 1 \end{pmatrix} = \begin{pmatrix} h_2^{*\prime} \\ -h_2^{*\prime} \end{pmatrix}.$$

Using W_1, only one W-matrix for above SGDD can be constructed (vide Eq. (5.2.6)) and its transpose is given by

$$W_1^{*\prime} = \begin{pmatrix} 1 & -1 & 1 & -1 & 0 & 0 \\ 0 & 0 & -1 & 1 & 1 & -1 \\ -1 & 1 & 0 & 0 & -1 & 1 \end{pmatrix} = \begin{pmatrix} h_2^{*\prime} & h_2^{*\prime} & 0' \\ 0' & -h_2^{*\prime} & h_2^{*\prime} \\ -h_2^{*\prime} & 0' & -h_2^{*\prime} \end{pmatrix}.$$

Again, there exists a W-matrix for the BIBD (cf. Chap. 4) which is given by

$$W_{(1)} = \begin{pmatrix} 1 & 0 & -1 \\ -1 & 1 & 0 \\ 0 & -1 & 1 \end{pmatrix}.$$

Using $W_{(1)}$, one more W-matrix for the SGDD can be constructed through Kronecker product and its transpose is given by

$$W_{(1)}^{*\prime} = W_{(1)}' \otimes 1_2' = \begin{pmatrix} 1_2' & -1_2' & 0' \\ 0' & 1_2' & -1_2' \\ -1_2' & 0' & 1_2' \end{pmatrix} = \begin{pmatrix} 1 & 1 & -1 & -1 & 0 & 0 \\ 0 & 0 & 1 & 1 & -1 & -1 \\ -1 & -1 & 0 & 0 & 1 & 1 \end{pmatrix}.$$

It is easy to check that $W_{(1)}^*$ is orthogonal to W_1^*.

Example 5.2.2 Consider a BIBD with parameters $v^* = 4$, $b^* = 24$, $r^* = 12$, $k^* = 2$, $\lambda^* = 4$ (this is obtained by repeating BIBD(4, 6, 3, 2, 1) 4 times) with the transpose of the incidence matrix

$$N^{*\prime} = 1_4 \otimes \begin{pmatrix} 1 & 1 & 0 & 0 \\ 0 & 1 & 1 & 0 \\ 0 & 0 & 1 & 1 \\ 1 & 0 & 0 & 1 \\ 1 & 0 & 1 & 0 \\ 0 & 1 & 0 & 1 \end{pmatrix}.$$

The SGDD with parameters $v = 16$, $b = 24$, $r = 12$, $k = 8$, $\lambda_1 = 12$, $\lambda_2 = 4$, $m = 4$, $n = 4$ is obtained by replacing each treatment of the BIBD with $n = 4$ treatments. The transpose of the incidence matrix $\mathbf{N}^{16 \times 24}$ of SGDD can be written as

$$\mathbf{N}' = (\mathbf{N}^{*\prime} \otimes \mathbf{1}_4') = \mathbf{1}_4 \otimes \begin{pmatrix} \mathbf{1}_4' & \mathbf{1}_4' & \mathbf{0}' & \mathbf{0}' \\ \mathbf{0}' & \mathbf{1}_4' & \mathbf{1}_4' & \mathbf{0}' \\ \mathbf{0}' & \mathbf{0}' & \mathbf{1}_4' & \mathbf{1}_4' \\ \mathbf{1}_4' & \mathbf{0}' & \mathbf{0}' & \mathbf{1}_4' \\ \mathbf{1}_4' & \mathbf{0}' & \mathbf{1}_4' & \mathbf{0}' \\ \mathbf{0}' & \mathbf{1}_4' & \mathbf{0}' & \mathbf{1}_4' \end{pmatrix} = \left(\mathbf{N}_1'^{\,24 \times 4}, \mathbf{N}_2'^{\,24 \times 4}, \mathbf{N}_3'^{\,24 \times 4}, \mathbf{N}_4'^{\,24 \times 4} \right).$$

$\mathbf{H}_4$ and $\mathbf{H}_{12}$ are written as follows:

$$\mathbf{H}_4 = \begin{pmatrix} 1 & 1 & 1 & 1 \\ 1 & -1 & 1 & -1 \\ 1 & -1 & -1 & 1 \\ 1 & 1 & -1 & -1 \end{pmatrix} = \begin{pmatrix} \mathbf{h}_1 \\ \mathbf{h}_2 \\ \mathbf{h}_3 \\ \mathbf{h}_4 \end{pmatrix} = \left(\mathbf{h}_1^*, \mathbf{h}_2^*, \mathbf{h}_3^*, \mathbf{h}_4^* \right), \qquad (5.2.7)$$

$$\mathbf{H}_{12} = \begin{pmatrix} 1 & 1 & -1 & 1 & -1 & -1 & -1 & 1 & 1 & 1 & -1 & 1 \\ 1 & -1 & 1 & 1 & -1 & 1 & -1 & -1 & -1 & 1 & 1 & 1 \\ 1 & 1 & -1 & 1 & 1 & -1 & 1 & -1 & -1 & -1 & 1 & 1 \\ 1 & 1 & 1 & -1 & 1 & 1 & -1 & 1 & -1 & -1 & -1 & 1 \\ 1 & -1 & -1 & -1 & 1 & 1 & 1 & -1 & 1 & 1 & -1 & 1 \\ 1 & -1 & 1 & -1 & -1 & 1 & 1 & 1 & -1 & 1 & 1 & 1 \\ 1 & -1 & -1 & -1 & -1 & -1 & -1 & -1 & -1 & -1 & -1 & -1 \\ 1 & 1 & 1 & -1 & 1 & -1 & -1 & -1 & 1 & 1 & 1 & -1 \\ 1 & 1 & 1 & 1 & -1 & 1 & 1 & -1 & 1 & -1 & -1 & -1 \\ 1 & -1 & 1 & 1 & 1 & -1 & 1 & 1 & -1 & 1 & -1 & -1 \\ 1 & -1 & -1 & 1 & 1 & 1 & -1 & 1 & 1 & -1 & 1 & -1 \\ 1 & 1 & -1 & -1 & 1 & 1 & 1 & 1 & -1 & 1 & 1 & -1 \end{pmatrix} = \left(\mathbf{h}_1^{**}, \mathbf{h}_2^{**}, \ldots, \mathbf{h}_{11}^{**}, \mathbf{h}_{12}^{**} \right),$$

Now we define the matrix: $\mathbf{U}_{i,j}^{(12 \times 4)} = \mathbf{h}_i^* \otimes \mathbf{h}_j^{**\prime}$; $\forall i = 2, 3, 4$; $j = 2, 3, \ldots, 12$.

It can easily be checked that these 33 $\mathbf{U}_{i,j}$'s give the optimum $\mathbf{W}$-matrices for an RBD(4, 12). We write $\mathbf{U}_{2,1} = \mathbf{W}^{(1)}, \mathbf{U}_{2,2} = \mathbf{W}^{(2)}, \ldots, \mathbf{U}_{4,11} = \mathbf{W}^{(33)}$ respectively. Let us consider

$$\begin{aligned} \mathbf{W}^{(1)} &= \mathbf{h}_2^* \otimes \mathbf{h}_2^{**\prime} \\ &= (1, -1, 1, -1)' \otimes (1, -1, 1, 1, -1, -1, -1, 1, 1, -1, -1, 1) \\ &= \begin{pmatrix} 1 & -1 & 1 & 1 & -1 & -1 & -1 & 1 & 1 & -1 & -1 & 1 \\ -1 & 1 & -1 & -1 & 1 & 1 & 1 & -1 & -1 & 1 & 1 & -1 \\ 1 & -1 & 1 & 1 & -1 & -1 & -1 & 1 & 1 & -1 & -1 & 1 \\ -1 & 1 & -1 & -1 & 1 & 1 & 1 & -1 & -1 & 1 & 1 & -1 \end{pmatrix} \\ &= (\mathbf{a}, -\mathbf{a}, \mathbf{a}, \mathbf{a}, -\mathbf{a}, -\mathbf{a}, -\mathbf{a}, \mathbf{a}, \mathbf{a}, -\mathbf{a}, -\mathbf{a}, \mathbf{a}), \end{aligned}$$

where $\mathbf{a} = \mathbf{h}_2^*$ is of order 4×1 (vide (5.2.7)).

By putting the elements of $\mathbf{W}^{(1)}$ in the non-zero positions of each $\mathbf{N}_i$ ($i = 1, 2, 3, 4$), $\mathbf{W}_1^*$ is obtained and its transpose is written as

$$\mathbf{W}_1^{*\prime} = \left(\mathbf{W}_{11}^{*\prime}, \mathbf{W}_{21}^{*\prime}, \mathbf{W}_{31}^{*\prime}, \mathbf{W}_{41}^{*\prime}\right), \tag{5.2.8}$$

where

$\mathbf{W}_{11}^{*} = (a, 0, 0, -a, a, 0, a, 0, 0, -a, -a, 0, -a, 0, 0, a, a, 0, -a, 0, 0, -a, a, 0)$
$\mathbf{W}_{21}^{*} = (a, -a, 0, 0, 0, a, a, -a, 0, 0, 0, -a, -a, a, 0, 0, 0, a, -a, -a, 0, 0, 0, a)$
$\mathbf{W}_{31}^{*} = (0, a, -a, 0, a, 0, 0, a, -a, 0, -a, 0, 0, -a, a, 0, a, 0, 0, -a, -a, 0, a, 0)$
$\mathbf{W}_{41}^{*} = (0, 0, a, -a, 0, a, 0, 0, a, -a, 0, -a, 0, 0, -a, a, 0, a, 0, 0, -a, -a, 0, a).$

In this way, by using the remaining 32 $\mathbf{W}$-matrices of RBD(4, 12), we get another 32 $\mathbf{W}_j^*$'s, $j = 2, 3, \ldots, 33$.

Now by taking the Khatri-Rao product of the $\mathbf{h}_2$ of (5.2.7) and the matrix $\mathbf{W}_1^*$ in (5.2.8), we get $\mathbf{W}_{2,1}^{**}$ whose transpose is

$$\mathbf{W}_{2,1}^{**\prime} = \mathbf{h}_2 \otimes \mathbf{W}_1^{*\prime} = \left(\mathbf{W}_{11}^{*\prime}, -\mathbf{W}_{21}^{*\prime}, \mathbf{W}_{31}^{*\prime}, -\mathbf{W}_{41}^{*\prime}\right).$$

Similarly, by taking the Khatri-Rao product of $(\mathbf{h}_1, \mathbf{W}_1^*)$, $(\mathbf{h}_3, \mathbf{W}_1^*)$ and $(\mathbf{h}_4, \mathbf{W}_1^*)$, respectively, three other optimum $\mathbf{W}$-matrices, i.e. $\mathbf{W}_{1,1}^{**}$, $\mathbf{W}_{3,1}^{**}$, $\mathbf{W}_{4,1}^{**}$ can be constructed. As before, by using different rows of $\mathbf{H}_4$ and other 32 $\mathbf{W}_j^*$'s, we get additional 128 optimum $\mathbf{W}$-matrices for the said SGDD.

Moreover, there exist three optimum $\mathbf{W}$-matrices for the BIBD and these are constructed by the method described in Chap. 4 and are given by

$$(\mathbf{h}_1 \otimes \mathbf{U}^*)'; \quad (\mathbf{h}_2 \otimes \mathbf{U}^*)'; \quad (\mathbf{h}_3 \otimes \mathbf{U}^*)';$$

where

$$\mathbf{U}^* = \begin{pmatrix} 1 & -1 & 0 & 0 \\ 0 & 1 & -1 & 0 \\ 0 & 0 & 1 & -1 \\ 1 & 0 & 0 & -1 \\ 1 & 0 & -1 & 0 \\ 0 & 1 & 0 & -1 \end{pmatrix}.$$

Using these three $\mathbf{W}$-matrices of BIBD, we can construct three more optimum $\mathbf{W}$-matrices for the SGDD as described in Remark 5.2.4 and it is easy to see that these three are orthogonal to previous 132 optimum $\mathbf{W}$-matrices. So we get 135 optimum $\mathbf{W}$-matrices in all for the said SGDD.

Example 5.2.3 Consider a BIBD with parameters $v^* = 4$, $b^* = 8$, $r^* = 6$, $k^* = 3$, $\lambda^* = 4$ with the transpose of the incidence matrix

$$\mathbf{N}^{*\prime} = \begin{pmatrix} 1 \\ 1 \end{pmatrix} \otimes \begin{pmatrix} 1 & 1 & 1 & 0 \\ 0 & 1 & 1 & 1 \\ 1 & 0 & 1 & 1 \\ 1 & 1 & 0 & 1 \end{pmatrix}.$$

An SGDD with parameters $v = 8$, $b = 8$, $r = 6$, $k = 6$, $\lambda_1 = 6$, $\lambda_2 = 4$, $m = 4$, $n = 2$ is obtained by replacing each treatment of the BIBD with $n = 2$ treatments. The transpose of the incidence matrix $\mathbf{N}$ of SGDD can be written as

$$\mathbf{N}' = \mathbf{N}^{*\prime} \otimes (1,\ 1).$$

$\mathbf{H}_2$ is written as

$$\mathbf{H}_2 = \begin{pmatrix} 1 & 1 \\ 1 & -1 \end{pmatrix} = (\mathbf{h}_1^* \ \mathbf{h}_2^*).$$

It follows that the matrix $\mathbf{W}_1$ given below is the transpose of a $\mathbf{W}$-matrix for an RBD(2, 6):

$$\mathbf{W}_1' = \begin{pmatrix} 1 & -1 \\ 1 & -1 \\ 1 & -1 \\ -1 & 1 \\ -1 & 1 \\ -1 & 1 \end{pmatrix} = \begin{pmatrix} \mathbf{h}_2^{*\prime} \\ \mathbf{h}_2^{*\prime} \\ \mathbf{h}_2^{*\prime} \\ -\mathbf{h}_2^{*\prime} \\ -\mathbf{h}_2^{*\prime} \\ -\mathbf{h}_2^{*\prime} \end{pmatrix}.$$

Proceeding in the lines of Theorem 5.2.1, we construct

$$\mathbf{W}_1^{*\prime} = (\mathbf{W}_{11}^{*\prime}, \mathbf{W}_{21}^{*\prime}, \mathbf{W}_{31}^{*\prime}, \mathbf{W}_{41}^{*\prime}),$$

where

$$\mathbf{W}_{11}^{*\prime} = \begin{pmatrix} \mathbf{h}_2^{*\prime} \\ \mathbf{0}' \\ \mathbf{h}_2^{*\prime} \\ \mathbf{h}_2^{*\prime} \\ -\mathbf{h}_2^{*\prime} \\ \mathbf{0}' \\ -\mathbf{h}_2^{*\prime} \\ -\mathbf{h}_2^{*\prime} \end{pmatrix} ; \ \mathbf{W}_{21}^{*\prime} = \begin{pmatrix} \mathbf{h}_2^{*\prime} \\ \mathbf{h}_2^{*\prime} \\ \mathbf{0}' \\ \mathbf{h}_2^{*\prime} \\ -\mathbf{h}_2^{*\prime} \\ -\mathbf{h}_2^{*\prime} \\ \mathbf{0}' \\ -\mathbf{h}_2^{*\prime} \end{pmatrix} ; \ \mathbf{W}_{31}^{*\prime} = \begin{pmatrix} \mathbf{h}_2^{*\prime} \\ \mathbf{h}_2^{*\prime} \\ \mathbf{h}_2^{*\prime} \\ \mathbf{0}' \\ -\mathbf{h}_2^{*\prime} \\ -\mathbf{h}_2^{*\prime} \\ -\mathbf{h}_2^{*\prime} \\ \mathbf{0}' \end{pmatrix} ; \ \mathbf{W}_{41}^{*\prime} = \begin{pmatrix} \mathbf{0}' \\ \mathbf{h}_2^{*\prime} \\ \mathbf{h}_2^{*\prime} \\ \mathbf{h}_2^{*\prime} \\ \mathbf{0}' \\ -\mathbf{h}_2^{*\prime} \\ -\mathbf{h}_2^{*\prime} \\ -\mathbf{h}_2^{*\prime} \end{pmatrix}.$$

and using $\mathbf{W}_1^*$ and the second row of $\mathbf{H}_4$ of (5.2.7), one optimum $\mathbf{W}$-matrix for the SGDD can be constructed where its transpose is

$$\mathbf{W}_{11}^{**\prime} = \mathbf{h}_2 \odot \mathbf{W}_1^{*\prime} = (\mathbf{W}_{11}^{*\prime}, -\mathbf{W}_{21}^{*\prime}, \mathbf{W}_{31}^{*\prime}, -\mathbf{W}_{41}^{*\prime}),$$

Similarly by taking the Khatri-Rao product of $(\mathbf{W}_1^*, \mathbf{h}_1)$, $(\mathbf{W}_1^*, \mathbf{h}_3)$ and $(\mathbf{W}_1^*, \mathbf{h}_4)$, three other optimum $\mathbf{W}$-matrices can be constructed for the above SGDD, where $\mathbf{h}_1$, $\mathbf{h}_3$ and $\mathbf{h}_4$ are, respectively, the first row, third row and 4th row of $\mathbf{H}_4$ in (5.2.7).

5.2.2 Semi-Regular Group Divisible Design (SRGDD) Set-Up

According to Bose et al. (1953), it is known that the existence of an SRGDD with parameters $v = mn$, $b = n^2\lambda_2$, $r = n\lambda_2$, $k = m$, $\lambda_1 = 0$, λ_2, m, n implies the existence of an orthogonal array, $OA(n^2\lambda_2, m, n, 2)$ and conversely. The definition of an orthogonal array (cf. Raghavarao 1971, p. 10) is given below:

Definition 5.2.4 A $k \times N$ matrix $\mathbf{A}$ with entries from a set of s (≥ 2) elements is called an orthogonal array of size N, k constraints, s levels, strength t, and index λ if any $t \times N$ sub-matrix of $\mathbf{A}$ contains all possible $t \times 1$ column vectors with same frequency λ. Such an array is denoted by $OA(N, k, s, t)$.

In this case, using the properties of orthogonal array (cf. Raghavarao 1971) one can find the optimum covariate designs which is stated in the following theorem.

Theorem 5.2.2 *Let the existence of an $OA(n^2\lambda_2, m, n, 2)$ and the existence of Hadamard matrices of order n and $m_1 = k(2 \leq m_1 < m)$ be assumed. Then $(n-1)(k-1)m_2$ optimum $\mathbf{W}$-matrices can be constructed for an SRGDD with parameters $v = m_1 n$, $b = n^2\lambda_2$, $r = n\lambda_2$, $k = m_1$, $\lambda_1 = 0$, λ_2, m_1, n where $m_1 + m_2 = m$ and $m_2 > 2$.*

Proof Let the orthogonal array $OA(n^2\lambda_2, m, n, 2)$ be denoted by the matrix $\mathbf{A}$ with $n^2\lambda_2$ columns and m rows. The n symbols in the pth row of the orthogonal array are denoted as $(p-1)n+1$, $(p-1)n+2, \ldots, pn$; $p = 1, 2, \ldots, m$. Let it be partitioned into two sub-matrices $\mathbf{A}_1$ and $\mathbf{A}_2$, i.e. $\left(\frac{\mathbf{A}_1}{\mathbf{A}_2}\right)$ where $\mathbf{A}_1$ corresponds to first m_1 rows and $\mathbf{A}_2$ corresponds to last m_2 ($m_2 = m - m_1$) rows of $\mathbf{A}$. Using $\mathbf{A}_1$, an SRGDD with parameters $v = m_1 n$, $b = n^2\lambda_2$, $r = n\lambda_2$, $k = m_1$, m_1, n, $\lambda_1 = 0$, λ_2, where $m_1 + m_2 = m$ and $m_2 > 2$ can be constructed, where the $n^2\lambda_2$ columns of $\mathbf{A}_1$ give the $b = n^2\lambda_2$ blocks of the SRGDD. Let a Hadamard matrix of order n be written as

$$\mathbf{H}_n = [\mathbf{h}_1, \mathbf{h}_2, \ldots, \mathbf{h}_{n-1}, \mathbf{1}]. \tag{5.2.9}$$

Again let the n symbols in each row of $\mathbf{A}_2$ be replaced by $(h_{j1}, h_{j2}, \ldots, h_{jn})$, where h_{ji}'s are the elements of $\mathbf{h}_j$, the jth column of $\mathbf{H}_n$, $j = 1, 2, \ldots, (n-1)$ and the new array $\mathbf{A}_2^{*\prime}(j) = (\mathbf{a}_1^{*\prime}(j), \mathbf{a}_2^{*\prime}(j), \ldots, \mathbf{a}_{m_2}^{*\prime}(j))$ thus obtained is still an orthogonal array of strength 2, but with the two symbols $+1$ and -1 in each row. Let the incidence matrix of the SRGDD corresponding to the orthogonal array $\mathbf{A}_1$ be denoted as $\mathbf{N}^{v \times b}$ with the jth column as, $\mathbf{n}_j = (n_{1j}, n_{2j}, \ldots, n_{vj})'$, $n_{ij} = 0$ or 1; $1 \leq i \leq v$, $1 \leq j \leq b$. The non-zero elements of each column of $\mathbf{N}$ (containing k non-zero elements) are replaced by the k elements (± 1) of $\mathbf{h}_u^*$, the uth column of $\mathbf{H}_k$, $u = 1, 2, \ldots, (k-1)$ in that order and thus $\mathbf{N}_u^*$ is obtained with the jth column as $\mathbf{n}_j^*(u) = (n_{1j}^*(u), n_{2j}^*(u), \ldots, n_{vj}^*(u))'$, $j = 1, 2, \ldots, b$. Obviously the element $n_{ij}^*(u)$ assumes one of the three distinct values $+1$ or -1 or 0. Now, a matrix $\mathbf{W}(j, u, q)$ is obtained by taking the Khatri-Rao product of $\mathbf{a}_q^*(j)$ and $\mathbf{N}_u^*$. A matrix $\mathbf{W}(j, u, q)$ is written as

$$\mathbf{W}(j, u, q) = \mathbf{a}_q^*(j) \odot \mathbf{N}_u^* = \begin{pmatrix} a_{1q}^*(j) \\ \vdots \\ a_{bq}^*(j) \end{pmatrix} \odot \left(\mathbf{n}_1^*(u) \ldots \mathbf{n}_b^*(u) \right) \qquad (5.2.10)$$

It is easy to see that the $\mathbf{W}(j, u, q)$, $q = 1, 2, \ldots, m_2$, $u = 1, 2, \ldots, (k-1)$, $j = 1, 2, \ldots, (n-1)$ matrices given by (5.2.10) satisfy the condition (3.1.5). Thus the theorem follows. $\qquad \square$

Example 5.2.4 Consider SRGDD with parameters $v = 8$, $b = 8$, $r = 4$, $k = 4$, $m_1 \doteq 4$, $n = 2$, $\lambda_1 = 0$, $\lambda_2 = 1$ which is obtained from OA(8, 7, 2, 2) as follows:

Let $\mathbf{A} = OA(8, 7, 2, 2)$ where

$$\mathbf{A}' = \begin{pmatrix} 1 & 3 & 5 & 7 & 9 & 11 & 13 \\ 2 & 3 & 6 & 7 & 10 & 11 & 14 \\ 1 & 4 & 6 & 7 & 9 & 12 & 14 \\ 2 & 4 & 5 & 7 & 10 & 12 & 13 \\ 1 & 3 & 5 & 8 & 10 & 12 & 14 \\ 2 & 3 & 6 & 8 & 9 & 12 & 13 \\ 1 & 4 & 6 & 8 & 10 & 11 & 13 \\ 2 & 4 & 5 & 8 & 9 & 11 & 14 \end{pmatrix} = (\mathbf{A}_1 | \mathbf{A}_2).$$

Using $\mathbf{A}_1$, the SRGDD with above parameters is obtained and the incidence matrix $\mathbf{N}$ corresponding to the design is written as in the form of its transpose

$$\mathbf{N}' = \begin{pmatrix} 1 & 0 & 1 & 0 & 1 & 0 & 1 & 0 \\ 0 & 1 & 1 & 0 & 0 & 1 & 1 & 0 \\ 1 & 0 & 0 & 1 & 0 & 1 & 1 & 0 \\ 0 & 1 & 0 & 1 & 1 & 0 & 1 & 0 \\ 1 & 0 & 1 & 0 & 1 & 0 & 0 & 1 \\ 0 & 1 & 1 & 0 & 0 & 1 & 0 & 1 \\ 1 & 0 & 0 & 1 & 0 & 1 & 0 & 1 \\ 0 & 1 & 0 & 1 & 1 & 0 & 0 & 1 \end{pmatrix}.$$

$\mathbf{H}_2$ and $\mathbf{H}_4$ are written as

$$\mathbf{H}_2 = \begin{pmatrix} 1 & 1 \\ 1 & -1 \end{pmatrix} = (\mathbf{h}_1, \mathbf{1}); \quad \mathbf{H}_4 = \begin{pmatrix} 1 & 1 & 1 & 1 \\ -1 & -1 & 1 & 1 \\ 1 & -1 & -1 & 1 \\ -1 & 1 & -1 & 1 \end{pmatrix} = (\mathbf{h}_1^*, \mathbf{h}_2^*, \mathbf{h}_3^*, \mathbf{1}).$$

Replacing the elements in the columns of A_2 by those of h_1, $A_2^*(1)$ can be written as

$$A_2^*(1) = \begin{pmatrix} 1 & -1 & 1 & -1 & -1 & 1 & -1 & 1 \\ 1 & 1 & -1 & -1 & -1 & -1 & 1 & 1 \\ 1 & -1 & -1 & 1 & -1 & 1 & 1 & -1 \end{pmatrix} = (a_1^*(1), a_2^*(1), a_3^*(1)).$$

If the non-zero elements of each row of N are replaced by the four elements (± 1) of first column h_1^* of H_4 in that order, then the transpose of N_1^* is obtained as

$$
\begin{array}{c}
\text{Block} \\
\downarrow
\end{array}
\begin{array}{c}
\text{Treatment} \rightarrow \\
1\ 2\ 3\ \ 4\ 5\ 6\ \ 7\ \ 8
\end{array}
$$

$$
N_1^{*\prime} = \begin{array}{c} 1 \\ 2 \\ 3 \\ 4 \\ 5 \\ 6 \\ 7 \\ 8 \end{array}
\begin{pmatrix}
1 & 0 & -1 & \ 0 & 1 & 0 & -1 & \ 0 \\
0 & 1 & -1 & \ 0 & 0 & 1 & -1 & \ 0 \\
1 & 0 & \ 0 & -1 & 0 & 1 & -1 & \ 0 \\
0 & 1 & \ 0 & -1 & 1 & 0 & -1 & \ 0 \\
1 & 0 & -1 & \ 0 & 1 & 0 & \ 0 & -1 \\
0 & 1 & -1 & \ 0 & 0 & 1 & \ 0 & -1 \\
1 & 0 & \ 0 & -1 & 0 & 1 & \ 0 & -1 \\
0 & 1 & \ 0 & -1 & 1 & 0 & \ 0 & -1
\end{pmatrix}
$$

Now using the Khatri-Rao product between the first column $a_1^*(1)$ of $A_2^*(1)$ and N_1^*, the following the transpose of the W-matrix can be constructed as

$$
\begin{pmatrix} 1 \\ -1 \\ 1 \\ -1 \\ -1 \\ 1 \\ -1 \\ 1 \end{pmatrix} \odot N_1^{*\prime} =
\begin{pmatrix}
1 & 0 & -1 & \ 0 & 1 & 0 & -1 & \ 0 \\
0 & -1 & 1 & \ 0 & 0 & -1 & 1 & \ 0 \\
1 & 0 & 0 & -1 & 0 & 1 & -1 & \ 0 \\
0 & -1 & 0 & 1 & -1 & 0 & 1 & \ 0 \\
-1 & 0 & 1 & 0 & -1 & 0 & 0 & 1 \\
0 & 1 & -1 & 0 & 0 & 1 & 0 & -1 \\
-1 & 0 & 0 & 1 & 0 & -1 & 0 & 1 \\
0 & 1 & 0 & -1 & 1 & 0 & 0 & -1
\end{pmatrix} = W'(1, 1, 1).
$$

Note that $W'(1, 1, 1)$ matches with $N_1^{*\prime}$. In this way, altogether 9 optimum W-matrices can be constructed for different choices of columns of $A_2^*(1)$ and first three columns of H_4 (excluding h_1).

Remark 5.2.5 It follows from Theorem 5.2.2 that the maximum number of W-matrices that can be constructed depends on the maximum value of $m_2(m_1 - 1)(n - 1)$ where $m_1 > 0$, $m_2 > 0$, $m_1 + m_2 = m$ and each of m_1, n is such that H_{m_1} and H_n exist.

Remark 5.2.6

(a) If n is even but H_n does not exist, then it is possible to construct $(k - 1)m_2$ optimum W-matrices for the SRGD with the above parameters by using a vector of the form $(1'_{\frac{n}{2}}, -1'_{\frac{n}{2}})'$ in place of the columns of H_n.

(b) Similarly if k is even but $\mathbf{H}_k$ does not exist, then it is possible to construct $(n-1)m_2$ optimum $\mathbf{W}$-matrices for the SRGD with the above parameters by using a vector of the form $(\mathbf{1}'_{\frac{k}{2}}, -\mathbf{1}'_{\frac{k}{2}})'$ in place of the columns of $\mathbf{H}_k$.

(c) Again if $n,\ k$ are both even but $\mathbf{H}_n$ and $\mathbf{H}_k$ do not exist, then it is possible to construct m_2 optimum $\mathbf{W}$-matrices for the SRGD with the above parameters by using two vectors of the form $(\mathbf{1}'_{\frac{n}{2}}, -\mathbf{1}'_{\frac{n}{2}})'$ and $(\mathbf{1}'_{\frac{k}{2}}, -\mathbf{1}'_{\frac{k}{2}})'$ in place of the columns of $\mathbf{H}_n$ and $\mathbf{H}_k$.

5.2.3 Regular Group Divisible (RGD) Design Set-Up

It is known that if from a BIBD with parameters v^*, b^*, r^*, k^*, $\lambda^* = 1$, all the r^* blocks in which a particular treatment occurs are deleted, then a RGD design with parameters $v = v^* - 1$, $b = b^* - r^*$, $r = r^* - 1$, $k = k^*$, $\lambda_1 = 0$, $\lambda_2 = 1$, $m = r^*$, $n = k^* - 1$ can be obtained (Bose et al. (1953); also see Raghavarao (1971)). It is difficult to construct covariate design optimally for such GD design obtained from arbitrary BIBD with the parameters v^*, b^*, r^*, k^*, $\lambda^* = 1$. However, for some series of BIBDs, it is possible to provide optimum covariate designs. Let the series of BIBD designs with parameters:

$$b^* = (4t+1)(3t+1), v^* = 4(3t+1), r^* = 4t+1, k^* = 4, \lambda^* = 1 \quad (5.2.11)$$

be considered with the initial blocks:

$$\left(x_1^{2i}, x_1^{2t+2i}, x_2^{\alpha+2i}, x_2^{\alpha+2t+2i}\right)$$
$$\left(x_2^{2i}, x_2^{2t+2i}, x_3^{\alpha+2i}, x_3^{\alpha+2t+2i}\right) \quad (5.2.12)$$
$$\left(x_3^{2i}, x_3^{2t+2i}, x_1^{\alpha+2i}, x_1^{\alpha+2t+2i}\right)$$
$$(0_1, 0_2, 0_3, \infty); \quad i = 0, 1, \ldots, t-1,$$

where $4t + 1$ is prime or prime power and x is a primitive root of $GF(4t + 1)$; 1, 2, 3 are the three symbols attached to x, α is an odd integer and ∞ the invariant treatment symbol (cf. Bose 1939). If the initial block containing treatment symbol ∞ in (5.2.12) is deleted and others are developed, then an RGD design with parameters:

$$b = 3t(4t+1), v = 3(4t+1), r = 4t, k = 4, \lambda_1 = 0, \lambda_2 = 1, m = 4t+1, n = 3 \quad (5.2.13)$$

is obtained. The $(4t+1)$ groups obtained by developing $(0_1, 0_2, 0_3)$ over $GF(4t+1)$ give the association scheme for the above RGD design. The following theorem provides optimum covariate designs for the series with parameters given in (5.2.13).

Theorem 5.2.3 *If $\mathbf{H}_t$ exists, then $3t$ optimum $\mathbf{W}$-matrices can be constructed for the RGD design with parameters given in (5.2.13).*

Proof Let the $3t$ initial blocks other than $(0_1, 0_2, 0_3, \infty)$ of (5.2.12) be divided into t sets of 3 blocks each, the ith set being

$$S_{i+1} = \left(\left(x_1^{2i}, x_1^{2t+2i}, x_2^{\alpha+2i}, x_2^{\alpha+2t+2i} \right), \left(x_2^{2i}, x_2^{2t+2i}, x_3^{\alpha+2i}, x_3^{\alpha+2t+2i} \right) \right.$$
$$\left. \left(x_3^{2i}, x_3^{2t+2i}, x_1^{\alpha+2i}, x_1^{\alpha+2t+2i} \right) \right), \quad i = 0, 1, \ldots, t - 1.$$

Also, let each of the initial blocks of S_{i+1} be displayed in the form of column vectors of the incidence matrix and development of these initial blocks will give rise to the sub-incidence matrix $\mathbf{N}_i$ of order $3(4t + 1) \times 3(4t + 1)$, once more we restrict to the transpose matrix where

$$\mathbf{N}_i' = \begin{pmatrix} \mathbf{N}_1^{(i)\prime} & \mathbf{N}_2^{(i)\prime} & \mathbf{0} \\ \mathbf{0} & \mathbf{N}_1^{(i)\prime} & \mathbf{N}_2^{(i)\prime} \\ \mathbf{N}_2^{(i)\prime} & \mathbf{0} & \mathbf{N}_1^{(i)\prime} \end{pmatrix}.$$

It is easy to see that $\mathbf{N}_1^{(i)}$ and $\mathbf{N}_2^{(i)}$ matrices corresponding to two portions of the initial blocks of S_i, are obtained by cyclically permuting the column vectors of each of the matrices. For $j = 1, 2$, the two non-zero positions of the first column of $\mathbf{N}_j^{(i)}$ is replaced by $+1$ and -1 successively and then this column is permuted cyclically in the same way as $\mathbf{N}_j^{(i)}$ was obtained. The resultant matrix is denoted by $\mathbf{W}_j^{(i)}$. By replacing the $\mathbf{N}_j^{(i)}$ by $\mathbf{W}_j^{(i)}$ in $\mathbf{N}_i$'s one would get a matrix $\mathbf{W}_{i1}$ of order $3(4t + 1) \times 3(4t + 1)$ whose transpose can be displayed as

$$\mathbf{W}_{i1}' = \begin{pmatrix} \mathbf{W}_1^{(i)\prime} & \mathbf{W}_2^{(i)\prime} & \mathbf{0} \\ \mathbf{0} & \mathbf{W}_1^{(i)\prime} & \mathbf{W}_2^{(i)\prime} \\ \mathbf{W}_2^{(i)\prime} & \mathbf{0} & \mathbf{W}_1^{(i)\prime} \end{pmatrix}.$$

Then two other matrices, viz. $\mathbf{W}_{i2}$ and $\mathbf{W}_{i3}$ are constructed from $\mathbf{W}_{i1}$ and $\mathbf{N}_i$, respectively, where their transpose matrices are respectively

$$\mathbf{W}_{i2}' = \begin{pmatrix} \mathbf{W}_1^{(i)\prime} & -\mathbf{W}_2^{(i)\prime} & \mathbf{0} \\ \mathbf{0} & \mathbf{W}_1^{(i)\prime} & -\mathbf{W}_2^{(i)\prime} \\ -\mathbf{W}_2^{(i)\prime} & \mathbf{0} & \mathbf{W}_1^{(i)\prime} \end{pmatrix}, \quad \mathbf{W}_{i3}' = \begin{pmatrix} \mathbf{N}_1^{(i)\prime} & -\mathbf{N}_2^{(i)\prime} & \mathbf{0} \\ \mathbf{0} & \mathbf{N}_1^{(i)\prime} & -\mathbf{N}_2^{(i)\prime} \\ -\mathbf{N}_2^{(i)\prime} & \mathbf{0} & \mathbf{N}_1^{(i)\prime} \end{pmatrix}.$$

It can be easily checked that these three matrices satisfy the condition (3.1.5) for each $i, i = 1, 2, \ldots, t$. If Hadamard matrix $\mathbf{H}_t = (h_{ml})$ exists, the number of $\mathbf{W}$-matrices can be increased t times. The $3t$ optimum $\mathbf{W}$-matrices can be constructed and their transpose can be displayed as

$$\mathbf{W}_j'(m) = \begin{pmatrix} h_{m1} \\ h_{m2} \\ \vdots \\ h_{mt} \end{pmatrix} \odot \begin{pmatrix} \mathbf{W}_{1j}' \\ \mathbf{W}_{2j}' \\ \vdots \\ \mathbf{W}_{tj}' \end{pmatrix} \quad \forall m = 1, 2, \ldots, t; \ j = 1, 2, 3. \quad (5.2.14)$$

It can be easily seen that these $3t$ matrices given in (5.2.14) satisfy the condition (3.1.5) and give optimum $\mathbf{W}$-matrices. $\qquad\qquad\square$

Example 5.2.5 With $t = 1$ the RGD design with parameters $v = 15, b = 15$, $r = 4, k = 4, \lambda_1 = 0, \lambda_2 = 1, m = 5, n = 3$ is considered and the initial blocks forming the single set are $\{(1_1, 4_1, 2_2, 3_2), (1_2, 4_2, 2_3, 3_3), (1_3, 4_3, 2_1, 3_1)\}$ and the groups of the association scheme are generated from $(0_1, 0_2, 0_3)$. The transpose of the incidence matrix of this design

$$\mathbf{N}' = \left(\begin{array}{ccccc|ccccc|ccccc} 0&1&0&0&1 & 0&0&1&1&0 & 0&0&0&0&0 \\ 1&0&1&0&0 & 0&0&0&1&1 & 0&0&0&0&0 \\ 0&1&0&1&0 & 1&0&0&0&1 & 0&0&0&0&0 \\ 0&0&1&0&1 & 1&1&0&0&0 & 0&0&0&0&0 \\ 1&0&0&1&0 & 0&1&1&0&0 & 0&0&0&0&0 \\ \hline 0&0&0&0&0 & 0&1&0&0&1 & 0&0&1&1&0 \\ 0&0&0&0&0 & 1&0&1&0&0 & 0&0&0&1&1 \\ 0&0&0&0&0 & 0&1&0&1&0 & 1&0&0&0&1 \\ 0&0&0&0&0 & 0&0&1&0&1 & 1&1&0&0&0 \\ 0&0&0&0&0 & 1&0&0&1&0 & 0&1&1&0&0 \\ \hline 0&0&1&1&0 & 0&0&0&0&0 & 0&1&0&0&1 \\ 0&0&0&1&1 & 0&0&0&0&0 & 1&0&1&0&0 \\ 1&0&0&0&1 & 0&0&0&0&0 & 0&1&0&1&0 \\ 1&1&0&0&0 & 0&0&0&0&0 & 0&0&1&0&1 \\ 0&1&1&0&0 & 0&0&0&0&0 & 1&0&0&1&0 \end{array}\right) = \begin{pmatrix} \mathbf{N}_1^{(i)\prime} & \mathbf{N}_2^{(i)\prime} & \mathbf{0} \\ \mathbf{0} & \mathbf{N}_1^{(i)\prime} & \mathbf{N}_2^{(i)\prime} \\ \mathbf{N}_2^{(i)\prime} & \mathbf{0} & \mathbf{N}_1^{(i)\prime} \end{pmatrix}$$

with

$$\mathbf{N}_i^{(1)\prime} = \begin{pmatrix} 0&1&0&0&1 \\ 1&0&1&0&0 \\ 0&1&0&1&0 \\ 0&0&1&0&1 \\ 1&0&0&1&0 \end{pmatrix}, \quad \mathbf{N}_2^{(2)\prime} = \begin{pmatrix} 0&0&1&1&0 \\ 0&0&0&1&1 \\ 1&0&0&0&1 \\ 1&1&0&0&0 \\ 0&1&1&0&0 \end{pmatrix}.$$

The transpose matrices of the three optimum $\mathbf{W}$-matrices are respectively

$$\mathbf{W}_{11}' = \begin{pmatrix} \mathbf{W}_1^{(1)\prime} & \mathbf{W}_2^{(1)\prime} & \mathbf{0} \\ \mathbf{0} & \mathbf{W}_1^{(1)\prime} & \mathbf{W}_2^{(1)\prime} \\ \mathbf{W}_2^{(1)\prime} & \mathbf{0} & \mathbf{W}_1^{(1)\prime} \end{pmatrix}.$$

The column headers above $\mathbf{N}'$: $0_1\ 1_1\ 2_1\ 3_1\ 4_1 \quad 0_2\ 1_2\ 2_2\ 3_2\ 4_2 \quad 0_3\ 1_3\ 2_3\ 3_3\ 4_3$

$$\mathbf{W}'_{12} = \begin{pmatrix} \mathbf{W}_1^{(1)\prime} & -\mathbf{W}_2^{(1)\prime} & \mathbf{0} \\ \mathbf{0} & \mathbf{W}_1^{(1)\prime} & -\mathbf{W}_2^{(1)\prime} \\ -\mathbf{W}_2^{(1)\prime} & \mathbf{0} & \mathbf{W}_1^{(1)\prime} \end{pmatrix}, \quad \mathbf{W}'_{13} = \begin{pmatrix} \mathbf{N}_1^{(1)\prime} & -\mathbf{N}_2^{(1)\prime} & \mathbf{0} \\ \mathbf{0} & \mathbf{N}_1^{(1)\prime} & -\mathbf{N}_2^{(1)\prime} \\ -\mathbf{N}_2^{(1)\prime} & \mathbf{0} & \mathbf{N}_1^{(1)\prime} \end{pmatrix}.$$

where,

$$\mathbf{W}_1^{(1)\prime} = \begin{pmatrix} 0 & 1 & 0 & 0 & -1 \\ -1 & 0 & 1 & 0 & 0 \\ 0 & -1 & 0 & 1 & 0 \\ 0 & 0 & -1 & 0 & 1 \\ 1 & 0 & 0 & -1 & 0 \end{pmatrix}, \quad \mathbf{W}_2^{(1)\prime} = \begin{pmatrix} 0 & 0 & 1 & -1 & 0 \\ 0 & 0 & 0 & 1 & -1 \\ -1 & 0 & 0 & 0 & 1 \\ 1 & -1 & 0 & 0 & 0 \\ 0 & 1 & -1 & 0 & 0 \end{pmatrix}$$

and $\mathbf{N}_1^{(1)}$ and $\mathbf{N}_2^{(1)}$ as above.

Appendix

A list of OCDs for suitable subclasses of GDDs, viz. SGDDs, SRGDDs and RGDDs divided as singular (S), semi-regular (SR), regular (R) is given below. These are extracted from the catalogue prepared by Clatworthy (1973) and amenable to construction of OCDs. See Dutta et al. (2009, 2010) in this context. In the constructional method column, T stands for Theorem and R for Remark (Tables 5.1, 5.2 and 5.3).

Table 5.1 OCDs in SGDDs

Sl. no.	Design no.	v	b	r	k	λ_1	λ_2	m	n	v^*	b^*	r^*	k^*	λ^*	t	$t+t_1$	Method of construction
1	S1	6	3	2	4	2	1	3	2	3	3	2	2	1	1	2	T 5.2.1(i), R 5.2.4
2	S2	6	6	4	4	4	2	3	2	3	6	4	2	2	2	5	T 5.2.1(i), R 5.2.4
3	S3	6	9	6	4	6	3	3	2	3	9	6	2	3	1	2	T 5.2.1(i), R 5.2.4
4	S4	6	12	8	4	8	4	3	2	3	12	8	2	4	4	11	T 5.2.1(i), R 5.2.4
5	S5	6	15	10	4	10	5	3	2	3	15	10	2	5	1	2	T 5.2.1(i), R 5.2.4
6	S7	8	12	6	4	6	2	4	2	4	12	6	2	2	1	5	T 5.2.1(iv), R 5.2.4
7	S9	10	10	4	4	4	1	5	2	5	10	4	2	1	2	5	T 5.2.1(i), R 5.2.4
8	S10	10	20	8	4	8	2	5	2	5	20	8	2	2	4	11	T 5.2.1(i), R 5.2.4
9	S12	12	30	10	4	10	2	6	2	6	30	10	2	2	1	3	R 5.2.3, R 5.2.4
10	S13	14	21	6	4	6	1	7	2	7	21	6	2	1	1	2	T 5.2.1(i), R 5.2.4
11	S15	18	36	8	4	8	1	9	2	9	36	8	2	1	4	11	T 5.2.1(i), R 5.2.4
12	S17	22	55	10	4	10	1	11	2	11	55	10	2	1	1	2	T 5.2.1(i), R 5.2.4
13	S19	8	8	6	6	6	4	4	2	4	8	6	3	4	0	4	T 5.2.1(iv), R 5.2.4

(continued)

Table 5.1 (continued)

Sl. no.	Design no.	v	b	r	k	λ_1	λ_2	m	n	v^*	b^*	r^*	k^*	λ^*	t	$t+t_1$	Method of construction
14	S21	9	3	2	6	2	1	3	3	3	3	2	2	1	1	1	R 5.2.4
15	S22	9	6	4	6	4	2	3	3	3	6	4	2	2	2	2	R 5.2.4
16	S23	9	9	6	6	6	3	3	3	3	9	6	2	3	1	1	R 5.2.4
17	S24	9	12	8	6	8	4	3	3	3	12	8	2	4	4	4	R 5.2.4
18	S25	9	15	10	6	10	5	3	3	3	15	10	2	5	1	1	R 5.2.4
19	S26	10	10	6	6	6	3	5	2	5	10	6	3	3	0	1	T 5.2.1(i)
20	S29	12	12	6	6	6	2	4	3	4	12	6	2	2	1	1	R 5.2.4
21	S31	12	20	10	6	10	4	6	2	6	20	10	3	4	0	2	R 5.2.3
22	S33	14	14	6	6	6	2	7	2	7	14	6	3	2	0	1	T 5.2.1(i)
23	S35	15	10	4	6	4	1	5	3	5	10	4	2	1	2	2	R 5.2.4
24	S36	15	20	8	6	8	2	5	3	5	20	8	2	2	4	4	R 5.2.4
25	S37	18	12	4	6	4	1	9	2	9	12	4	3	1	0	3	T 5.2.1(i)
26	S39	18	24	8	6	8	2	9	2	9	24	8	3	2	0	7	T 5.2.1(i)
27	S40	18	30	10	6	10	2	6	3	6	30	10	2	2	1	1	R 5.2.4
28	S42	21	21	6	6	6	1	7	3	7	21	6	2	1	1	1	R 5.2.4
29	S44	26	26	6	6	6	1	13	2	13	26	6	3	1	0	1	T 5.2.1(i)
30	S45	27	36	8	6	8	1	9	3	9	36	8	2	1	4	4	R 5.2.4
31	S48	33	55	10	6	10	1	11	3	11	55	10	2	1	1	1	R 5.2.4
32	S50	42	70	10	6	10	1	21	2	21	70	10	3	1	0	1	T 5.2.1(i)

(continued)

Table 5.1 (continued)

Sl. no.	Design no.	v	b	r	k	λ_1	λ_2	m	n	v^*	b^*	r^*	k^*	λ^*	t	$t+t_1$	Method of construction
33	S51	10	5	4	8	4	3	5	2	5	5	4	4	3	3	6	T 5.2.1(i), R 5.2.4
34	S52	10	10	8	8	8	6	5	2	5	10	8	4	6	6	13	T 5.2.1(i), R 5.2.4
35	S53	12	3	2	8	2	1	3	4	3	3	2	2	1	1	4	T 5.2.1(i), R 5.2.4
36	S54	12	6	4	8	4	2	3	4	3	6	4	2	2	2	11	T 5.2.1(i), R 5.2.4
37	S55	12	9	6	8	6	3	3	4	3	9	6	2	3	1	12	T 5.2.1(i), R 5.2.4
38	S56	12	12	8	8	8	4	3	4	3	12	8	2	4	4	25	T 5.2.1(i), R 5.2.4
39	S57	12	15	10	8	10	5	3	4	3	15	10	2	5	1	20	T 5.2.1(i), R 5.2.4
40	S58	12	15	10	8	10	6	6	2	6	15	10	4	6	1	3	R 5.2.3
41	S59	14	7	4	8	4	2	7	2	7	7	4	4	2	3	6	T 5.2.1(i), R 5.2.4
42	S60	14	14	8	8	8	4	7	2	7	14	8	4	4	6	13	T 5.2.1(i), R 5.2.4
43	S62	16	12	6	8	6	2	4	4	4	12	6	2	2	1	45	T 5.2.1(iii), R 5.2.4
44	S65	18	18	8	8	8	3	9	2	9	18	8	4	3	6	13	T 5.2.1(i), R 5.2.4
45	S66	20	10	4	8	4	1	5	4	5	10	4	2	1	2	11	T 5.2.1(i), R 5.2.4

(continued)

Table 5.1 (continued)

Sl. no.	Design no.	v	b	r	k	λ_1	λ_2	m	n	v^*	b^*	r^*	k^*	λ^*	t	$t+t_1$	Method of construction
46	S67	20	15	6	8	6	2	10	2	10	15	6	4	2	0	2	R 5.2.3
47	S68	20	20	8	8	8	2	5	4	5	20	8	2	2	4	25	T 5.2.1(i), R 5.2.4
48	S70	24	30	10	8	10	2	6	4	6	30	10	2	2	1	39	R 5.2.2(ii)
49	S71	26	13	4	8	4	1	13	2	13	13	4	4	1	3	6	T 5.2.1(i), R 5.2.4
50	S72	26	26	8	8	8	2	13	2	13	26	8	4	2	6	13	T 5.2.1(i), R 5.2.4
51	S73	28	21	6	8	6	1	7	4	7	21	6	2	1	1	12	T 5.2.1(i), R 5.2.4
52	S76	32	40	10	8	10	2	16	2	16	40	10	4	2	1	17	T 5.2.1(iv), R 5.2.4
53	S77	36	36	8	8	8	1	9	4	9	36	8	2	1	4	25	T 5.2.1(i), R 5.2.4
54	S79	44	55	10	8	10	1	11	4	11	55	10	2	1	1	20	T 5.2.1(i), R 5.2.4
55	S80	50	50	8	8	8	1	25	2	25	50	8	4	1	6	13	T 5.2.1(i), R 5.2.4
56	S99	12	12	10	10	10	8	6	2	6	12	10	5	8	0	2	R 5.2.3
57	S100	15	3	2	10	2	1	3	5	3	3	2	2	1	1	1	R 5.2.4
58	S101	15	6	4	10	4	2	3	5	3	6	4	2	2	2	2	R 5.2.4
59	S102	15	9	6	10	6	3	3	5	3	9	6	2	3	1	1	R 5.2.4

(continued)

Table 5.1 (continued)

Sl. no.	Design no.	v	b	r	k	λ_1	λ_2	m	n	v^*	b^*	r^*	k^*	λ^*	t	$t+t_1$	Method of construction
60	S103	15	12	8	10	8	4	3	5	3	12	8	2	4	4	4	R 5.2.4
61	S104	15	15	10	10	10	5	3	5	3	15	10	2	5	1	1	R 5.2.4
62	S105	18	18	10	10	10	5	9	2	9	18	10	5	5	0	1	T 5.2.1(i)
63	S107	20	12	6	10	6	2	4	5	4	12	6	2	2	1	1	R 5.2.4
64	S111	22	22	10	10	10	4	11	2	11	22	10	5	4	0	1	T 5.2.1(i)
65	S112	25	10	4	10	4	1	5	5	5	10	4	2	1	2	2	R 5.2.4
66	S113	25	20	8	10	8	2	5	5	5	20	8	2	2	4	4	R 5.2.4
67	S115	30	30	10	10	10	2	6	5	6	30	10	2	2	1	1	R 5.2.4
68	S116	35	21	6	10	6	1	7	5	7	21	6	2	1	1	1	R 5.2.4
69	S119	42	42	10	10	10	2	21	2	21	42	10	5	2	0	1	T 5.2.1(i)
70	S120	45	36	8	10	8	1	9	5	9	36	8	2	1	4	4	R 5.2.4
71	S121	50	30	6	10	6	1	25	2	25	30	6	5	1	0	1	T 5.2.1(i)
72	S123	55	55	10	10	10	1	11	5	11	55	10	2	1	1	1	R 5.2.4
73	S124	82	82	10	10	10	1	41	2	41	82	10	5	1	0	1	T 5.2.1(i)

$(v, b, r, k, \lambda_1, \lambda_2, m, n)$ denote the parameters of the PBIBD and $(v^*, b^*, r^*, k^*, \lambda^*)$ denote the parameters of the BIBD used to construct the PBIBD. t denotes the number of optimum **W**-matrices and t_1 denotes the additional number of **W**-matrices mentioned in Remark 5.2.4

Table 5.2 OCDs in SRGDDs

		v	b	r	k	λ_1	λ_2	m	n	$OA(n^2\lambda_2, m, n, 2)$	m_2	$c = (n-1)(k-1)m_2$ (cf. T 5.2.2)
74	SR1	4	4	2	2	0	1	2	2	$OA(4, 3, 2, 2)$	1	1
75	SR2	4	8	4	2	0	2	2	2	$OA(8, 7, 2, 2)$	5	5
76	SR3	4	12	6	2	0	3	2	2	$OA(12, 11, 2, 2)$	9	9
77	SR4	4	16	8	2	0	4	2	2	$OA(16, 15, 2, 2)$	13	13
78	SR5	4	20	10	2	0	5	2	2	$OA(20, 19, 2, 2)$	17	17
79	SR9	8	16	4	2	0	1	2	4	$OA(16, 5, 4, 2)$	1	3
80	SR10	8	32	8	2	0	2	2	4	$OA(32, 10, 4, 2)$	8	24
81	SR13	12	36	6	2	0	1	2	6	$OA(36, 7, 6, 2)$	5	5
82	SR15	16	64	8	2	0	1	2	8	$OA(64, 9, 8, 2)$	6	42
83	SR17	20	100	10	2	0	1	2	10	$OA(100, 11, 10, 2)$	9	9
84	SR36	8	8	4	4	0	2	4	2	$OA(8, 7, 2, 2)$	3	9
85	SR37	8	12	6	4	0	3	4	2	$OA(12, 11, 2, 2)$	7	21
86	SR39	8	16	8	4	0	4	4	2	$OA(16, 15, 2, 2)$	11	33
87	SR40	8	20	10	4	0	5	4	2	$OA(20, 19, 2, 2)$	15	45

(continued)

Table 5.2 (continued)

		v	b	r	k	λ_1	λ_2	m	n	$OA(n^2\lambda_2, m, n, 2)$	m_2	$c = (n-1)(k-1)m_2$ (cf. T 5.2.2)
88	SR44	16	16	4	4	0	1	4	4	OA(16, 5, 4, 2)	1	9
89	SR45	16	32	8	4	0	2	4	4	OA(32, 10, 4, 2)	6	54
90	SR49	32	64	8	4	0	1	4	8	OA(64, 9, 8, 2)	5	105
91	SR51	40	100	10	4	0	1	4	10	OA(100, 11, 10, 2)	7	21
92	SR66	12	8	4	6	0	2	6	2	OA(8, 7, 2, 2)	1	1
93	SR67	12	12	6	6	0	3	6	2	OA(12, 11, 2, 2)	5	5
94	SR69	12	16	8	6	0	4	6	2	OA(16, 15, 2, 2)	9	9
95	SR70	12	20	10	6	0	5	6	2	OA(20, 19, 2, 2)	13	13
96	SR74	24	32	8	6	0	2	6	4	OA(32, 10, 4, 2)	4	12
97	SR78	48	64	8	6	0	1	6	8	OA(64, 9, 8, 2)	3	21
98	SR91	16	12	6	8	0	3	8	2	OA(12, 11, 2, 2)	3	21
99	SR92	16	16	8	8	0	4	8	2	OA(16, 15, 2, 2)	7	49
100	SR93	16	20	10	8	0	2	8	4	OA(32, 19, 2, 2)	11	77
101	SR95	32	32	8	8	0	2	8	4	OA(32, 10, 4, 2)	2	42
102	SR97	64	64	8	8	0	1	8	8	OA(64, 9, 8, 2)	1	49
103	SR106	20	12	6	10	0	3	10	2	OA(12, 11, 2, 2)	1	1
104	SR107	20	16	8	10	0	4	10	2	OA(16, 15, 2, 2)	5	5
105	SR108	20	20	10	10	0	5	10	2	OA(20, 19, 2, 2)	9	9

Table 5.3 OCDs in RGDDs

		v	b	r	k	λ_1	λ_2	m	n	t	$3t$	Method of construction
106	R114	15	15	4	4	0	1	5	3	1	3	Example 5.2.5
107	R129	27	54	8	4	0	1	9	3	2	6	T 5.2.3

References

Bose RC (1939) On the construction of balanced incomplete block designs. Ann Eugen 9:353–399

Bose RC, Shimamoto T (1952) Classification and analysis of partially balanced incomplete block designs with two-associate classes. J Am Stat Assoc 47:151–184

Bose RC, Shrikhande SS, Bhattacharya KN (1953) On the construction of group divisible incomplete block designs. Ann Math Stat 24:167–195

Clatworthy WH (1973) Tables of two-associate class partially balanced designs. US Department of Commerce, National Bureau of Standards, Washington

Dutta G, Das P, Mandal NK (2009) Optimum covariate designs in partially balanced incomplete block (PBIB) design set-ups. J Stat Plan Inference 139:2823–2835

Dutta G, Das P, Mandal NK (2010) Tables for optimum covariate designs in PBIBD set-ups. J Indian Soc Agric Stat 64:375–389

Nandi HK, Adhikari B (1966) On the definition of Bose-Shimamoto cyclical association scheme. Calcutta Stat Assoc Bull 15:165–168

Raghavarao D (1971) Constructions and combinatorial problems in design of experiments. Wiley, New York

Rao CR (1973) Linear statistical inference and its applications. Wiley, New York

Vartak MN (1954) On an application of Kronecker product of matrices to statistical designs. Ann Math Stat 26:420–438

Zelen M (1954) A note on partially balanced designs. Ann Math Stat 25:599–602

Chapter 6
OCDs in Binary Proper Equireplicate Block Design Set-Up

6.1 Introduction

In Chaps. 4 and 5, we have considered OCDs in the set-ups of BIBDs and PBIBDs, which belong to the class of BPEBDs. It was observed earlier that the constructions of OCDs on BIBDs and PBIBDs depend heavily on the method of constructions of these designs and also that the designs having cyclic nature were more suitable for constructing OCDs. Dutta et al. (2010) investigated the problem of construction of OCDs for the general class of BPEBDs where b is a multiple of v. The cyclic designs with 'full sets', a number of BIBDS, PBIBDs and a host of other designs belong to this class of PBEBDs and consequently the construction of OCDs on these set-ups will follow from the general method. The only restriction that the designs have to follow is that b should be a multiple of v.

The cyclic designs with 'partial sets' do not have b as a multiple of v but as these are BPEBDs we have considered the set-ups as a related discussion. In this chapter, we mainly concentrate on Dutta et al. (2010) and describe the construction of OCDs described therein.

6.2 BPEBDs with $b = mv$

It can be noticed that it is difficult to construct OCDs for any arbitrary block design. The procedures depend heavily on the methods of construction of the corresponding block designs and often optimum W-matrices are searched for designs which are mainly constructed through the method of differences. But now we shall describe a technique for constructing OCDs in BPEBDs with $b = mv$, $m =$ positive integer, which does not depend on the method of construction and hence can be widely applied to a large class of commonly used block designs. The following lemma and theorem will help us in the construction of OCDs in such set-ups.

© Springer India 2015
P. Das et al., *Optimal Covariate Designs*, DOI 10.1007/978-81-322-2461-7_6

Lemma 6.2.1 *Let* C *be a* $k \times b$ *matrix with* v *elements* $t_1, t_2, \ldots, t_v$ *where* $b = mv$, $m =$ *a positive integer, such that each element occurs at most once in each column and an equal number of times in the whole matrix* C. *Then from* C *we can construct a* $v \times b$ *matrix* A *with* $(k + 1)$ *symbols* $a_1, a_2, \ldots, a_k$ *and* 0 *such that each of the non-null symbols occurs once and only once in each of the b columns and m times in each of the v rows of* A.

Proof From the properties of the matrix C it can be easily seen that the columns can be identified with the b blocks of a BPEBD d with constant block size k and with v treatments $t_1, t_2, \ldots, t_v$. We know from Agrawal (1966) that for a BPEBD with $b = mv$, the k treatments in the b blocks of d can always be arranged such that each treatment occurs m times in each of the k positions in the blocks. We denote such an arrangement by a $k \times b$ matrix B. From the above matrix B, we can construct a $v \times b$ matrix A by putting the element a_l in its (i, j)th cell if t_i occurs in the lth row and jth column of B, $l = 1, 2, \ldots, k$, $i = 1, 2, \ldots, b$, $j = 1, 2, \ldots, v$. Other positions are filled in with zeros. Obviously it follows from the property of B that each of $a_1, a_2, \ldots, a_k$ occurs once and only once in each of the b columns of A. As every treatment occurs m times in each of the k rows of B, it is evident that each of the symbols $a_1, a_2, \ldots, a_k$ occurs m times in each row of A. Thus the lemma is proved.   □

Remark 6.2.1 It may sometimes be challenging to construct a B mentioned above. But if a BPEBD with $b = mv$ has a cyclic solution, it is very straightforward to construct the B-matrix. When the block design with $b = mv$ does not have a cyclic solution, the construction of B seems to be difficult and a trial and error method is used to get the desired configuration, whose existence is guaranteed by Lemma 6.2.1.

Now we prove the main theorem.

Theorem 6.2.1 *For any BPEBD* $d(v, b, r, k)$ *with* $b = mv$; m (≥ 1) *a positive integer,* $(k - 1)$ *optimum* W-*matrices can be constructed provided* H_k, *a Hadamard matrix of order* k, *exists.*

Proof We write the matrix H_k as

$$H_k = (1, h_1, h_2, \ldots, h_{k-1}) \qquad (6.2.1)$$

From a BPEBD $d(v, b, r, k)$, we can always, by Lemma 6.2.1, construct a $v \times b$ matrix A where each of $a_1, a_2, \ldots, a_k$ occurs m times in each row and once in each column. We identify the k elements of h_i with the symbols $a_1, a_2, \ldots, a_k$ and replace these symbols in A with their identified elements of h_i; $i = 1, 2, \ldots, k$. Thus we get $(k - 1)$ matrices $W_1, W_2, \ldots, W_{k-1}$ corresponding to $h_1, h_2, \ldots, h_{k-1}$ respectively. From the properties of the matrix A and those of H_k, it easily follows that the W_i's satisfy the optimality condition (3.1.5).   □

Example 6.2.1 Let us consider the symmetric BIBD with parameters $v = b = 7$, $r = k = 4, \lambda = 2$ constructed heuristically by Nandi (1946). The blocks are: (1, 2, 3, 4), (1, 2, 5, 6), (1, 3, 6, 7), (1, 4, 5, 7), (2, 3, 5, 7), (2, 4, 6, 7), (3, 4, 5, 6). The **B**- and **A**-matrices of Lemma 6.2.1 can respectively be written as

$$\mathbf{B} = \begin{pmatrix} 1 & 2 & 3 & 4 & 5 & 7 & 6 \\ 2 & 1 & 6 & 5 & 7 & 4 & 3 \\ 3 & 5 & 1 & 7 & 2 & 6 & 4 \\ 4 & 6 & 7 & 1 & 3 & 2 & 5 \end{pmatrix} \quad \text{and} \quad \mathbf{A} = \begin{pmatrix} a_1 & a_2 & a_3 & a_4 & 0 & 0 & 0 \\ a_2 & a_1 & 0 & 0 & a_3 & a_4 & 0 \\ a_3 & 0 & a_1 & 0 & a_4 & 0 & a_2 \\ a_4 & 0 & 0 & a_1 & 0 & a_2 & a_3 \\ 0 & a_3 & 0 & a_2 & a_1 & 0 & a_4 \\ 0 & a_4 & a_2 & 0 & 0 & a_3 & a_1 \\ 0 & 0 & a_4 & a_3 & a_2 & a_1 & 0 \end{pmatrix}.$$

Consider

$$\mathbf{H}_4 = \begin{pmatrix} 1 & 1 & 1 & 1 \\ 1 & -1 & -1 & 1 \\ 1 & 1 & -1 & -1 \\ 1 & -1 & 1 & -1 \end{pmatrix} = (\mathbf{1}, \mathbf{h}_1, \mathbf{h}_2, \mathbf{h}_3).$$

Using Lemma 6.2.1, we construct the following three **W**-matrices by using the identification $\mathbf{a} = \mathbf{h}_1$, $\mathbf{a} = \mathbf{h}_2$ and $\mathbf{a} = \mathbf{h}_3$ respectively, where $\mathbf{a}' = (a_1, a_2, a_3, a_4)$ and they are as follows:

$$\mathbf{W}_1 = \begin{pmatrix} 1 & -1 & 1 & -1 & 0 & 0 & 0 \\ -1 & 1 & 0 & 0 & 1 & -1 & 0 \\ 1 & 0 & 1 & 0 & -1 & 0 & -1 \\ -1 & 0 & 0 & 1 & 0 & -1 & 1 \\ 0 & 1 & 0 & -1 & 1 & 0 & -1 \\ 0 & -1 & -1 & 0 & 0 & 1 & 1 \\ 0 & 0 & -1 & 1 & -1 & 1 & 0 \end{pmatrix}, \mathbf{W}_2 = \begin{pmatrix} 1 & -1 & -1 & 1 & 0 & 0 & 0 \\ -1 & 1 & 0 & 0 & -1 & 1 & 0 \\ -1 & 0 & 1 & 0 & 1 & 0 & -1 \\ 1 & 0 & 0 & 1 & 0 & -1 & -1 \\ 0 & -1 & 0 & -1 & 1 & 0 & 1 \\ 0 & 1 & -1 & 0 & 0 & -1 & 1 \\ 0 & 0 & 1 & -1 & -1 & 1 & 0 \end{pmatrix},$$

$$\mathbf{W}_3 = \begin{pmatrix} 1 & 1 & -1 & -1 & 0 & 0 & 0 \\ 1 & 1 & 0 & 0 & -1 & -1 & 0 \\ -1 & 0 & 1 & 0 & -1 & 0 & 1 \\ -1 & 0 & 0 & 1 & 0 & 1 & -1 \\ 0 & -1 & 0 & 1 & 1 & 0 & -1 \\ 0 & -1 & 1 & 0 & 0 & -1 & 1 \\ 0 & 0 & -1 & -1 & 1 & 1 & 0 \end{pmatrix}.$$

It is easy to observe that $\mathbf{W}_i$-matrices satisfy the optimality condition (3.1.5).

Remark 6.2.2 If k is even, then it follows from Theorem 6.2.1 that at least one optimum **W**-matrix can always be constructed by identifying the **a**'s with $(\mathbf{1}'_{\frac{k}{2}}, -\mathbf{1}'_{\frac{k}{2}})$.

In the following theorem, we shall see that the number of optimum **W**-matrices can be increased substantially if the BPEBD obeys an additional condition of k-resolvability. Now we give the definition of α-resolvability of a design (Raghavarao 1971, p. 59).

Definition 6.2.1 A BPEBD with number of treatments $= v$, number of blocks $= b$, number of replications of each treatment $= r$ and block size $= k$ is said to be α-resolvable if the sets can be grouped into t classes $S_1, S_2, \ldots, S_t$, each with β sets, such that in each class every symbol is replicated α times.

We then have

$$v\alpha = k\beta, \quad b = t\beta, \quad r = t\alpha.$$

Thus a k-resolvable BPEBD with $b = mv$ requires that the $b = mv$ blocks can be partitioned into m sets $S_1, S_2, \ldots, S_m$ each of which contains v blocks such that each of the v treatments occurs k times in each $S_i, i = 1, 2, \ldots, m$.

Theorem 6.2.2 *For a k-resolvable BPEBD with $b = mv$, it is possible to construct $m(k - 1)$ optimum **W**-matrices, provided $\mathbf{H}_k$ and $\mathbf{H}_m$ exist.*

Proof As the design is k-resolvable, then Lemma 6.2.1 is applicable to the blocks of each S_i and from these v blocks a matrix $\mathbf{A}_i^{v \times v}$ can be constructed where $\mathbf{A}_i$ contains each of the symbols $a_1, a_2, \ldots, a_k$ once and only once in each row and in each column, $i = 1, 2, \ldots, m$. It is also to be noted that

$$\mathbf{A} = (\mathbf{A}_1, \mathbf{A}_2, \ldots, \mathbf{A}_m) \tag{6.2.2}$$

is the $\mathbf{A}$-matrix of Lemma 6.2.1 corresponding to the $b = mv$ blocks of BPEBD where each of the symbols occurs m times in each row and just once in each column of $\mathbf{A}$.

According to the method described in Theorem 6.2.1, we can construct a matrix $\mathbf{W}_{ji}$ from $\mathbf{A}_j$ by identifying $a_1, a_2, \ldots, a_k$ with the elements of $\mathbf{h}_i$, the ith column of $\mathbf{H}_k$ in (6.2.1). By juxtaposing $\mathbf{W}_{ji}, j = 1, 2, \ldots, m$, for fixed i, we obtain a matrix $\mathbf{W}_i$, where

$$\mathbf{W}_i = (\mathbf{W}_{1i}, \mathbf{W}_{2i}, \ldots, \mathbf{W}_{mi}), \quad i = 1, 2, \ldots, (k - 1). \tag{6.2.3}$$

Varying i in (6.2.3), we get $(k - 1)$ matrices $\mathbf{W}_1, \mathbf{W}_2, \ldots, \mathbf{W}_{k-1}$ which are optimum $\mathbf{W}$-matrices for the BPEBD. As the BPEBD is k-resolvable and $\mathbf{H}_m$ exists, we can increase the number of optimal $\mathbf{W}$-matrices. By taking the Khatri-Rao product among $\mathbf{h}_j^* = (h_{j1}^*, h_{j2}^*, \ldots, h_{jm}^*)'$, the jth column of $\mathbf{H}_m$ and $\mathbf{W}_i$ of (6.2.3), $m(k - 1)$ matrices $\mathbf{W}_{ji}^*$ can be constructed, where

$$\mathbf{W}_{ji}^* = \mathbf{h}_j^{*'} \odot \mathbf{W}_i = \left(h_{j1}^* \mathbf{W}_{1i}, h_{j2}^* \mathbf{W}_{2i}, \ldots, h_{jm}^* \mathbf{W}_{mi} \right),$$
$$\forall i = 1, 2, \ldots, (k - 1); \ j = 1, 2, \ldots, m. \tag{6.2.4}$$

It is easy to verify that $\mathbf{W}_{ji}^*$'s satisfy the condition (3.1.5) and hence give $m(k - 1)$ optimum $\mathbf{W}$-matrices for the k-resolvable BPEBD. $\qquad\square$

Example 6.2.2 Let us consider the following 2-resolvable BIBD with parameters $v = 5, b = 10, r = 4, k = 2, \lambda = 1$ where the blocks can be represented in the form of a matrix $\mathbf{B}$ of order 2×10 as

$$\mathbf{B} = \begin{pmatrix} 1\,2\,3\,4\,5 & 1\,2\,3\,4\,5 \\ 2\,3\,4\,5\,1 & 3\,4\,5\,1\,2 \end{pmatrix} = (\mathbf{B}_1, \mathbf{B}_2)$$

Now the **A**-matrix of order 10×5 can be constructed as

$$
\begin{array}{c}
Tr. \\
\downarrow
\end{array}
\begin{array}{c}
\qquad Bl. \rightarrow \\
1\ \ 2\ \ 3\ \ 4\ \ 5\ \ 6\ \ 7\ \ 8\ \ 9\ \ 10
\end{array}
$$

$$
\mathbf{A} = \begin{array}{c} 1 \\ 2 \\ 3 \\ 4 \\ 5 \end{array}
\begin{pmatrix}
a_1 & 0 & 0 & 0 & a_2 & a_1 & 0 & 0 & a_2 & 0 \\
a_2 & a_1 & 0 & 0 & 0 & 0 & a_1 & 0 & 0 & a_2 \\
0 & a_2 & a_1 & 0 & 0 & a_2 & 0 & a_1 & 0 & 0 \\
0 & 0 & a_2 & a_1 & 0 & 0 & a_2 & 0 & a_1 & 0 \\
0 & 0 & 0 & a_2 & a_1 & 0 & 0 & a_2 & 0 & a_1
\end{pmatrix} = (\mathbf{A}_1, \mathbf{A}_2)
$$

Considering the column $(1, -1)'$ of $\mathbf{H}_2$ and identifying 1 with a_1 and -1 with a_2, one **W**-matrix can be constructed by using Theorem 6.2.1 as

$$
\mathbf{W}_1' = \begin{array}{c} \\ \end{array}
\begin{array}{ccccc}
1 & 2 & 3 & 4 & 5
\end{array}
\left(
\begin{array}{ccccc}
1 & -1 & 0 & 0 & 0 \\
0 & 1 & -1 & 0 & 0 \\
0 & 0 & 1 & -1 & 0 \\
0 & 0 & 0 & 1 & -1 \\
-1 & 0 & 0 & 0 & 1 \\
\hline
1 & 0 & -1 & 0 & 0 \\
0 & 1 & 0 & -1 & 0 \\
0 & 0 & 1 & 0 & -1 \\
-1 & 0 & 0 & 1 & 0 \\
0 & -1 & 0 & 0 & 1
\end{array}
\right) = \begin{pmatrix} \mathbf{W}_{11}' \\ \mathbf{W}_{21}' \end{pmatrix}.
$$

Since this is a resolvable design and $m = 2$, when $\mathbf{H}_2$ exists, two optimum **W**-matrices can be constructed by using Theorem 6.2.2 as

$$\mathbf{W}_{11}^* = \mathbf{1}_2' \odot (\mathbf{W}_{11}, \mathbf{W}_{21}); \quad \mathbf{W}_{21}^* = (1, -1) \odot (\mathbf{W}_{11}, \mathbf{W}_{21}) = (\mathbf{W}_{11} - \mathbf{W}_{21}).$$

Remark 6.2.3 Let $\mathbf{H}_k$ exist and m (>2) be even, then $2(k-1)$ optimum **W**-matrices can be obtained for a resolvable BPEBD by using $(k-1)$ columns (except the column of all 1's) of $\mathbf{H}_k$ and using $\mathbf{1}_m$ and $(\mathbf{1}_{\frac{m}{2}}', -\mathbf{1}_{\frac{m}{2}}')'$ for the two choices of orthogonal vectors in the Khatri-Rao product in Theorem 6.2.2.

Remark 6.2.4 If both of k (>2) and m (>2) are even, then two optimum **W**-matrices can be constructed for a resolvable BPEBD by using the two pairs of vectors $((\mathbf{1}_{\frac{k}{2}}', -\mathbf{1}_{\frac{k}{2}}')', \mathbf{1}_m)$ and $((\mathbf{1}_{\frac{k}{2}}', -\mathbf{1}_{\frac{k}{2}}')', (\mathbf{1}_{\frac{m}{2}}', -\mathbf{1}_{\frac{m}{2}}')')$ for the columns of $\mathbf{H}_k$ and $\mathbf{H}_m$ respectively in Theorem 6.2.2.

Remark 6.2.5 Let $\mathbf{H}_m$ exist and k (>2) be even. Then following Theorem 6.2.2, m optimum **W**-matrices can be constructed for a resolvable BPEBD by using m columns of $\mathbf{H}_m$ and $(\mathbf{1}_{\frac{k}{2}}', -\mathbf{1}_{\frac{k}{2}}')'$ as a column of fictitious $\mathbf{H}_k$.

Remark 6.2.6 If t optimum **W**-matrices exist for any BPEBD, then the same number of **W**-matrices exist for the dual of that design where the blocks of the original design play the role of the treatments. This is because an optimum **W**-matrix for the dual design is also optimum for the original design.

6.3 Cyclic Designs

Cyclic designs are BPEBDs obtained by developing $m(\geq 1)$ initial blocks where the v treatments are the elements of a module M. All cyclic designs belong to the class of PBIBDs with at most $\frac{v}{2}$ associate classes. Many incomplete block designs may be set out as cyclic designs.

If there are v treatments denoted by $0, 1, \ldots, v-1$ which are elements of a module M, and are arranged in blocks of size k so that each treatment is replicated r times, then the cyclic design with these parameters is denoted by $C(v, k, r)$. Given any initial block, another block is generated by adding α (mod v) to each treatment of the initial block where $\alpha \in M$. If all the v blocks thus obtained from the given initial block are all distinct then this set of blocks is said to form a 'full set'. If v and k are relatively prime to each other then the v blocks generated from an initial block always give a full set with parameters $(v, k = r)$. On the other hand if v and k have a common divisor d, then for every value of d, there always exists at least one initial block where all the v blocks generated from an initial block are not distinct; only $\frac{v}{d}$ of them are distinct. This set of blocks forms a 'partial set' with parameters $(v, k, r = \frac{v}{d})$. Full or partial sets can be used singly or in combination to construct cyclic designs. For a detailed study, one is referred to John et al. (1972) and John (1987).

According to John (1987), for given v and k, the $\binom{v}{k}$ distinct blocks can be set out in a number of cyclic sets where the sets are either 'full sets' consisting of v blocks each or are 'partial sets'. If v and k are relatively prime, then all the sets are 'full sets'. On the other hand, if v and k are not relatively prime then 'partial sets' consisting of $\frac{v}{d}$ distinct blocks arise, where d is any common divisor of v and k. For example, for $v = 7$ and $k = 3$ the $35 = \binom{7}{3}$ all possible distinct blocks can be set out in five cyclic 'full sets' each of 7 blocks and the five initial blocks can be taken as $(0, 1, 2), (0, 1, 3), (0, 1, 4), (0, 1, 5), (0, 2, 4)$ mod 7. On the other hand, for $v = 8$ and $k = 4$, the $70 = \binom{8}{4}$ all possible blocks can not be divided into all 'full sets' as 'full sets' because 8 does not divide 70. Moreover as 4 and 2 are common factors of $v = 8$ and $k = 4$, there should be 'partial sets', one containing 4 blocks and another containing 2 blocks. The seventy distinct blocks can be set out in 8 'full sets' of 8 blocks each; one half-set of four blocks viz. $(0, 1, 4, 5), (1, 2, 5, 6), (2, 3, 6, 7), (3, 4, 7, 0)$ and one quarter-set of two blocks given by $(0, 2, 4, 6), (1, 3, 5, 7)$.

If there exists a partial set consisting of $\frac{k}{d}$ blocks in a cyclic design, then we see that each treatment is replicated $\frac{k}{d}$ times in these blocks. So the number of covariates to be accommodated in a cyclic design depends on whether the sets are full or partial and also on the number of sets. When cyclic designs consist of 'full sets' only, then a systematic way for assigning values to the covariates can be developed. However, when a design contains 'partial sets', it is difficult to specify the number of covariates to be accommodated beforehand. Some examples of cyclic designs containing 'partial sets' are considered in Dutta et al. (2010) where they have provided a solution for OCDs through an ad hoc method.

It is to be noted that Das et al. (2003) and Dutta (2004) proposed OCDs on some series of BIBD's which belonged to the class of cyclic designs. Moreover, all irreducible BIBDs can also be obtained by cyclically developing some sets of initial blocks. So we can cover all these designs and a lot of other designs under a general technique described in the following section.

6.3.1 Cyclic Designs Containing 'Full Sets' Only

It is proved in Theorem 6.2.2 that, for resolvable BPEBDs the number of covariates can be increased over the number of covariates for ordinary BPEBD with the same parameters. It can be easily noted that cyclic designs with m initial blocks giving m full sets of blocks always give resolvable BPEBDs. The particular case when the resolvable BPEBDs are cyclic designs, construction can be done more easily by exploiting the circular nature of the blocks. The precise statement follows.

Theorem 6.3.1 *Let a cyclic design with parameters $v, b = mv, r = mk, k$ be obtained by developing m initial blocks each of size k and also let $\mathbf{H}_k$ and $\mathbf{H}_m$ exist. Then $m(k-1)$ optimum $\mathbf{W}$-matrices can be constructed.*

Proof Let $\mathbf{H}_k$ and $\mathbf{H}_m$ be written respectively in the following form:

$$\mathbf{H}_k = (\mathbf{1}, \mathbf{h}_1, \mathbf{h}_2, \ldots, \mathbf{h}_{k-1}) \text{ and } \mathbf{H}_m = (\mathbf{h}_1^*, \mathbf{h}_2^*, \ldots, \mathbf{h}_m^*)$$

where

$$\mathbf{h}_i^{*\prime} = (h_{i1}^*, h_{i2}^*, \ldots, h_{im}^*); \quad i = 1, 2, \ldots, m.$$

Also let the m initial blocks of the design be displayed in the form of column vectors in the incidence matrix of that design. The k non-zero elements of the qth initial block are replaced by the k elements of $\mathbf{h}_j$ in that order and are cyclically permuted to get a $v \times v$ matrix $\mathbf{W}_{jq}$. Again from $\mathbf{h}_i^*$ and $\mathbf{W}_{jq}, q = 1, 2, \ldots, m$, we construct a $v \times b$ matrix $\mathbf{W}_{ji}^*$ by applying Khatri-Rao product, i.e.

$$\mathbf{W}_{ji}^* = \mathbf{h}_i^{*\prime} \odot \mathbf{W}_j = \left(h_{i1}^* \mathbf{W}_{j1}, \ h_{i2}^* \mathbf{W}_{j2}, \ \ldots, \ h_{im}^* \mathbf{W}_{jm} \right); \tag{6.3.1}$$

where

$$\mathbf{W}_j = (\mathbf{W}_{j1}, \mathbf{W}_{j2}, \ldots, \mathbf{W}_{jm}), \quad j = 1, 2, \ldots, (k-1). \tag{6.3.2}$$

By varying j and i, $m(k-1)$ such $\mathbf{W}_{ji}^*$-matrices can be obtained. It can easily be seen that these matrices satisfy the condition (3.1.5) and are optimum. Thus the theorem follows.   □

Note 6.3.1 It is to be noted that the cyclic design with 'full sets' is k-resolvable. So $m(k-1)$ optimum $\mathbf{W}$-matrices could have been constructed following Theorem 6.2.2. But as the design possesses cyclic nature, OCDs can be constructed more easily through Theorem 6.3.1.

Example 6.3.1 Let a cyclic design with parameters $v = 13, b = 26, r = 8, k = 4$ with the initial blocks as $(1, 4, 12, 13), (1, 4, 10, 13)$ mod 13, be constructed. Let $\mathbf{H}_4$ and $\mathbf{H}_2$ be written as

$$\mathbf{H}_4 = \begin{pmatrix} 1 & 1 & 1 & 1 \\ 1 & -1 & -1 & 1 \\ 1 & 1 & -1 & -1 \\ 1 & -1 & 1 & -1 \end{pmatrix} = (\mathbf{1}, \mathbf{h}_1, \mathbf{h}_2, \mathbf{h}_3); \quad \mathbf{H}_2 = \begin{pmatrix} 1 & 1 \\ 1 & -1 \end{pmatrix} = (\mathbf{h}_1^*, \mathbf{h}_2^*).$$

We identify the elements of $\mathbf{h}_1$ with the non-zero elements of the first column of $\mathbf{N}_1$, the part of the incidence matrix corresponding to the first initial block. Then we permute cyclically this column in the same way as $\mathbf{N}_1$ was obtained and get matrix $\mathbf{W}_{11}$. In the same way, by identifying the elements of $\mathbf{h}_1$ with non-zero elements of the first column of $\mathbf{N}_2$, the part of the incidence matrix corresponding to the second initial block and cyclically permuting it, we get $\mathbf{W}_{12}$. These matrices can be visualized as

$$\begin{array}{c} \text{Treatments} \rightarrow \\ 1 \;\; 2 \;\; 3 \;\;\; 4 \;\; 5 \;\; 6 \;\; 7 \;\; 8 \;\; 9 \;\; 10 \;\; 11 \;\; 12 \;\; 13 \end{array}$$
$$\mathbf{W}_{11}' = \begin{pmatrix} 1 & 0 & 0 & -1 & 0 & 0 & 0 & 0 & 0 & 0 & 0 & 1 & -1 \\ \multicolumn{13}{c}{\text{and cyclic permutations}} \end{pmatrix};$$

$$\begin{array}{c} \text{Treatments} \rightarrow \\ 1 \;\; 2 \;\; 3 \;\;\; 4 \;\; 5 \;\; 6 \;\; 7 \;\; 8 \;\; 9 \;\; 10 \;\; 11 \;\; 12 \;\; 13 \end{array}$$
$$\mathbf{W}_{12}' = \begin{pmatrix} 1 & 0 & 0 & -1 & 0 & 0 & 0 & 0 & 1 & 0 & 0 & -1 \\ \multicolumn{12}{c}{\text{and cyclic permutations}} \end{pmatrix}.$$

Then by Theorem 6.3.1, six optimum $\mathbf{W}$-matrices for this design can be constructed by using $\mathbf{h}_1, \mathbf{h}_2, \mathbf{h}_3$ and the two columns of $\mathbf{H}_2$. For instance, if $\mathbf{h}_1$ and $\mathbf{h}_1^*$ are used, then from (6.3.1), the two $\mathbf{W}$-matrices $\mathbf{W}_{11}^*, \mathbf{W}_{12}^*$ are given by

$$\mathbf{W}_{11}^* = \mathbf{h}_1^{*\prime} \odot \mathbf{W}_1 = (1, 1) \odot (\mathbf{W}_{11}, \mathbf{W}_{12}) = (\mathbf{W}_{11}, \mathbf{W}_{12})$$

and
$$\mathbf{W}_{12}^* = \mathbf{h}_2' * \odot \mathbf{W}_1 = \left(1, -1\right) \odot \left(\mathbf{W}_{11}, \mathbf{W}_{12}\right) = \left(\mathbf{W}_{11}, -\mathbf{W}_{12}\right)$$

Incidentally, it is seen that the cyclic design, discussed in Chap.5, is a two-associate class PBIBD with parameters $v = 13, b = 26, r = 8, k = 4, \lambda_1 = 1$, $\lambda_2 = 3, n_1 = n_2 = 6$. The first associates of the treatment i are $(i + 2, i + 5, i + 6, i + 7, i + 8, i + 11)$ (mod 13).

Below we make some remarks regarding methods of construction of optimum W-matrices where at least one of $\mathbf{H}_k$ and $\mathbf{H}_m$ does not exist so that Theorem 6.3.1 can not be applied.

Remark 6.3.1 If $\mathbf{H}_k$ exists but m is an odd integer, so that $\mathbf{H}_m$ does not exist, then $(k - 1)$ optimum W-matrices can be obtained for the said cyclic design and they are given by $\mathbf{1}_m \odot \mathbf{W}_j, j = 1, 2, \ldots, k - 1$, where $\mathbf{W}_j$ is obtained from (6.3.2).

Remark 6.3.2 If none of $\mathbf{H}_k$ and $\mathbf{H}_m$ exists and $k \equiv 2 \pmod 4$ and m is an odd integer, then one optimum W-matrix can be constructed as $\mathbf{1}_m \odot \mathbf{W}^*$, where $\mathbf{W}^*$ is a matrix analogous to $\mathbf{W}_j$ of equation (6.3.2) obtained by using $(\mathbf{1}_{\frac{k}{2}}', -\mathbf{1}_{\frac{k}{2}}')'$ for the cyclical permutation in the incidence matrix.

Remark 6.3.3 Let $\mathbf{H}_k$ exist and $m \equiv 2 \pmod 4, m > 2$. In this case, $2(k-1)$ optimum W-matrices given by $\mathbf{h}^{**} \odot \mathbf{W}_j$ can be obtained, where $\mathbf{h}^{**}$ is $\mathbf{1}_m$ or $(\mathbf{1}_{\frac{m}{2}}', -\mathbf{1}_{\frac{m}{2}}')'$.

Remark 6.3.4 If each of k, m is of the form 2 (mod 4) so that none of $\mathbf{H}_k$ and $\mathbf{H}_m$ exists, then 2 optimum W-matrices can be constructed as $\mathbf{h}^{**} \odot \mathbf{W}^*$ where $\mathbf{h}^{**}$ is $\mathbf{1}_m$ or $(\mathbf{1}_{\frac{m}{2}}', -\mathbf{1}_{\frac{m}{2}}')'$ and $\mathbf{W}^*$ is the same as in Remark 6.3.2.

Remark 6.3.5 If $\mathbf{H}_m$ exists but $\mathbf{H}_k$ does not where $k \equiv 2 \pmod 4, k > 2$, then m optimum W-matrices can be constructed as $\mathbf{h}_i^* \odot \mathbf{W}^*, i = 1, 2, \ldots, m$ where $\mathbf{W}^*$ is the same as Remark 6.3.2.

6.3.2 Cyclic Designs Containing Some Partial Sets

It was mentioned earlier that it is difficult to propose a systematic method for finding OCDs for cyclic designs containing 'partial sets'. It should be noted that the number of optimum W-matrices depends on the properties of the 'partial sets' and consequently on the nature of the columns of $\mathbf{H}_k$ whose elements are used to replace the non-zero elements in the blocks of the incidence matrix. We consider the following example illustrating an ad hoc method which depends on the nature of the partial set.

Example 6.3.2 Consider the irreducible BIBD with parameters $v = 6, b = \binom{6}{4} = 15$; $r = \binom{5}{3} = 10, k = 4, \lambda = \binom{4}{2} = 6$. The design can be obtained from the three initial blocks: [(0, 1, 2, 3), (0, 2, 3, 4), (0, 2, 3, 5)] mod 6, where the first two give 'full sets' containing six distinct blocks each and the last one gives a 'partial set' containing only three distinct blocks. We consider $\mathbf{H}_4$ as

$$\mathbf{H}_4 = \begin{pmatrix} 1 & 1 & 1 & 1 \\ 1 & -1 & -1 & 1 \\ 1 & 1 & -1 & -1 \\ 1 & -1 & 1 & -1 \end{pmatrix} = (\mathbf{1}, \mathbf{h}_1, \mathbf{h}_2, \mathbf{h}_3). \tag{6.3.3}$$

It is to be noted that, as in Theorem 6.3.1, all the three columns $\mathbf{h}_1$, $\mathbf{h}_2$ and $\mathbf{h}_3$ of (6.3.3) cannot be used in each of the three subsets of blocks obtained by developing cyclically the three initial blocks. The last three blocks obtained from the third initial block are 'partially cyclic'; only $\mathbf{h}_1$, $\mathbf{h}_2$ can be used to construct $\mathbf{W}$-matrices but $\mathbf{h}_3$ cannot be used as it will not lead to zero column-sums. Using $\mathbf{h}_1$ and $\mathbf{h}_2$, we get two $\mathbf{W}$-matrices, namely $\mathbf{W}'_{(1)}$ and $\mathbf{W}'_{(2)}$ respectively by applying the method described in Theorem 6.3.1:

Treatments $\rightarrow$

$$\mathbf{W}'_{(1)} = \begin{array}{cccccc} 0 & 1 & 2 & 3 & 4 & 5 \\ \begin{pmatrix} 1 & -1 & 1 & -1 & 0 & 0 \\ 0 & 1 & -1 & 1 & -1 & 0 \\ 0 & 0 & 1 & -1 & 1 & -1 \\ -1 & 0 & 0 & 1 & -1 & 1 \\ 1 & -1 & 0 & 0 & 1 & -1 \\ -1 & 1 & -1 & 0 & 0 & 1 \\ \hline 1 & 0 & -1 & 1 & -1 & 0 \\ 0 & 1 & 0 & -1 & 1 & -1 \\ -1 & 0 & 1 & 0 & -1 & 1 \\ 1 & -1 & 0 & 1 & 0 & -1 \\ -1 & 1 & -1 & 0 & 1 & 0 \\ 0 & -1 & 1 & -1 & 0 & 1 \\ \hline 1 & 0 & -1 & 1 & 0 & -1 \\ -1 & 1 & 0 & -1 & 1 & 0 \\ 0 & -1 & 1 & 0 & -1 & 1 \end{pmatrix} \end{array} \; ; \;\; \mathbf{W}'_{(2)} = \begin{array}{cccccc} 0 & 1 & 2 & 3 & 4 & 5 \\ \begin{pmatrix} 1 & -1 & -1 & 1 & 0 & 0 \\ 0 & 1 & -1 & -1 & 1 & 0 \\ 0 & 0 & 1 & -1 & -1 & 1 \\ 1 & 0 & 0 & 1 & -1 & -1 \\ -1 & 1 & 0 & 0 & 1 & -1 \\ -1 & -1 & 1 & 0 & 0 & 1 \\ \hline 1 & 0 & -1 & -1 & 1 & 0 \\ 0 & 1 & 0 & -1 & -1 & 1 \\ 1 & 0 & 1 & 0 & -1 & -1 \\ -1 & 1 & 0 & 1 & 0 & -1 \\ -1 & -1 & 1 & 0 & 1 & 0 \\ 0 & -1 & -1 & 1 & 0 & 1 \\ \hline 1 & 0 & -1 & -1 & 0 & 1 \\ -1 & 1 & 0 & 1 & -1 & 0 \\ 0 & -1 & 1 & 0 & 1 & -1 \end{pmatrix} \end{array}$$

Treatments $\rightarrow$

6.3.3 Cyclic Designs Where Each Element Corresponds to a Number of Symbols

Here any treatment is denoted by α_j where α is any element of the module $M = (0, 1, \ldots, m)$ and j is one of the n symbols $1, 2, \ldots, n$. The following is an example of a design which is obtained by the classical method of difference (cf. Bose 1939) where each symbol of the module $M = (0, 1, 2, 3, 4)$ corresponds to two symbols

1 and 2. If the blocks can be grouped into sets which have cycles, then optimum W-matrices can be constructed by exploiting this property. The method is illustrated through the following example.

Example 6.3.3 Consider the GD design with parameters $v = 10, b = 20, r = 8$, $k = 4, \lambda_1 = 0, \lambda_2 = 3, m = 5, n = 2$ with the initial group $(0_1, 0_2)$ mod 5. The initial blocks are $(0_1, 1_2, 2_2, 4_2), (0_2, 1_1, 2_1, 4_1), (0_1, 2_2, 3_2, 4_2), (0_2, 2_1, 3_1, 4_1)$ mod 5. We divide the four initial blocks into two sets viz. $S_1 = \{(0_1, 1_2, 2_2, 4_2),$ $(0_1, 2_2, 3_2, 4_2)\}$ and $S_2 = \{(0_2, 1_1, 2_1, 4_1), (0_2, 2_1, 3_1, 4_1)\}$. In the first five columns of the incidence matrix corresponding to the initial block $(0_1, 1_2, 2_2, 4_2)$ of S_1, the non-zero elements are replaced by the elements of $\mathbf{h}_1$ of (6.3.3) and in the last five columns corresponding to $(0_1, 2_2, 3_2, 4_2)$ of S_1, the non-zero elements are replaced by those of $-\mathbf{h}_1$. We denote this matrix, of order 10×10, by $\mathbf{U}_1^{(1)}$. In the same way by using $\mathbf{h}_1$ and $-\mathbf{h}_1$ in the two initial blocks of S_2 we get a matrix $\mathbf{U}_1^{(2)}$. $\mathbf{U}_1^{(1)}$ and $\mathbf{U}_1^{(2)}$ are given by

$$
\mathbf{U}_1^{(1)} = \begin{array}{c} Tr. \\ \downarrow \end{array} \begin{array}{c} 0_1 \\ 1_1 \\ 2_1 \\ 3_1 \\ 4_1 \\ 0_2 \\ 1_2 \\ 2_2 \\ 3_2 \\ 4_2 \end{array} \left(\begin{array}{ccccc|ccccc} 1 & 0 & 0 & 0 & 0 & -1 & 0 & 0 & 0 & 0 \\ 0 & 1 & 0 & 0 & 0 & 0 & -1 & 0 & 0 & 0 \\ 0 & 0 & 1 & 0 & 0 & 0 & 0 & -1 & 0 & 0 \\ 0 & 0 & 0 & 1 & 0 & 0 & 0 & 0 & -1 & 0 \\ 0 & 0 & 0 & 0 & 1 & 0 & 0 & 0 & 0 & -1 \\ \hline 0 & -1 & 0 & 1 & -1 & 0 & 1 & -1 & 1 & 0 \\ -1 & 0 & -1 & 0 & 1 & 0 & 0 & 1 & -1 & 1 \\ 1 & -1 & 0 & -1 & 0 & 1 & 0 & 0 & 1 & -1 \\ 0 & 1 & -1 & 0 & -1 & -1 & 1 & 0 & 0 & 1 \\ -1 & 0 & 1 & -1 & 0 & 1 & -1 & 1 & 0 & 0 \end{array} \right),
$$

$$
\mathbf{U}_1^{(2)} = \begin{array}{c} Tr. \\ \downarrow \end{array} \begin{array}{c} 0_1 \\ 1_1 \\ 2_1 \\ 3_1 \\ 4_1 \\ 0_2 \\ 1_2 \\ 2_2 \\ 3_2 \\ 4_2 \end{array} \left(\begin{array}{ccccc|ccccc} 0 & 1 & 0 & -1 & 1 & 0 & -1 & 1 & -1 & 0 \\ 1 & 0 & 1 & 0 & -1 & 0 & 0 & -1 & 1 & -1 \\ -1 & 1 & 0 & 1 & 0 & -1 & 0 & 0 & -1 & 1 \\ 0 & -1 & 1 & 0 & 1 & 1 & -1 & 0 & 0 & -1 \\ 1 & 0 & -1 & 1 & 0 & -1 & 1 & -1 & 0 & 0 \\ \hline -1 & 0 & 0 & 0 & 0 & 1 & 0 & 0 & 0 & 0 \\ 0 & -1 & 0 & 0 & 0 & 0 & 1 & 0 & 0 & 0 \\ 0 & 0 & -1 & 0 & 0 & 0 & 0 & 1 & 0 & 0 \\ 0 & 0 & 0 & -1 & 0 & 0 & 0 & 0 & 1 & 0 \\ 0 & 0 & 0 & 0 & -1 & 0 & 0 & 0 & 0 & 1 \end{array} \right).
$$

From $\mathbf{U}_1^{(1)}$ and $\mathbf{U}_2^{(2)}$, we construct two optimum W-matrices as

$$
\mathbf{W}^{(1,1)} = (\mathbf{U}_1^{(1)}, \mathbf{U}_1^{(2)}), \quad \mathbf{W}^{(1,2)} = (\mathbf{U}_1^{(1)}, -\mathbf{U}_1^{(2)}).
$$

Similarly we can get four more optimum W-matrices viz. $\mathbf{W}^{(2,1)}, \mathbf{W}^{(2,2)}, \mathbf{W}^{(3,1)}$ and $\mathbf{W}^{(3,2)}$ by using $\mathbf{h}_2$ and $\mathbf{h}_3$.

6.4 t-Fold BPEBDs

In Das et al. (2003) it was been seen that optimum $\mathbf{W}$-matrices could be constructed for BIBDs with repeated blocks. In this section, we propose to extend the result to any BPEBD with repeated blocks. Thus the method will be applicable to BIBDs, PBIBDs and a lot of others with repeated blocks. Also it has been seen that in our method the number of optimum $\mathbf{W}$-matrices can substantially be increased. Precise statement follows.

Theorem 6.4.1 *Let t repetitions of the blocks of a BPEBD $d(v, b, r, k)$ be considered where $b = mv$. Also let $\mathbf{H}_k$ and $\mathbf{H}_t$ exist. Then we can construct $t(k - 1)$ optimum $\mathbf{W}$-matrices for the t-fold design BPEBD $d(v, bt, rt, k)$.*

Proof As $\mathbf{H}_k$ exists, according to the method described in Theorem 6.2.1, it is possible to construct $(k - 1)$ optimum $\mathbf{W}$-matrices for the lth BPEBD $d(v, b, r, k)$ with $b = mv, m$, a positive integer, where the nth optimum $\mathbf{W}$-matrix for the lth BPEBD be denoted as $\mathbf{U}_l^{(n)}, l = 1, 2, \ldots, t, n = 1, 2, \ldots, k - 1$.

For fixed n, the optimum $\mathbf{W}$-matrix for the whole design considering all the folds together is given by

$$\mathbf{U}^{(n)} = (\mathbf{U}_1^{(n)}, \mathbf{U}_2^{(n)}, \ldots, \mathbf{U}_t^{(n)}).$$

By assumption, $\mathbf{H}_t$ exists and it is written as

$$\mathbf{H}_t = (\mathbf{h}_1^*, \mathbf{h}_2^*, \ldots, \mathbf{h}_{t-1}^*, \mathbf{h}_t^*) \tag{6.4.1}$$

where $\mathbf{h}_t^* = (1, 1, \ldots, 1)'$.

Then as the t-fold BPEBD is resolvable, then by Theorem 6.2.2

$$\mathbf{U}_{jn} = \mathbf{h}_j^{*\prime} \odot (\mathbf{U}_1^{(n)}, \mathbf{U}_2^{(n)}, \ldots, \mathbf{U}_t^{(n)}), \quad j = 1, 2, \ldots, t; \ n = 1, 2, \ldots, k - 1 \tag{6.4.2}$$

gives $t(k - 1)$ optimum $\mathbf{W}$-matrices for the t-fold BPEBD $d(v, tb, tr, k)$. □

Corollary 6.4.1 *Let t repetitions of the blocks of a BPEBD $d(v, b, r, k)$ be considered where $b = mv$ where $b = mv = p^h + 1$, m, h are positive integers, $p^h \equiv 1 \pmod 4$ and p is a prime odd number. Also let $\mathbf{H}_k$ and $\mathbf{H}_t$ exist. Then we can construct $(t - 1)(b - 1)(k - 2)$ optimum $\mathbf{W}$-matrices in addition to $t(k - 1)$ optimum $\mathbf{W}$-matrices constructed earlier.*

Proof Additional $(t - 1)(b - 1)(k - 2)$ optimum $\mathbf{W}$-matrices for the BPEBD $d(v, bt, rt, k)$ can be constructed through the following steps:

Step 1:
Let, s be an odd prime power and let $\alpha_0, \alpha_1, \ldots, \alpha_{s-1}$ denote the elements of GF(s). Consider, an $s \times s$ matrix $\mathbf{Q} = (q_{ij})$, where $q_{ij} = \chi(\alpha_i - \alpha_j), i, j = 0, 1, \ldots, (s - 1)$ and χ is the Legendre function satisfying

$$\chi(\beta) = 1 \quad \text{if } \beta \text{ is a quadratic residue in GF(s)}$$
$$= 0 \quad \text{if } \beta = 0$$
$$= -1 \text{ otherwise.}$$

This map satisfies $\chi(\beta_1\beta_2) = \chi(\beta_1)\chi(\beta_2)$. It is well known that $\mathbf{Q}$ satisfies the following properties (cf. Hedayat et al. (1999), p. 150):

a. $\mathbf{Q1}_s = \mathbf{0}, \quad \mathbf{1}'_s\mathbf{Q} = \mathbf{0}'$
b. $\mathbf{QQ}' = s\mathbf{I}_s - \mathbf{J}_s$
c. $\mathbf{Q}$ is symmetric if $s \equiv 1 \pmod 4$, skew−symmetric if $s \equiv 3 \pmod 4$.

Under condition of the theorem it follows that $b \equiv 2 \pmod 4$ and $b - 1$ is an odd prime power and so a symmetric matrix $\mathbf{Q} = (q_{ij})$ of order $(b - 1) \times (b - 1)$ exists.

Let $\mathbf{u}_{li}^{(n)}$ be the ith row of the nth optimum $\mathbf{W}$-matrix $\mathbf{U}_l^{(n)}$ for the lth BPEBD $d(v, b, r, k)$, $n = 1, 2, \ldots, k - 1$, $l = 1, 2, \ldots, t$. Using the rows of $\mathbf{U}_l^{(n)}$ and the elements of $\mathbf{Q}^{(b-1)\times(b-1)}$, we define a matrix $\mathbf{A}_l(n, n')$ of order $b \times (b - 1)v$ where the ith row is the partitioned into $(b - 1)$ sub-vectors $\mathbf{a}_{ij}^{(l)}(n, n')$ of order $1 \times v$, as

$$\mathbf{a}_{ij}^{(l)}(n, n') = \mathbf{u}_{lj}^{(n)} \quad \text{if } i = j = 1, 2, \ldots, b - 1$$
$$= q_{ij}\mathbf{u}_{lj}^{(n')} \quad \text{if } i \neq j = 1, 2, \ldots, b - 1, \ n \neq n' = 1, 2, \ldots, k - 1$$
$$= -\mathbf{u}_{lb}^{(n)} \quad \text{if } i = b, \ j = 1, 2, \ldots, (b - 1).$$

Similarly we can define another matrix $\mathbf{B}_l(n, n')$ of order $b \times (b - 1)v$ where sub-vector $\mathbf{b}_{ij}^{(l)}(n, n')$ in $\mathbf{B}_l(n, n')$ stands for $\mathbf{a}_{ij}^{(l)}(n, n')$ of the $\mathbf{A}_l(n, n')$-matrix. Actually

$$\mathbf{b}_{ij}^{(l)}(n, n') = -\mathbf{u}_{lj}^{(n')} \quad \text{if } i = j = 1, 2, \ldots, b - 1$$
$$= q_{ij}\mathbf{u}_{lj}^{(n)} \quad \text{if } i \neq j = 1, 2, \ldots, b - 1, \ n \neq n' = 1, 2, \ldots, k - 1$$
$$= \mathbf{u}_{lb}^{(n')} \quad \text{if } i = b, \ j = 1, 2, \ldots, (b - 1).$$

This is to be noted that $\mathbf{u}_{li}^{(n)}\mathbf{u}_{li}^{(n)'} = k =$ block size and $\mathbf{u}_{li}^{(n)}\mathbf{u}_{li}^{(n')'} = 0$ for all l, i, $n \neq n'$. Now we construct the matrix $\mathbf{C}_l(n, n')$ of order $b \times 2(b - 1)v$ as follows:

$$\mathbf{C}_l(n, n') = (\mathbf{A}_l(n, n') : \mathbf{B}_l(n, n')).$$

Let $\mathbf{C}_{lj}(n, n')$ denote the jth set of v columns of $\mathbf{C}_l(n, n')$, $j = 1, 2, \ldots, 2(b - 1)$, i.e.

$$\mathbf{C}_{lj}(n, n') = \begin{pmatrix} \mathbf{a}_{1j}^{(l)}(n, n') \\ \mathbf{a}_{2j}^{(l)}(n, n') \\ \vdots \\ \mathbf{a}_{bj}^{(l)}(n, n') \end{pmatrix}, \quad \mathbf{C}_{l,b-1+j}(n, n') = \begin{pmatrix} \mathbf{b}_{1j}^{(l)}(n, n') \\ \mathbf{b}_{2j}^{(l)}(n, n') \\ \vdots \\ \mathbf{b}_{bj}^{(l)}(n, n') \end{pmatrix}, \quad j = 1, 2, \ldots, b - 1.$$

$$(6.4.3)$$

Using the properties (b) and (c) of the $\mathbf{Q}$-matrix, we can easily check that $\mathbf{1}'_b[\mathbf{C}_{lj}(n, n')$ $*\mathbf{C}_{lj'}(n, n')]\mathbf{1}_v = 0 \, \forall \, j \neq j' = 1, 2, \ldots, 2(b-1)$, where '*' denotes the Hadamard product.

Step 3:
For fixed $n \neq n'$, let us define the following $v \times bt$ matrix as

$$\mathbf{W}_{pi}(n, n') = \mathbf{h}_p^{*'} \odot \left(\mathbf{C}'_{1i}(n, n'),\, \mathbf{C}'_{2i}(n, n'),\, \ldots,\, \mathbf{C}'_{ti}(n, n')\right);$$
$$p = 1, 2, \ldots, (t-1),\ i = 1, 2, \ldots, 2(b-1). \qquad (6.4.4)$$

We show below that $\mathbf{W}'_{lpi}(n, n')$ gives an optimum $\mathbf{W}$-matrix for a BPEBD $d(v, bt, rt, k)$.

From the properties of $\mathbf{C}_{li}(n, n')$ and $\mathbf{h}_p^*$ it can be proved that

(i) $\mathbf{1}'_{tb}\mathbf{W}_{pi}(n, n') = \mathbf{0}' \, \forall i, \ p$ as $\mathbf{1}_{tb} = \mathbf{1}_t \otimes \mathbf{1}_b$
(ii) $\mathbf{u}_{li}^{(n)'}\mathbf{1}_v = 0 \, \forall i = 1, 2, \ldots, b, n = 1, 2, \ldots, (k-1), l = 1, 2, \ldots, t,$
(iii) $\mathbf{1}'_{mb}[\mathbf{W}_{pi}(n, n') * \mathbf{W}_{pj}(n, n')]\mathbf{1}_v = 0 \, \forall l = 1, 2, \ldots, (t-1); i \neq j = 1, 2, \ldots,$
      $2(b-1),$

which imply conditions $\mathbf{C}_1$, $\mathbf{C}_2$ and $\mathbf{C}_3$ of (3.1.5) respectively.

Again for $(n, n') \neq (n'', n''')$, $\mathbf{W}_{pi}(n, n')$ and $\mathbf{W}_{pi}(n'', n''')$ are orthogonal in the sense that sum of the elements of the Hadamard product of the above matrices is zero. Thus we have only $\frac{k-2}{2}$ such distinct pairs of (n, n'). So using these $\frac{k-2}{2}$ distinct pairs for each l and i, j, we can generate $\frac{2(t-1)(b-1)(k-2)}{2}$ i.e. $(t-1)(b-1)(k-2)$ optimum $\mathbf{W}$-matrices for the BPEBD $d(v, bt, rt, k)$. We can also easily check that these $\mathbf{W}$-matrices are orthogonal to the $t(k-1)$ optimum $\mathbf{W}$-matrices of (6.4.2).

So in all, we get $t(k-1) + (t-1)(b-1)(k-2)$ optimum $\mathbf{W}$-matrices for the BPEBD $d(v, bt, rt, k)$. $\square$

Example 6.4.1 We consider 2-fold of the BIBD($v = 9$, $b = 18$, $r = 8$, $k = 4$, $\lambda = 3$) with the initial blocks (x^0, x^2, x^4, x^6) and (x, x^3, x^5, x^7), where x is a primitive root of GF(3^2).

We write $\mathbf{H}_4$ as

$$\mathbf{H}_4 = \begin{pmatrix} 1 & 1 & 1 & 1 \\ 1 & -1 & -1 & 1 \\ 1 & 1 & -1 & -1 \\ 1 & -1 & 1 & -1 \end{pmatrix} = (\mathbf{1}, \mathbf{h}_1, \mathbf{h}_2, \mathbf{h}_3).$$

Applying Theorem 6.2.1 we construct $\mathbf{U}_l^{(1)}$ for the lth fold of the design by using $\mathbf{h}_1$ as

$$
\mathbf{U}_l^{(1)\ 18\times 9} =
\begin{pmatrix}
0 & 1 & x & 2x+1 & 2x+2 & 2 & 2x & x+2 & x+1 \\
0 & 1 & 0 & -1 & 0 & 1 & 0 & -1 & 0 \\
1 & 0 & -1 & 0 & -1 & 1 & 0 & 0 & 0 \\
0 & -1 & 0 & 0 & -1 & 0 & 0 & 1 & 1 \\
-1 & 0 & 0 & 0 & 1 & 0 & 1 & -1 & 0 \\
0 & -1 & -1 & 1 & 0 & 0 & 1 & 0 & 0 \\
1 & 1 & 0 & 0 & 0 & 0 & -1 & 0 & -1 \\
0 & 0 & 0 & 1 & 1 & -1 & 0 & 0 & -1 \\
-1 & 0 & 1 & -1 & 0 & 0 & 0 & 0 & 1 \\
0 & 0 & 1 & 0 & 0 & -1 & -1 & 1 & 0 \\
0 & 0 & 1 & 0 & -1 & 0 & 1 & 0 & -1 \\
0 & 0 & 0 & 1 & 0 & 0 & -1 & -1 & 1 \\
1 & 0 & 0 & -1 & 0 & -1 & 1 & 0 & 0 \\
0 & 1 & -1 & 0 & 0 & -1 & 0 & 0 & 1 \\
-1 & 0 & 0 & 0 & 1 & 0 & 1 & 0 & -1 \\
0 & 0 & -1 & -1 & 1 & 0 & 0 & 1 & 0 \\
1 & -1 & 1 & 0 & 0 & 0 & 0 & -1 & 0 \\
0 & -1 & 0 & 0 & 1 & 1 & -1 & 0 & 0 \\
-1 & 1 & 0 & 1 & -1 & 0 & 0 & 0 & 0
\end{pmatrix}
=
\begin{pmatrix}
\mathbf{u}_{l,1}^{(1)} \\
\mathbf{u}_{l,2}^{(1)} \\
\mathbf{u}_{l,3}^{(1)} \\
\mathbf{u}_{l,4}^{(1)} \\
\mathbf{u}_{l,5}^{(1)} \\
\mathbf{u}_{l,6}^{(1)} \\
\mathbf{u}_{l,7}^{(1)} \\
\mathbf{u}_{l,8}^{(1)} \\
\mathbf{u}_{l,9}^{(1)} \\
\mathbf{u}_{l,10}^{(1)} \\
\mathbf{u}_{l,11}^{(1)} \\
\mathbf{u}_{l,12}^{(1)} \\
\mathbf{u}_{l,13}^{(1)} \\
\mathbf{u}_{l,14}^{(1)} \\
\mathbf{u}_{l,15}^{(1)} \\
\mathbf{u}_{l,16}^{(1)} \\
\mathbf{u}_{l,17}^{(1)} \\
\mathbf{u}_{l,18}^{(1)}
\end{pmatrix}
; l = 1, 2.
$$

Similarly using $\mathbf{h}_2$ we get $\mathbf{U}_l^{(2)}$ as

$$
\mathbf{U}_l^{(2)\ 18\times 9} =
\begin{pmatrix}
0 & 1 & x & 2x+1 & 2x+2 & 2 & 2x & x+2 & x+1 \\
0 & 1 & 0 & -1 & 0 & -1 & 0 & 1 & 0 \\
-1 & 0 & 1 & 0 & -1 & 1 & 0 & 0 & 0 \\
0 & -1 & 0 & 0 & 1 & 0 & 0 & -1 & 1 \\
1 & 0 & 0 & 0 & 1 & 0 & -1 & -1 & 0 \\
0 & 1 & -1 & -1 & 0 & 0 & 1 & 0 & 0 \\
1 & -1 & 0 & 0 & 0 & 0 & -1 & 0 & 1 \\
0 & 0 & 0 & 1 & -1 & 1 & 0 & 0 & -1 \\
-1 & 0 & 1 & 1 & 0 & 0 & 0 & 0 & -1 \\
0 & 0 & -1 & 0 & 0 & -1 & 1 & 1 & 0 \\
0 & 0 & 1 & 0 & -1 & 0 & -1 & 0 & 1 \\
0 & 0 & 0 & -1 & 0 & 0 & -1 & 1 & 1 \\
0 & 1 & -1 & 0 & 0 & 1 & 0 & 0 & -1 \\
1 & 0 & 0 & 0 & 0 & 1 & 0 & -1 & -1 \\
0 & 0 & 1 & -1 & -1 & 0 & 0 & 1 & 0 \\
1 & -1 & -1 & 0 & 0 & 0 & 0 & -1 & 0 \\
0 & -1 & 0 & 0 & 1 & -1 & 1 & 0 & 0 \\
-1 & -1 & 0 & 1 & 1 & 0 & 0 & 0 & 0
\end{pmatrix}
=
\begin{pmatrix}
\mathbf{u}_{l,1}^{(2)} \\
\mathbf{u}_{l,2}^{(2)} \\
\mathbf{u}_{l,3}^{(2)} \\
\mathbf{u}_{l,4}^{(2)} \\
\mathbf{u}_{l,5}^{(2)} \\
\mathbf{u}_{l,6}^{(2)} \\
\mathbf{u}_{l,7}^{(2)} \\
\mathbf{u}_{l,8}^{(2)} \\
\mathbf{u}_{l,9}^{(2)} \\
\mathbf{u}_{l,10}^{(2)} \\
\mathbf{u}_{l,11}^{(2)} \\
\mathbf{u}_{l,12}^{(2)} \\
\mathbf{u}_{l,13}^{(2)} \\
\mathbf{u}_{l,14}^{(2)} \\
\mathbf{u}_{l,15}^{(2)} \\
\mathbf{u}_{l,16}^{(2)} \\
\mathbf{u}_{l,17}^{(2)} \\
\mathbf{u}_{l,18}^{(2)}
\end{pmatrix}
; l = 1, 2.
$$

In the same way, using $\mathbf{h}_3$ we can construct $\mathbf{U}_l^{(3)}$. Since $b - 1 = 17$, so we can construct $\mathbf{Q}$-matrix of order 17×17. From (6.4.3) we write the $\mathbf{C}_l(n, n')$-matrix by using the appropriate elements of $\mathbf{Q}$. To save space, below we show $\mathbf{C}_{l,1}(1, 2)$, $\mathbf{C}_{l,2}$, $\mathbf{C}_{l,18}(1, 2)$ and $\mathbf{C}_{l,19}(1, 2)$ only.

$$\mathbf{C}_{l,1}(1, 2) = (\mathbf{u}_{l,1}^{(1)\prime}, \mathbf{u}_{l,2}^{(2)\prime}, -\mathbf{u}_{l,3}^{(2)\prime}, \mathbf{u}_{l,4}^{(2)\prime}, -\mathbf{u}_{l,5}^{(2)\prime}, \mathbf{u}_{l,6}^{(2)\prime}, -\mathbf{u}_{l,7}^{(2)\prime}, \mathbf{u}_{l,8}^{(2)\prime}, -\mathbf{u}_{l,9}^{(2)\prime}, \mathbf{u}_{l,10}^{(2)\prime}, -\mathbf{u}_{l,11}^{(2)\prime},$$
$$\mathbf{u}_{l,12}^{(2)\prime}, -\mathbf{u}_{l,13}^{(2)\prime}, \mathbf{u}_{l,14}^{(2)\prime}, -\mathbf{u}_{l,15}^{(2)\prime}, \mathbf{u}_{l,16}^{(2)\prime}, -\mathbf{u}_{l,17}^{(2)\prime}, -\mathbf{u}_{l,18}^{(1)\prime})\prime,$$
$$\mathbf{C}_{l,2}(1, 2) = (\mathbf{u}_{l,1}^{(2)\prime}, \mathbf{u}_{l,2}^{(1)\prime}, \mathbf{u}_{l,3}^{(2)\prime}, \mathbf{u}_{l,4}^{(2)\prime}, \mathbf{u}_{l,5}^{(2)\prime}, -\mathbf{u}_{l,6}^{(2)\prime}, \mathbf{u}_{l,7}^{(2)\prime}, -\mathbf{u}_{l,8}^{(2)\prime}, -\mathbf{u}_{l,9}^{(2)\prime}, \mathbf{u}_{l,10}^{(2)\prime}, \mathbf{u}_{l,11}^{(2)\prime},$$
$$-\mathbf{u}_{l,12}^{(2)\prime}, -\mathbf{u}_{l,13}^{(2)\prime}, -\mathbf{u}_{l,14}^{(2)\prime}, -\mathbf{u}_{l,15}^{(2)\prime}, \mathbf{u}_{l,16}^{(2)\prime}, -\mathbf{u}_{l,17}^{(2)\prime}, -\mathbf{u}_{l,18}^{(1)\prime})\prime,$$
$$\mathbf{C}_{l,18}(1, 2) = (-\mathbf{u}_{l,1}^{(2)\prime}, \mathbf{u}_{l,2}^{(1)\prime}, -\mathbf{u}_{l,3}^{(1)\prime}, \mathbf{u}_{l,4}^{(1)\prime}, -\mathbf{u}_{l,5}^{(1)\prime}, -\mathbf{u}_{l,6}^{(1)\prime}, -\mathbf{u}_{l,7}^{(1)\prime}, \mathbf{u}_{l,8}^{(1)\prime}, -\mathbf{u}_{l,9}^{(1)\prime}, \mathbf{u}_{l,10}^{(1)\prime},$$
$$-\mathbf{u}_{l,11}^{(1)\prime}, \mathbf{u}_{l,12}^{(1)\prime}, -\mathbf{u}_{l,13}^{(1)\prime}, \mathbf{u}_{l,14}^{(1)\prime}, -\mathbf{u}_{l,15}^{(1)\prime}, \mathbf{u}_{l,16}^{(1)\prime}, -\mathbf{u}_{l,17}^{(1)\prime}, \mathbf{u}_{l,18}^{(2)\prime})\prime,$$
$$\mathbf{C}_{l,19}(1, 2) = (\mathbf{u}_{l,1}^{(1)\prime}, -\mathbf{u}_{l,2}^{(2)\prime}, \mathbf{u}_{l,3}^{(1)\prime}, \mathbf{u}_{l,4}^{(1)\prime}, \mathbf{u}_{l,5}^{(1)\prime}, -\mathbf{u}_{l,6}^{(1)\prime}, \mathbf{u}_{l,7}^{(1)\prime}, -\mathbf{u}_{l,8}^{(1)\prime}, -\mathbf{u}_{l,9}^{(1)\prime}, \mathbf{u}_{l,10}^{(1)\prime}, \mathbf{u}_{l,11}^{(1)\prime},$$
$$-\mathbf{u}_{l,12}^{(1)\prime}, -\mathbf{u}_{l,13}^{(1)\prime}, -\mathbf{u}_{l,14}^{(1)\prime}, -\mathbf{u}_{l,15}^{(1)\prime}, \mathbf{u}_{l,16}^{(1)\prime}, -\mathbf{u}_{l,17}^{(1)\prime}, \mathbf{u}_{l,18}^{(2)\prime})\prime; l = 1, 2.$$

As $t = 2$, $\mathbf{H}_2$ exists and is written as

$$\mathbf{H}_2 = \begin{pmatrix} 1 & 1 \\ -1 & 1 \end{pmatrix} = (\mathbf{h}_1^*, \mathbf{h}_2^*).$$

From (6.4.4),

$$\mathbf{W}_{1,1}(1, 2)^{36 \times 9} = \mathbf{h}_1^* \odot \begin{pmatrix} \mathbf{C}_{1,1}(1, 2) \\ \mathbf{C}_{2,1}(1, 2) \end{pmatrix} = \begin{pmatrix} \mathbf{C}_{1,1}(1, 2) \\ -\mathbf{C}_{2,1}(1, 2) \end{pmatrix}$$

can be constructed.

Similarly other $\mathbf{W}$-matrices such as $\mathbf{W}_{1,2}(1, 2)$, $\mathbf{W}_{1,18}(1, 2)$, and $\mathbf{W}_{1,19}(1, 2)$ can be constructed. In this way we can construct $(t-1)(b-1)(k-2) = 34$ $\mathbf{W}$-matrices for BIBD(9, 36, 16, 4, 6) which is a 2-fold of the BIBD(9, 18, 8, 4, 3). Again from (6.4.2) 6 additional optimum $\mathbf{W}$-matrices can be obtained. So in all, we get 40 optimum $\mathbf{W}$-matrices for BIBD(9, 36, 16, 4, 6).

So far we have we assume $b = mv$ for the BPEBDs. Now we try to construct optimum $\mathbf{W}$-matrices for t-fold BPEBD $d(v, b, r, k)$, where $b \neq mv$, $m(\geq 1)$, a positive integer.

Theorem 6.4.2 *Suppose a cyclic BPEBD $d(v = b, r = k)$ exists. Again, if $\mathbf{H}_k$, $\mathbf{H}_t$ and $\mathbf{Q}$ of order $(b-1) \times (b-1)$ exist, then we can construct $(t-1)(b-1)(k-2) + (t-1)(k-1)$ optimum $\mathbf{W}$-matrices for the t-fold of these BPEBD $d(v = b, r = k)$.*

Proof As $\mathbf{H}_k$ and $\mathbf{H}_t$ exist, $(t-1)(k-1)$ optimum $\mathbf{W}$-matrices can be constructed using the cyclic property of the incidence matrix as in Theorem 6.3.1 and the additional number $(t-1)(b-1)(k-2)$ optimum $\mathbf{W}$-matrices can be constructed by using properties of $\mathbf{Q}$-matrix as in proof of Corollary 6.4.1. □

Appendix

As mentioned earlier, BIBDs and PBIBDs form an important sub-class of BPEBDs and the lists of BIBDs and PBIBDs are readily available (Tables 6.1 and 6.2). So, for OCDs in BPEBD set-up with $b = mv$ we have considered BIBDs with $b = mv$ where $v, b \leq 100$, $r, k \leq 15$ from the list given in Raghavarao (1971) and the GDDs with $b = mv$ and $r \leq 10$, $k \leq 10$ from the same catalogue of Clatworthy (1973). In this connection, we have to mention that in Chap. 5, we have also prepared tables for OCDs in GDDs set-up. However the readers need not get confused between the tables of the two respective chapters. Here we can construct some more OCDs in GDDs set-up and these are separated by using a '$*$' mark for additional designs, which are not in Tables 5.1, 5.2 and 5.3 of Chap. 5, in the 11th column in Table 6.2. A number of BIBDs and GDDs with $b = mv$ are cyclic designs (Table 6.3).

These designs have not been considered separately in cyclic design class. A separate list for the cyclic BIBDs having partial cycles has only been considered. Here c denotes the number of optimum **W**'s for BPEBD. Other parameters have usual significance.

Table 6.1 OCDs in BIBD with $b = mv$

Sl. no.	Parameters					c	Design no.	Method of construction
	v	b	r	k	λ			
1	5	10	4	2	1	2	3	Theorem 6.2.2
2	5	5	4	4	3	3	4	Theorem 6.2.1
3	7	7	4	4	2	3	11	Theorem 6.2.1
4	7	21	6	2	1	1	12	Theorem 6.2.1
5	7	7	6	6	5	1	13	Theorem 6.2.1
6	9	36	8	2	1	4	18	Theorem 6.2.2
7	9	18	8	4	3	6	19	Theorem 6.2.2
8	9	9	8	8	7	7	21	Theorem 6.2.1
9	11	11	6	6	3	1	30	Theorem 6.2.1
10	11	55	10	2	1	1	31	Theorem 6.2.1
11	11	11	10	10	9	1	32	Theorem 6.2.1
12	13	13	4	4	1	3	37	Theorem 6.2.1
13	13	26	12	6	5	2	40	Remark 6.2.5
14	15	15	8	8	4	7	44	Theorem 6.2.1
15	16	16	6	6	2	1	47	Remark 6.2.2
16	16	16	10	10	6	1	49	Remark 6.2.2
17	19	19	10	10	5	1	56	Remark 6.2.2
18	19	57	12	4	2	3	57	Theorem 6.2.1
19	21	42	12	6	3	2	61	Remark 6.2.5

(continued)

Table 6.1 (continued)

Sl. no.	Parameters					c	Design no.	Method of construction
	v	b	r	k	λ			
20	23	23	12	12	6	11	Dual of 64	Theorem 6.2.1
21	23	50	8	4	1	6	66	Theorem 6.2.2
22	27	27	14	14	7	1	Dual of 71	Remark 6.2.2
23	31	31	6	6	1	1	75	Remark 6.2.2
24	31	31	10	10	3	1	76	Remark 6.2.2
25	45	45	12	12	3	11	85	Theorem 6.2.1
26	57	57	8	8	1	7	87	Theorem 6.2.1
27	91	91	10	10	1	1	91	Remark 6.2.2

Table 6.2 OCDs in GDDs

Sl. no.	v	b	r	k	λ_1	λ_2	m	n	c	Design no.	Method of construction
1	6	3	2	4	2	1	3	2	2	S1	Remark 6.2.6
2	6	6	4	4	4	2	3	2	3	S2	Theorem 6.2.1
3	6	12	8	4	8	4	3	2	6	S4	Theorem 6.2.2
4	10	10	4	4	4	1	5	2	3	S9	Theorem 6.2.1
5	10	20	8	4	8	2	5	2	6	S10	Theorem 6.2.2
6	18	36	8	4	8	1	9	2	6	S15	Theorem 6.2.2
7	8	8	6	6	6	4	4	2	1	S19	Remark 6.2.2
8	9	3	2	6	2	1	3	3	1	S21	Remark 6.2.2
9	9	9	6	6	6	3	3	3	1	S23	Remark 6.2.2
10	10	10	6	6	6	3	5	2	1	S26	Remark 6.2.2
11	12	12	6	6	6	2	4	3	1	S29	Remark 6.2.2
12	14	14	6	6	6	2	7	2	1	S33	Remark 6.2.2
13	21	21	6	6	6	1	7	3	1	S42	Remark 6.2.2
14	26	26	6	6	6	1	13	2	1	S44	Remark 6.2.2
15	10	5	4	8	4	3	5	2	6	S51	Remark 6.2.6
16	10	10	8	8	8	6	5	2	7	S52	Theorem 6.2.2
17	12	3	2	8	2	1	3	4	4	S53	Remark 6.2.6
18	12	6	4	8	4	2	3	4	6	S54	Remark 6.2.6
19	12	12	8	8	8	4	3	4	7	S56	Theorem 6.2.2
20	14	7	4	8	4	2	7	2	6	S59	Remark 6.2.6
21	14	14	8	8	8	4	7	2	7	S60	Theorem 6.2.2
22	18	18	8	8	8	3	9	2	7	S65	Theorem 6.2.2
23	20	10	4	8	4	1	5	4	6	S66	Remark 6.2.6

(continued)

Table 6.2 (continued)

Sl. no.	v	b	r	k	λ_1	λ_2	m	n	c	Design no.	Method of construction
24	20	20	8	8	8	2	5	4	7	S68	Theorem 6.2.2
25	26	13	4	8	4	1	13	2	6	S71	Remark 6.2.6
26	26	26	8	8	8	2	13	2	7	S72	Theorem 6.2.2
27	36	36	8	8	8	1	9	4	7	S77	Theorem 6.2.2
28	50	50	8	8	8	1	25	2	7	S80	Theorem 6.2.2
29	12	12	10	10	10	8	6	2	1	S99	Remark 6.2.2
30	15	3	2	10	2	1	3	5	1	S100	Remark 6.2.6
31	15	15	10	10	10	5	3	5	1	S104	Remark 6.2.2
32	18	18	10	10	10	5	9	2	1	S105	Remark 6.2.2
33	22	22	10	10	10	4	11	2	1	S111	Remark 6.2.2
34	30	30	10	10	10	2	6	5	1	S115	Remark 6.2.2
35	42	42	10	10	10	2	21	2	1	S119	Remark 6.2.2
36	55	55	10	10	10	1	11	5	1	S123	Remark 6.2.2
37	82	82	10	10	10	1	41	2	1	S124	Remark 6.2.2
38	4	4	2	2	0	1	2	2	1	SR1	Theorem 6.2.1
39	4	8	4	2	0	2	2	2	2	SR2	Theorem 6.2.2
40	4	12	6	2	0	3	2	2	1	SR3	Theorem 6.2.1
41	4	16	8	2	0	4	2	2	4	SR4	Theorem 6.2.2
42	4	20	10	2	0	5	2	2	1	SR5	Theorem 6.2.1
43	6	18	6	2	0	2	2	3	1	SR7*	Theorem 6.2.1
44	8	16	4	2	0	1	2	4	2	SR9	Theorem 6.2.2
45	8	32	8	2	0	2	2	4	4	SR10	Theorem 6.2.2
46	10	50	10	2	0	2	2	5	1	SR12*	Theorem 6.2.1
47	12	36	6	2	0	1	2	6	1	SR13	Theorem 6.2.1
48	16	64	8	2	0	1	2	8	4	SR15	Theorem 6.2.2
49	20	100	10	2	0	1	2	10	1	SR17	Theorem 6.2.1
50	8	8	4	4	0	2	4	2	3	SR36	Theorem 6.2.1
51	8	16	8	4	0	4	4	2	6	SR39	Theorem 6.2.2
52	16	16	4	4	0	1	4	4	3	SR44	Theorem 6.2.1
53	16	32	8	4	0	2	4	4	6	SR45	Theorem 6.2.2
54	32	64	8	4	0	1	4	8	6	SR49	Theorem 6.2.2
55	12	12	6	6	0	3	6	2	1	SR67	Remark 6.2.2
56	12	12	6	6	2	3	3	4	1	SR68*	Remark 6.2.2
57	18	18	6	6	0	2	6	3	1	SR72*	Remark 6.2.2
58	16	16	8	8	0	4	8	2	7	SR92	Theorem 6.2.1
59	32	32	8	8	0	2	8	4	7	SR95	Theorem 6.2.1

(continued)

Table 6.2 (continued)

Sl. no.	v	b	r	k	λ_1	λ_2	m	n	c	Design no.	Method of construction
60	64	64	8	8	0	1	8	8	7	SR97	Theorem 6.2.1
61	20	20	10	10	0	5	10	2	1	SR108	Remark 6.2.2
62	4	8	4	2	2	1	2	2	2	R1*	Theorem 6.2.2
62	4	8	4	2	2	1	2	2	2	R1*	Theorem 6.2.2
63	4	12	6	2	4	1	2	2	1	R4*	Theorem 6.2.1
64	4	16	8	2	6	1	2	2	4	R8*	Theorem 6.2.2
65	4	16	8	2	4	2	2	2	4	R9*	Theorem 6.2.2
66	4	16	8	2	2	3	2	2	4	R10*	Theorem 6.2.2
67	4	20	10	2	8	1	2	2	1	R14*	Theorem 6.2.1
68	4	20	10	2	6	2	2	2	1	R15*	Theorem 6.2.1
69	4	20	10	2	4	3	2	2	1	R16*	Theorem 6.2.1
70	4	20	10	2	2	4	2	2	1	R17*	Theorem 6.2.1
71	6	12	4	2	0	1	3	2	2	R18*	Theorem 6.2.2
72	6	18	6	2	2	1	3	2	1	R19*	Theorem 6.2.1
73	2	24	8	2	4	1	3	2	4	R22*	Theorem 6.2.2
74	6	24	8	2	0	2	3	2	4	R23*	Theorem 6.2.2
75	6	24	8	2	1	2	2	3	4	R24*	Theorem 6.2.2
76	6	30	10	2	6	1	3	2	1	R28*	Theorem 6.2.1
77	8	24	6	2	0	1	4	2	1	R29*	Theorem 6.2.1
78	8	32	8	2	2	1	4	2	4	R30*	Theorem 6.2.2
79	8	40	10	2	2	1	2	4	1	R32*	Theorem 6.2.1
80	8	40	10	2	4	1	4	2	1	R33*	Theorem 6.2.1
81	9	27	6	2	0	1	3	3	1	R34*	Theorem 6.2.1
82	9	45	10	2	2	1	3	3	1	R35*	Theorem 6.2.1
83	10	40	8	2	0	1	5	2	4	R36*	Theorem 6.2.2
84	10	50	10	2	2	1	5	2		R37*	Theorem 6.2.1
85	12	48	8	2	0	1	3	4	4	R38*	Theorem 6.2.2
86	12	16	10	2	0	1	6	2	1	R40*	Theorem 6.2.1
87	15	75	10	2	0	1	3	5	1	R41*	Theorem 6.2.1
88	6	6	4	4	3	2	2	3	3	R94*	Theorem 6.2.1
89	6	12	8	4	6	4	2	3	6	R95*	Theorem 6.2.2
90	6	12	8	4	4	5	3	2	6	R96*	Theorem 6.2.2
91	8	16	8	4	4	3	2	4	6	R98*	Theorem 6.2.2
92	8	16	8	4	6	3	4	2	6	R99*	Theorem 6.2.2
93	9	9	4	4	3	1	3	3	3	R104*	Theorem 6.2.1
94	9	18	8	4	6	2	3	3	6	R105*	Theorem 6.2.2

(continued)

Table 6.2 (continued)

Sl. no.	v	b	r	k	λ_1	λ_2	m	n	c	Design no.	Method of construction
95	10	20	8	4	0	3	5	2	6	R106	Theorem 6.2.2
96	12	12	4	4	2	1	6	2	3	R109*	Theorem 6.2.1
97	12	24	8	4	4	2	6	2	6	R110*	Theorem 6.2.2
98	14	14	4	4	0	1	7	2	3	R112*	Theorem 6.2.1
99	14	28	8	4	0	2	7	2	6	R113*	Theorem 6.2.2
100	15	15	4	4	0	1	5	3	3	R114*	Theorem 6.2.1
101	15	30	8	4	6	1	5	3	6	R115*	Theorem 6.2.2
102	15	30	8	4	0	2	5	3	6	R116*	Theorem 6.2.2
103	15	30	8	4	1	2	3	5	6	R117*	Theorem 6.2.2
104	16	32	8	4	4	1	4	4	6	R120*	Theorem 6.2.2
105	26	52	8	4	0	1	13	2	6	R128*	Theorem 6.2.2
106	27	54	8	4	0	1	9	3	6	R129*	Theorem 6.2.2
107	28	56	8	4	0	1	7	4	6	R130*	Theorem 6.2.2
108	10	10	6	6	5	2	2	5	1	R166*	Remark 6.2.2
109	15	15	6	6	5	1	3	5	1	R168*	Remark 6.2.2
110	27	27	6	6	3	1	9	3	1	R170*	Remark 6.2.2
111	28	28	6	6	2	1	7	4	1	R171*	Remark 6.2.2
112	12	12	8	8	6	5	6	2	7	R186*	Theorem 6.2.1
113	14	14	8	8	7	2	2	7	7	R187*	Theorem 6.2.1
114	21	21	8	8	7	1	3	7	7	R188*	Theorem 6.2.1
115	24	24	8	8	4	2	4	6	7	R189*	Theorem 6.2.1
116	48	48	8	8	4	1	12	4	7	R190*	Theorem 6.2.1
117	63	63	8	8	0	1	9	7	7	R191*	Theorem 6.2.1
118	12	12	10	10	9	8	4	3	1	R203*	Remark 6.2.2
119	14	14	10	10	8	6	2	7	1	R204*	Remark 6.2.2
120	14	14	10	10	6	7	7	2	1	R205*	Remark 6.2.2
121	18	18	10	10	9	2	2	9	1	R206*	Remark 6.2.2
122	27	27	10	10	9	1	3	9	1	R207*	Remark 6.2.2
123	32	32	10	10	6	2	4	8	1	R208*	Remark 6.2.2
124	75	75	10	10	5	1	15	5	1	R209*	Remark 6.2.2

Table 6.3 OCDs in cyclic designs

Sl. no.	Parameters					Solution	c	Method of construction
	v	b	r	k	λ			
1	6	15	10	4	6	Two 'full sets' of blocks each and the initial blocks: $[(0, 1, 2, 3), (0, 2, 3, 4)]$ mod 6; the initial blocks $(0, 2, 3, 5)$ mod 6	2	Analogous to Example 6.3.2
2	19	37	12	4	2	Difference set: $(0, x^0, x^6, x^{12})$; $(0, x^1, x^7, x^{1}3)$; $(0, x^2, x^8, x^{14})$; x is a primitive root of GF(19)	3	Remark 6.3.1
3	22	77	14	4	2	Difference set: $(x_1^0, x_1^3, x_2^\alpha, x_2^{\alpha+3})$; $(x_1^1, x_1^4, x_2^{\alpha+1}, x_2^{\alpha+4})$; $(x_1^2, x_1^5, x_2^{\alpha+2}, x_2^{\alpha+5})$; $(x_2^0, x_2^3, x_3^\alpha, x_3^{\alpha+3})$; $(x_2^1, x_2^4, x_3^{\alpha+1}, x_3^{\alpha+4})$ $(x_2^2, x_2^5, x_3^{\alpha+2}, x_3^{\alpha+5})$; $(x_3^0, x_3^3, x_1^\alpha, x_1^{\alpha+3})$; $(x_3^1, x_3^4, x_1^{\alpha+1}, x_1^{\alpha+4})$; $(x_3^2, x_3^5, x_1^{\alpha+2}, x_1^{\alpha+5})$; $(\infty, 0_1, 0_2, 0_3)$; $(\infty, 0_1, 0_2, 0_3)$; x is a primitive root of GF(7)	2	Analogous to Example 6.3.2 and Example 6.3.3

References

Agrawal H (1966) Some generalizations of distinct representatives with applications to statistical designs. Ann Math Stat 37:525–526

Bose RC (1939) On the construction of balanced incomplete block designs. Ann Eugen. 9:353–399

Clatworthy WH (1973) Tables of two-associate class partially balanced designs. U.S., Department of Commerce, National Bureau of Standards

Das K, Mandal NK, Sinha BK (2003) Optimal experimental designs with covariates. J Stat Plan Inference 115:273–285

Dutta G (2004) Optimum choice of covariates in BIBD set-up. Calcutta Stat Assoc Bull 55:39–55

Dutta G, Das P, Mandal NK (2009) Optimum covariate designs in partially balanced incomplete block (PBIB) design set-ups. J Stat Plan Inference 139:2823–2835

Dutta G, Das P, Mandal NK (2010) Optimum covariate designs in binary proper equi-replicate block design set-up. Discrete Math 310:1037–1049

Hedayat AS, Sloane NJA, Stufken J (1999) Orthogonal arrays: theory and applications. Springer, New York

John JA (1987) Cyclic designs. Chapman & Hall, New York

John JA, Wolock FW, David HA (1972) Cyclic designs, vol 62. Applied mathematics series. National Bureau of Standards

Nandi HK (1946) Enumeration of nonisomorphic solutions of balanced incomplete block designs. Sankhyā 7:305–312

Raghavarao D (1971) Constructions and combinatorial problems in design of experiments. Wiley, New York

Chapter 7
OCDs in Balanced Treatment Incomplete Block Design Set-Up

7.1 Introduction

Here we deal with the balanced treatment incomplete block (BTIB) design set-up with $p + 1$ treatments and c covariates and investigate the problem of most efficient estimation of the covariate parameters in BTIB design set-ups. As a useful class of designs for testing test treatments against control, Bechhofer and Tamhane (1981) introduced BTIB design. Suppose in a test-control design d, the treatments are indexed by $c_0, 1, \ldots, v$ with c_0 denoting the control treatment and $1, 2, \ldots, v$ denoting the $v(\geq 2)$ test treatments. Let k denote the common size of each block, and b denote the number of blocks available for experimentation. Thus $n = kb$ is the total number of experimental units. According to Bechhofer and Tamhane (1981), the design d is called a BTIB design if

(a) d is incomplete, i.e. no block contains all the $v + 1$ treatments,
(b) $\lambda_{c_0 i} = \lambda_{c_0}$, $i = 1, 2, \ldots, v$ and $\lambda_{i_1 i_2} = \lambda$, $i_1 \neq i_2 = 1, 2, \ldots, v$, where

$$\lambda_{uu'} = \sum_{j=1}^{b} n_{uj} n_{u'j}, \quad u \neq u' = c_0, 1, \ldots, v \text{ and } n_{ij} \text{ denotes the number of times}$$

the ith treatment appears in the jth block, $i = c_0, 1, \ldots, v$, $j = 1, 2, \ldots, b$.

According to Bechhofer and Tamhane (1981), a BTIB design neither needs to satisfy the condition that $r_i = \sum_{j=1}^{b} n_{ij}$, $(1 \leq i \leq v)$, the number of replications of the ith test treatment are all equal nor, does it require to be binary in the test/control treatments. But as mentioned earlier Dutta and Das (2013) considered only those BTIB designs which were constructed in Bechhofer and Tamhane (1981) and Das et al. (2005) where the designs had all $r_i = r$.

© Springer India 2015
P. Das et al., *Optimal Covariate Designs*, DOI 10.1007/978-81-322-2461-7_7

So here the discussion is restricted to the set-up of BTIB design with parameters v, b, k, r, r_{c_0}, λ, λ_{c_0} which is denoted by BTIB(v, b, k, r, r_{c_0}, λ, λ_{c_0}), where $r_{c_0} = \sum_{j=1}^{b} n_{c_0 j}$ is the replication of the control treatment.

Let y_{ijl} be the response and $z_{ijl}^{(t)}$ be the value of the tth covariate when the treatment i is applied to the unit l of block j, $i = c_0, 1, \ldots, v$, $j = 1, 2, \ldots, b$, $l = 1, 2, \ldots, n_{ij}(n_{ij} = 0, 1, 2, \ldots)$, $t = 1, 2, \ldots, c$. The model which we work with is

$$y_{ijl} = \mu + \tau_i + \beta_j + \sum_{t=1}^{c} \gamma_t z_{ijl}^{(t)} + e_{ijl}, \tag{7.1.1}$$

where μ is the general mean, τ_i is the effect of treatment i, β_j the effect of block j, $\gamma_1, \gamma_2, \ldots, \gamma_c$ are the regression coefficients associated with the c covariates Z_1, $Z_2, \ldots, Z_c$ respectively and e_{ijl} is the observational error corresponding to y_{ijl}. As usual, the random errors $\{e_{ijl}\}$ are assumed to be uncorrelated and homoscedastic with common variance σ^2. As in other chapters it is assumed that the values of each covariate are in the interval $[-1, 1]$, i.e.

$$z_{ijl}^{(t)} \in [-1, 1], \; i = c_0, 1, \ldots, v; \; j = 1, 2, \ldots, b; \; l = 1, 2, \ldots, n_{ij}; \; t = 1, 2, \ldots, c. \tag{7.1.2}$$

In matrix notation Model (7.1.1) can be represented as

$$(\mathbf{Y}, \; \mu \mathbf{1}_n + \mathbf{X}_1 \tau + \mathbf{X}_2 \beta + \mathbf{Z}\gamma, \; \mathbf{I}_n \sigma^2) \tag{7.1.3}$$

where $\mathbf{Y} = (\ldots, y_{ijl}, \ldots)'$ is the vector of observations of order $n \times 1$, τ, β and γ correspond, respectively, to the vectors of treatment effects, block effects and the covariate effects; $\mathbf{X}_1$ and $\mathbf{X}_2$ are, respectively, the design matrices of treatment effects and block effects and $\mathbf{Z} = ((z_{ijl}^{(t)}))$ is the design matrix corresponding to the covariate effects. $\mathbf{1}_n$ is a vector of order n with all elements unity and $\mathbf{I}_n$ is the identity matrix of order n.

With reference to the model (7.1.3), it is evident that for the estimation of the covariate effects orthogonally to the treatment and block effects we will impose the conditions as stated in (3.1.3) and the regression parameters are estimated with maximum efficiency if additionally, (3.1.4) holds.

7.2 OCDs and W-Matrices

With respect to Model (7.1.3), γ is estimated most efficiently if $\mathbf{Z}$-matrix satisfies conditions (3.1.3) and (3.1.4). The choice of the $\mathbf{Z}$-matrix is usually difficult under the most general block design set-up. As mentioned in Chap. 4 that in the binary design set-up Das et al. (2003) had represented each column of the $\mathbf{Z}$-matrix by a matrix $\mathbf{W}$, where the rows of $\mathbf{W}$ corresponded to the treatments and the columns of $\mathbf{W}$ corresponded to the blocks. This brought in some ease in construction of the $\mathbf{Z}$-matrix. But a BTIB design need not be binary; however, in the BTIB designs which we are considering here, the portion for the test treatments is binary, but the control treatment may occur more than once in a block. So to represent the columns of an optimum $\mathbf{Z}$-matrix by $\mathbf{W}$-matrices, we convert the incidence matrix of a BTIB design to one which is binary.

$$\text{Let } \mathbf{N} = \left(\mathbf{n}'_C, \mathbf{N}'_T \right)'$$

be the incidence matrix of the BTIB design, where $\mathbf{n}_C^{1 \times b}$ indicates the incidence vector of the control treatment and $\mathbf{N}_T$ indicates the incidence matrix of the test treatments. For our convenience in construction of OCDs, we replace the incidence vector $\mathbf{n}_C$ of the control treatment by a $n_{c_0} \times b$ matrix $\mathbf{N}_C^*$ with elements 0 and 1 where the jth column of $\mathbf{N}_C^*$ contains $n_{c_{0j}}$ unities in some order and $(n_{c_0} - n_{c_{0j}})$ zeros in other places, $j = 1, 2, \ldots, b$, n_{c_0} being the maximum of $n_{c_{01}}, n_{c_{02}}, \ldots, n_{c_{0b}}$. To fix ideas and to illustrate the technique we will use a definite order of 1s and 0s where 1's are followed by 0's. Therefore, the incidence matrix $\mathbf{N}$ can be written in a transformed form with the elements 0 and 1 as

$$\mathbf{N}^{*(n_{c_0}+v) \times b} = \left(\mathbf{N}_C^{*'}, \mathbf{N}_T' \right)' \qquad (7.2.1)$$

where $\mathbf{N}_C^{*(n_{c_0} \times b)}$ is actually the incidence matrix corresponding to the control treatment in the binary form.

Now we can make a correspondence between the elements of any column of $\mathbf{Z}$ with the positive entries of $\mathbf{N}^*$. Also, as the other entries of $\mathbf{N}^*$ are zeros and the z-values are ± 1, we can get a matrix $\mathbf{W}^{(t)}$ from $\mathbf{N}^*$ by replacing n_{ij}^*'s by $\pm n_{ij}^*$ according to the values of tth column of $\mathbf{Z}$. The $\mathbf{W}^{(t)}$-matrix precisely represents the tth column of $\mathbf{Z}$. Note from (7.2.1) that $\mathbf{W}^{(t)}$ can be accordingly partitioned as

$$\mathbf{W}^{(t) \ (n_{c_0}+v) \times b} = \left(\mathbf{W}_C^{(t)'}, \mathbf{W}_T^{(t)'} \right)'. \qquad (7.2.2)$$

Here optimum $\mathbf{W}$-matrices are being constructed from $\mathbf{N}^*$, the incidence matrix of a BTIB design, by putting $+1$ or -1 in the non-zero positions of every row and every column of $\mathbf{N}^*$. From the definition of the $\mathbf{W}$-matrix it follows that the conditions (3.1.3) and (3.1.4) change to the following:

C_1. $\mathbf{W}^{(t)}$-matrix has all column sums equal to zero;

C_2. $\mathbf{W}_T^{(t)}$-matrix has all row sums equal to zero;

C_3. The grand total of all the entries in the Hadamard product of $\mathbf{W}^{(t)}$ and $\mathbf{W}^{(t')}$ *is equal to* $n\delta_{tt'}$, $1 \le t \ne t' \le c$.

$$(7.2.3)$$

We may note in passing that (i) $\mathbf{W}_C^{(t)} = (\pm 1, 0)$ does not posses any such property of its row total (ii) however, it is trivially true that $\sum \sum w_C^{(t)}(i, j)$ is equal to zero.

Definition 7.2.1 With respect to model (7.1.3), the $\mathbf{W}$-matrices corresponding to the c covariates are said to be optimum if they satisfy the condition (7.2.3).

Remark 7.2.1 It is to be noted that if $c = 1$, only the conditions C_1 and C_2 are to be satisfied by the $\mathbf{W}$-matrix to be optimum.

7.3 Optimum Covariate Designs

As it has already been mentioned earlier that the construction of OCDs depends on the methods of construction of the corresponding BTIB designs, we divide this section into three subsections according to the methods of construction.

7.3.1 BTIB Design Obtained from Generator Designs

Following (Bechhofer and Tamhane 1981), we define *generator* designs which are BTIB designs with v test treatments and b blocks of size k each such that no proper subsets of blocks can give rise to a BTIB design. Suppose that there are s_0 generator designs $D_1, D_2, \ldots, D_{s_0}$ (say) and let $\lambda^{(i)}, \lambda_{c_0}^{(i)}$ be the frequency parameters associated with D_i and let b_i be the number of blocks required by D_i $(i = 1, 2, \ldots, s_0)$. Then a BTIB design $D = \bigcup_{i=1}^{s_0} f_i D_i$ obtained by forming unions of $f_i > 0$ replications of D_i has the frequency parameters $\lambda = \sum_{i=1}^{s_0} f_i \lambda^{(i)}$, $\lambda_{c_0} = \sum_{i=1}^{s_0} f_i \lambda_{c_0}^{(i)}$ and $b = \sum_{i=1}^{s_0} f_i b_i$ blocks cf. Bechhofer and Tamhane (1981). We consider the generator designs constructed by Bechhofer and Tamhane (1981) and construct OCDs for these BTIB designs.

(i) For each $v \geq 2, k = 2$ there are exactly two generator designs and these are

$$D_1 = \left\{ \begin{array}{cccc} c_0 & c_0 & \cdots & c_0 \\ 1 & 2 & \cdots & v \end{array} \right\}, \quad D_2 = \left\{ \begin{array}{cccc} 1 & 1 & \cdots & v-1 \\ 2 & 3 & \cdots & v \end{array} \right\}. \quad (7.3.1)$$

From these generator designs, implementable BTIB designs of the type $D = f_1 D_1 \cup f_2 D_2$ can be constructed for $f_1, f_2 > 0$. When $f_1 = f_2 = f$ (say), the corresponding design parameters for D are

$$v, \; b = f(v(v+1)/2), \; k = 2, \; r = r_{c_0} = fv, \; \lambda_{c_0} = f, \; \lambda = f. \quad (7.3.2)$$

For the construction of OCDs for BTIB design with v even and $f = 1$ the following lemma will be helpful.

Lemma 7.3.1 *Let v (≥ 2) be even and $f = 1$ in the parameters (7.3.2). Then a* **W**-*matrix for the BTIB design with parameters v, $b = v(v + 1)/2$, $k = 2$, $r = r_{c_0} = v$, $\lambda = 1$, $\lambda_{c_0} = 1$ obtained by* $D = D_1 \cup D_2$, *can always be constructed.*

Proof The incidence matrix $\mathbf{N}^*$ of the design D can be written as

$$\mathbf{N}^* = \begin{pmatrix} \mathbf{1}'_v & \mathbf{0}' & \mathbf{0}' & \cdots & \mathbf{0}' & \mathbf{0}' \\ \mathbf{I}_v & \mathbf{A}_1^* & \mathbf{A}_2^* & \cdots & \mathbf{A}_{(v-2)/2}^* & \mathbf{B} \end{pmatrix} \tag{7.3.3}$$

where $\mathbf{A}_i^{*\prime}$ of order $v \times v$ is obtained from cyclic permutation of the following row

$$\begin{matrix} 1 \; 2 \, \ldots \, i \;\; i+1 \; i+2 \, \ldots \, v \\ (\, 1 \; 0 \ldots 0 \;\; 1 \quad\; 0 \quad\; \ldots 0 \,); \;\; i = 1, 2, \ldots, (v-2)/2 \end{matrix} \tag{7.3.4}$$

and

$$\mathbf{B} = \left(\mathbf{I}_{v/2}, \; \mathbf{I}_{v/2} \right)'. \tag{7.3.5}$$

Corresponding to $\mathbf{A}_i'$ we construct a matrix $\mathbf{W}_{2,i}^{*\prime}$ by cyclical permutation of the vector

$$\begin{matrix} 1 \; 2 \, \ldots \, i \;\; i+1 \; i+2 \, \ldots \, v \\ (\, 1 \; 0 \ldots 0 \;\; -1 \quad\; 0 \quad\; \ldots 0 \,); \;\; i = 1, 2, \ldots, (v-2)/2 \end{matrix} \tag{7.3.6}$$

obtained from (7.3.4) by replacing the non-null elements by 1 and -1 respectively, $i = 1, 2, \ldots, (v-2)/2$. Thus we get the following matrix

$$\mathbf{W}_2^{* \; v \times v(v-2)/2} = \left(\mathbf{W}_{2,1}^*, \; \mathbf{W}_{2,2}^*, \ldots, \mathbf{W}_{2,(v-2)/2}^* \right) \tag{7.3.7}$$

after juxtaposition of $\mathbf{W}_{2,i}^*$'s. Again corresponding to the matrices $\mathbf{I}_v$ of (7.3.3) and $\mathbf{B}$ of (7.3.5) we define the two following matrices

$$\mathbf{W}_1^{* \; v \times v} = \begin{pmatrix} -\mathbf{I}_{v/2} & \mathbf{0} \\ \mathbf{0} & \mathbf{I}_{v/2} \end{pmatrix}, \;\; \mathbf{W}_3^{* \; v \times v/2} = \begin{pmatrix} \mathbf{I}_{v/2} \\ -\mathbf{I}_{v/2} \end{pmatrix}. \tag{7.3.8}$$

Using $\mathbf{W}_1^*$, $\mathbf{W}_2^*$ and $\mathbf{W}_3^*$ in $\mathbf{N}^*$ of (7.3.3), the following $\mathbf{W}$-matrix of order $(v + 1) \times v(v + 1)/2$

$$\mathbf{W}^* = \left(\begin{array}{c|c|c} \mathbf{1}'_{v/2} \;\; -\mathbf{1}'_{v/2} & \mathbf{0}' & \mathbf{0}' \\ \hline \mathbf{W}_1^* & \mathbf{W}_2^* & \mathbf{W}_3^* \end{array} \right) \tag{7.3.9}$$

can be seen to satisfy conditions C_1–C_3 of (7.2.3). $\qquad\qquad\qquad \square$

Example 7.3.1 Take $v = 4$. The incidence matrix $\mathbf{N}^*$ of the design when $v = 4$ looks like

$$\mathbf{N}^* = \begin{pmatrix} 1\ 1\ 1\ 1 & 0\ 0\ 0\ 0 & 0\ 0 \\ 1\ 0\ 0\ 0 & 1\ 0\ 0\ 1 & 1\ 0 \\ 0\ 1\ 0\ 0 & 1\ 1\ 0\ 0 & 0\ 1 \\ 0\ 0\ 1\ 0 & 0\ 1\ 1\ 0 & 1\ 0 \\ 0\ 0\ 0\ 1 & 0\ 0\ 1\ 1 & 0\ 1 \end{pmatrix} = \begin{pmatrix} \mathbf{1}_4' & \mathbf{0}' & \mathbf{0}' \\ \hline \mathbf{I}_4 & \mathbf{A}_1^* & \mathbf{B} \end{pmatrix} \qquad (7.3.10)$$

From (7.3.6), $\mathbf{W}_{2,1}^{*\prime}$ can be written as follows:

$$\mathbf{W}_{2,1}^{*\prime} = \begin{pmatrix} 1 & -1 & 0 & 0 \\ 0 & 1 & -1 & 0 \\ 0 & 0 & 1 & -1 \\ -1 & 0 & 0 & 1 \end{pmatrix}.$$

Therefore from (7.3.9), $\mathbf{W}^*$-matrix is given by

$$\mathbf{W}^* = \left(\begin{array}{cc|cc|cccc|cc} 1 & 1 & -1 & -1 & 0 & 0 & 0 & 0 & 0 & 0 \\ \hline -1 & 0 & 0 & 0 & 1 & 0 & 0 & -1 & 1 & 0 \\ 0 & -1 & 0 & 0 & -1 & 1 & 0 & 0 & 0 & 1 \\ \hline 0 & 0 & 1 & 0 & 0 & -1 & 1 & 0 & -1 & 0 \\ 0 & 0 & 0 & 1 & 0 & 0 & -1 & 1 & 0 & -1 \end{array} \right). \qquad (7.3.11)$$

The following theorem gives method of construction of OCDs for the general form of the above designs.

Theorem 7.3.1 *Let v be even and a Hadamard matrix of order f exist. Then f optimum $\mathbf{W}$-matrices for the series of BTIB designs with parameters given in (7.3.2) can be constructed.*

Proof Incidence matrix of the design $(f\mathrm{D}_1 \cup f\mathrm{D}_2)$ is actually f replications of $\mathbf{N}^*$ of (7.3.3) and hence can be written as

$$\mathbf{N}^{**\ (v+1)\times fv(v+1)/2} = \mathbf{1}_f' \bigotimes \mathbf{N}^* \qquad (7.3.12)$$

where $\mathbf{N}^*$ is defined in (7.3.3) and $\bigotimes$ denotes Kornecker product. By assumption, Hadamard matrix of order f exists and is written as

$$\mathbf{H}_f = (\mathbf{h}_1,\ \mathbf{h}_2, ..., \mathbf{h}_f). \qquad (7.3.13)$$

Now we construct the matrix $\mathbf{W}_i^{**}$ as follows

$$\mathbf{W}_i^{**} = \mathbf{h}_i \bigotimes \mathbf{W}^*; \quad i = 1, 2, ..., f \qquad (7.3.14)$$

where $\mathbf{W}^*$ is defined in (7.3.9). We can easily check that $\mathbf{W}_i^{**}$'s satisfy all properties of optimum $\mathbf{W}$-matrices. $\square$

Example 7.3.2 Let $v = 4$, $f = 4$. Considering $\mathbf{N}^*$ from (7.3.10)

$$\mathbf{N}^{**} = \mathbf{1}'_f \bigotimes \begin{pmatrix} 1\,1\,1\,1 & 0\,0\,0\,0\,0\,0 \\ 1\,0\,0\,0 & 1\,0\,0\,1\,1\,0 \\ 0\,1\,0\,0 & 1\,1\,0\,0\,0\,1 \\ 0\,0\,1\,0 & 0\,1\,1\,0\,1\,0 \\ 0\,0\,0\,1 & 0\,0\,1\,1\,0\,1 \end{pmatrix} \qquad (7.3.15)$$

$\mathbf{H}_4$, a Hadamard matrix of order 4, is written as

$$\mathbf{H}_4 = \begin{pmatrix} 1 & 1 & 1 & 1 \\ 1 & -1 & -1 & 1 \\ 1 & 1 & -1 & -1 \\ 1 & -1 & 1 & -1 \end{pmatrix} = (\mathbf{h}_1,\ \mathbf{h}_2,\ \mathbf{h}_3,\ \mathbf{h}_4) \qquad (7.3.16)$$

where $\mathbf{h}_l$ is the lth column of $\mathbf{H}_4$. Four optimum $\mathbf{W}$-matrices can be constructed as follows:

$$\mathbf{W}_1^{**} = \mathbf{h}'_1 \bigotimes \mathbf{W}^* = (\mathbf{W}^*,\ \mathbf{W}^*,\ \mathbf{W}^*,\ \mathbf{W}^*);\quad \mathbf{W}_2^{**} = \mathbf{h}'_2 \bigotimes \mathbf{W}^*$$
$$= (\mathbf{W}^*,\ -\mathbf{W}^*,\ \mathbf{W}^*,\ -\mathbf{W}^*);$$

$$\mathbf{W}_3^{**} = \mathbf{h}'_3 \bigotimes \mathbf{W}^* = (\mathbf{W}^*,\ -\mathbf{W}^*,\ -\mathbf{W}^*,\ \mathbf{W}^*);\quad \mathbf{W}_4^{**} = \mathbf{h}'_4 \bigotimes \mathbf{W}^*$$
$$= (\mathbf{W}^*,\ \mathbf{W}^*,\ -\mathbf{W}^*,\ -\mathbf{W}^*);$$

where $\mathbf{W}^*$ is given in (7.3.11).

(ii) Following Bechhofer and Tamhane (1981), for given $(v,\ k)$ and $k \geq 3$, let a generator design $\mathbf{D}_m$ have $m + 1$ plots assigned to the control treatment in each block; the v test treatments be assigned to the remaining $k_m = (k - m - 1)$-plots of the b_m blocks $(0 \leq m \leq k - 2)$ of the design in such away that they form a BIBD. The incidence matrix of $\mathbf{D}_m$ can be transformed into $\mathbf{N}^*$ of expression (7.2.1), where $\mathbf{N}_C^*$ looks like the incidence matrix of a RBD with $(m+1)$ treatments arranged in b_m blocks. This is denoted by RBD($m + 1$, b_m). $\mathbf{N}_T$ is incidence matrix of the BIBD with parameters v, b_m, r_m, $k_m = k - m - 1$, λ_m which is denoted by BIBD(v, b_m, r_m, k_m, λ_m). Here three cases have been considered viz $m = 0$, $m =$ even and $m =$ odd and in each of the three cases, OCDs can be constructed for generator designs.

Case 1: When $m = 0$ one plot in each block is assigned to the control treatment and the test treatments in the blocks each of size $k_0 = (k - 1)$ form a BIBD($v, b_0, r_0, k_0, \lambda_0$). Let $\mathbf{N}$ be the incidence matrix of the BTIB design $\mathbf{D}_0$ with parameters v, $b = b_0$, $r = r_0$, $k = k_0 + 1$, $\lambda = \lambda_0$, $\lambda_{c_0} = r$ and it can be written as,

$$\mathbf{N} = (\mathbf{1}_{b_0},\ \mathbf{N}'_T)' \qquad (7.3.17)$$

where $\mathbf{N}_T$ is the incidence matrix of a BIBD(v, b_0, r_0, k_0, λ_0). It is convenient for the construction of OCD for D_0, if $\mathbf{N}_T$ is the incidence matrix of a k_0-resolvable BIBD with $b_0 = sv$, $s \geq 1$ being an integer. This requires that the $b_0 = sv$ blocks can be partitioned into s sets $T_1, T_2, \ldots, T_s$ each of which contains v blocks such that each of the v treatments occurs k_0 times in each T_i, $i = 1, 2, \ldots, s$. By exploiting the properties of k_0-resolvable BIBD, it is possible to construct OCD for D_0. Precise statement follows:

Theorem 7.3.2 *If a k_0-resolvable BIBD(v, $b_0 = sv$, $r_0 = sk_0$, $k_0 = k - 1$, λ_0) exists, then it is possible to construct $sk_0/2$ optimum $\mathbf{W}$-matrices for the generator design D_0 with parameters v, $b = sv = b_0$, k, $r = sk_0 = r_0$, $r_{c_0} = sv$, $\lambda = \lambda_0$, $\lambda_{c_0} = sk_0$, provided $\mathbf{H}_{k_0+1}$ and $\mathbf{H}_{s/2}$ exist.*

Proof As the BIBD is k_0-resolvable, then Lemma 6.2.1 of Chap. 6 is applicable to the blocks of each T_i. According to Lemma 6.2.1 of Chap. 6, the k_0 treatments in the v blocks of T_i can always be arranged such that each treatment occurs exactly once in each of the k_0 positions in the blocks and this arrangement is denoted by a $k_0 \times v$ matrix $\mathbf{B}_i$ and from the matrix $\mathbf{B}_i$, it is possible to construct a $v \times v$ matrix $\mathbf{A}_i^{v \times v}$ by putting an element a_l in its (m, q) th cell if mth treatment occurs in the lth row and q th column of $\mathbf{B}_i$, $l = 1, 2, \ldots, k_0$, m, $q = 1, 2, \ldots, v$. Other positions are filled in with zeros. It is easily seen that $\mathbf{A}_i$ contains each of the symbols $a_1, a_2, \ldots, a_{k_0}$ once and only once in each row and in each column, $i = 1, 2, \ldots, s$. Now a matrix $\mathbf{A}_i^*$ of order $(v + 1) \times 2v$ is defined by pairing the $\mathbf{A}_i$'s as follows

$$\mathbf{A}_i^* = \begin{pmatrix} \mathbf{1}_v' & \mathbf{1}_v' \\ \mathbf{A}_{2i-1} & \mathbf{A}_{2i} \end{pmatrix}; \quad \forall i = 1, 2, \ldots, s/2. \tag{7.3.18}$$

It is given that $\mathbf{H}_k$, a Hadamard matrix of order k, exists. Let it be written as:

$$\mathbf{H}_{k_0+1} = \begin{pmatrix} 1 & \mathbf{1}_{k_0}' \\ \mathbf{1}_{k_0} & \mathbf{H}_{k_0}^* \end{pmatrix}. \tag{7.3.19}$$

where

$$\mathbf{H}_{k_0}^* = (\mathbf{h}_1^*, \mathbf{h}_2^*, \ldots, \mathbf{h}_{k_0}^*) = \text{core matrix of } \mathbf{H}_{k_0+1}, \tag{7.3.20}$$

$\mathbf{h}_j^*$ is the jth column of $\mathbf{H}_{k_0}^*$. Now a matrix $\mathbf{W}_{j,i}$ can be constructed from $\mathbf{A}_i^*$ by identifying the symbols $a_1, a_2, \ldots, a_{k_0}$ of $\mathbf{A}_{2i-1}$ and $\mathbf{A}_{2i}$ with the elements of $\mathbf{h}_j^*$ and $-\mathbf{h}_j^*$ respectively and also replacing first row of $\mathbf{A}_i^*$ by $(\mathbf{1}_v', -\mathbf{1}_v')$. By juxtaposing $\mathbf{W}_{j,i}$, $i = 1, 2, \ldots, s/2$, for fixed j, we obtain a matrix $\mathbf{W}_j$, where

$$\mathbf{W}_j = (\mathbf{W}_{j,1}, \mathbf{W}_{j,2}, \ldots, \mathbf{W}_{j,s/2}); \quad j = 1, 2, \ldots, k_0. \tag{7.3.21}$$

Varying j in (7.3.21), k_0 optimum $\mathbf{W}$-matrices, $\mathbf{W}_1, \mathbf{W}_2, \ldots, \mathbf{W}_{k_0}$, are obtained. Again as $\mathbf{H}_{s/2}$ exists, it is possible to increase the number of optimum $\mathbf{W}$-matrices. By taking the Khatri–Rao product of $\mathbf{h}_i = (h_{i1}, h_{i2}, \ldots, h_{i,s/2})$, the ith row of $\mathbf{H}_{s/2}$, with $\mathbf{W}_j$ of (7.3.21), $sk_0/2$ matrices $\mathbf{W}_{ji}$ can be constructed, where

$$\mathbf{W}_{ji} = \mathbf{h}_i \odot \mathbf{W}_j = (h_{i1}\mathbf{W}_{j,1}, \; h_{i2}\mathbf{W}_{j,2}, \ldots, h_{i,s/2}\mathbf{W}_{j,s/2}),$$
$$\forall i = 1, 2, \ldots, s/2; \; j = 1, 2, \ldots, k_0. \tag{7.3.22}$$

It is easy to verify that $\mathbf{W}_{ji}$'s satisfy condition (7.2.3). □

Remark 7.3.1 If $s = 2$, then k_0 optimum **W**-matrices, $\mathbf{W}_1, \mathbf{W}_2,\ldots,\mathbf{W}_{k_0}$, can be constructed whenever $\mathbf{H}_k$ exists.

This is illustrated by considering the following example:

Example 7.3.3 Let us consider the following 3-resolvable BIBD(5, 10, 6, 3, 3) obtained by cyclical development of two initial blocks (0, 1, 2) and (0, 1, 3) constructed from the module $M = (0, 1, 2, 3, 4)$. The blocks with the treatments renamed as (1, 2, 3, 4, 5) can be represented in the form of a matrix **B** of order 3×10 as

$$\mathbf{B} = \begin{pmatrix} 1\,2\,3\,4\,5 & 1\,2\,3\,4\,5 \\ 2\,3\,4\,5\,1 & 2\,3\,4\,5\,1 \\ 3\,4\,5\,1\,2 & 4\,5\,1\,2\,3 \end{pmatrix} = (\mathbf{B}_1, \mathbf{B}_2).$$

Now the **A**-matrix of order 5×10 can be constructed from **B** as

$$\mathbf{A} = \begin{pmatrix} a_1 & 0 & 0 & a_3 & a_2 & a_1 & 0 & a_3 & 0 & a_2 \\ a_2 & a_1 & 0 & 0 & a_3 & a_2 & a_1 & 0 & a_3 & 0 \\ a_3 & a_2 & a_1 & 0 & 0 & 0 & a_2 & a_1 & 0 & a_3 \\ 0 & a_3 & a_2 & a_1 & 0 & a_3 & 0 & a_2 & a_1 & 0 \\ 0 & 0 & a_3 & a_2 & a_1 & 0 & a_3 & 0 & a_2 & a_1 \end{pmatrix} = (\mathbf{A}_1, \mathbf{A}_2).$$

The core matrix $\mathbf{H}_3^*$ obtained from Hadamard matrix $\mathbf{H}_4$ of (7.3.16) is given

$$\mathbf{H}_3^* = \begin{pmatrix} -1 & -1 & 1 \\ 1 & -1 & -1 \\ -1 & 1 & -1 \end{pmatrix} = (\mathbf{h}_1^*, \mathbf{h}_2^*, \mathbf{h}_3^*). \tag{7.3.23}$$

Considering $\mathbf{h}_1^*$ for $\mathbf{A}_1$ and then identifying -1 with a_1, 1 with a_2 and -1 with a_3 of $\mathbf{A}_1$ and similarly identifying the elements of $-\mathbf{h}_1^*$ with those of $\mathbf{A}_2$, $\mathbf{W}_1$ can be constructed by using Theorem 7.3.2 as

$$\mathbf{W}_1 = \begin{pmatrix} 1 & 1 & 1 & 1 & 1 & -1 & -1 & -1 & -1 & -1 \\ -1 & 0 & 0 & -1 & 1 & 1 & 0 & 1 & 0 & -1 \\ 1 & -1 & 0 & 0 & -1 & -1 & 1 & 0 & 1 & 0 \\ -1 & 1 & -1 & 0 & 0 & 0 & -1 & 1 & 0 & 1 \\ 0 & -1 & 1 & -1 & 0 & 1 & 0 & -1 & 1 & 0 \\ 0 & 0 & -1 & 1 & -1 & 0 & 1 & 0 & -1 & 1 \end{pmatrix}. \tag{7.3.24}$$

Similarly $\mathbf{W}_2$ and $\mathbf{W}_3$ can be constructed by using $\mathbf{h}_2^*$ and $\mathbf{h}_3^*$ respectively. It is easy to see that $\mathbf{W}_1$, $\mathbf{W}_2$ and $\mathbf{W}_3$ satisfy condition (7.2.3).

Remark 7.3.2 If $c^* = sk_0/2$ optimum **W**-matrices for the generator design D_0 with parameters v, $b = sv$, k, $r = sk_0$, $r_{c_0} = sv$, $\lambda = \lambda_0$, $\lambda_{c_0} = sk_0$ exist, then it is possible to construct c^*u optimum **W**-matrices for the BTIB$(v$, $b = svu$, k, $r = suk_0$, $r_{c_0} = suv$, $\lambda = u\lambda_0$, $\lambda_{c_0} = suk_0)$, which is obtained by repeating D_0 u-times, provided $\mathbf{H}_u$ exists.

Case 2: m=even and $\mathbf{N}^*$, the incidence matrix of D_m which is a BTIB design with parameters v, $b = b_m$, k, $r = r_m$, $r_{c_0} = (m+1)b_m$, $\lambda = \lambda_m$, $\lambda_{c_0} = (m+1)r_m$ can be written as

$$\mathbf{N}^* = (\mathbf{J}'^{(m+1)\times b_m}, \ \mathbf{N}_T')' = (\mathbf{J}^{b_m\times m}, \ \mathbf{N}_0^{*\prime})' \qquad (7.3.25)$$

where $\mathbf{N}_0^*$ is the incidence matrix of D_0, a BTIB design with parameters v, $b = b_m$, $k - m$, $r = r_m$, $r_c = b_m$, $\lambda = \lambda_m$, $\lambda_{c_0} = r_m$. $\mathbf{J}^{m\times b_m}$, a matrix of order $m \times b_m$ with all elements unity, can be considered as the incidence matrix of an RBD$(m$, $b_m)$. From Theorem 7.3.2 it follows that if the BIBD in D_0 is k_m-resolvable, then OCDs can be constructed for D_0. By combining the optimum **W**-matrices for D_0 with those for RBD$(m$, $b_m)$ (cf. Chap. 3), it is possible to construct OCDs for D_m. The results are stated in the following theorem.

Theorem 7.3.3 *If u_1 optimum* **W***-matrices exist for D_0 with parameters v, $b = b_m$, $k_m + 1 = k - m$, $r = r_m$, $r_{c_0} = b_m$, $\lambda = \lambda_m$, $\lambda_{c_0} = r_m$ and u_2 optimum* **W***-matrices exist for RBD$(m$, $b_m)$, then $u = \min\{u_1, u_2\}$ optimum* **W***s for D_m with parameters v, $b = b_m$, k, $r = r_m$, $r_{c_0} = (m+1)b_m$, $\lambda = \lambda_m$, $\lambda_{c_0} = (m+1)r_m$ can be obtained.*

Proof Let the u_1 optimum **W**-matrices for D_0 and u_2 optimum **W**-matrices for RBD$(m$, $b_m)$ be denoted as $\mathbf{W}_{0,1}$, $\mathbf{W}_{0,2},...,\mathbf{W}_{0,u_1}$ and $\mathbf{W}_{m,RBD}^{(1)}, \mathbf{W}_{m,RBD}^{(2)},...,$ $\mathbf{W}_{m,RBD}^{(u_2)}$ respectively. It can be seen that the following u matrices

$$\mathbf{W}_{m,i} = \left(\mathbf{W}_{m,RBD}^{(i)\prime}, \ \mathbf{W}_{0,i}'\right)', \quad i = 1, 2, \ldots, u \qquad (7.3.26)$$

satisfy the condition C and hence u optimum **W**-matrices are obtained for D_m. $\square$

Example 7.3.4 Consider D_2 with parameters $v = 5$, $b_2 = 10$, $k = 6$, $r = 6$, $r_{c_0} = 30$, $\lambda = 3$, $\lambda_{c_0} = 18$ and the incidence matrix of D_2 is

$$\mathbf{N} = \begin{pmatrix} 3\,3\,3\,3\,3 & 3\,3\,3\,3\,3 \\ 1\,0\,0\,1\,1 & 1\,0\,1\,0\,1 \\ 1\,1\,0\,0\,1 & 1\,1\,0\,1\,0 \\ 1\,1\,1\,0\,0 & 0\,1\,1\,0\,1 \\ 0\,1\,1\,1\,0 & 1\,0\,1\,1\,0 \\ 0\,0\,1\,1\,1 & 0\,1\,0\,1\,1 \end{pmatrix}.$$

Here 3 optimum **W**s exist for D_0 (see Example 7.3.3) but one optimum **W**-matrix can be constructed for RBD(2, 10) which is given by

$$\mathbf{W}_{2,RBD}^{(1)} = \begin{pmatrix} 1 & 1 & 1 & 1 & 1 & -1 & -1 & -1 & -1 & -1 \\ -1 & -1 & -1 & -1 & -1 & 1 & 1 & 1 & 1 & 1 \end{pmatrix}.$$

Therefore one optimum $\mathbf{W}$ can be constructed for D_2 by using any one of 3 optimum $\mathbf{W}$s for D_0 given in (7.3.24) and $\mathbf{W}_{2,RBD}^{(1)}$ as follows:

$$\left(\begin{array}{ccccc|ccccc} 1 & 1 & 1 & 1 & 1 & -1 & -1 & -1 & -1 & -1 \\ -1 & -1 & -1 & -1 & -1 & 1 & 1 & 1 & 1 & 1 \\ 1 & 1 & 1 & 1 & 1 & -1 & -1 & -1 & -1 & -1 \\ -1 & 0 & 0 & -1 & 1 & 1 & 0 & 1 & 0 & -1 \\ 1 & -1 & 0 & 0 & -1 & -1 & 1 & 0 & 1 & 0 \\ -1 & 1 & -1 & 0 & 0 & 0 & -1 & 1 & 0 & 1 \\ 0 & -1 & 1 & -1 & 0 & 1 & 0 & -1 & 1 & 0 \\ 0 & 0 & -1 & 1 & -1 & 0 & 1 & 0 & -1 & 1 \end{array} \right).$$

Case 3: When $m = $ odd, then optimum $\mathbf{W}$-matrices for D_m are obtained in the same manner pairing the optimum $\mathbf{W}$-matrices.

Theorem 7.3.4 *Suppose u_1 and u_2 optimum $\mathbf{W}$-matrices exist for RBD$((m+1), b_m)$ and BIBD$(v, b_m, r_m, k_m = k - m - 1, \lambda_m)$ respectively. Then $u = \min\{u_1, u_2\}$ optimum $\mathbf{W}$-matrices for D_m with parameters v, $b = b_m$, k, $r = r_m$, $r_{c_0} = (m+1)b_m$, $\lambda = \lambda_m$, $\lambda_{c_0} = (m+1)r_m$ can be constructed.*

Proof Let the $\mathbf{W}_{RBD}^{(1)}$, $\mathbf{W}_{RBD}^{(2)}$,...,$\mathbf{W}_{RBD}^{(u_1)}$ be u_1 optimum $\mathbf{W}$-matrices of RBD and the $\mathbf{W}_{BIBD}^{(1)}$, $\mathbf{W}_{BIBD}^{(2)}$,...,$\mathbf{W}_{BIBD}^{(u_2)}$ be u_2 optimum $\mathbf{W}$-matrices of BIBD. Then u optimum $\mathbf{W}$ matrices of D_m can be constructed as follows:

$$\mathbf{W}_i = (\mathbf{W}_{RBD}^{(i)\prime}, \mathbf{W}_{BIBD}^{(i)\prime})', \quad i = 1, 2, ..., u. \tag{7.3.27}$$

It can be easily checked that $\mathbf{W}_i$s satisfy the condition (7.2.3). □

Remark 7.3.3 For given (v, k), it is possible to construct optimum $\mathbf{W}$s for D_0, D_1,..., D_{k-2} respectively by using Theorems 7.3.2–7.3.4 and by imposing suitable conditions. The generator design D_{k-1} contains no control treatment; it is an RBD or a BIBD with the v test treatments if $v = k$ or $v > k$ respectively. Optimum $\mathbf{W}$s for D_{k-1} are the same as the optimum $\mathbf{W}$s for the corresponding RBD or BIBD as the case may be. Hence the optimum $\mathbf{W}$s for the combined BTIB design, $D = \bigcup_{i=1}^{k-1} f_i D_i$ with at least one $f_i > 0$ $(i = 1, 2, ..., k-2)$ can be obtained by suitably using the $\mathbf{W}$s of the generator designs D_m $(0 \le m \le k-1)$. But it is difficult to say beforehand how many $\mathbf{W}$s exist for D since the number of optimum $\mathbf{W}$-matrices depends on the parameters of the generator designs and the number of generator designs used.

Remark 7.3.4 Below we describe the construction of OCDs for a BTIB design which looks similar to that described in Remark 7.3.3, but the constructional method described therein is not applicable since the irreducible BIBD used here is not necessarily resolvable. Let G_i be the set of $\binom{v}{i}$ blocks formed by choosing all possible i

treatments from a set of v test treatments and then by augmenting with $(v - i)$ repetitions of the control treatment c_0, $i = 1, 2, \ldots, v$. It easily follows that $\bigcup_{i=1}^{v} G_i = G$ forms a BTIB design with parameters v, $b = 2^v - 1$, $k = v$, $r = 2^{v-1}$, $r_{c_0} = p(2^{v-1} - 1)$, $\lambda = 2^{v-2}$ and $\lambda_{c_0} = (v - 1)2^{v-2}$. For $v = 4$, $k = 4$, the construction of OCDs for such BTIB design is illustrated bellow.

$$G_1 = \begin{Bmatrix} c_0 & c_0 & c_0 & c_0 \\ c_0 & c_0 & c_0 & c_0 \\ c_0 & c_0 & c_0 & c_0 \\ 3 & 4 & 1 & 2 \end{Bmatrix}, \quad G_2 = \begin{Bmatrix} c_0 & c_0 & c_0 & c_0 & c_0 & c_0 \\ c_0 & c_0 & c_0 & c_0 & c_0 & c_0 \\ 1 & 2 & 3 & 4 & 1 & 2 \\ 2 & 3 & 4 & 1 & 3 & 4 \end{Bmatrix},$$

$$G_3 = \begin{Bmatrix} c_0 & c_0 & c_0 & c_0 \\ 1 & 2 & 3 & 4 \\ 2 & 3 & 4 & 1 \\ 3 & 4 & 1 & 2 \end{Bmatrix}, \quad G_4 = \begin{Bmatrix} 1 \\ 2 \\ 3 \\ 4 \end{Bmatrix}.$$

Using $\mathbf{h}_2$, $\mathbf{h}_3$ and $\mathbf{h}_4$ of (7.3.15), 3 optimum $\mathbf{W}$s can be constructed by exploiting the inherent cyclic nature of the G_i's where

$$\mathbf{W}_1 = \left(\mathbf{W}_C^{(1)'}, \mathbf{W}_T^{(1)'} \right)' =$$

	G_1				G_2						G_3				G_4
Control Tr. ↓	0	0	0	0	0	0	0	0	0	0	1	−1	1	−1	0
	0	0	0	0	1	1	1	1	−1	−1	−1	1	−1	1	0
	1	1	1	1	−1	−1	−1	−1	−1	−1	1	−1	1	−1	0
Test Tr. ↓	−1	0	−1	1	1	0	0	−1	1	0	−1	0	0	0	1
	1	−1	0	−1	−1	1	0	0	0	1	0	1	0	0	−1
	−1	1	−1	0	0	−1	1	0	1	0	0	0	−1	0	1
	0	−1	1	−1	0	0	−1	1	0	1	0	0	0	1	−1

,

$$\mathbf{W}_2 = \left(\mathbf{W}_C^{(2)'}, \mathbf{W}_T^{(2)'} \right)' =$$

	G_1				G_2						G_3				G_4
Control Tr. ↓	0	0	0	0	0	0	0	0	0	0	1	1	1	1	0
	0	0	0	0	−1	1	1	−1	1	1	−1	−1	1	1	0
	1	1	1	1	−1	−1	−1	−1	−1	−1	−1	−1	−1	−1	0
Test Tr. ↓	−1	0	1	−1	1	0	0	1	−1	0	1	0	0	0	−1
	−1	−1	0	1	1	−1	0	0	0	−1	0	1	0	0	1
	1	−1	−1	0	0	1	−1	0	1	0	0	0	−1	0	1
	0	1	−1	−1	0	0	1	1	0	1	0	0	0	−1	−1

,

$$\mathbf{W}_3 = \left(\mathbf{W}_C^{(3)\prime}, \ \mathbf{W}_T^{(3)\prime} \right)^\prime =$$

$$
\begin{array}{c}
\\
\text{Control} \\
\text{Tr.} \\
\downarrow \\
\hline
\text{Test} \\
\text{Tr.} \\
\downarrow
\end{array}
\begin{array}{c}
\overset{G_1}{} \quad | \quad \overset{G_2}{} \quad | \quad \overset{G_3}{} \ | \ \overset{G_4}{} \\
\left(
\begin{array}{cccc|cccccc|cccc|c}
0 & 0 & 0 & 0 & 0 & 0 & 0 & 0 & 0 & 0 & -1 & 1 & -1 & 1 & 0 \\
0 & 0 & 0 & 0 & 1 & -1 & -1 & 1 & -1 & -1 & -1 & 1 & 1 & -1 & 0 \\
1 & 1 & 1 & 1 & -1 & -1 & -1 & -1 & 1 & 1 & 1 & -1 & 1 & -1 & 0 \\
\hline
1 & 0 & -1 & -1 & -1 & 0 & 0 & 1 & -1 & 0 & 1 & 0 & 0 & 0 & 1 \\
-1 & 1 & 0 & -1 & 1 & 1 & 0 & 0 & 0 & -1 & 0 & -1 & 0 & 0 & 1 \\
-1 & -1 & 1 & 0 & 0 & 1 & 1 & 0 & 1 & 0 & 0 & 0 & -1 & 0 & -1 \\
0 & -1 & -1 & 1 & 0 & 0 & 1 & -1 & 0 & 1 & 0 & 0 & 0 & 1 & -1
\end{array}
\right)
\end{array}.
$$

7.3.2 BTIB Designs Obtained from BIBDs

In this section, we consider two constructional methods of BTIB designs. Method (i) was given by Bechhofer and Tamhane (1981) and Method (ii) was described in Das et al. (2005). Both are based on BIBDs.

(i) From Bechhofer and Tamhane (1981), it is known that from a BIBD(v^*, b, r, k, λ) where $v^* > v$, a BTIB design with parameters v, b, k, r, $r_{c_0} = (v^* - v)r$, λ, $\lambda_{c_0} = (v^* - v)\lambda$ can be obtained by replacing the treatments $v + 1$, $v + 2, \ldots, v^*$ by the control treatment. Here v should be such that each of the new blocks after replacement contains at least one test treatment. The optimum $\mathbf{W}$-matrices for BTIB design can be constructed by using optimum $\mathbf{W}$s for the corresponding BIBD. The following theorem gives the results precisely:

Theorem 7.3.5 *If c^* optimum $\mathbf{W}$-matrices exist for BIBD(v^*, b, r, k, λ), then an equal number of optimum $\mathbf{W}$ matrices for BTIB design with parameters v, b, k, r, $r_{c_0} = (v^* - v)r$, λ, $\lambda_{c_0} = (v^* - v)\lambda$ can be constructed.*

Proof of the theorem follows from the fact that any optimum $\mathbf{W}$-matrix for the BIBD remains optimum for the corresponding BTIB design as the latter is obtained from the first one by just renaming of the $(v^* - v)$ treatments.

The method will be clear from an example. Consider a symmetric BIBD ($v^* = b = 7, r = k = 4, \lambda = 2$). Here 3 optimum $\mathbf{W}$s for the BIBD can be constructed (cf. Chap. 4) and these are denoted by $\mathbf{W}_{1,BIBD}$, $\mathbf{W}_{2,BIBD}$ and $\mathbf{W}_{3,BIBD}$. These $\mathbf{W}$s provide three OCDs for the BTIB design. Take $v = 5$ and Treatments 6 and 7 of the BIBD are relabeled by the control treatment c of BTIB design. Using $\mathbf{W}_{1,BIBD}$, an optimum $\mathbf{W}$-matrix for the BTIB design can be constructed as

$$
\mathbf{W}'_{1,BIBD} =
\begin{array}{c}
\text{Treatment} \longrightarrow \\
\begin{array}{cc}
\text{Test} & \quad\;\; \text{Control}
\end{array}
\end{array}
$$

$$
\mathbf{W}'_{1,BIBD} =
\left(
\begin{array}{ccccc|cc}
1 & 2 & 3 & 4 & 5 & 6 & 7 \\
\hline
1 & 0 & 0 & -1 & 0 & 1 & -1 \\
-1 & 1 & 0 & 0 & -1 & 0 & 1 \\
1 & -1 & 1 & 0 & 0 & -1 & 0 \\
0 & 1 & -1 & 1 & 0 & 0 & -1 \\
-1 & 0 & 1 & -1 & 1 & 0 & 0 \\
0 & -1 & 0 & 1 & -1 & 1 & 0 \\
0 & 0 & -1 & 0 & 1 & -1 & 1
\end{array}
\right) = \mathbf{W}'_{1,BTIBD}.
$$

Similarly from $\mathbf{W}_{2,BIBD}$ and $\mathbf{W}_{3,BIBD}$, two more optimum $\mathbf{W}$s for BTIB design can be obtained.

Remark 7.3.5 It is to be noted that the number of optimum $\mathbf{W}$s for the BTIB design considered above does not depend on the numbers of either the test treatments or the control treatments used but depends only on the existence of optimum $\mathbf{W}$-matrices for the corresponding BIBD as the same $\mathbf{W}$-matrixe for BIBD is used for the BTIB design.

(ii) Consider a BIBD, d_0, with the parameters v^*, b^*, r^*, k^*, λ^*. Replace a given set of i ($0 \leq i \leq v^* - 2$) of the treatments in d_0 by the control treatment and call the resultant design $BIB_i(v^*, b^*, k^*)$. Finally, each block of the design $BIB_i(v^*, b^*, k^*)$ is augmented by $u \geq 0$ replications of the control treatment, such that $(i, u) \neq (0, 0)$. Denote this design by d. Then, it is easy to see that d is a BTIB design with parameters $v = v^* - i$, $b = b^*$, $k = k^* + u$, $r = r^*$, $r_{c_0} = ir^* + b^*u$, $\lambda = \lambda^*$, $\lambda_{c_0} = i\lambda^* + r^*u$, $0 \leq i \leq v^* - 2$, $u \geq 0$. For convenience, the design d is denoted by $BIB_i(v^*, b^*, k^*, u)$. Note that a $BIB_0(v^*, b^*, k^*, u)$ is a BTIB of the R-type while a $BIB_1(v^*, b^*, k^*, u)$ is a BTIB of the S-type. For a definition of R- and S-type BTIB designs see Hedayat and Majumdar (1984). OCDs for such BTIB design can be obtained by the following theorem:

Theorem 7.3.6 *Suppose that c_1^* and c_2^* optimum $\mathbf{W}$-matrices exist for RBD(b^*, u) and BIBD$(v^*, b^*, r^*, k^*, \lambda^*)$ respectively. Then $c^{**} = min\{c_1^*, c_2^*\}$ optimum $\mathbf{W}$-matrices exist for BTIB$(v = v^* - i, b = b^*, k = k^* + u, r = r^*, r_{c_0} = ir^* + b^*u, \lambda = \lambda^*, \lambda_{c_0} = i\lambda^* + r^*u)$ for $i = 0, 1, \ldots, v^* - 2$.*

Proof The proof is simple and follows along the lines of Theorems 7.3.4 and 7.3.5. $\qquad\square$

7.3.3 BTIB Designs Obtained from Group Divisible (GD) Designs

In this section, again we deal with two methods of construction of BTIB designs from GD designs where the first method was illustrated in Bechhofer and Tamhane (1981) and the lastone in Das et al. (2005).

(i) According to Bechhofer and Tamhane (1981), BTIB design can be constructed from group divisible (GD) design with u treatments and b blocks of size k. The association scheme of such a GD design can be obtained by representing the treatments in the form of an $m \times w$ array (with $mw = u$). Any two treatments in the same row of the array are first associates, and those in the different rows are second associates. For OCDs in GD design set-up one is referred to Chap. 5. Suppose that $m \geq k$. One can take $v = m$ and relabel the entries in each of $n_1 > 0$ columns of the array by $1, 2, \ldots, v$ and entries in the remaining $n_2 = w - n_1 > 0$ columns by the control treatment c, thus obtaining a BTIB design. We shall only consider those BTIB designs, no block of which contains the control treatment entirely. OCDs can be constructed for such BTIB design and precise statement follows:

Theorem 7.3.7 *If c^{***} optimum W-matrices exist for a GD design then an equal number of optimum W-matrices exists for the BTIB design obtained from the GD-PBIBD.*

Proof of the theorem is simple. The method is explained through an example by considering the singular group divisible (SGD) design S2 ($u = 6$, $b = 6$, $r = 4$, $k = 4$, $\lambda_1 = 4$, $\lambda_2 = 2$) in Clatworthy's table (1973), page 83, given by

$$blocks \longrightarrow$$

1	2	3	4	5	6
4	5	6	1	2	3
2	3	1	5	6	4
5	6	4	2	3	1

with the following association scheme:

$$\begin{Bmatrix} 1 & 4 \\ 2 & 5 \\ 3 & 6 \end{Bmatrix}.$$

By relabeling the treatments 4, 5 and 6 by c_0's, a BTIB design can be obtained where no block contains the control treatment entirely. The blocks are:

$$blocks \longrightarrow$$

b_1	b_2	b_3	b_4	b_5	b_6
1	2	1	1	2	1
2	3	3	2	3	3
c_0	c_0	c_0	c_0	c_0	c_0
c_0	c_0	c_0	c_0	c_0	c_0

For the given SGD design, 5 optimum W-matrices (cf. Chap. 5) can be constructed and using these W-matrices 5 optimum Ws for the BTIB design can also be constructed by identifying the control treatment in different blocks with the original treaments 4, 5, and 6. One of the **W**-matrix is given by

$$
\mathbf{W}_1 =
\begin{array}{c}
\text{control Tr.} \\
\downarrow \\
\\
\text{Test Tr.} \\
\downarrow
\end{array}
\left(
\begin{array}{cccccc}
0 & -1 & 1 & 0 & 1 & -1 \\
-1 & 1 & 0 & 1 & -1 & 0 \\
-1 & 0 & 1 & 1 & 0 & -1 \\
0 & 1 & -1 & 0 & -1 & 1 \\
1 & -1 & 0 & -1 & 1 & 0 \\
1 & 0 & -1 & -1 & 0 & 1
\end{array}
\right) = (\mathbf{W}_C^{(1)\prime}, \ \mathbf{W}_T^{(1)\prime})'.
$$

Remark 7.3.6 In Chap. 5, we provide a list of optimum **W**s for a large number of GD designs. These optimum **W**s may be used to generate the optimum **W**s for the BTIB designs, obtainable from these GD designs.

(ii) Consider two GD designs d_1 and d_2. Suppose that the parameters of d_i are v, b_i, r_i, k_i, λ_{1i}, λ_{2i}, $i = 1, 2$. Assume that $k_2 > k_1$ and $\lambda_{11} + \lambda_{12} = \lambda_{21} + \lambda_{22} = \lambda$. Then the design obtained by taking the union of the blocks of d_1 and d_2, after adding the control treatment $k_2 - k_1$ times to the blocks of d_1, is a BTIB design (cf. Das et al. 2005) with the following parameters:

$$
v, \ b = b_1 + b_2, \ k = k_2, \ r = r_1 + r_2, \ r_{c_0} = b(k_2 - k_1), \ \lambda, \ \lambda_{c_0} = r_1(k_2 - k_1).
$$
(7.3.28)

Following theorem describes the construction of OCDs for such BTIB designs.

Theorem 7.3.8 *Suppose c_i^{**} and c_3^{**} optimum **W**-matrices exist for the GD designs with parameters v, b_i, r_i, k_i, λ_{1i}, λ_{2i}, $i = 1, 2$, $k_2 > k_1$ and $\lambda_{11} + \lambda_{12} = \lambda_{21} + \lambda_{22} = \lambda$ and the RBD($k_2 - k_1$, b) respectively, then it is possible to construct $c^{****} = min\{c_1^{**}, \ c_2^{**}, \ c_3^{**}\}$ optimum **W**-matrices for the BTIB design with the parameters given in (7.3.28).*

Proof The proof is simple and hence omitted.   □

Remark 7.3.7 Theorem 7.3.8 also holds for any two 2-associate PBIB designs with same association scheme. For details, we refer the original paper of Dutta and Das (2013).

Remark 7.3.8 In this chapter, construction of OCDs in BTIB design set-up has been considered and a large number of commonly used BTIB designs is covered. It is expected that these designs will serve practical purposes to a large extent. As the results are of varied nature, a summary of the BTIB design set-ups and the conditions of existence of OCDs, etc. is presented in the following table which may be helpful for ready reference (Tables 7.1, 7.2 and 7.3).

Table 7.1 BTIB designs obtained from generator designs

Design	Conditions	No. of optimum W-matrices
$D_0 \cup D_1$ (Lemma 7.3.1)	$m = 0$, $k = 2$, $v =$ even	1
$f(D_0 \cup D_1)$ (Theorem 7.3.1)	$m = 0$, $k = 2$, $v =$ even and $\mathbf{H}_f$ exists	f
D_0 (Theorem 7.3.2)	$k > 2$, $m = 0$, $(k-1)-$ resolvable BIBD$(v, sv, s(k-1), k-1, \lambda_0)$ $\mathbf{H}_k$ and $\mathbf{H}_{s/2}$ exist	$s(k-1)/2$
D_m (Theorem 7.3.3)	$k > 2$, $m(> 0) =$ even, u_2 OCDs for RBD(m, b_m) and u_1 OCDs for $D_0(v, b_m, k-m, r_m, b_m, \lambda_m, \lambda_c = r_m)$ exist	$u = \min\{u_1, u_2\}$
D_m (Theorem 7.3.4)	$k > 2$, $m(> 0) =$ odd, u_1 OCDs for RBD$(m+1, b_m)$ and u_2 OCDs for BIBD$(v, b_m, r_m, k-m-1, \bar{\lambda}_m)$ exist	$u = \min\{u_1, u_2\}$

Table 7.2 BTIB designs obtained from BIBDs

Design	Conditions	No. of optimum **W**-matrices
BTIB design mentioned in Theorem 7.3.5	Existence of c^* OCDs for BIBD$(v^*, b^*, r^*, k^*, \lambda^*)$	c^*
BTIB design mentioned in Theorem 7.3.6	Existence of c_1^* OCDs for RBD(b^*, u) and c_2^* OCDs for BIBD$(v^*, b^*, r^*, k^*, \lambda^*)$	$c^{**} = \min\{c_1^*, c_2^*\}$

Table 7.3 BTIB designs obtained from GD designs

Design	Conditions	No. of optimum **W**-matrices
BTIB design mentioned in Theorem 7.3.7	Existence of c^{***} OCDs for GD design	c^{***}
BTIB design mentioned in Theorem 7.3.8	Existence of c_i^{**} OCDs for GDDs$(v, b_i, r_i, k_i, \lambda_{1i}, \lambda_{2i})$, $i = 1, 2, k_2 > k_1, \lambda_{11} + \lambda_{12} = \lambda_{21} + \lambda_{22}$ and existence of c_3^{**} OCDs for RBD$(k_2 - k_1, b)$	$c^{****} = \min\{c_1^{**}, c_2^{**}, c_3^{**}\}$

References

Bechhofer RE, Tamhane AC (1981) Incomplete block designs for comparing treatments with a control: general theory. Technometrics 23:45–57

Clatworthy WH (1973) Tables of two-associate class partially balanced designs. National Bureau of Standards, US Department of Commerce

Das A, Dey A, Kageyama S, Sinha K (2005) A-efficient balanced treatment incomplete block designs. Aust J Comb 32:243–252

Das K, Mandal NK, Sinha Bikas K (2003) Optimal experimental designs for models with covariates. J Stat Plan Inference 115:273–285

Dutta G, Das P (2013) Optimum designs for estimation of regression parameters in a balanced treatment incomplete block design set-up. J Stat Plan Inference 143:1203–1214

Hedayat AS, Majumdar D (1984) A-optimal incomplete block designs for control-test treatment comparisons. Technometrics 26:363–370

Chapter 8
Miscellaneous Other Topics and Issues

8.1 OCDs in the Crossover Designs

The problem of optimal choice of covariates in the set-up of crossover design or repeated measurement design (RMD) has been considered by Dutta and SahaRay (2013). A crossover design is used in an experiment in which a unit is exposed to various treatments over different periods. In such an experiment, t treatments are assigned to n experimental units each of which receives one treatment during each of the p periods. Such designs are very often used in many industrial, agricultural and biological experiments. Under the traditional model, it is assumed that each treatment assigned to an experimental unit (e.u.) has a direct effect on the e.u. in the same period and also carryover effects (residual effects) in the subsequent periods. Efficient estimation and testing of the direct effects as well as residual effects are of interest to practitioners from an application point of view. The reader is referred to Stufken (1996) and Bose and Dey (2009) for a review on this topic. In practice, situations arise when controllable covariates are used conveniently in this set-up to control the experimental error. For example, in treating arthritis pain or prevention of heart disease, the duration of daily exercise or walking plays a role, besides the effects of medicines. Thus the duration of exercise or walking can be viewed as a controllable covariate when formulating an appropriate model for the study of the effects of different medicines in such cases. So the problem arises to propose appropriate designs which will allow most efficient estimation of these covariate effects on the response. The aim is to address this issue dealing with c covariates for some classes of strongly balanced and balanced crossover designs which are known to be universally optimal for the estimation of direct treatment effects and residual treatment effects in an appropriate class of competing designs.

© Springer India 2015
P. Das et al., *Optimal Covariate Designs*, DOI 10.1007/978-81-322-2461-7_8

8.1.1 Preliminary Definitions and Notations

We assume t treatments, denoted by $1, 2, \ldots, t$ are to be compared using n e.u.s over p periods. Let $\Omega_{t,n,p}$ denote the class of such crossover designs. A design $d \in \Omega_{t,n,p}$ is *uniform on the periods* if each treatment is assigned to an equal number of subjects in each period. A design $d \in \Omega_{t,n,p}$ is *uniform on the subjects* if each treatment is assigned equally often to each subject. A design is said to be *uniform* if it is uniform on the periods and uniform on the subjects. A crossover design is said to be *balanced*, if no treatment is immediately preceded by itself and each treatment is immediately preceded by every other treatment equally often. A crossover design is called *strongly balanced* if each treatment is immediately preceded by every treatment (including itself) equally often.

Here, we deal with a covariate model allowing c covariates under the crossover design set-up. Let $d(i, j)$ denote the treatment assigned by $d \in \Omega_{t,n,p}$ in the ith period to the jth e.u.; $i = 1, 2, \ldots, p$, $j = 1, 2, \ldots, n$. The model of response for the observation y_{ij} with $z_{ij}^{(l)}$, the value of the lth covariate Z_l received in the ith period on the jth experimental unit is given by

$$y_{ij} = \mu + \alpha_i + \beta_j + \tau_{d(i,j)} + \rho_{d(i-1,j)} + \sum_{l=1}^{c} \gamma_l z_{ij}^{(l)} + e_{ij}, \qquad (8.1.1)$$

where μ is the general mean, α_i is the ith period effect, β_j is the jth experimental unit effect, $\tau_{d(i,j)}$ is the direct effect due to treatment $d(i, j)$, $\rho_{d(i-1,j)}$ is the first order residual effect of treatment $d(i - 1, j)$ with $\rho_{d(0,j)} = 0$ for all $j = 1, 2, \ldots, n$; γ_l is the regression coefficient associated with the lth covariate, $l = 1, 2, \ldots, c$. As usual, the random errors $\{e_{ij}\}'s$ are assumed to be uncorrelated and homoscedastic with the common variance σ^2.

Writing the observations unit wise, in matrix notation the above model can be represented as

$$(\mathbf{Y}, \ \mu\mathbf{1}_{np} + \mathbf{X}_1\alpha + \mathbf{X}_2\beta + \mathbf{X}_3\tau + \mathbf{X}_4\rho + \mathbf{Z}\gamma, \ \mathbf{I}_{np}\sigma^2) \qquad (8.1.2)$$

where $\mathbf{Y}$ is the observation vector of order $np \times 1$, α, β, τ, ρ and γ correspond, respectively, to the vectors of period effects, experimental unit effects, direct effects, first-order residual effects and the covariate effects; $\mathbf{X}_1$, $\mathbf{X}_2$, $\mathbf{X}_3$, $\mathbf{X}_4$ and $\mathbf{Z}$ denote, respectively, the part of the design matrix corresponding to the period effects, experimental unit effects, direct effects, first-order residual effects and covariate effects, $\mathbf{1}_{np}$ is a vector of all ones of order np and $\mathbf{I}_{np}$ is the identity matrix of order np.

In model (8.1.2) each of the covariates Z_l's, $l = 1, 2, \ldots, c$ is assumed to be a controllable non-stochastic variable. Applying a location scale transformation of the original limits of the values of the covariates, without loss of generality, it is assumed that the np values $z_{ij}^{(l)}$'s taken by the lth covariate Z_l can vary within the interval $[-1, 1]$, i.e.

$$z_{ij}^{(l)} \in [-1, 1], \ i = 1, 2, \ldots, p; \ j = 1, 2, \ldots, n; \ l = 1, 2, \ldots, c. \quad (8.1.3)$$

With reference to model (8.1.2), it is evident that orthogonal estimation of the ANOVA effects and the covariate effects is possible whenever the following conditions:

$$\mathbf{X}_i' \mathbf{Z} = \mathbf{0}, \ \forall i = 1, 2, 3, 4 \quad (8.1.4)$$

are satisfied. Further, the covariate effects are estimated with the maximum efficiency if and only if (cf. Pukelsheim 1993)

$$\mathbf{Z}' \mathbf{Z} = np\mathbf{I}_c. \quad (8.1.5)$$

Therefore, optimal estimation of each of the covariate effects is possible while the estimates of the ANOVA effects remain unaltered, if and only if $\mathbf{Z}$ satisfies the conditions (8.1.4) and (8.1.5) simultaneously. In the sequel, any Hadamard matrix of order R is written as

$$\mathbf{H}_R = (\mathbf{h}_1^{(R)}, \ldots, \mathbf{h}_R^{(R)}). \quad (8.1.6)$$

For a Hadamard matrix in the *seminormal* form we assume, without loss of generality, $\mathbf{h}_R^{(R)}$ to be $\mathbf{1}$.

Note that under model (8.1.2) for any $d \in \Omega_{t,n,p}$, $\mathbf{X}_1 = \mathbf{I}_p \otimes \mathbf{1}_n$ and $\mathbf{X}_2 = \mathbf{1}_p \otimes \mathbf{I}_n$. Thus for d, conditions (8.1.4) and (8.1.5) are equivalent to the following conditions:

$$
\left.
\begin{aligned}
&(i) \ \ z_{ij}^{(l)} && = \pm 1 \ \forall \ i = 1, 2, \ldots, p; \ j = 1, 2, \ldots, n; \ l = 1, 2, \ldots, c, \\
&(ii) \ \sum_{i=1}^{p} z_{ij}^{(l)} && = \ 0 \ \forall \ j = 1, 2, \ldots, n; \ l = 1, 2, \ldots, c, \\
&(iii) \ \sum_{j=1}^{n} z_{ij}^{(l)} && = \ 0 \ \forall \ i = 1, 2, \ldots, p; \ l = 1, 2, \ldots, c, \\
&(iv) \sum_{(i,j):d(i,j)=k} z_{ij}^{(l)} && = \ 0 \ \forall \ k = 1, 2, \ldots, t; \ l = 1, 2, \ldots, c, \\
&(v) \sum_{(i,j):d(i-1,j)=k} z_{ij}^{(l)} && = \ 0 \ \forall \ k = 1, 2, \ldots, t; \ l = 1, 2, \ldots, c, \\
&(vi) \sum_{i=1}^{p} \sum_{j=1}^{n} z_{ij}^{(l)} z_{ij}^{(l')} && = \ 0 \ \forall \ l \neq l' = 1, 2, \ldots, c.
\end{aligned}
\right\}
$$

$$(8.1.7)$$

Thus to obtain an OCD for any $d \in \Omega_{t,n,p}$ it is required to construct the $\mathbf{Z}$-matrix satisfying the conditions laid down in (8.1.7). In general, for any arbitrary d this problem of construction is combinatorially intractable. Dutta and SahaRay (2013) handled this issue of construction by adopting the technique used by Das et al. (2003) where *each column* of the $\mathbf{Z}$-matrix can be recast to a $\mathbf{W}$-matrix. Using this idea, the lth column of $\mathbf{Z}$-matrix, a vector of order $np \times 1$ is represented in the form of the matrix $\mathbf{W}^{(l)}$ of order $p \times n$, where the columns correspond to the experimental units in the order $1, 2, \ldots, n$ and the rows correspond to the periods in the order $1, 2, \ldots, p$.

To elucidate the idea, the lth column of $\mathbf{Z}$-matrix is written as $\mathbf{W}^{(l)}$-matrix in the following way:

$$\mathbf{W}^{(l)} = \begin{pmatrix} z_{11}^{(l)} & z_{12}^{(l)} & \cdots & z_{1n}^{(l)} \\ z_{21}^{(l)} & z_{22}^{(l)} & \cdots & z_{2n}^{(l)} \\ & \vdots & & \\ z_{p1}^{(l)} & z_{p2}^{(l)} & \cdots & z_{pn}^{(l)} \end{pmatrix}, \quad l = 1, 2, \ldots, c. \tag{8.1.8}$$

The requirement of the $\mathbf{Z}$-matrix satisfying the conditions (ii) and (iii) of (8.1.7) is equivalent to having zero row sums and zero column sums for each row and each column of $\mathbf{W}^{(l)}$, $l = 1, 2, \ldots, c$. To visualize the conditions (iv) and (v) of (8.1.7) in terms of the $\mathbf{W}$-matrix we define two more matrices of order $p \times n$ as follows:

$$\mathbf{V}_1 = \begin{pmatrix} d(1, 1) & d(1, 2) & \cdots & d(1, n) \\ d(2, 1) & d(2, 2) & \cdots & d(2, n) \\ & \vdots & & \\ d(p, 1) & d(p, 2) & \cdots & d(p, n) \end{pmatrix}, \quad \mathbf{V}_2 = \begin{pmatrix} 0 & 0 & \cdots & 0 \\ d(1, 1) & d(1, 2) & \cdots & d(1, n) \\ & \vdots & & \\ d(p-1, 1) & d(p-1, 2) & \cdots & d(p-1, n) \end{pmatrix}.$$
$$\tag{8.1.9}$$

Recalling that $d(i, j)$ denotes the treatment assigned to the jth unit in the ith period of $d \in \Omega_{t,n,p}$, $i = 1, 2, \ldots, p$, $j = 1, 2, \ldots, n$, it is now easy to verify that the requirement of the lth column of the $\mathbf{Z}$-matrix satisfying the conditions (iv) and (v) of (8.1.7) is equivalent to the requirement of the sums of $z_{ij}^{(l)}$'s corresponding to the same treatment to be equal to zero after superimposition of $\mathbf{W}^{(l)}$ on $\mathbf{V}_1$ and $\mathbf{V}_2$ respectively, $l = 1, 2, \ldots, c$.

Thus the necessary and sufficient conditions in terms of the elements of $\mathbf{W}^{(l)}$, $l = 1, 2, \ldots, c$ for the existence of an OCD are summed up as follows:

(C_1) Each of the elements of $\mathbf{W}^{(l)}$ is either $+1$ or -1;

(C_2) $\mathbf{W}^{(l)}$-matrix has all row sums equal to zero;

(C_3) $\mathbf{W}^{(l)}$-matrix has all column sums equal to zero;

(C_4) After superimposing $\mathbf{W}^{(l)}$ on $\mathbf{V}_1$, for every treatment as specified in $\mathbf{V}_1$, the sum of the elements of $\mathbf{W}^{(l)}$ corresponding to the same treatment is equal to zero;

(C_5) After superimposing $\mathbf{W}^{(l)}$ on $\mathbf{V}_2$, for every treatment as specified in $\mathbf{V}_2$ the sum of the elements of $\mathbf{W}^{(l)}$ corresponding to the same treatment is equal to zero;

(C_6) The grand total of all entries in the Hadamard product of $\mathbf{W}^{(l)}$ and $\mathbf{W}^{(l')}$ is equal to np or zero depending on $l = l'$ or $l \neq l'$ respectively.

$$\tag{8.1.10}$$

It is worthwhile to note that a covariate design $\mathbf{Z}$ for c covariates is equivalent to c $\mathbf{W}$-matrices which are convenient to work with.

Definition 8.1.1 With respect to model (8.1.2), the c $\mathbf{W}$-matrices corresponding to the c covariates are said to be optimum if they satisfy the conditions laid down in (8.1.10).

Remark 8.1.1 It is to be noted that if $c = 1$, only the conditions C_1–C_5 of (8.1.10) are to be satisfied by the $\mathbf{W}$-matrix for an OCD to exist.

Definition 8.1.2 The maximum number of covariates cannot exceed the error degrees of freedom for the ANOVA part of a given set-up.

Here, our aim is to construct an OCD. In other words optimum $\mathbf{W}$-matrices, with as many $\mathbf{W}$-matrices as possible for a crossover design which is uniform strongly balanced or strongly balanced, uniform on the periods and uniform on the units in the first $p - 1$ periods or uniform balanced. The construction of $\mathbf{W}$-matrices is much dependent on the particular method of construction of the underlying basic crossover design. We will denote by c^* the maximum value of c, the number of covariates in a given context as attained by a given method of construction. In reality, a limited number of covariates turn out to be useful. Thus given the choice of c^* optimum $\mathbf{W}$-matrices, the experimenter has the flexibility of selecting the optimum values of the required number of covariates from a large pool of possible options, appropriate to the experimental situation and availability of the resources.

8.1.2 Main Results

Here the construction of $\mathbf{W}$-matrices satisfying (8.1.10) for different series of strongly balanced and balanced crossover designs obtained through different constructional methods are given. We briefly discuss the method of construction of the underlying basic crossover design to understand the construction of optimum $\mathbf{W}$-matrices as their interdependency has already been pointed out.

Strongly Balanced Crossover Design Set-up in $\Omega_{t, \lambda_1 t^2, \lambda_2 t}$
It has been shown in Stufken (1996) that a uniform strongly balanced crossover design d^* in $\Omega_{t,n,p}$ is universally optimal for the estimation of direct treatment effects and residual treatment effects and can always be constructed using latin squares and orthogonal arrays whenever $n = \lambda_1 t^2$ and $p = \lambda_2 t$ for integers $\lambda_1 \geq 1$ and $\lambda_2 \geq 2$. We start with this particular method of construction of d^* assuming $\lambda_1 = 1$ and obtain an OCD. The construction of OCD with $n = \lambda_1 t^2$, $\lambda_1 > 1$ will be taken up later.

Let $\mathbf{A}_t$ be an orthogonal array, denoted by OA $(t^2, 3, t, 2)$ with entries from $S = \{1, 2, \ldots, t\}$. Such an orthogonal array can easily be obtained from a latin square $L = ((l_{ij}))$ of order t, $t \geq 2$ as follows:

$$
\mathbf{A}_t : \begin{array}{ccc} \overbrace{1\;1\;\ldots\;1}^{t\ \text{times}} & \overbrace{2\;2\;\ldots\;2}^{t\ \text{times}} & \ldots\quad \overbrace{t\;t\;\ldots\;t}^{t\ \text{times}} \\ 1\;2\;\ldots\;t & 1\;2\;\ldots\;t & \ldots\quad 1\;2\;\ldots t \\ l_{11}l_{12}\ldots l_{1t} & l_{21}l_{22}\ldots l_{2t} & \ldots\quad l_{t1}l_{t2}\ldots l_{tt} \end{array}. \tag{8.1.11}
$$

Let $\mathbf{B}_t$ be an orthogonal array OA $(t^2,\ 2,\ t,\ 2)$, obtained from $\mathbf{A}_t$ by deleting the third row in $\mathbf{A}_t$. For $i \in \{1, 2, \ldots, t-1\}$ let $\mathbf{A}_i = \mathbf{A}_t + i$ and $\mathbf{B}_i = \mathbf{B}_t + i$, where i is added to each element of $\mathbf{A}_t$ or $\mathbf{B}_t$, modulo t. Let the two arrays $\mathbf{A}$ and $\mathbf{B}$ of order $3t \times t^2$ and $2t \times t^2$, respectively, be defined as

$$
\mathbf{A} = \begin{pmatrix} \mathbf{A}_1 \\ \mathbf{A}_2 \\ \vdots \\ \mathbf{A}_t \end{pmatrix}, \quad \mathbf{B} = \begin{pmatrix} \mathbf{B}_1 \\ \mathbf{B}_2 \\ \vdots \\ \mathbf{B}_t \end{pmatrix}. \tag{8.1.12}
$$

With $\lambda_2 \geq 2$, writing $\lambda_2 = 3\delta_1 + 2\delta_2$ for non-negative integers δ_1 and δ_2, the $p \times t^2$ array d^* defined by

$$
d^* = (\mathbf{A}', \ldots, \mathbf{A}',\ \mathbf{B}', \ldots, \mathbf{B}')' \tag{8.1.13}
$$

consisting of δ_1 copies of $\mathbf{A}$ and δ_2 copies of $\mathbf{B}$ is a uniform strongly balanced crossover design in $\Omega_{t,n,p}$.

We now present the actual construction of OCD, in other words optimum $\mathbf{W}$-matrices for d^* in $\Omega_{t,t^2,p}$ under a variety of choices of t accommodating the maximum number of covariates as attained by the given method of construction.

Case 1: $t = 0(mod\ 4)$

The following theorem relates to an OCD for d^* in $\Omega_{t,t^2,3t}$.

Theorem 8.1.1 *Suppose $\mathbf{H}_t$, $\mathbf{H}_{3t}$ and further $s(\geq 2)$ mutually orthogonal latin squares (MOLS) of order t exist. Let d^* in $\Omega_{t,t^2,3t}$ be constructed as described in (8.1.13). Then there exists a set of $(3t-1)(t-1)(s-1)$ optimum $\mathbf{W}$-matrices $d^* \in \Omega_{t,t^2,3t}$.*

Proof Without loss of generality we assume that $\mathbf{H}_t$ and $\mathbf{H}_{3t}$ are in the *seminormal* form. Let $L_1,\ L_2, \ldots, L_s$ be s MOLS of order t, based on the symbols $1, 2, \ldots, t$. Suppose L_s is used for constructing $\mathbf{A}_t$ in (8.1.11) and $L_s^{(q)} = L_s + q$, where q is added to each element of L_s modulo t, is used to develop the third row of $\mathbf{A}_q$, $q = 1, 2, \ldots, t-1$ in (8.1.12) to give rise to d^* in $\Omega_{t,t^2,3t}$ as described in (8.1.13). Now we proceed to construct the optimum $\mathbf{W}$-matrices for d^* in $\Omega_{t,t^2,3t}$ as follows.

In each of the L_i, $i = 1, 2, \ldots, s-1$, replace the symbols $1, 2, \ldots, t$ by the elements of $\mathbf{h}_j^{(t)}$ in order, for $j = 1, 2, \ldots, t-1$. Let $\mathbf{d}_m^{ij'}$ denote the replaced mth row of L_i, $m = 1, 2, \ldots, t$ written with the symbols of $\mathbf{h}_j^{(t)}$. Now juxtaposing side by side these t rows, we obtain a row vector $\mathbf{D}_{ij}'$ of order t^2 given by

$$\mathbf{D}'_{ij} = \left(\mathbf{d}_1^{ij'} : \mathbf{d}_2^{ij'} : \ldots : \mathbf{d}_t^{ij'}\right).\qquad(8.1.14)$$

Now we construct $\mathbf{W}_{ijf}$ of order $3t \times t^2$ as follows:

$$\mathbf{W}^{(l)} = \mathbf{W}_{ijf} = \mathbf{h}_f^{(3t)} \otimes \mathbf{D}'_{ij};$$
$$i = 1, 2, \ldots, s-1, \quad j = 1, \ldots, t_{\bullet} - 1, \quad f = 1, \ldots, 3t-1,$$
$$l = (i-1)(t-1)(3t-1) + (j-1)(3t-1) + f.$$

$$(8.1.15)$$

Using the properties of latin square, Hadamard matrices and the fact that L_i, $i = 1, 2, \ldots, s-1$ is orthogonal with $L_s^{(q)}$, $q = 1, 2, \ldots, t-1$, defined above, it is not hard to see that $\mathbf{W}^{(l)}$'s satisfy the conditions of (8.1.10) and the maximum number of covariates in the given context attained by the method of construction is $c^* = (3t-1)(t-1)(s-1)$. □

An illustration of the above method of construction with $t = 4$ follows.

Example 8.1.1 $t = 4, d^* \in \Omega_{4,16,12}$

$$L_1 = \begin{pmatrix} 1 & 2 & 3 & 4 \\ 2 & 1 & 4 & 3 \\ 3 & 4 & 1 & 2 \\ 4 & 3 & 2 & 1 \end{pmatrix}, \quad L_2 = \begin{pmatrix} 1 & 2 & 3 & 4 \\ 3 & 4 & 1 & 2 \\ 4 & 3 & 2 & 1 \\ 2 & 1 & 4 & 3 \end{pmatrix}, \quad L_3 = \begin{pmatrix} 1 & 2 & 3 & 4 \\ 4 & 3 & 2 & 1 \\ 2 & 1 & 4 & 3 \\ 3 & 4 & 1 & 2 \end{pmatrix}$$

and

$$\mathbf{A} = \begin{pmatrix}
2 & 2 & 2 & 3 & 3 & 3 & 3 & 4 & 4 & 4 & 4 & 1 & 1 & 1 & 1 \\
2 & 3 & 4 & 1 & 2 & 3 & 4 & 1 & 2 & 3 & 4 & 1 & 2 & 3 & 4 & 1 \\
2 & 3 & 4 & 1 & 1 & 4 & 3 & 2 & 3 & 2 & 1 & 4 & 4 & 1 & 2 & 3 \\
3 & 3 & 3 & 3 & 4 & 4 & 4 & 4 & 1 & 1 & 1 & 1 & 2 & 2 & 2 & 2 \\
3 & 4 & 1 & 2 & 3 & 4 & 1 & 2 & 3 & 4 & 1 & 2 & 3 & 4 & 1 & 2 \\
3 & 4 & 1 & 2 & 2 & 1 & 4 & 3 & 4 & 3 & 2 & 1 & 1 & 2 & 3 & 4 \\
4 & 4 & 4 & 4 & 1 & 1 & 1 & 1 & 2 & 2 & 2 & 2 & 3 & 3 & 3 & 3 \\
4 & 1 & 2 & 3 & 4 & 1 & 2 & 3 & 4 & 1 & 2 & 3 & 4 & 1 & 2 & 3 \\
4 & 1 & 2 & 3 & 3 & 2 & 1 & 4 & 1 & 4 & 3 & 2 & 2 & 3 & 4 & 1 \\
1 & 1 & 1 & 1 & 2 & 2 & 2 & 2 & 3 & 3 & 3 & 3 & 4 & 4 & 4 & 4 \\
1 & 2 & 3 & 4 & 1 & 2 & 3 & 4 & 1 & 2 & 3 & 4 & 1 & 2 & 3 & 4 \\
1 & 2 & 3 & 4 & 4 & 3 & 2 & 1 & 2 & 1 & 4 & 3 & 3 & 4 & 1 & 2
\end{pmatrix}.$$

The forms of $\mathbf{H}_4$ and $\mathbf{H}_{12}$ for our use are

$$
\mathbf{H}_4 = \begin{pmatrix} -1 & 1 & -1 & 1 \\ 1 & -1 & -1 & 1 \\ -1 & -1 & 1 & 1 \\ 1 & 1 & 1 & 1 \end{pmatrix}
$$

(8.1.16)

$$
\mathbf{H}_{12} = \begin{pmatrix}
1 & 1 & 1 & 1 & 1 & 1 & 1 & 1 & 1 & 1 & 1 & 1 \\
-1 & 1 & -1 & 1 & 1 & 1 & -1 & -1 & -1 & 1 & -1 & 1 \\
-1 & -1 & 1 & -1 & 1 & 1 & 1 & -1 & -1 & -1 & 1 & 1 \\
1 & -1 & -1 & 1 & -1 & 1 & 1 & 1 & -1 & -1 & -1 & 1 \\
-1 & 1 & -1 & -1 & 1 & -1 & 1 & 1 & 1 & -1 & -1 & 1 \\
-1 & -1 & 1 & -1 & -1 & 1 & -1 & 1 & 1 & 1 & -1 & 1 \\
-1 & -1 & -1 & 1 & -1 & -1 & 1 & -1 & 1 & 1 & 1 & 1 \\
1 & -1 & -1 & -1 & 1 & -1 & -1 & 1 & -1 & 1 & 1 & 1 \\
1 & 1 & -1 & -1 & -1 & 1 & -1 & -1 & 1 & -1 & 1 & 1 \\
1 & 1 & 1 & -1 & -1 & -1 & 1 & -1 & -1 & 1 & -1 & 1 \\
-1 & 1 & 1 & 1 & -1 & -1 & -1 & 1 & -1 & -1 & 1 & 1 \\
1 & -1 & 1 & 1 & 1 & -1 & -1 & -1 & 1 & -1 & -1 & 1
\end{pmatrix}
$$

(8.1.17)

Now using $\mathbf{h}_1^{(4)}$, the first column of $\mathbf{H}_4$ and L_1, we construct $\mathbf{D}'_{11}$ as

$$
\mathbf{D}'_{11} = \begin{pmatrix} -1 & 1 & -1 & 1 : & 1 & -1 & 1 & -1 : & -1 & 1 & -1 & 1 : & 1 & -1 & 1 & -1 \end{pmatrix}.
$$

Hence using $\mathbf{h}_1^{(12)}$, the first column of $\mathbf{H}_{12}$, $\mathbf{W}^{(1)} = \mathbf{W}_{111} = \mathbf{h}_1^{(12)} \otimes (\mathbf{d}_1^{11}, \ \mathbf{d}_2^{11},$
$\mathbf{d}_3^{11}, \ \mathbf{d}_4^{11})$ takes the form

$$
\mathbf{W}^{(1)} = \begin{pmatrix}
-1 & 1 & -1 & 1 & 1 & -1 & 1 & -1 & -1 & 1 & -1 & 1 & 1 & -1 & 1 & -1 \\
1 & -1 & 1 & -1 & -1 & 1 & -1 & 1 & 1 & -1 & 1 & -1 & -1 & 1 & -1 & 1 \\
1 & -1 & 1 & -1 & -1 & 1 & -1 & 1 & 1 & -1 & 1 & -1 & -1 & 1 & -1 & 1 \\
-1 & 1 & -1 & 1 & 1 & -1 & 1 & -1 & -1 & 1 & -1 & 1 & 1 & -1 & 1 & -1 \\
1 & -1 & 1 & -1 & -1 & 1 & -1 & 1 & 1 & -1 & 1 & -1 & -1 & 1 & -1 & 1 \\
1 & -1 & 1 & -1 & -1 & 1 & -1 & 1 & 1 & -1 & 1 & -1 & -1 & 1 & -1 & 1 \\
1 & -1 & 1 & -1 & -1 & 1 & -1 & 1 & 1 & -1 & 1 & -1 & -1 & 1 & -1 & 1 \\
-1 & 1 & -1 & 1 & 1 & -1 & 1 & -1 & -1 & 1 & -1 & 1 & 1 & -1 & 1 & -1 \\
-1 & 1 & -1 & 1 & 1 & -1 & 1 & -1 & -1 & 1 & -1 & 1 & 1 & -1 & 1 & -1 \\
-1 & 1 & -1 & 1 & 1 & -1 & 1 & -1 & -1 & 1 & -1 & 1 & 1 & -1 & 1 & -1 \\
1 & -1 & 1 & -1 & -1 & 1 & -1 & 1 & 1 & -1 & 1 & -1 & -1 & 1 & -1 & 1 \\
-1 & 1 & -1 & 1 & 1 & -1 & 1 & -1 & -1 & 1 & -1 & 1 & 1 & -1 & 1 & -1
\end{pmatrix}.
$$

Similarly 65 more choices of the optimum $\mathbf{W}$-matrices can be constructed.

Remark 8.1.2 In practice in the above situation the experimenter has the flexibility to choose the values of the required number of optimum covariates from the set of 66 possible optimum choices.

Remark 8.1.3 In particular for $t = 4$, three more optimum $\mathbf{W}^{(l)}$ for d^* in $\Omega_{4,16,12}$ can be constructed by trial and error method as follows:

$$
\mathbf{W}^{(67)} = \begin{pmatrix}
1'_4 & -1'_4 & 1'_4 & -1'_4 \\
1'_4 & -1'_4 & 1'_4 & -1'_4 \\
1'_4 & -1'_4 & 1'_4 & -1'_4 \\
-1'_4 & 1'_4 & -1'_4 & 1'_4 \\
-1'_4 & 1'_4 & -1'_4 & 1'_4 \\
-1'_4 & 1'_4 & -1'_4 & 1'_4 \\
-1'_4 & 1'_4 & -1'_4 & 1'_4 \\
-1'_4 & 1'_4 & -1'_4 & 1'_4 \\
-1'_4 & 1'_4 & -1'_4 & 1'_4 \\
1'_4 & -1'_4 & 1'_4 & -1'_4 \\
1'_4 & -1'_4 & 1'_4 & -1'_4 \\
1'_4 & -1'_4 & 1'_4 & -1'_4
\end{pmatrix}, \quad
\mathbf{W}^{(68)} = \begin{pmatrix}
1'_4 & -1'_4 & 1'_4 & -1'_4 \\
-1'_4 & 1'_4 & -1'_4 & 1'_4 \\
1'_4 & -1'_4 & 1'_4 & -1'_4 \\
-1'_4 & 1'_4 & -1'_4 & 1'_4 \\
1'_4 & -1'_4 & 1'_4 & -1'_4 \\
1'_4 & -1'_4 & 1'_4 & -1'_4 \\
-1'_4 & 1'_4 & -1'_4 & 1'_4 \\
1'_4 & -1'_4 & 1'_4 & -1'_4 \\
1'_4 & -1'_4 & 1'_4 & -1'_4 \\
1'_4 & -1'_4 & 1'_4 & -1'_4 \\
-1'_4 & 1'_4 & -1'_4 & 1'_4 \\
-1'_4 & 1'_4 & -1'_4 & 1'_4
\end{pmatrix},
$$

$$
\mathbf{W}^{(69)} = \begin{pmatrix}
1'_4 & -1'_4 & -1'_4 & 1'_4 \\
1'_4 & -1'_4 & -1'_4 & 1'_4 \\
1'_4 & -1'_4 & -1'_4 & 1'_4 \\
-1'_4 & 1'_4 & 1'_4 & -1'_4 \\
-1'_4 & 1'_4 & 1'_4 & -1'_4 \\
-1'_4 & 1'_4 & 1'_4 & -1'_4 \\
1'_4 & -1'_4 & -1'_4 & 1'_4 \\
1'_4 & -1'_4 & -1'_4 & 1'_4 \\
1'_4 & -1'_4 & -1'_4 & 1'_4 \\
-1'_4 & 1'_4 & 1'_4 & -1'_4 \\
-1'_4 & 1'_4 & 1'_4 & -1'_4 \\
-1'_4 & 1'_4 & 1'_4 & -1'_4
\end{pmatrix}.
$$

Theorem 8.1.2 *Suppose* $\mathbf{H}_t$, *and further* $s (\geq 2)$ *mutually orthogonal latin squares (MOLS) of order* t *exist. Let* d^* *in* $\Omega_{t,t^2,2t}$ *be constructed as described in* (8.1.13). *Then there exists a set of* $(2t - 1)(t - 1)s$ *optimum* $\mathbf{W}$-*matrices* $d^* \in \Omega_{t,t^2,2t}$.

Proof The proof is along the similar lines of the proof of Theorem 3.1. Note that $d^* \in \Omega_{t,t^2,2t}$ as described in (8.1.13) can be constructed without requiring to use L_s. So L_s can also be used to construct the row vector $\mathbf{D}'_{ij}$ (8.1.14) of order t^2 as before, $i = 1, 2, \ldots, s$; $j = 1, 2, \ldots, t - 1$. Since $\mathbf{H}_t$ and hence $\mathbf{H}_{2t}$ exist, assuming both of these in the *seminormal* form, we construct $\mathbf{W}^{(l)}$ of order $2t \times t^2$ as follows:

$$
\mathbf{W}^{(l)} = \mathbf{W}_{ijf} = \mathbf{h}_f^{(2t)} \otimes \left(\mathbf{d}_1^{ij'} : \mathbf{d}_2^{ij'} : \ldots : \mathbf{d}_t^{ij'} \right);
$$

$$
i = 1, 2, \ldots, s, \quad j = 1, \ldots, t - 1, \quad f = 1, \ldots, 2t - 1,
$$

$$
l = (i - 1)(t - 1)(2t - 1) + (j - 1)(2t - 1) + f
$$

with $c^* = (2t - 1)(t - 1)s$ in the given context. $\qquad \square$

Remark 8.1.4 For $t = 4$, four more optimum $\mathbf{W}^{(l)}$ for d^* in $\Omega_{4,16,8}$ can be constructed by trial and error method as described below:

$$
\mathbf{W}^{(64)} = \begin{pmatrix}
1'_4 & -1'_4 & 1'_4 & -1'_4 \\
1'_4 & -1'_4 & 1'_4 & -1'_4 \\
-1'_4 & 1'_4 & -1'_4 & 1'_4 \\
-1'_4 & 1'_4 & -1'_4 & 1'_4 \\
-1'_4 & 1'_4 & -1'_4 & 1'_4 \\
-1'_4 & 1'_4 & -1'_4 & 1'_4 \\
1'_4 & -1'_4 & 1'_4 & -1'_4 \\
1'_4 & -1'_4 & 1'_4 & -1'_4
\end{pmatrix}, \quad
\mathbf{W}^{(65)} = \begin{pmatrix}
1'_4 & -1'_4 & 1'_4 & -1'_4 \\
-1'_4 & 1'_4 & -1'_4 & 1'_4 \\
-1'_4 & 1'_4 & -1'_4 & 1'_4 \\
1'_4 & -1'_4 & 1'_4 & -1'_4 \\
-1'_4 & 1'_4 & -1'_4 & 1'_4 \\
1'_4 & -1'_4 & 1'_4 & -1'_4 \\
1'_4 & -1'_4 & 1'_4 & -1'_4 \\
-1'_4 & 1'_4 & -1'_4 & 1'_4
\end{pmatrix},
$$

$$
\mathbf{W}^{(66)} = \begin{pmatrix}
1'_4 & -1'_4 & -1'_4 & 1'_4 \\
1'_4 & -1'_4 & -1'_4 & 1'_4 \\
-1'_4 & 1'_4 & 1'_4 & -1'_4 \\
-1'_4 & 1'_4 & 1'_4 & -1'_4 \\
1'_4 & -1'_4 & -1'_4 & 1'_4 \\
1'_4 & -1'_4 & -1'_4 & 1'_4 \\
-1'_4 & 1'_4 & 1'_4 & -1'_4 \\
-1'_4 & 1'_4 & 1'_4 & -1'_4
\end{pmatrix}, \quad
\mathbf{W}^{(67)} = \begin{pmatrix}
1'_4 & -1'_4 & -1'_4 & 1'_4 \\
-1'_4 & 1'_4 & 1'_4 & -1'_4 \\
-1'_4 & 1'_4 & 1'_4 & -1'_4 \\
1'_4 & -1'_4 & -1'_4 & 1'_4 \\
1'_4 & -1'_4 & -1'_4 & 1'_4 \\
-1'_4 & 1'_4 & 1'_4 & -1'_4 \\
-1'_4 & 1'_4 & 1'_4 & -1'_4 \\
1'_4 & -1'_4 & -1'_4 & 1'_4
\end{pmatrix}.
$$

Case 2: $t \equiv 2 \,(mod\ 4), t \neq 2, 6$

It is clear that $\mathbf{H}_t$ does not exist but if s MOLS of order t exist then $(s - 1)$ optimum $\mathbf{W}$-matrices can be constructed for $d^* \in \Omega_{t,t^2,3t}$ (vide 8.1.13) using the same steps followed in the proof of Theorem 3.1 and the vector $\mathbf{a}_1 = \left(1'_{\frac{t}{2}}, -1'_{\frac{t}{2}}\right)'$ and $\mathbf{a}_2 = \left(1'_{\frac{3t}{2}}, -1'_{\frac{3t}{2}}\right)'$ instead of the columns of $\mathbf{H}_t$ and $\mathbf{H}_{3t}$ respectively. Similarly if $\mathbf{H}_{2t}$ exists, $(2t - 1)s$ optimum $\mathbf{W}$-matrices can be constructed for $d^* \in \Omega_{t,t^2,2t}$ (vide 8.1.13) following the same steps of Theorem 8.1.2 using the vector $\mathbf{a}_1 = \left(1'_{\frac{t}{2}}, -1'_{\frac{t}{2}}\right)'$ instead of the columns of $\mathbf{H}_t$.

Case 3: $t = 2$

Since a pair of MOLS does not exist for $t = 2$, the methods discussed in earlier cases do not apply here to construct an OCD. We adopt trial and error method to construct optimum $\mathbf{W}$-matrices.

Theorem 8.1.3 *Let d_1^* in $\Omega_{2,4,6}$ and d_2^* in $\Omega_{2,4,4}$ be constructed as described in (8.1.13). Then there exist 2 optimum $\mathbf{W}$-matrices for each of d_1^* and d_2^*.*

Proof Recalling (8.1.12) it is easy to see that d_1^* and d_2^* given below represent the strongly balanced design in $\Omega_{2,4,6}$ and $\Omega_{2,4,4}$ respectively.

$$
d_1^* : \begin{pmatrix} 2 & 2 & 1 & 1 \\ 2 & 1 & 2 & 1 \\ 2 & 1 & 1 & 2 \\ 1 & 1 & 2 & 2 \\ 1 & 2 & 1 & 2 \\ 1 & 2 & 2 & 1 \end{pmatrix}, \quad d_2^* = \begin{pmatrix} 2 & 2 & 1 & 1 \\ 2 & 1 & 2 & 1 \\ 1 & 1 & 2 & 2 \\ 1 & 2 & 1 & 2 \end{pmatrix}. \tag{8.1.18}
$$

Optimum **W**-matrices denoted by $\mathbf{W}_1^*$ and $\mathbf{W}_2^*$ for d_1^* and $\mathbf{W}_1^{**}$ and $\mathbf{W}_2^{**}$ for d_2^*, respectively, can be constructed as

$$
\mathbf{W}_1^* = \begin{pmatrix} 1 & 1 & -1 & -1 \\ 1 & -1 & -1 & 1 \\ -1 & 1 & 1 & -1 \\ -1 & -1 & 1 & 1 \\ -1 & 1 & 1 & -1 \\ 1 & -1 & -1 & 1 \end{pmatrix}, \quad \mathbf{W}_2^* = \begin{pmatrix} 1 & -1 & 1 & -1 \\ 1 & 1 & -1 & -1 \\ 1 & -1 & 1 & -1 \\ -1 & 1 & -1 & 1 \\ -1 & -1 & 1 & 1 \\ -1 & 1 & -1 & 1 \end{pmatrix}
$$

$$
\mathbf{W}_1^{**} = \begin{pmatrix} 1 & 1 & -1 & -1 \\ -1 & 1 & -1 & 1 \\ -1 & -1 & 1 & 1 \\ 1 & -1 & 1 & -1 \end{pmatrix}, \quad \mathbf{W}_2^{**} = \begin{pmatrix} 1 & -1 & -1 & 1 \\ -1 & 1 & 1 & -1 \\ 1 & -1 & -1 & 1 \\ -1 & 1 & 1 & -1 \end{pmatrix}.
$$

□

Case 4: $t = 6$

It is known that for $t = 6$ a pair of MOLS does not exist and hence we take up the construction of OCD in this case separately.

We start with a uniform strongly balanced crossover design $d^* \in \Omega_{6,36,18}$ constructed (vide 8.1.13) using the latin square L (say) given by

$$
L = \begin{pmatrix} 1 & 2 & 3 & 4 & 5 & 6 \\ 2 & 1 & 4 & 3 & 6 & 5 \\ 6 & 5 & 1 & 2 & 3 & 4 \\ 5 & 6 & 2 & 1 & 4 & 3 \\ 4 & 3 & 6 & 5 & 2 & 1 \\ 3 & 4 & 5 & 6 & 1 & 2 \end{pmatrix}. \tag{8.1.19}
$$

Theorem 8.1.4 *Let d_1^* in $\Omega_{6,36,18}$ and d_2^* in $\Omega_{6,36,12}$ be constructed (vide 8.1.13) using L of (8.1.19). Then there exist an optimum **W**-matrix for d_1^* and 11 optimum **W** matrices for d_2^*.*

Proof Let **D** be a matrix of order 6×6 with elements ± 1 as follows:

$$
\mathbf{D} = \begin{pmatrix}
1 & 1 & 1 & -1 & -1 & -1 \\
1 & 1 & -1 & -1 & -1 & 1 \\
1 & -1 & -1 & -1 & 1 & 1 \\
-1 & -1 & -1 & 1 & 1 & 1 \\
-1 & -1 & 1 & 1 & 1 & -1 \\
-1 & 1 & 1 & 1 & -1 & -1
\end{pmatrix} = \begin{pmatrix}
\mathbf{d}_1' \\
\mathbf{d}_2' \\
\mathbf{d}_3' \\
\mathbf{d}_4' \\
\mathbf{d}_5' \\
\mathbf{d}_6'
\end{pmatrix}. \tag{8.1.20}
$$

It is to be noted that the row sums and column sums of $\mathbf{D}$ are zero. Moreover superimposing $\mathbf{D}$ on L, it can be seen that for each symbol in L, the sum of the corresponding elements of $\mathbf{D}$ is also zero. Thus an optimum $\mathbf{W}$-matrix for d_1^* in $\Omega_{6,36,18}$ (vide 8.1.13) using L of (8.1.19) can be formed taking $\mathbf{a} = \left(\mathbf{1}_9', -\mathbf{1}_9'\right)'$ and the rows of matrix $\mathbf{D}$ as

$$
\mathbf{W}^{(1)} = \mathbf{a} \otimes \left(\mathbf{d}_1' : \mathbf{d}_2' :, \mathbf{d}_3' : \mathbf{d}_4' : \mathbf{d}_5' : \mathbf{d}_6'\right).
$$

But for d_2^* in $\Omega_{6,36,12}$ (vide 8.1.12), 11 optimum $\mathbf{W}$-matrices can be formed using $\mathbf{H}_{12}$ of (8.1.17) as follows:

$$
\mathbf{W}^{(l)} = \mathbf{h}_l^{(12)} \otimes \left(\mathbf{d}_1' : \mathbf{d}_2' : \mathbf{d}_3' : \mathbf{d}_4' : \mathbf{d}_5' : \mathbf{d}_6'\right), \quad l = 1, 2, \ldots, 11.
$$

□

So far we have discussed the construction of optimum $\mathbf{W}$-matrices for uniform strongly balanced crossover design d_1^* in $\Omega_{t,t^2,3t}$ and d_2^* in $\Omega_{t,t^2,2t}$ separately. Let c_1^* and c_2^* denote the maximum number of optimum $\mathbf{W}$-matrices for d_1^* and d_2^* respectively in the given context. Now we will consider the construction of optimum $\mathbf{W}$-matrices for a strongly balanced crossover design d^* in $\Omega_{t,t^2,p}$ (vide 8.1.13) where $p = (3\delta_1 + 2\delta_2)t$ for non-negative integers δ_1 and δ_2. Write

$$
d^* = \left[d_1^{*\prime}, \ldots, d_1^{*\prime}, d_2^{*\prime}, \ldots, d_2^{*\prime}\right]' \tag{8.1.21}
$$

taking δ_1 copies of d_1^* and δ_2 copies of d_2^*.

Define

$$
\delta_0 = \min\{\delta_1, \delta_2\} \text{ and } c_0 = \min\{c_1^*, c_2^*\}. \tag{8.1.22}
$$

Corollary 8.1.1 *Suppose $\mathbf{H}_{\delta_1}$ and $\mathbf{H}_{\delta_2}$ exist. Let d^* in $\Omega_{t,t^2,p}$ be constructed as described in (8.1.21) for $p = (3\delta_1 + 2\delta_2)t$, δ_1, $\delta_2 \geq 0$, non-negative integers. Then there exists a set of $\delta_0 c_0$ optimum $\mathbf{W}$-matrices for d^* where δ_0 and c_0 are defined in (8.1.22).*

Proof Let the c_1^* optimum $\mathbf{W}$-matrices for d_1^* be denoted by $\mathbf{W}_1^*, \ldots, \mathbf{W}_{c_1^*}^*$ and the c_2^* optimum $\mathbf{W}$-matrices for d_2^* be denoted by $\mathbf{W}_1^{**}, \ldots, \mathbf{W}_{c_2^*}^{**}$. Then it can be easily seen that $\mathbf{W}^{(l)}$ defined as

$$\mathbf{W}^{(l)} = \mathbf{W}_{ij} = \begin{pmatrix} \mathbf{W}^*_{ij} \\ \mathbf{W}^{**}_{ij} \end{pmatrix}; \text{ where } \mathbf{W}^*_{ij} = \mathbf{h}_i^{(\delta_1)} \otimes \mathbf{W}^*_j \text{ and } \mathbf{W}^{**}_{ij} = \mathbf{h}_i^{(\delta_2)} \otimes \mathbf{W}^{**}_j$$

(8.1.23)

$i = 1, 2, \ldots, \delta_0$, $j = 1, 2, \ldots, c_0$, $l = c_0(i-1) + j$, are the required $\mathbf{W}$-matrices for d^*. □

Remark 8.1.5 Note that $\mathbf{H}_{\delta_1}$ and $\mathbf{H}_{\delta_2}$ are not necessarily assumed to be in the semi-normal form. Thus $\mathbf{h}_i^{(\delta_1)}$ and $\mathbf{h}_i^{(\delta_2)}$ can as well be of the form of a vector all ones.

Remark 8.1.6 It is not hard to see that the set of $\delta_0 c_0$ $\mathbf{W}$-matrices in Corollary 8.1.1 is not unique.

Remark 8.1.7 The construction of optimum $\mathbf{W}$-matrices for a strongly balanced design d^* in $\Omega_{t,\lambda_1 t^2, p}$ for $\lambda_1 > 1$ can easily be obtained by taking the Kronecker product of the rows of $\mathbf{H}_{\lambda_1}$ and the corresponding optimum $\mathbf{W}$-matrix of $\Omega_{t,t^2,p}$ whenever $\mathbf{H}_{\lambda_1}$ exists. In case of non-existence of $\mathbf{H}_{\lambda_1}$ for λ_1 even, the role of the rows of $\mathbf{H}_{\lambda_1}$ above can be taken by the vectors $\mathbf{1}'_{\lambda_1}$ and $(\mathbf{1}'_{\frac{\lambda_1}{2}}, -\mathbf{1}'_{\frac{\lambda_1}{2}})'$. In case of λ_1 odd, the vector of all ones serves the purpose.

Case 5: t odd

Whenever t is odd, it is easy to verify that an OCD for a uniform strongly balanced crossover design d^* in $\Omega_{t,t^2,p}$ as described in (8.1.13) does not exist as Condition C_2 of (8.1.10) is not attainable. Let a uniform strongly balanced crossover design $d^{**} \in \Omega_{t,\lambda_1 t^2, p}$ be defined as

$$d^{**} = \mathbf{1}'_{\lambda_1} \otimes d^*$$

(8.1.24)

for some positive integer λ_1. The following theorem relates to the construction of OCD for d^{**}.

Theorem 8.1.5 *Suppose* $\mathbf{H}_{\lambda_1 t}$, $\mathbf{H}_p$ *and a pair of mutually orthogonal latin squares of order t exist. Let d^{**} be defined as in (8.1.24). Then there exists a set of* $(\lambda_1 t - 1)(p - 1)$ *optimum $\mathbf{W}$-matrices for d^{**}.*

Proof Suppose L_1 and L_2 are pairwise orthogonal latin squares of order t and L_2 has been used in (8.1.12) and (8.1.13) to construct a uniform strongly balanced crossover design d^* in $\Omega_{t,t^2,p}$. Now we proceed to construct the optimum $\mathbf{W}$-matrices for d^{**}. Assuming $\mathbf{H}_{\lambda_1 t}$ and $\mathbf{H}_p$ in the *seminormal* form, for each $i = 1, 2, \ldots, \lambda_1 t - 1$, partitioning $\mathbf{h}_i^{(\lambda_1 t)}$ into λ_1 parts as

$$\mathbf{h}_i^{(\lambda_1 t)} = \left(\mathbf{h}_{i1}^{(\lambda_1 t)\prime}, \ldots, \mathbf{h}_{ij}^{(\lambda_1 t)\prime}, \ldots, \mathbf{h}_{i\lambda_1}^{(\lambda_1 t)\prime} \right)'$$

(8.1.25)

we construct a row vector $\mathbf{D}^{*\prime}_{ij}$ of order t^2 considering L_1 and $\mathbf{h}_{ij}^{(\lambda_1 t)}$, for every fixed $j \in \{1, 2, \ldots, \lambda_1\}$, following the steps as described in Theorem 3.1. Thus

$$\mathbf{D}_{ij}^{*\prime} = \left(\mathbf{d}_1^{*ij\prime}, \, \mathbf{d}_2^{*ij\prime}, \ldots, \mathbf{d}_t^{*ij\prime}\right). \tag{8.1.26}$$

Now we construct $\mathbf{W}_{if}^{(j)}$ of order $p \times t^2$ as follows:

$$\mathbf{W}_{if}^{(j)} = \mathbf{h}_f^{(p)} \otimes \left(\mathbf{d}_1^{*ij\prime}, \, \mathbf{d}_2^{*ij\prime}, \ldots, \mathbf{d}_t^{*ij\prime}\right)'; \ \ i = 1, 2, \ldots, \lambda_1 t - 1, \ f = 1, 2, \ldots, p - 1. \tag{8.1.27}$$

Finally $\mathbf{W}^{(l)}$ matrix of order $p \times \lambda_1 t^2$ is given by:

$$\mathbf{W}^{(l)} = (\mathbf{W}_{if}^{(1)}, \ldots, \mathbf{W}_{if}^{(j)}, \ldots, \mathbf{W}_{if}^{(\lambda_1)}),$$
$$i = 1, 2, \ldots, \lambda_1 t - 1, \ f = 1, 2, \ldots, p - 1, \ l = (i - 1)(p - 1) + f.$$

It can be easily checked that these $\mathbf{W}^{(l)}$'s are the required optimum $\mathbf{W}$-matrices for d^{**} in $\Omega_{t, \lambda_1 t^2, p}$ and $c^* = (\lambda_1 t - 1)(p - 1)$ in this given context. $\square$

Remark 8.1.8 If for p even, $\mathbf{H}_p$ does not exist, then $\mathbf{a} = \left(\mathbf{1}_{\frac{p}{2}}', -\mathbf{1}_{\frac{p}{2}}'\right)'$ can be used instead of $\mathbf{h}_f^{(p)}$ in the above theorem.

Strongly Balanced Crossover Design Set-Up in $\Omega_{t, \lambda_1 t, \lambda_2 t + 1}$

It has been shown in Stufken (1996) that a strongly balanced crossover design that is uniform on the periods and uniform on the units in the first $p - 1$ periods is universally optimal for the estimation of direct treatment effects as well as residual treatment effects in $\Omega_{t, n, p}$. We now take up the construction of OCD for such design whenever t is odd and λ_1 is even, as otherwise an OCD fails to exist.

Whenever t is odd, a uniform balanced design d_0^* exists in $\Omega_{t, 2t, t}$, which is obtained by juxtaposing two special latin squares of order t side by side (cf. Bose and Dey 2009; Williams 1949). A strongly balanced design $\tilde{d}^{**}$ obtained by repeating the last period of d_0^* is uniform on the periods and uniform on the units in the first t periods (cf. Cheng and Wu 1980). Now for some positive integer λ, taking λ copies of this design let a strongly balanced design $\tilde{d}^*$ in $\Omega_{t, 2\lambda t, t+1}$ be constructed as

$$\tilde{d}^* = \mathbf{1}_\lambda' \otimes \tilde{d}^{**} \tag{8.1.28}$$

Theorem 8.1.6 *Suppose $\mathbf{H}_{2\lambda}$ exists. Let $\tilde{d}^*$ be defined as in (8.1.28). Then there exists a set of $2\lambda - 1$ optimum $\mathbf{W}$-matrices for $\tilde{d}^*$.*

Proof Assuming $\mathbf{H}_{2\lambda}$ in the *seminormal* form, the optimum $\mathbf{W}^{(l)}$-matrix for $\tilde{d}^*$ in $\Omega_{t, 2\lambda t, t+1}$ can be constructed as:

$$\mathbf{W}^{(l)} = \mathbf{a}^* \otimes \mathbf{h}_l^{(2\lambda)} \otimes \mathbf{1}_t', \ l = 1, 2, \ldots, 2\lambda - 1,$$

where $\mathbf{a}^* = \left(\mathbf{1}_{\frac{t+1}{2}}', -\mathbf{1}_{\frac{t+1}{2}}'\right)'$. $\square$

It has been shown in Stufken (1996) that the above idea of (Cheng and Wu 1980) to construct a strongly balanced design from a uniform balanced design can be extended to cover $p = \lambda_2 t + 1$. The required uniform balanced design d_0^* in $\Omega_{t, \lambda_1 t, \lambda_2 t}$ is a $\lambda_2 \times \lambda_1$ array of special latin square of order t. We refer to Stufken (1996) and Bose and Dey (2009) for the details of the construction. Now repeating the last period of this uniformly balanced design, we get a strongly balanced design $\tilde{d}^*$ in $\Omega_{t, \lambda_1 t, \lambda_2 t + 1}$ which is uniform on the periods and uniform on the units in the first $p - 1$ periods. The following theorem deals with the construction of OCD for this $\tilde{d}^*$.

Corollary 8.1.2 *Suppose* $\mathbf{H}_{\lambda_2 t + 1}$ *and* $\mathbf{H}_{\lambda_1}$ *exist. Then there exists a set of* $\lambda_2 t (\lambda_1 - 1)$ *optimum* $\mathbf{W}$*-matrices for a strongly balanced* $\tilde{d}^*$ *in* $\Omega_{t, \lambda_1 t, \lambda_2 t + 1}$.

Proof It is readily verified that assuming $\mathbf{H}_p$ and $\mathbf{H}_{\lambda_1}$ in the *seminormal* form,

$$\mathbf{W}^{(l)} = \mathbf{W}_{ij} = \mathbf{h}_i^{(\lambda_2 t + 1)} \otimes \mathbf{h}_j^{(\lambda_1)\prime} \otimes \mathbf{1}_t',$$

$$i = 1, 2, \ldots, \lambda_2 t, \ j = 1, 2, \ldots, \lambda_1 - 1, \ l = (\lambda_1 - 1)(i - 1) + j \quad (8.1.29)$$

are the required optimum $\mathbf{W}$-matrices. $\square$

Balanced Crossover Design Set-Up

In this section we consider the construction of OCD for Williams square (1949) and Patterson (1952) designs as the basic designs which are uniform balanced crossover design with appropriate parameters.

It is known that for all even values of t, a uniform balanced design d_0^* in $\Omega_{t,t,t}$ exists which is a balanced latin square and is referred to as a Williams Square in the literature. There does not exist any optimum $\mathbf{W}$-matrix for d_0^* in $\Omega_{t,t,t}$ as $t - 1$ being odd, Condition C_5 is not attainable. Let for some positive integer λ, a uniform balanced crossover design be constructed as

$$d_0^{**} = \mathbf{1}_\lambda' \otimes d_0^*. \quad (8.1.30)$$

We next deal with the construction of optimum $\mathbf{W}$-matrices for d_0^{**} in $\Omega_{t, \lambda t, t}$.

Theorem 8.1.7 *Suppose* $\mathbf{H}_t$ *and* $\mathbf{H}_\lambda$ *exist. Then there exist* $(t - 1)^2 (\lambda - 1)$ *optimum* $\mathbf{W}$*-matrices for* d_0^{**} *in* $\Omega_{t, \lambda t, t}$ *as defined in (8.1.30).*

Proof Assuming $\mathbf{H}_t$ and $\mathbf{H}_\lambda$ in the *seminormal* form

$$\mathbf{W}^{(l)} = \mathbf{W}_{ijf} = \mathbf{h}_f^{(\lambda)\prime} \otimes \mathbf{h}_i^{(t)} \otimes \mathbf{h}_j^{(t)\prime}; \ \ i, j = 1, 2, \ldots, t - 1, \ f = 1, 2, \ldots, \lambda - 1,$$

$$l = (i - 1)(\lambda - 1)(t - 1) + (j - 1)(\lambda - 1) + f \quad (8.1.31)$$

are the required optimum $\mathbf{W}$-matrices for d_0^{**} in $\Omega_{t, \lambda t, t}$. $\square$

Remark 8.1.9 If $\mathbf{H}_t$ does not exist but $\mathbf{H}_\lambda$ exists then a set of $\lambda - 1$ optimum $\mathbf{W}$-matrices for d_0^* can be constructed as

$$\mathbf{W}_l^* = \mathbf{h}_l^{(\lambda)'} \otimes \mathbf{a}^* \otimes \mathbf{a}^{*'}, \ l = 1, 2, \ldots, \lambda$$

where $\mathbf{a}^* = \left(\mathbf{1}_{t/2}', -\mathbf{1}_{t/2}'\right)'$.

Remark 8.1.10 An OCD for a uniform balanced crossover design in $\Omega_{t,t,t}$ or $\Omega_{t,2t,t}$ cannot be constructed for t odd.

A popular choice of balanced crossover design is the one given by Patterson (1952) for $p \leq t$, as this often involves a moderate number of subjects while keeping the number of periods small. For t a prime or prime power, consider $\{L_i\}, i = 1, 2, \ldots, t - 1$, a complete set of MOLS of order t where L_{i+1} can be obtained by cyclically permuting the last $t - 1$ rows of L_i. Then the $t \times t(t-1)$ array P given by

$$P = (L_1, L_2, \ldots, L_{t-1}). \tag{8.1.32}$$

yields a Patterson design in $\Omega_{t,t(t-1),t}$. Now, on deleting any $t - p$ rows of P one gets a Patterson design in $\Omega_{t,t(t-1),p}$ with $p < t$ (cf. Bose and Dey 2009; Patterson 1952). The construction of optimum $\mathbf{W}$-matrices for a Patterson design in $\Omega_{t,t(t-1),p}$ is very much dependent on the existence of the optimum $\mathbf{W}$-matrices for a randomized block design (RBD) (cf. Chap. 3).

Now we consider the following theorem which gives the optimum $\mathbf{W}$-matrices for a Patterson design.

Theorem 8.1.8 *If there exists a set of c* $\mathbf{W}$*-matrices of order* $p \times (t - 1)$ *for an RBD* $(p, t - 1)$, *then there exists a set of c optimum* $\mathbf{W}$*-matrices for a Patterson design in* $\Omega_{t,t(t-1),p}$.

Proof The optimum $\mathbf{W}$-matrices for the Patterson design in $\Omega_{t,t(t-1),p}$ can be obtained by replacing 1 by $\mathbf{1}_t'$ and -1 by $-\mathbf{1}_t'$ in the $\mathbf{W}$-matrices of RBD $(p, (t - 1))$. □

For t prime of the form $4u + 3$, where u is a positive integer, a Patterson design exists in $\Omega_{t,2t,(t+1)/2}$ which is formed by juxtaposing two RBDs$((t + 1)/2, t)$ side by side. For details of the method of construction we refer to Patterson (1952).

Theorem 8.1.9 *Suppose* $\mathbf{H}_{(t+1)/2}$ *exists. Then there exists a set of* $(t - 1)/2$ *optimum* $\mathbf{W}$*-matrices for a Patterson design in* $\Omega_{t,2t,(t+1)/2}$.

Proof Assuming $\mathbf{H}_{(t+1)/2}$ in the *seminormal* form,

$$\mathbf{W}^{(l)} = \mathbf{h}_l^{((t+1)/2)} \otimes (1, \ -1) \otimes \mathbf{1}_t'; \ l = 1, 2, \ldots, (t - 1)/2 \tag{8.1.33}$$

□

8.2 OCDs in Multi-factor Set-Ups

Rao et al. (2003) proposed optimum covariate designs (OCD) through mixed orthogonal arrays for set-ups involving at most two factors where the effects for the qualitative factors and those of the quantitative controllable covariates were orthogonally estimable. In essence, completely randomised designs and randomized block designs were studied in Rao et al. (2003). Dutta and Das (2013) extended these results and proposed OCDs for the m-factor set-ups where the factorial effects involving at most t ($\leq m$) factors and those of the covariates are orthogonally estimable. It is seen that for such model specifications, optimum designs can be obtained through extended mixed orthogonal arrays (EMOA, Dutta et al. 2009) which reduce to mixed orthogonal arrays for the particular set-ups of Rao et al. (2003).

In this section, we will introduce extended mixed orthogonal arrays and cite some applications. In the process, we will deal with the following simple illustrative examples:

(i) RBD with $b = v = 4$ and two observations per cell; (ii) LSD of order 4; (iii) Graeco LSD of order 4; (iv) LSD with 2 observations per cell; (v) LSD of order 6.

8.2.1 Model and Optimality Conditions

Let $F_1, F_2, \ldots, F_m$ be m factors with $s_1, s_2, \ldots, s_m$ levels, respectively ($s_i \geq 2$, $1 \leq i \leq m$), and $Z^{(1)}, Z^{(2)}, \ldots, Z^{(c)}$ denote c covariates. Also, let n combinations be chosen from all possible $v = \prod_{\alpha=1}^{m} s_\alpha$ level combinations and Ω denote the set of n chosen level combinations. For a level combination $(j_1, j_2, \ldots, j_m)$ of Ω, let $(y_{j_1 j_2 \ldots j_m}, z^{(1)}_{j_1 j_2 \ldots j_m}, z^{(2)}_{j_1 j_2 \ldots j_m}, \ldots, z^{(c)}_{j_1 j_2 \ldots j_m})$ denote the vector of observation and the values assumed by the covariates. As mentioned earlier, we assume the location-scale transformed version of the covariate values, viz $|z^{(l)}_{j_1 j_2 \ldots j_m}| \leq 1$ for all $(j_1, j_2, \ldots, j_m) \in \Omega$ and $l = 1, 2, \ldots, c$. We also assume that the level combinations in Ω are so chosen that the interactions, involving at most t factors ($1 \leq t \leq m$), are orthogonally estimable and all the effects involving $(t + 1)$ and higher order interactions are negligible and contribute to the error.

The reader may note that we are now in the framework of a factorial design of a very general nature. In the above we are referring to asymmetric factorial design. The definition of main effects and interaction effects are very standard and excellent expository article of Bose (1947) provides all the basic results in this direction (also see Gupta and Mukerjee 1989 and Kshirsagar 1983).

The following linear model is assumed (cf. Kshirsagar 1983)

$$y_{j_1 j_2 \ldots j_m} = \mu + \sum_{1 \leq i_1 \leq m} \theta^{i_1}_{j_{i_1}} + \sum_{1 \leq i_1 < i_2 \leq m} \theta^{i_1 i_2}_{j_{i_1} j_{i_2}} + \cdots + \sum_{1 \leq i_1 < i_2 < \ldots < i_t \leq m} \theta^{i_1 i_2 \ldots i_t}_{j_{i_1} j_{i_2} \ldots j_{i_t}}$$

$$+ \sum_{l=1}^{c} \gamma_l z_{j_1 j_2 \ldots j_m}^{(l)} + e_{j_1 j_2 \ldots j_m}, \tag{8.2.1}$$

where, $(j_1, j_2, \ldots, j_m) \in \Omega$, $\theta_{j_{i_1}}^{i_1}$ is the effect due to j_{i_1}th level of F_{i_1}, $\theta_{j_{i_1} j_{i_2}}^{i_1 i_2}$ is the interaction between j_{i_1}th level of F_{i_1} and j_{i_2}th level of F_{i_2} and so on, $1 \leq t \leq m$. Further, γ_l is the regression coefficient for the lth concomitant variable $Z^{(l)}$, $l = 1, 2, \ldots, c$. The restrictions on the factorial effects are the following:

$$\sum_{j_{i_\alpha}=1}^{s_{i_\alpha}} \theta_{j_{i_1} j_{i_2} \ldots j_{i_u}}^{i_1 i_2 \ldots i_u} = 0 \ \ \forall j_{i_1}, \ j_{i_2}, \ldots, j_{i_u} (\neq j_{i_\alpha}), \ 1 \leq i_1 < i_2 < \ldots < i_u \leq m, \ 1 \leq u \leq t.$$

In matrix notations, the model (8.2.1) can be rewritten as

$$(\mathbf{Y}, \ \mathbf{X}\theta + \mathbf{Z}\gamma, \ \sigma^2 \mathbf{I}_n), \tag{8.2.2}$$

where $\mathbf{X}$ and $\mathbf{Z}$ are suitably defined.

Therefore, for the model (8.2.2), the condition (3.1.3) for estimating the γ-components orthogonally to the ANOVA effects reduces to

$$\mathbf{Z}'\mathbf{X} = \mathbf{0}. \tag{8.2.3}$$

Further, from (3.1.3) and (3.1.4) it follows that the most efficient estimation of γ-components independently of the ANOVA effects is possible whenever, in addition to (8.2.3), we can also ascertain

$$\mathbf{Z}'\mathbf{Z} = n\mathbf{I}_c. \tag{8.2.4}$$

The condition (8.2.4) implies that $z_{j_1 j_2 \ldots j_m}^{(l)} = \pm 1 \ \forall l$ and $\displaystyle\sum_{(j_1, j_2, \ldots, j_m) \in \Omega} z_{j_1 j_2 \ldots j_m}^{(l)} z_{j_1 j_2 \ldots j_m}^{(l')}$
$= n \delta_{ll'} \ \forall \ 1 \leq l \neq l' \leq c$, where $\delta_{ll'} = 1 (0)$ when $l = l' \ (l \neq l')$.

Let us consider a fixed set of t factors viz. $F_1, F_2, \ldots, F_t$. Also, let $\mathbf{X}^{12 \ldots t}$ denote the coefficient matrix of order $n \times (\prod_{i=1}^{t} s_i)$, corresponding to the factorial effects of the factors $F_1, F_2, \ldots, F_t$. It is not difficult to verify that $\mathbf{X}^{12 \ldots t}$ is a $(0, 1)$-matrix and the condition $\mathbf{X}^{12 \ldots t\prime} \mathbf{Z} = \mathbf{0}$ implies that $\mathbf{Z}$ is also orthogonal to any design matrix corresponding to any sub-set of the factors $F_1, F_2, \ldots, F_t$. Again, it must be noted that conditions such as above need to be satisfied for any choice of t factors out of m factors.

Thus we get the following theorem.

Theorem 8.2.1 *With respect to the linear model (8.2.2) for an m-factor set-up, the following conditions:*

(i) $z_{j_1 j_2 \ldots j_m}^{(l)} = \pm 1 \ \forall \ (j_1, j_2, \ldots, j_m) \in \Omega, \ l = 1, 2, \ldots, c;$

(ii) $\sum'' z_{j_1 j_2 \ldots j_m}^{(l)} = 0;$ *the summation* $\sum''$ *is taken over all those level combinations in* Ω *which contain any given level combination for the t factors* $F_{i_1}, F_{i_2}, \ldots, F_{i_t}, \ 1 \le i_1 < i_2 < \ldots < i_t \le m;$

(iii) $\displaystyle\sum_{(j_1, j_2, \ldots, j_m) \in \Omega} z_{j_1 j_2 \ldots j_m}^{(l)} z_{j_1 j_2 \ldots j_m}^{(l')} = n \delta_{ll'} \ \forall \ l, \ l' = 1, 2, \ldots, c,$ *where* $\delta_{ll'} = 1(0)$ *when* $l = l'(l \ne l'),$

are necessary and sufficient for the optimal estimation of each of the covariate effects γ_l*'s, with the minimum variance Var* $(\widehat{\gamma_l}) = \frac{\sigma^2}{n} \ \forall \ l = 1, 2, \ldots, c.$

From data analysis point view, to attain simplicity and optimality, it is desirable that a fractional factorial design should be such that all t and less factor effects would be orthogonally estimable with balance. This requires that the fraction denoted by **A**, should be an MOA $(n, \ s_1 \times s_2 \times \cdots \times s_t, \ u), u = \min\{2t, \ m\}$ (cf. Dutta et al. 2009). To construct an OCD on this set-up, we should search for **z** vectors with elements ± 1 such that condition (8.2.3) for orthogonality to the design matrix is satisfied. This, in effect, implies that the elements, viz. ± 1 of any **z** vector should occur orthogonally to any choice of t rows of **A**, i.e. all the level combinations for the choice of any t rows from **A** and any one row of **Z** should occur an equal number of times.

It thus transpires that a systematic study of OA, MOA and EMOA introduced below, can be profitably utilized for construction of OCDs in factorial design contexts.

8.2.2 Extended Mixed Orthogonal Array (EMOA) and Construction of OCDs

We describe a new type of array, called extended mixed orthogonal arrays (EMOA) introduced in Dutta et al. (2009) in connection with OCDs in the set-ups of split- and strip-plot designs. The definition of EMOA is as follows.

Definition 8.2.1 Let us consider a $k \times n$ array where the k rows corresponding to the k factors be divided into p sets $S_1, S_2, \ldots, S_p$. The ith set S_i contains $k_i \ (\ge 2)$ factors $F_{i1}, F_{i2}, \ldots, F_{ik_i}$, with $\displaystyle\sum_{i=1}^{p} k_i = k$, where F_{ij} has $s_{ij} \ (\ge 2)$ levels. The array is said to be an extended mixed orthogonal array (EMOA) if

(i) for the choice of any $d_i \ (\ge 2)$ factors from S_i, all possible level combinations of these d_i factors occur equally often (say λ; λ may depend on the selected factors), $i = 1, 2, \ldots, p;$

(ii) for the choice of any d sets $(d \ge 2)$, say $S_{i_1}, S_{i_2}, \ldots, S_{i_d}$, the level combinations arising out of any t_{i_1} factors from S_{i_1}, any t_{i_2} factors from $S_{i_2}, \ldots,$ any t_{i_d} factors from S_{i_d}, where $1 \le t_{i_j} \le d_{i_j}, 1 \le i_1 < i_2 < \cdots < i_d \le p$, occur equally often (say μ times; μ may depend on the selected factors).

Such an array is denoted by EMOA $[n, \ k, \ \prod_{i=1}^{p}\prod_{j=1}^{k_i} s_{ij}, (d_1, \ d_2, \ldots, d_p),$ $(d; \ t_1, \ t_2, \ldots, t_p)]$. The frequency parameters λ and μ can be obtained from the parameters already included in the notation above. An EMOA $[n, \ k, \ \prod_{i=1}^{p}\prod_{j=1}^{k_i} s_{ij}, (d_1, \ d_2, \ldots, d_p), \ (d; \ t_1, \ t_2, \ldots, t_p)]$ is also an EMOA $[n, \ k,$ $\prod_{i=1}^{p}\prod_{j=1}^{k_i} s_{ij}, \ (d_1', \ d_2', \ldots, d_p'), \ (d'; \ t_1', \ t_2', \ldots, t_p')]$, where $d_i' < d_i, t_i' < t_i, \forall i$ and $d' < d$. It is to be noted that a compound orthogonal array (cf. Hedayat et al. 1999, p. 230) and the array proposed by Chakravarti (1956) can also be seen as EMOAs with particular parameters.

Remark 8.2.1 If $d_i = t$ and u_i be a non-negative integer such that $1 \leq u_i \leq t_i$, $1 \leq i \leq p$ satisfying $\sum_{j=1}^{q} u_{i_j} = t$, where $2 \leq q \leq d$ and $1 \leq i_1 < i_2 < \ldots < i_q \leq p$, then the EMOA $[n, \ k,$ $\prod_{i=1}^{p}\prod_{j=1}^{k_i} s_{ij}, \ (d_1, \ d_2, \ldots, d_p), \ (d; \ t_1, \ t_2, \ldots, t_p)]$ is an MOA $(n, \ \prod_{i=1}^{p}\prod_{j=1}^{k_i} s_{ij}, \ t)$. Also it follows that an EMOA can always be looked upon as a MOA of strength 2.

Remark 8.2.2 From the above discussions it follows that the OCD on the factorial set-up under consideration can be displayed in the form of an array; the chosen n level combinations of the m factors form n columns of a $m \times n$ matrix denoted by $\mathbf{A}$ and the z-values of the c covariates form c rows of a $c \times n$ matrix denoted by $\mathbf{B}$. This $(m + c) \times n$ array $\begin{pmatrix} \mathbf{A} \\ \mathbf{B} \end{pmatrix}$ is such that $\mathbf{A}$ forms an MOA$(n, \ s_1 \times s_2 \times \cdots \times s_m, \ u)$, $u = \min\{2t, \ m\}$, with elements in the ith row as the levels of $F_i, i = 1, 2, \ldots, m$ and $\mathbf{B}$ forms an OA $(n, \ c, \ 2, \ 2)$ with elements $+1$ or -1 in each row. From the discussion after Theorem 8.2.1, it follows that all the level combinations for the choice of any t rows from $\mathbf{A}$ and any one row from $\mathbf{B}$ occur an equal number of times. This array is actually an EMOA $[n, \ m + c, \ s_1 \times s_2 \times \cdots \times s_m \times 2^c, \ (u, \ 2), \ (2; \ t, \ 1)]$.

Thus we get the following theorem.

Theorem 8.2.2 *The existence of an EMOA* $[n, \ m + c, \ s_1 \times s_2 \times \cdots \times s_m \times 2^c, \ (u, \ 2), \ (2; \ t, \ 1)]$ *implies the existence of an OCD in a multi-factor set-up where all the main effects and interactions up to t-factors are orthogonally estimable, $u = \min\{2t, m\}$.*

Below we cite an example of an EMOA for clear understanding of the concepts and definitions.

Example 8.2.1 Let us consider the following orthogonal array **D**, with parameters
$(16, 5, 4, 2)$

$$\begin{pmatrix} 1 & 1 & 1 & 1 & 2 & 2 & 2 & 2 & 3 & 3 & 3 & 3 & 4 & 4 & 4 & 4 \\ 1 & 2 & 3 & 4 & 1 & 2 & 3 & 4 & 1 & 2 & 3 & 4 & 1 & 2 & 3 & 4 \\ 1 & 2 & 3 & 4 & 2 & 1 & 4 & 3 & 3 & 4 & 1 & 2 & 4 & 3 & 2 & 1 \\ 1 & 2 & 3 & 4 & 3 & 4 & 1 & 2 & 4 & 3 & 2 & 1 & 2 & 1 & 4 & 3 \\ 1 & 2 & 3 & 4 & 4 & 3 & 2 & 1 & 2 & 1 & 4 & 3 & 3 & 4 & 1 & 2 \end{pmatrix} \qquad (8.2.5)$$

and another orthogonal array **B** with parameters $(4, 3, 2, 2)$

$$\mathbf{B} = \begin{pmatrix} 1 & -1 & 1 & -1 \\ 1 & -1 & -1 & 1 \\ 1 & 1 & -1 & -1 \end{pmatrix}. \qquad (8.2.6)$$

Replacing the level j in the first row of **D** by the jth column of **B**, $j = 1, 2, 3$ and 4, we construct an array $\mathbf{B}_1$ of order 3×16. Now let $\mathbf{D}_1$ be the 4×16 array obtained from **D** after ignoring the first row. Then the 7×32 array **C** obtained as

$$\mathbf{C} = \begin{pmatrix} \mathbf{B}_1 & \overline{\mathbf{B}}_1 \\ \mathbf{D}_1 & \mathbf{D}_1 \end{pmatrix}$$

is an EMOA $[32, 7, 2^3 \times 4^4, (3, 2), (2; 3, 1)]$, where $\overline{\mathbf{B}}_1$ is the array obtained from $\mathbf{B}_1$ by interchanging -1 and 1. Let, S_1 denote the set of three rows corresponding to the B's and S_2 denote the set of four rows corresponding to $\mathbf{D}_1$'s. Then see that each level combination arising out of three rows of S_1 occurs four times, while any level combination arising out of any two rows of S_2 occurs twice. So $\lambda_{123} = 4$ while $\lambda_{i_1 i_2} = 2, 4 \leq i_1, i_2 \leq 7$. Again, for the choice of the three rows of S_1 and any row from S_2 all possible level combinations occur just once. So $\mu_{123 i_1} = 1, 4 \leq i_1 \leq 7$.

Some modified versions of this array have been used in the construction of OCDs in the examples considered below.

8.2.3 Examples of OCDs

We undertake several examples for construction of OCDs in simple experimental set-ups. Subsequently, in the sections to follow, we develop general results.

Example 8.2.2 RBD with $b = v = 4$ and 2 observations per cell.

Consider $\mathbf{H}_{16}$ and denote the columns of $\mathbf{H}_{16}$ by the vectors $\mathbf{z}^{(1)}, \mathbf{z}^{(2)}, \dots, \mathbf{z}^{(16)}$.

Let $\mathbf{y}^{(1)}$ denote the 16×1 observation vector arising out of the RBD involving the first observation in each cell. Similarly, we also have $\mathbf{y}^{(2)}$ available as the second observation vector across the 16 cells. It does not matter if the observations are laid down row-wise or column-wise.

Let $\mathbf{z}^{(1)}$, $\mathbf{z}^{(2)}, \ldots, \mathbf{z}^{(16)}$ be associated with $\mathbf{y}^{(1)}$ vector and let their 'negations' occupy the respective positions in $\mathbf{y}^{(2)}$.

It is noted that $\mathbf{U} =$ vector of average of the two observations in each cell has, for its expectation, exclusively the terms involving the general mean μ, the block effect parameter(s) and the treatment parameter(s) and these are free from the covariate parameter(s) represented by the $\mathbf{z}$-components. On the other hand, $\mathbf{V} =$ vector of differences [divided by 2] has, for its expectation, terms involving only the covariate parameters and these are free from the 'design parameters'. These covariate parameters have associated with them the corresponding $\mathbf{z}$ vectors. Since the $\mathbf{z}$ vectors are mutually orthogonal with elements (± 1), we are in a position to optimally accommodate 16 covariates.

Remark 8.2.3 This approach is definitely very transparent and one can see how orthogonalization of the two observations within each cell has resulted into separation of the two sets of parameters: design parameters and covariates parameter.

Remark 8.2.4 It is not clear if this approach might lead to the possibility of including any more covariates optimally. Towards an affirmative answer for this we take a look at the RBD with $v = b = 4$ with one observation per cell. From Das et al. (2003) and Rao et al. (2003) it is known that for this set-up with single observation per cell there are nine $\mathbf{z}$ vectors that can be accommodated optimally. Denote these vectors of order 16×1 by $\mathbf{z}_1^*, \mathbf{z}_2^*, \ldots, \mathbf{z}_9^*$. It is now enough to repeat these $\mathbf{z}^*$ at both the positions in each cell. These provide additional 9 covariates, besides the 16 outlined above, thereby giving a total of 25 covariates, the maximum number that can be achieved with 32 observations and $v = b = 4$.

Below we demonstrate an equivalent but unified method of arriving at the same result by means of EMOA to ascertain the existence of an OCD with 25 covariates optimally included.

Consider the array $\mathbf{D}$ defined in (8.2.5). By replacing the levels 1, 2, 3, 4 in the kth row of $\mathbf{D}$ by the elements of the jth row of $\mathbf{B}$ defined in (8.2.6) successively, $j = 1, 2, 3$, we can construct an array $\mathbf{C}_k$ of order $3 \times 16, k = 1, 2, 3, 4, 5$. Now let $\mathbf{A}_1$ be the 2×16 array obtained from $\mathbf{D}$ after ignoring the last three rows. $\mathbf{A}_1$ provides the set-up for the RBD. Then the 27×32 array $\mathbf{E}$ is obtained as

$$\mathbf{E} = \begin{pmatrix} \mathbf{A}_1 & \mathbf{A}_1 \\ \mathbf{1}'_{16} & -\mathbf{1}'_{16} \\ \mathbf{C}_1 & -\mathbf{C}_1 \\ \mathbf{C}_2 & -\mathbf{C}_2 \\ \mathbf{C}_3 & -\mathbf{C}_3 \\ \mathbf{C}_4 & -\mathbf{C}_4 \\ \mathbf{C}_5 & -\mathbf{C}_5 \\ \mathbf{C}_3 & \mathbf{C}_3 \\ \mathbf{C}_4 & \mathbf{C}_4 \\ \mathbf{C}_5 & \mathbf{C}_5 \end{pmatrix} = (\mathbf{E}^{R_1}, \mathbf{E}^{R_2}).$$

It is readily verified that **E** is an EMOA [32, 27, $4^2 \times 2^{25}$, (2, 2), (2; 1, 1)] which provides 25 optimal covariates. The matrix **E** is displayed below in the partitioned form with the observations separately shown in two cells.

Cell position 1

$$
\mathbf{E}^{R_1} =
$$

```
1 1 1 1 2 2 2 2 3 3 3 3 4 4 4 4
1 2 3 4 1 2 3 4 1 2 3 4 1 2 3 4
───────────────────────────────────────
 1  1  1  1  1  1  1  1  1  1  1  1  1  1  1  1
 1  1  1  1 -1 -1 -1 -1  1  1  1  1 -1 -1 -1 -1
 1  1  1  1 -1 -1 -1 -1 -1 -1 -1 -1  1  1  1  1
 1  1  1  1  1  1  1  1 -1 -1 -1 -1 -1 -1 -1 -1
 1 -1  1 -1  1 -1  1 -1  1 -1  1 -1  1 -1  1 -1
 1 -1 -1  1  1 -1 -1  1  1 -1 -1  1  1 -1 -1  1
 1  1 -1 -1  1  1 -1 -1  1  1 -1 -1  1  1 -1 -1
 1 -1  1 -1 -1  1 -1  1  1 -1  1 -1 -1  1 -1  1
 1 -1 -1  1 -1  1  1 -1  1  1 -1  1 -1  1 -1  1
 1  1 -1 -1  1  1 -1 -1 -1 -1  1  1 -1 -1  1  1
 1 -1  1 -1  1 -1  1 -1 -1  1 -1  1 -1  1 -1  1
 1 -1 -1  1 -1  1  1 -1  1 -1 -1  1  1 -1  1 -1
 1  1 -1 -1 -1 -1  1  1 -1 -1  1  1  1  1 -1 -1
 1 -1  1 -1 -1  1 -1  1  1 -1  1 -1  1 -1  1 -1
 1 -1 -1  1  1 -1 -1  1 -1  1  1 -1  1  1 -1 -1
 1  1 -1 -1 -1 -1  1  1  1  1 -1 -1 -1 -1  1  1
 1 -1  1 -1 -1  1 -1  1 -1  1 -1  1  1 -1  1 -1
 1 -1 -1  1  1 -1 -1  1  1 -1  1 -1 -1  1  1 -1
 1  1 -1 -1 -1 -1  1  1  1  1 -1 -1 -1 -1  1  1
 1 -1  1 -1 -1  1 -1  1 -1  1 -1  1 -1  1 -1  1
 1 -1 -1  1 -1  1  1 -1 -1  1  1 -1  1 -1  1 -1
 1  1 -1 -1 -1 -1  1  1 -1 -1  1  1  1  1 -1 -1
 1 -1  1 -1 -1 -1  1  1 -1 -1  1  1  1  1 -1 -1
 1 -1  1 -1 -1 -1  1 -1  1 -1  1 -1  1  1 -1  1
 1 -1 -1  1  1 -1 -1  1 -1  1  1 -1 -1  1  1 -1
 1  1 -1 -1 -1 -1  1  1  1  1 -1 -1 -1 -1  1  1
```

Cell position 2

$$
\mathbf{E}^{R_2} =
\begin{array}{cccccccccccccccc}
1 & 1 & 1 & 1 & 2 & 2 & 2 & 2 & 3 & 3 & 3 & 3 & 4 & 4 & 4 & 4 \\
1 & 2 & 3 & 4 & 1 & 2 & 3 & 4 & 1 & 2 & 3 & 4 & 1 & 2 & 3 & 4 \\
\hline
-1 & -1 & -1 & -1 & -1 & -1 & -1 & -1 & -1 & -1 & -1 & -1 & -1 & -1 & -1 & -1 \\
-1 & -1 & -1 & -1 & 1 & 1 & 1 & 1 & -1 & -1 & -1 & -1 & 1 & 1 & 1 & 1 \\
-1 & -1 & -1 & -1 & 1 & 1 & 1 & 1 & 1 & 1 & 1 & 1 & -1 & -1 & -1 & -1 \\
-1 & -1 & -1 & -1 & -1 & -1 & -1 & -1 & 1 & 1 & 1 & 1 & 1 & 1 & 1 & 1 \\
-1 & 1 & -1 & 1 & -1 & 1 & -1 & 1 & -1 & 1 & -1 & 1 & -1 & 1 & -1 & 1 \\
-1 & 1 & 1 & -1 & -1 & 1 & 1 & -1 & -1 & 1 & 1 & -1 & -1 & 1 & 1 & -1 \\
-1 & -1 & 1 & 1 & -1 & -1 & 1 & 1 & -1 & -1 & 1 & 1 & -1 & -1 & 1 & 1 \\
-1 & 1 & -1 & 1 & 1 & -1 & 1 & -1 & 1 & -1 & 1 & -1 & 1 & -1 & 1 & -1 \\
-1 & 1 & 1 & -1 & 1 & -1 & -1 & 1 & 1 & -1 & -1 & 1 & 1 & -1 & -1 & 1 \\
-1 & -1 & 1 & 1 & -1 & -1 & 1 & 1 & 1 & 1 & -1 & -1 & 1 & 1 & -1 & -1 \\
-1 & 1 & -1 & 1 & 1 & -1 & 1 & -1 & 1 & -1 & 1 & -1 & 1 & -1 & 1 & -1 \\
-1 & 1 & 1 & -1 & 1 & -1 & 1 & -1 & 1 & -1 & 1 & -1 & 1 & -1 & 1 & -1 \\
-1 & -1 & 1 & 1 & 1 & -1 & -1 & 1 & -1 & -1 & -1 & -1 & 1 & 1 & 1 & 1 \\
-1 & 1 & -1 & 1 & 1 & -1 & 1 & -1 & 1 & -1 & 1 & -1 & 1 & -1 & 1 & -1 \\
-1 & 1 & 1 & -1 & -1 & 1 & 1 & -1 & 1 & -1 & -1 & 1 & 1 & -1 & -1 & 1 \\
-1 & -1 & 1 & 1 & 1 & -1 & -1 & -1 & -1 & 1 & 1 & 1 & 1 & -1 & -1 & -1 \\
1 & -1 & 1 & -1 & -1 & 1 & -1 & 1 & 1 & -1 & 1 & -1 & 1 & -1 & 1 \\
1 & -1 & -1 & 1 & -1 & 1 & 1 & -1 & -1 & 1 & 1 & -1 & 1 & -1 & -1 & 1 \\
1 & 1 & -1 & -1 & 1 & 1 & -1 & -1 & -1 & -1 & 1 & 1 & -1 & -1 & 1 & 1 \\
1 & -1 & 1 & -1 & 1 & -1 & 1 & -1 & -1 & 1 & -1 & 1 & -1 & 1 & -1 & 1 \\
1 & -1 & -1 & 1 & -1 & 1 & 1 & -1 & 1 & -1 & -1 & 1 & -1 & 1 & 1 & -1 \\
1 & 1 & -1 & -1 & -1 & -1 & 1 & 1 & -1 & -1 & 1 & 1 & 1 & 1 & -1 & -1 \\
1 & -1 & 1 & -1 & -1 & 1 & -1 & 1 & -1 & 1 & -1 & 1 & 1 & -1 & 1 & -1 \\
1 & -1 & -1 & 1 & 1 & -1 & -1 & 1 & -1 & 1 & 1 & -1 & -1 & 1 & 1 & -1 \\
1 & 1 & -1 & -1 & -1 & -1 & 1 & 1 & 1 & 1 & -1 & -1 & -1 & -1 & 1 & 1 \\
\end{array}
$$

Remark 8.2.5 In the structure of the EMOA we can readily identify the two sets of z-vectors arising out of the first Method. In $\mathbf{D}_1$, the rows of the three components $[\mathbf{C}_3, \mathbf{C}_3]$; $[\mathbf{C}_4, \mathbf{C}_4]$; $[\mathbf{C}_5, \mathbf{C}_5]$ represent the nine $\mathbf{z}^*$s vectors of Remark 8.2.4 while the rest are identified as the $\mathbf{z}$ vectors.

Remark 8.2.6 Here we note that

$$
\mathbf{E}_1^{R_1} = \begin{pmatrix} \mathbf{A}_1 \\ \hline \mathbf{C}_3 \\ \mathbf{C}_4 \\ \mathbf{C}_5 \end{pmatrix}
$$

is an EMOA $[16, 11, 4^2 \times 2^9, (2, 2), (2; 1, 1)]$ and thus from Theorem 8.2.2 it follows that this EMOA provides an OCD for RBD set-up with 4 blocks and 4 treatments with single observation per cell, i.e. a standard RBD with $b = v = 4$ (this in agreement with Dutta et al. 2009; Rao et al. 2003). Here we accommodate the maximum possible number of 9 covariates optimally.

Example 8.2.3 LSD of order 4 with provision for formation of six z vectors
Define

$$\mathbf{E}^{\text{LSD}} = \begin{pmatrix} \mathbf{A}_2 \\ \mathbf{C}_4 \\ \mathbf{C}_5 \end{pmatrix}.$$

It is observed that $\mathbf{E}^{\text{LSD}}$ is an EMOA $[16, 9, 4^3 \times 2^6, (2, 2), (2; 1, 1)]$, where $\mathbf{A}_2$ is the 3×16 array obtained from $\mathbf{D}$ after ignoring the last two rows. $\mathbf{A}_2$ gives the set-up the 4×4 LSD. We can easily infer from Theorem 8.2.2 that this EMOA gives an OCD for LSD set-up with 4 rows, 4 columns and 4 treatments. $\mathbf{E}^{\text{LSD}}$ is displayed as follows:

$$
\mathbf{E}^{\text{LSD}} =
\begin{array}{cccccccccccccccc}
1 & 1 & 1 & 1 & 2 & 2 & 2 & 2 & 3 & 3 & 3 & 3 & 4 & 4 & 4 & 4 \\
1 & 2 & 3 & 4 & 1 & 2 & 3 & 4 & 1 & 2 & 3 & 4 & 1 & 2 & 3 & 4 \\
1 & 2 & 3 & 4 & 2 & 1 & 4 & 3 & 3 & 4 & 1 & 2 & 4 & 3 & 2 & 1 \\
\hline
1 & -1 & 1 & -1 & 1 & -1 & 1 & -1 & -1 & 1 & -1 & 1 & -1 & 1 & -1 & 1 \\
1 & -1 & -1 & 1 & -1 & 1 & 1 & -1 & 1 & -1 & -1 & 1 & -1 & 1 & 1 & -1 \\
1 & 1 & -1 & -1 & -1 & -1 & 1 & 1 & -1 & -1 & 1 & 1 & 1 & 1 & -1 & -1 \\
1 & -1 & 1 & -1 & -1 & 1 & -1 & 1 & -1 & 1 & -1 & 1 & 1 & -1 & 1 & -1 \\
1 & -1 & -1 & 1 & 1 & -1 & -1 & 1 & -1 & 1 & 1 & -1 & -1 & 1 & 1 & -1 \\
1 & 1 & -1 & -1 & -1 & -1 & 1 & 1 & 1 & 1 & -1 & -1 & -1 & -1 & 1 & 1 \\
\end{array}
$$

Thus we construct OCD with 6 covariates which is the maximum possible number of covariates.

Example 8.2.4 Graeco LSD of order 4 with provision for formation of three z vectors.

Define

$$\mathbf{E}^{\text{GLSD}} = \begin{pmatrix} \mathbf{A}_3 \\ \mathbf{C}_5 \end{pmatrix}.$$

It is observed that $\mathbf{E}^{GLSD}$ is an EMOA $[16, 7, 4^4 \times 2^3, (2, 2), (2; 1, 1)]$, where $\mathbf{A}_3$ is the 4×16 array providing the set-up for Graeco LSD and is obtained from $\mathbf{D}$ after ignoring the last row. It easily follows from Theorem 8.2.2 that this EMOA is the OCD for Graeco LSD set-up. $\mathbf{E}^{GLSD}$ is displayed as follows:

$$
\mathbf{E}^{GLSD} =
\begin{array}{cccccccccccccccc}
1 & 1 & 1 & 1 & 2 & 2 & 2 & 2 & 3 & 3 & 3 & 3 & 4 & 4 & 4 & 4 \\
1 & 2 & 3 & 4 & 1 & 2 & 3 & 4 & 1 & 2 & 3 & 4 & 1 & 2 & 3 & 4 \\
1 & 2 & 3 & 4 & 2 & 1 & 4 & 3 & 3 & 4 & 1 & 2 & 4 & 3 & 2 & 1 \\
1 & 2 & 3 & 4 & 3 & 4 & 1 & 2 & 4 & 3 & 2 & 1 & 2 & 1 & 4 & 3 \\
\hline
1 & -1 & 1 & -1 & -1 & 1 & -1 & 1 & -1 & 1 & -1 & 1 & 1 & -1 & 1 & -1 \\
1 & -1 & -1 & 1 & 1 & -1 & -1 & 1 & -1 & 1 & 1 & -1 & -1 & 1 & 1 & -1 \\
1 & 1 & -1 & -1 & -1 & -1 & 1 & 1 & 1 & 1 & -1 & -1 & -1 & -1 & 1 & 1 \\
\end{array}
$$

Here we construct OCD with maximum possible number of 3 covariates.

Example 8.2.5 LSD of order 4 with 2 observations per cell.

By mimicking the arguments as in the case of an RBD of Example 8.2.2 with two observations per cell, we can immediately associate 6 covariates in an optimal manner since there are 6 error d.f. in the set-up of a latin square of order 4. These are analogous to the $\mathbf{z}^*$ vectors of Example 8.2.2. The remaining 16 $\mathbf{z}$ components are obtained by referring to $\mathbf{H}_{16}$ in the same way as was done there. The whole analysis can be carried out by referring to EMOA. This is explained below.

Define

$$
\mathbf{F} =
\left(
\begin{array}{c|c}
\mathbf{A}_2 & \mathbf{A}_2 \\
\hline
\mathbf{1}'_{16} & -\mathbf{1}'_{16} \\
\mathbf{C}_1 & -\mathbf{C}_1 \\
\mathbf{C}_2 & -\mathbf{C}_2 \\
\mathbf{C}_3 & -\mathbf{C}_3 \\
\mathbf{C}_4 & -\mathbf{C}_4 \\
\mathbf{C}_5 & -\mathbf{C}_5 \\
\mathbf{C}_4 & \mathbf{C}_4 \\
\mathbf{C}_5 & \mathbf{C}_5 \\
\end{array}
\right)
= (\mathbf{F}^{R_1}, \mathbf{F}^{R_2}),
$$

which is readily verified to be an EMOA $[32, 25, 4^3 \times 2^{22}, (2, 2), (2; 1, 1)]$. The matrix $\mathbf{F}$ is displayed below in the partitioned form with the observations separately shown in two cell positions.

Cell position 1

$$
\mathbf{F}^{R_1} =
\begin{array}{cccccccccccccccc}
1 & 1 & 1 & 1 & 2 & 2 & 2 & 2 & 3 & 3 & 3 & 3 & 4 & 4 & 4 & 4 \\
1 & 2 & 3 & 4 & 1 & 2 & 3 & 4 & 1 & 2 & 3 & 4 & 1 & 2 & 3 & 4 \\
1 & 2 & 3 & 4 & 2 & 1 & 4 & 3 & 3 & 4 & 1 & 2 & 4 & 3 & 2 & 1 \\
\hline
1 & 1 & 1 & 1 & 1 & 1 & 1 & 1 & 1 & 1 & 1 & 1 & 1 & 1 & 1 & 1 \\
1 & 1 & 1 & 1 & -1 & -1 & -1 & -1 & 1 & 1 & 1 & 1 & -1 & -1 & -1 & -1 \\
1 & 1 & 1 & 1 & -1 & -1 & -1 & -1 & -1 & -1 & -1 & -1 & 1 & 1 & 1 & 1 \\
1 & 1 & 1 & 1 & 1 & 1 & 1 & 1 & -1 & -1 & -1 & -1 & -1 & -1 & -1 & -1 \\
1 & -1 & 1 & -1 & 1 & -1 & 1 & -1 & 1 & -1 & 1 & -1 & 1 & -1 & 1 & -1 \\
1 & -1 & -1 & 1 & 1 & -1 & -1 & 1 & 1 & -1 & -1 & 1 & 1 & -1 & -1 & 1 \\
1 & 1 & -1 & -1 & 1 & 1 & -1 & -1 & 1 & 1 & -1 & -1 & 1 & 1 & -1 & -1 \\
1 & -1 & 1 & -1 & -1 & 1 & -1 & 1 & 1 & -1 & 1 & -1 & -1 & 1 & -1 & 1 \\
1 & -1 & -1 & 1 & -1 & 1 & 1 & -1 & 1 & -1 & -1 & 1 & -1 & 1 & 1 & -1 \\
1 & 1 & -1 & -1 & 1 & 1 & -1 & -1 & -1 & -1 & 1 & 1 & -1 & -1 & 1 & 1 \\
1 & -1 & 1 & -1 & 1 & -1 & 1 & -1 & -1 & 1 & -1 & 1 & -1 & 1 & -1 & 1 \\
1 & -1 & -1 & 1 & 1 & -1 & -1 & 1 & -1 & 1 & 1 & -1 & -1 & 1 & 1 & -1 \\
1 & 1 & -1 & -1 & -1 & -1 & 1 & 1 & -1 & -1 & 1 & 1 & 1 & 1 & -1 & -1 \\
1 & -1 & 1 & -1 & -1 & 1 & -1 & 1 & -1 & 1 & -1 & 1 & 1 & -1 & 1 & -1 \\
1 & -1 & -1 & 1 & -1 & 1 & 1 & -1 & -1 & 1 & 1 & -1 & 1 & -1 & -1 & 1 \\
1 & 1 & -1 & -1 & -1 & -1 & 1 & 1 & 1 & 1 & -1 & -1 & -1 & -1 & 1 & 1 \\
1 & -1 & 1 & -1 & 1 & -1 & 1 & -1 & -1 & 1 & -1 & 1 & 1 & -1 & 1 & -1 \\
1 & -1 & -1 & 1 & -1 & 1 & 1 & -1 & 1 & -1 & -1 & 1 & 1 & -1 & -1 & 1 \\
1 & 1 & -1 & -1 & -1 & -1 & 1 & 1 & 1 & 1 & -1 & -1 & 1 & 1 & -1 & -1 \\
1 & -1 & 1 & -1 & -1 & 1 & -1 & 1 & 1 & -1 & 1 & -1 & 1 & -1 & 1 & -1 \\
1 & -1 & -1 & 1 & -1 & 1 & 1 & -1 & 1 & -1 & -1 & 1 & -1 & 1 & 1 & -1 \\
1 & 1 & -1 & -1 & -1 & -1 & 1 & 1 & 1 & 1 & -1 & -1 & -1 & -1 & 1 & 1 \\
\end{array}
$$

Cell position 2

$$
\mathbf{F}^{R_2} =
\begin{array}{cccccccccccccccc}
1 & 1 & 1 & 1 & 2 & 2 & 2 & 2 & 3 & 3 & 3 & 3 & 4 & 4 & 4 & 4 \\
1 & 2 & 3 & 4 & 1 & 2 & 3 & 4 & 1 & 2 & 3 & 4 & 1 & 2 & 3 & 4 \\
1 & 2 & 3 & 4 & 2 & 1 & 4 & 3 & 3 & 4 & 1 & 2 & 4 & 3 & 2 & 1 \\
\hline
-1 & -1 & -1 & -1 & -1 & -1 & -1 & -1 & -1 & -1 & -1 & -1 & -1 & -1 & -1 & -1 \\
-1 & -1 & -1 & -1 & 1 & 1 & 1 & 1 & -1 & -1 & -1 & -1 & 1 & 1 & 1 & 1 \\
-1 & -1 & -1 & -1 & 1 & 1 & 1 & 1 & 1 & 1 & 1 & 1 & -1 & -1 & -1 & -1 \\
-1 & -1 & -1 & -1 & -1 & -1 & -1 & -1 & 1 & 1 & 1 & 1 & 1 & 1 & 1 & 1 \\
-1 & 1 & -1 & 1 & -1 & 1 & -1 & 1 & -1 & 1 & -1 & 1 & -1 & 1 & -1 & 1 \\
-1 & 1 & 1 & -1 & -1 & 1 & 1 & -1 & -1 & 1 & 1 & -1 & -1 & 1 & 1 & -1 \\
-1 & -1 & 1 & 1 & -1 & -1 & 1 & 1 & -1 & -1 & 1 & 1 & -1 & -1 & 1 & 1 \\
-1 & 1 & -1 & 1 & 1 & -1 & 1 & -1 & 1 & -1 & 1 & -1 & 1 & -1 & 1 & -1 \\
-1 & 1 & 1 & -1 & 1 & -1 & -1 & 1 & 1 & -1 & -1 & 1 & 1 & -1 & -1 & 1 \\
-1 & -1 & 1 & 1 & -1 & -1 & 1 & 1 & 1 & 1 & -1 & -1 & 1 & 1 & -1 & -1 \\
-1 & 1 & -1 & 1 & -1 & 1 & 1 & -1 & 1 & -1 & 1 & -1 & 1 & -1 & 1 & -1 \\
-1 & 1 & 1 & -1 & 1 & -1 & -1 & 1 & 1 & -1 & 1 & -1 & -1 & 1 & -1 & 1 \\
-1 & -1 & 1 & 1 & 1 & 1 & -1 & -1 & 1 & 1 & -1 & -1 & -1 & -1 & 1 & 1 \\
-1 & 1 & -1 & 1 & 1 & -1 & 1 & -1 & 1 & -1 & 1 & -1 & 1 & -1 & 1 & -1 \\
-1 & 1 & 1 & -1 & -1 & 1 & 1 & -1 & 1 & -1 & -1 & 1 & 1 & -1 & -1 & 1 \\
-1 & -1 & 1 & 1 & 1 & 1 & -1 & -1 & -1 & -1 & 1 & 1 & 1 & 1 & -1 & -1 \\
1 & -1 & 1 & -1 & 1 & -1 & 1 & -1 & 1 & -1 & 1 & -1 & 1 & -1 & 1 & -1 \\
1 & -1 & -1 & 1 & -1 & 1 & 1 & -1 & 1 & -1 & -1 & 1 & -1 & 1 & 1 & -1 \\
1 & 1 & -1 & -1 & -1 & -1 & 1 & 1 & -1 & -1 & 1 & 1 & 1 & 1 & -1 & -1 \\
1 & -1 & 1 & -1 & -1 & 1 & -1 & 1 & -1 & 1 & -1 & 1 & 1 & -1 & 1 & -1 \\
1 & -1 & -1 & 1 & 1 & -1 & -1 & 1 & 1 & -1 & -1 & 1 & 1 & -1 & -1 & 1 \\
1 & 1 & -1 & -1 & -1 & -1 & 1 & 1 & 1 & 1 & -1 & -1 & -1 & -1 & 1 & 1 \\
\end{array}
$$

Example 8.2.6 LSD of order 6.

It is very difficult to construct OCDs for latin square design when MOLS do not exist. However, using some special structure of latin square it is possible to construct at least one OCD in some case. We consider the following latin square (Sinha 2009, p. 224)

$$
L = \begin{pmatrix}
1 & 2 & 3 & 4 & 5 & 6 \\
2 & 1 & 4 & 3 & 6 & 5 \\
6 & 5 & 1 & 2 & 3 & 4 \\
5 & 6 & 2 & 1 & 4 & 3 \\
4 & 3 & 6 & 5 & 2 & 1 \\
3 & 4 & 5 & 6 & 1 & 2
\end{pmatrix} . \tag{8.2.7}
$$

Using **W**-matrix given in Sinha (2009), we can construct the following EMOA [36, $4, 6 \times 6 \times 6 \times 2^1, (2, 1), (2; 2, 1)$]:

$$
\begin{pmatrix}
1\,1\,1 & 1 & 1 & 1\,2\,2 & 2 & 2 & 2\,2\,3 & 3 & 3 & 3\,3\,3 \\
1\,2\,3 & 4 & 5 & 6\,1\,2 & 3 & 4 & 5\,6\,1 & 2 & 3 & 4\,5\,6 \\
1\,2\,3 & 4 & 5 & 6\,2\,1 & 4 & 3 & 6\,5\,6 & 5 & 1 & 2\,3\,4 \\
1\,1\,1 & -1 & -1 & -1\,1\,1 & -1 & -1 & -1\,1\,1 & -1 & -1 & -1\,1\,1
\end{pmatrix}
$$

$$
\begin{pmatrix}
4 & 4 & 4\,4\,4\,4 & 5 & 5\,5\,5\,5 & 5 & 6\,6\,6\,6 & 6 & 6 \\
1 & 2 & 3\,4\,5\,6 & 1 & 2\,3\,4\,5 & 6 & 1\,2\,3\,4 & 5 & 6 \\
5 & 6 & 2\,1\,4\,3 & 4 & 3\,6\,5\,2 & 1 & 3\,4\,5\,6 & 1 & 2 \\
-1 & -1 & -1\,1\,1\,1 & -1 & -1\,1\,1\,1 & -1 & -1\,1\,1\,1 & -1 & -1
\end{pmatrix},
$$

which gives an OCD with one covariate.

8.2.4 OCDs on Some General Set-Ups

Following are some examples of general nature and the OCDs thereon follow from direct application of Theorem 8.2.2. In all the results stated above and below, c denotes the number of covariates optimally included. This may be noted once and for all.

Generalization 1 (**main effects plan set-up**): Let **A** be an MOA(n, $s_1 \times s_2 \times \cdots \times s_m$, 2) giving an orthogonal main effects plan. Then, according to Theorem 8.2.2, the matrix $\begin{pmatrix} \mathbf{A} \\ \mathbf{B} \end{pmatrix}$ gives an OCD if $\begin{pmatrix} \mathbf{A} \\ \mathbf{B} \end{pmatrix}$ is an EMOA [n, $m + c$, $s_1 \times s_2 \times \cdots \times s_m \times 2^c$, (2, 2), (2; 1, 1)]. It follows that this EMOA is an MOA(n, $s_1 \times s_2 \times \cdots \times s_m \times 2^c$, 2).

Below we discuss a particular type of main effect plan obtained through hypergraecolatin square.

Hypergraecolatin square set-up: Let **A** be an $m \times s^2$ matrix giving an OA (s^2, m, s, 2) obtained from m mutually orthogonal *latin* squares (MOLS) of order s. The columns of **A** actually give the set-up of a hypergraecolatin square (cf. Raghavarao 1971). Then **B** gives an OCD in the above set-up if $\begin{pmatrix} \mathbf{A} \\ \mathbf{B} \end{pmatrix}$ = MOA (s^2, $s^m \times 2^c$, 2), $c \le (s-1)(s+1-m)$.

If $m = 3$, then **B** gives an OCD for an $s \times s$ latin square set-up (compare Example 8.2.4).

The following theorem states a method of getting OCDs for this set-up with a compromise on the error d.f. and pushing them to the covariates.

Theorem 8.2.3 *Suppose* $\mathbf{H}_s$ *and* $(m-2)$ *MOLS of order* s *with symbols* $1, 2, \ldots, s$ *exist,* $m \le s + 1$. *Let* $\mathbf{A} = OA$ (s^2, $m_1 + 2$, s, 2) *be constructed from* m_1 *MOLS out of the* $(m-2)$ $(=m_1 + m_2)$ *MOLS of order* s. *Then an OCD for the estimation of* $c = m_2(s-1)$ *regression coefficients in the set-up of an orthogonal main effects plan involving* $(m_1 + 2)$ *factors can be constructed from the remaining* m_2 *MOLS.*

Proof First we construct an orthogonal array, OA $(s^2, m, s, 2)$ using the $(m - 2)$ MOLS of order s (cf. Hedayat et al. 1999). Let this orthogonal array be denoted by the following matrix $\mathbf{E}$ in a partition form as

$$\mathbf{E} = \begin{pmatrix} \mathbf{A}_{(m_1+2) \times s^2} \\ \mathbf{D}_{m_2 \times s^2} \end{pmatrix}.$$

Here $\mathbf{D}$ is a resolvable orthogonal array of strength one (cf. Raghavarao 1971). A Hadamard matrix of order s is written as

$$\mathbf{H}_s = (\mathbf{h}_1, \mathbf{h}_2, \ldots, \mathbf{h}_{s-1}, \mathbf{1}). \tag{8.2.8}$$

Let the symbol i of $\mathbf{D}$, be replaced by h_{ji}, where h_{ji} is the ith element of the vector $\mathbf{h}_j$, $i = 1, 2, \ldots, s$, and a new $m_2 \times s^2$ array $\mathbf{B}^{(j)}$ is obtained from $\mathbf{D}$, $j = 1, 2, \ldots, s-1$. Note that $\mathbf{B}^{(j)}$ is an orthogonal array of strength 2 with the two symbols $+1$ and -1, $j = 1, 2, \ldots, s-1$. Next we construct the $m_2(s-1) \times s^2$ array $\mathbf{B}$ by the juxtaposition of $\mathbf{B}^{(1)}, \mathbf{B}^{(2)}, \ldots, \mathbf{B}^{(s-1)}$ row-wise, as

$$\mathbf{B} = \begin{pmatrix} \mathbf{B}^{(1)} \\ \mathbf{B}^{(2)} \\ \vdots \\ \mathbf{B}^{(s-1)} \end{pmatrix}. \tag{8.2.9}$$

We can easily check that $\begin{pmatrix} \mathbf{A} \\ \mathbf{B} \end{pmatrix}$ is an EMOA $[s^2, m_1+2+c, s^{m_1+2} \times 2^c, (2, 2), (2; 1, 1)]$ where $c = m_2(s - 1)$. So by Theorem 8.2.2, the result follows. $\square$

Remark 8.2.7 If $m_1 = 1$, then $\mathbf{B}$ gives an OCD for an $s \times s$ latin square design set-up and in this case $c = (s - 1)(m - 3)$.

Generalization 2 **(Set-up of m-way classification with single observation per cell)**: Let $\mathbf{A}$ be an $m \times v$ array containing all the $v = \prod_{i=1}^{m} s_i$ level combinations of the m factors. $\mathbf{A}$ is actually an MOA of strength m and all the factorial effects (v in number), together with the mean, are orthogonally estimable from the v observations. But as there is no error degrees of freedom left, no covariate can be accommodated. For this, according to the usual practice, we assume that the m-factor interactions are negligible and contribute to error. As $\mathbf{A}$ is an MOA with strength m, then all the factorial effects up to $(m-1)$-factor interactions are orthogonally estimable. So a $c \times v$ matrix $\mathbf{B}$ with elements ± 1, gives an OCD for the estimation of c regression coefficients if $\begin{pmatrix} \mathbf{A} \\ \mathbf{B} \end{pmatrix}$ is an EMOA $[v, m + c, s_1 \times s_2 \times \cdots \times s_m \times 2^c, (m, 2), (2; m - 1, 1)]$, where $c \leq \prod_{i=1}^{m} (s_i - 1)$.

Generalization 3 (**Set-up of m-way classification with r (> 1) observations per cell**): Let in the above set-up each level combination be repeated r (> 1) times in **A**. Then all the v factorial effects can be included in the model as the replications provide with the error and a matrix **B** satisfying $\begin{pmatrix} \mathbf{A} \\ \mathbf{B} \end{pmatrix}$ = EMOA $[vr, \ m+c, \ s_1 \times s_2 \times \cdots \times s_m \times 2^c, \ (m, \ 2), \ (2; \ m, \ 1)]$ will give an OCD for the estimation of c regression coefficients where $c \le v(r-1)$.

Generalizations 2 and 3 indicate how the OCDs can be obtained for this set-up through EMOAs with suitable parameters. Constructions of such EMOAs can be obtained by suitable adaptation of those for the MOAs given in Rao et al. (2003). The results are stated in the following theorems.

Theorem 8.2.4 *If $r = 1$ and,*

(i) *if there exists a Hadamard matrix of order $s_i (i = 1, \ldots, m)$, then an EMOA*

$$[v = \prod_{i=1}^{m} s_i, \ m+c, \ s_1 \times s_2 \times \cdots \times s_m \times 2^c, \ (m, \ 2), \ (2; \ m-1, 1)] \ exists,$$

where $c = \prod_{i=1}^{m}(s_i - 1)$;

(ii) *if Hadamard matrices of orders $s_1/2$, $2s_2$ and $s_i (i = 3, \ldots, m)$ exist, where s_2 is even, then an EMOA $[v = \prod_{i=1}^{m} s_i, \ m+c, \ s_1 \times s_2 \times \cdots \times s_m \times 2^c, \ (m, \ 2),$*

$(2; \ m-1, 1)]$ exists, where $c = \{(s_1 - 1)(s_2 - 1) - (s_2 - 2)\} \prod_{i=3}^{m}(s_i - 1)$;

(iii) *if Hadamard matrices of orders s_1 and $s_i (i = 3, \ldots, m)$ exist and $s_2 = 2$ (mod 4) and $(s_2 - 1)$ is a prime or prime power, then an EMOA $[v = \prod_{i=1}^{m} s_i, \ m+c, \ s_1 \times s_2 \times \cdots \times s_m \times 2^c, \ (m, \ 2), \ (2; \ m-1, 1)]$ exists, where*

$$c = \{(s_1 - 1)(s_2 - 1) - (s_2 - 2)\} \prod_{i=3}^{m}(s_i - 1).$$

Theorem 8.2.5 *If $r > 1$ and,*

(i) *if there exist Hadamard matrices of orders $v = \prod_{i=1}^{m} s_i$, r, then an EMOA $[vr, \ m+ c, \ s_1 \times s_2 \times \cdots \times s_m \times 2^c, \ (m, \ 2), \ (2; \ m, \ 1)]$ exists, where $c = v(r-1)$;*

(ii) *if Hadamard matrices of orders $v/2$, $v == \prod_{i=1}^{m} s_i$ and $2r$ exist, where r is even, then an EMOA $[vr, \ m+c, \ s_1 \times s_2 \times \cdots \times s_m \times 2^c, \ (m, \ 2), \ (2; \ m, \ 1)]$ exists, where $c = v(r-1)$;*

(iii) *if a Hadamard matrix of order* $v = \prod_{i=1}^{m} s_i$ *exists and* $r \equiv 2 \ (mod \ 4)$ *and* $(r - 1)$
is a prime or prime power, then an EMOA $[vr, \ m + c, \ s_1 \times s_2 \times \cdots \times s_m \times$
$2^c, \ (m, \ 2), \ (2; \ m, \ 1)]$ *exists, where* $c = v(r - 1)$. *Below we cite an example
of an EMOA for clear understanding of Theorems 8.2.4 and 8.2.5.*

Example 8.2.7 Let us consider a $4 \times 2 \times 2$ full factorial with one observation per
cell. Then EMOA $[16, 6, 4 \times 2 \times 2 \times 2^3, (2, 2), (2, 1, 1)]$ can be constructed as
follows:

$$
\begin{pmatrix}
1 & 1 & 1 & 1 & 2 & 2 & 2 & 2 & 3 & 3 & 3 & 3 & 4 & 4 & 4 & 4 \\
1 & 1 & 2 & 2 & 1 & 1 & 2 & 2 & 1 & 1 & 2 & 2 & 1 & 1 & 2 & 2 \\
1 & 2 & 1 & 2 & 1 & 2 & 1 & 2 & 1 & 2 & 1 & 2 & 1 & 2 & 1 & 2 \\
1 & -1 & -1 & 1 & 1 & -1 & 1 & 1 & -1 & 1 & -1 & -1 & 1 & -1 & 1 & 1 & -1 \\
1 & -1 & -1 & 1 & 1 & -1 & 1 & 1 & -1 & -1 & 1 & 1 & -1 & 1 & -1 & -1 & 1 \\
1 & -1 & -1 & 1 & 1 & 1 & -1 & -1 & 1 & -1 & 1 & 1 & -1 & -1 & 1 & 1 & -1
\end{pmatrix} = \begin{pmatrix} \mathbf{A}^{3 \times 16} \\ \mathbf{Z}'^{3 \times 16} \end{pmatrix},
$$

where $\mathbf{Z}^{16 \times 3} = (\mathbf{z}_1, \ \mathbf{z}_2, \ \mathbf{z}_3)$ and $\mathbf{z}_1' = (1, -1, 1, -1) \otimes (1, -1) \otimes (1, -1)$, $\mathbf{z}_2' = (1, -1, -1, 1) \otimes (1, -1) \otimes (1, -1)$, $\mathbf{z}_3' = (1, 1, -1, -1) \otimes (1, -1) \otimes (1, -1)$. Here
we accommodate 3 covariates optimally in the factorial set-up when all the main
effects and two-factor interactions are orthogonally estimable.

Again let us consider the $4 \times 2 \times 2$ full factorial with two observations per cell. Then
EMOA $[32, 22, 4 \times 2 \times 2 \times 2^{19}, (2, 2), (2, 1, 1)]$ can be constructed as follows:

$$
\begin{pmatrix}
\text{First set of} & \text{Second set of} \\
\text{16 observations} & \text{16 observations} \\
\hline
\mathbf{H}_{16} & -\mathbf{H}_{16}
\end{pmatrix}.
$$

Here we accommodate 16 covariates optimally in the factorial set-up with two observations per cell when all the main effects and interactions are orthogonally estimable.

Remark 8.2.8 (**RBD set-up as a particular case of Generalization 2**): Let $\mathbf{A}_{2 \times s_1 s_2}$
contain the all possible level combinations of an RBD with s_1 blocks and s_2 treatments. Then by Remark 8.2.1, $\mathbf{B}$, a $c \times s_1 s_2$ matrix with elements ± 1 gives an OCD
if $\begin{pmatrix} \mathbf{A} \\ \mathbf{B} \end{pmatrix}$ is an EMOA $[s_1 s_2, 2 + c, s_1 \times s_2 \times 2^c, (2, 2), (2; 1, 1)]$, which is actually
an MOA $(s_1 s_2, \ s_1 \times s_2 \times 2^c; \ 2)$. This is in full agreement with Rao et al. (2003)
(compare Example 3.4.1 of Chap. 3).

Remark 8.2.9 (**CRD set-up as a particular case of Generalization 3**): If in partic-
ular, $m = 1$ in the set-up of Remark 8.2.10, then a matrix $\mathbf{B}$, where $\begin{pmatrix} \mathbf{A} \\ \mathbf{B} \end{pmatrix} = \text{MOA}$
$(vr, \ v \times 2^c; \ 2)$ gives an OCD for the estimation of the regression coefficients under
the CRD set-up with v treatments. This is also in agreement with Rao et al. (2003)
(compare Example 3.4.1 of Chap. 3).

Remark 8.2.10 (**Incomplete block design set-up**): Let $m = 2$ and the columns of **A** give the set-up of an incomplete block design where the block and the treatment effects are non-orthogonally estimable. The same conditions (i)–(iii) of Theorem 8.2.1 apply for an OCD, but no general result similar to Theorem 8.2.2 can be proposed. The OCDs are difficult to construct here unless some patterns in the incidence matrices exist (cf. Chaps. 4, 5 and 6).

8.3 OCDs in Split-Plot and Strip-Plot Design Set-Ups

In the previous chapters we considered set-ups where the errors were assumed to be uncorrelated. In this section, we consider the problem of finding OCDs for the estimation of covariate parameters in the correlated set-ups of standard split-plot and strip-plot designs with the levels of the whole-plot factor laid out in r randomized blocks. An EMOA and Hadamard matrices play the key role for such construction.

8.3.1 Preliminaries

In the earlier chapters, we considered the set-up where the observations are uncorrelated. For the correlated model, the issue of finding the optimal covariate designs was considered by Dutta et al. (2009) which we discuss in the present section. For the general variance-covariance structure, it is difficult to construct the optimum **Z**-matrix retaining orthogonality with effects related to the ANOVA part. Dutta et al. (2009) dealt with standard split-plot and strip-plot design set-ups (cf. Cochran and Cox 1950) for which variance–covariance matrices have special structures that can be conveniently exploited to find the OCDs.

Consider the following non-stochastic controllable covariates model of a standard split-plot design set-up with the levels of the whole-plot factor (whole-plot treatments) in r randomized blocks (cf. Chakrabarti 1962)

$$(\mathbf{Y}, \ \mu\mathbf{1}_{rpq} + \mathbf{X}_1\alpha + \mathbf{X}_2\beta + \mathbf{X}_3\tau + \mathbf{X}_4\delta + \mathbf{Z}\gamma, \ \sigma^2\mathbf{\Sigma}) \tag{8.3.1}$$

where $\mathbf{Y} = (y_{111}, \ldots, y_{ijk}, \ldots, y_{rpq})'$ is the $rpq \times 1$ observation vector corresponding to the rpq level combinations of the three factors, viz. the block (R), the whole plot factor (A), and the sub-plot factor (B) arranged lexicographically; $\mathbf{X}_1, \mathbf{X}_2, \mathbf{X}_3$, $\mathbf{X}_4, \mathbf{Z}$ are the design matrices corresponding to the block effects vector $\alpha^{r \times 1}$, the whole-plot effects vector $\beta^{p \times 1}$, the sub-plot effects vector $\tau^{q \times 1}$, the whole-plot $\times$ sub-plot interaction effects vector $\delta^{pq \times 1}$ and the covariate effects $\gamma^{c \times 1}$ respectively. Obviously, $\mathbf{1}_{rpq}$ is the coefficient vector corresponding to the intercept term μ. It may be noted that $\mathbf{X}_{ij}$'s are (0,1) incidence matrices. $\mathbf{Z}$ is the matrix of covariate values. For convenience, we partition $\mathbf{X}_g$ ($g = 1, 2, 3, 4$) and $\mathbf{Z}$ as follows:

$$
\left.
\begin{aligned}
\mathbf{X}_g^{rpq \times n_g} &= \left(\mathbf{X}_{11}^{(g)\prime \; n_g \times q}, \ldots, \mathbf{X}_{1p}^{(g)\prime \; n_g \times q}, \ldots, \mathbf{X}_{ij}^{(g)\prime \; n_g \times q}, \ldots, \mathbf{X}_{rp}^{(g)\prime \; n_g \times q} \right)' \\
\mathbf{Z}^{rpq \times c} &= \left(\mathbf{Z}_{11}^{\prime \; c \times q}, \ldots, \mathbf{Z}_{1p}^{\prime \; c \times q}, \ldots, \mathbf{Z}_{ij}^{\prime \; c \times q}, \ldots, \mathbf{Z}_{rp}^{\prime \; c \times q} \right)'
\end{aligned}
\right\}
$$

$$(8.3.2)$$

where n_g stands for the number of parameters in the gth classification corresponding to the block, the whole-plot treatment and the sub-plot treatment, i.e. $n_g = r, p, q, pq$ for $g = 1, 2, 3, 4$ respectively. $\mathbf{X}_{ij}^{(1) \; q \times r}, \mathbf{X}_{ij}^{(2) \; q \times p}, \mathbf{X}_{ij}^{(3) \; q \times q}, \mathbf{X}_{ij}^{(4) \; q \times pq}$ and $\mathbf{Z}_{ij}^{q \times c}$ are the portions of the design matrices $\mathbf{X}_1, \mathbf{X}_2, \mathbf{X}_3, \mathbf{X}_4$ and $\mathbf{Z}$, respectively, corresponding to the observations of the ith block and the jth whole-plot treatment ($i = 1, 2, \ldots, r$; $j = 1, 2, \ldots, p$). Thus, if the structure of $\mathbf{X}_{ij}^{(1)}$ is investigated it is noted that in the ith column, 1 corresponds to each of the q observations on the q levels of the sub-factor B when R and A are fixed at i and j respectively. Other columns contain 0's only. We write $\mathbf{1}$ as $q \times 1$ vector with all elements unity, $\mathbf{e}_i$ as $q \times 1$ unit vector with 1 at the ith position, $\boldsymbol{\delta}_j = (\delta_{j1}, \delta_{j2}, \ldots, \delta_{jq})$, $j = 1, 2, \ldots, p$, jth vector of interactions of jth whole plot treatment with q sub-plot treatments, $j = 1, 2, \ldots, p$. With these notations we write the following $\mathbf{X}_{ij}^{(g)}$ matrices.

$$
\begin{array}{c}
\alpha_1 \; \alpha_2 \; \ldots \; \alpha_{i-1} \; \alpha_i \; \alpha_{i+1} \; \ldots \; \alpha_r \\[4pt]
\mathbf{X}_{ij}^{(1)} = \left(\begin{array}{ccccccccc} \mathbf{0} & \mathbf{0} & \ldots & \mathbf{0} & \mathbf{1} & \mathbf{0} & \ldots & \mathbf{0} \end{array} \right)^{q \times r} \quad \forall j;
\end{array}
$$

$$(8.3.3)$$

It is to be noted that the structure of $\mathbf{X}_{ij}^{(1)}$ is independent of j. In this way, we can write the other $\mathbf{X}_{ij}$-matrices as follows:

$$
\begin{array}{c}
\beta_1 \; \beta_2 \; \ldots \; \beta_{j-1} \; \beta_j \; \beta_{j+1} \; \ldots \; \beta_p \\[4pt]
\mathbf{X}_{ij}^{(2)} = \left(\begin{array}{ccccccccc} \mathbf{0} & \mathbf{0} & \ldots & \mathbf{0} & \mathbf{1} & \mathbf{0} & \ldots & \mathbf{0} \end{array} \right)^{q \times p} \quad \forall i;
\end{array}
$$

$$(8.3.4)$$

$$
\begin{array}{c}
\tau_1 \; \tau_2 \; \ldots \; \tau_q \\[4pt]
\mathbf{X}_{ij}^{(3)} = \left(\begin{array}{cccc} \mathbf{e}_1 & \mathbf{e}_2 & \ldots & \mathbf{e}_q \end{array} \right)^{q \times q} \quad \forall i, \; j;
\end{array}
$$

$$(8.3.5)$$

$$
\begin{array}{c}
\boldsymbol{\delta}_1 \; \boldsymbol{\delta}_2 \; \ldots \; \boldsymbol{\delta}_{j-1} \; \boldsymbol{\delta}_j \; \boldsymbol{\delta}_{j+1} \; \ldots \; \boldsymbol{\delta}_p \\[4pt]
\mathbf{X}_{ij}^{(4)} = \left(\begin{array}{ccccccccc} \mathbf{0} & \mathbf{0} & \ldots & \mathbf{0} & \mathbf{I}_q & \mathbf{0} & \ldots & \mathbf{0} \end{array} \right)^{q \times pq} \quad \forall i;
\end{array}
$$

$$(8.3.6)$$

Now we consider the following example which illustrates the above set-up and the representations.

Example 8.3.1 Let us take $r = 2$, $p = 2$, $q = 4$. Then $\mathbf{X}_1, \mathbf{X}_2, \mathbf{X}_3$ and $\mathbf{X}_4$ are written as follows.

$$
\mathbf{X}_1 = \begin{pmatrix} 1 & 0 \\ 1 & 0 \\ 1 & 0 \\ 1 & 0 \\ \hline 1 & 0 \\ 1 & 0 \\ 1 & 0 \\ 1 & 0 \\ \hline 0 & 1 \\ 0 & 1 \\ 0 & 1 \\ 0 & 1 \\ \hline 0 & 1 \\ 0 & 1 \\ 0 & 1 \\ 0 & 1 \end{pmatrix} = \begin{pmatrix} \mathbf{X}_{11}^{(1)4\times2} \\ \mathbf{X}_{12}^{(1)4\times2} \\ \mathbf{X}_{21}^{(1)4\times2} \\ \mathbf{X}_{22}^{(1)4\times2} \end{pmatrix}, \quad
\mathbf{X}_2 = \begin{pmatrix} 1 & 0 \\ 1 & 0 \\ 1 & 0 \\ 1 & 0 \\ \hline 0 & 1 \\ 0 & 1 \\ 0 & 1 \\ 0 & 1 \\ \hline 1 & 0 \\ 1 & 0 \\ 1 & 0 \\ 1 & 0 \\ \hline 0 & 1 \\ 0 & 1 \\ 0 & 1 \\ 0 & 1 \end{pmatrix} = \begin{pmatrix} \mathbf{X}_{11}^{(2)4\times2} \\ \mathbf{X}_{12}^{(2)4\times2} \\ \mathbf{X}_{21}^{(2)4\times2} \\ \mathbf{X}_{22}^{(2)4\times2} \end{pmatrix},
$$

$$
\mathbf{X}_3 = \begin{pmatrix} 1&0&0&0 \\ 0&1&0&0 \\ 0&0&1&0 \\ 0&0&0&1 \\ \hline 1&0&0&0 \\ 0&1&0&0 \\ 0&0&1&0 \\ 0&0&0&1 \\ \hline 1&0&0&0 \\ 0&1&0&0 \\ 0&0&1&0 \\ 0&0&0&1 \\ \hline 1&0&0&0 \\ 0&1&0&0 \\ 0&0&1&0 \\ 0&0&0&1 \end{pmatrix} = \begin{pmatrix} \mathbf{X}_{11}^{(3)4\times4} \\ \mathbf{X}_{12}^{(3)4\times4} \\ \mathbf{X}_{21}^{(3)4\times4} \\ \mathbf{X}_{22}^{(3)4\times4} \end{pmatrix}, \quad
\mathbf{X}_4 = \begin{pmatrix} 1&0&0&0&0&0&0&0 \\ 0&1&0&0&0&0&0&0 \\ 0&0&1&0&0&0&0&0 \\ 0&0&0&1&0&0&0&0 \\ \hline 0&0&0&0&1&0&0&0 \\ 0&0&0&0&0&1&0&0 \\ 0&0&0&0&0&0&1&0 \\ 0&0&0&0&0&0&0&1 \\ \hline 1&0&0&0&0&0&0&0 \\ 0&1&0&0&0&0&0&0 \\ 0&0&1&0&0&0&0&0 \\ 0&0&0&1&0&0&0&0 \\ \hline 0&0&0&0&1&0&0&0 \\ 0&0&0&0&0&1&0&0 \\ 0&0&0&0&0&0&1&0 \\ 0&0&0&0&0&0&0&1 \end{pmatrix} = \begin{pmatrix} \mathbf{X}_{11}^{(4)4\times8} \\ \mathbf{X}_{12}^{(4)4\times8} \\ \mathbf{X}_{21}^{(4)4\times8} \\ \mathbf{X}_{22}^{(4)4\times8} \end{pmatrix}.
$$

For a standard split-plot design, where intra-class correlation structure of the dispersion matrix is assumed, the elements of Σ-matrix (cf. Chakrabarti 1962) of (8.3.1) are given by

$$
\frac{1}{\sigma^2} Cov(y_{ijk}, y_{i',j',k'}) = \begin{cases} 1 \text{ if } i=i', \ j=j', \ k=k' \\ \rho \text{ if } i=i', \ j=j', \ k\neq k' \\ 0 \text{ otherwise,} \end{cases} \tag{8.3.7}
$$

and it can be expressed as

$$\Sigma = \mathbf{I}_{pr} \bigotimes \Sigma_1; \quad \Sigma_1 = (1 - \rho)\mathbf{I}_q + \rho\mathbf{J}_q \tag{8.3.8}$$

where ρ is the common intra-class correlation coefficient among the observations corresponding to the sub-plot treatments within the same whole-plot treatment in a block and $\mathbf{J}_u = \mathbf{1}_u\mathbf{1}_u'$ is the square matrix of order u with all elements unity. Following Cochran and Cox (1950), p. 220, we assume $\rho > 0$ as the observations corresponding to the different levels of the sub-plot treatments under the same level of the whole-plot treatment are expected to be positively correlated.

In this correlated set-up, we are concerned with the optimum choice of $\mathbf{Z}$ for the estimation of each of the regression parameters in the split-plot set-up with maximum accuracy in the sense of minimizing the variance of the best linear unbiased estimators of regression parameters retaining orthogonality with the estimators of the ANOVA effects.

The Optimality Conditions for the Split-Plot Design Set-Up
The information matrix for $\eta = (\mu, \ \alpha', \ \beta', \ \tau', \ \delta', \ \gamma')'$ in the split-plot design set-up (8.3.1) is given by

$$\mathbf{I}(\eta) = (\mathbf{X}, \ \mathbf{Z})'\Sigma^{-1}(\mathbf{X}, \ \mathbf{Z}). \tag{8.3.9}$$

where $\mathbf{X} = (\mathbf{1}, \ \mathbf{X}_1, \ \mathbf{X}_2, \ \mathbf{X}_3, \ \mathbf{X}_4)$. From (8.3.8), Σ^{-1} can be written as

$$\left.\begin{array}{l} \Sigma^{-1} = \mathbf{I}_{pr} \otimes \Sigma_1^{-1} \\ \Sigma_1^{-1} = \frac{1}{1-\rho}\left(\mathbf{I}_q - \frac{\rho}{1+(q-1)\rho}\mathbf{J}_q\right) \end{array}\right\} \tag{8.3.10}$$

It is evident from (8.3.9) that γ is estimable orthogonally to the ANOVA effects if and only if

$$\mathbf{X}_g'\Sigma^{-1}\mathbf{Z} = \mathbf{0}, \quad g = 1, \ 2, \ 3, \ 4, \tag{8.3.11}$$

where $\mathbf{X}_g$ is the design matrix of order $rpq \times n_g$ corresponding to the gth ANOVA effect described in (8.3.2), with $n_g = r, \ p, \ q, \ pq$ respectively for $g = 1, \ 2, \ 3, \ 4$. Using (8.3.10), the orthogonality conditions in (8.3.11) can be reduced to

$$\sum_{i=1}^{r}\sum_{j=1}^{p}\mathbf{X}_{ij}^{(g)'}\mathbf{Z}_{ij} - \frac{\rho}{1+(q-1)\rho}\sum_{i=1}^{r}\sum_{j=1}^{p}\mathbf{X}_{ij}^{(g)'}\mathbf{J}_q\mathbf{Z}_{ij} = \mathbf{0}, \tag{8.3.12}$$

which is satisfied if

$$\sum_{i=1}^{r}\sum_{j=1}^{p}\mathbf{X}_{ij}^{(g)'}\mathbf{Z}_{ij} = \mathbf{0},$$ (8.3.13)

and

$$\frac{\rho}{1+(q-1)\rho}\sum_{i=1}^{r}\sum_{j=1}^{p}\mathbf{X}_{ij}^{(g)'}\mathbf{J}_{q}\mathbf{Z}_{ij} = \mathbf{0}.$$ (8.3.14)

For $g = 1$, 2, 3 and 4, (8.3.3)–(8.3.6) imply that the left-hand side of (8.3.13) becomes, respectively, the $r \times c$ matrix

$$\left(\sum_{j=1}^{p}\sum_{l=1}^{q}z_{lm}^{(ij)}\right)_{i=1,2,\ldots,r,\ m=1,2,\ldots,c},$$ (8.3.15)

the $p \times c$ matrix

$$\left(\sum_{i=1}^{r}\sum_{l=1}^{q}z_{lm}^{(ij)}\right)_{j=1,2,\ldots,p,\ m=1,2,\ldots,c},$$ (8.3.16)

the $q \times c$ matrix

$$\left(\sum_{i=1}^{r}\sum_{j=1}^{p}z_{lm}^{(ij)}\right)_{l=1,2,\ldots,q,\ m=1,2,\ldots,c},$$ (8.3.17)

and the $pq \times c$ matrix

$$\left(\sum_{i=1}^{r}z_{lm}^{(ij)}\right)_{j=1,2,\ldots,p,\ l=1,2,\ldots,q,\ m=1,2,\ldots,c}.$$ (8.3.18)

Again for $g = 1$, 2, 3 and 4, (8.3.3)–(8.3.6) imply that the left-hand side of (8.3.14) becomes, respectively, the $r \times c$ matrix

$$q\left(\sum_{j=1}^{p}\sum_{l=1}^{q}z_{lm}^{(ij)}\right)_{i=1,2,\ldots,r,\ m=1,2,\ldots,c},$$ (8.3.19)

the $p \times c$ matrix

$$q\left(\sum_{i=1}^{r}\sum_{l=1}^{q} z_{lm}^{(ij)}\right)_{j=1,2,\ldots,p,\ m=1,2,\ldots,c}, \tag{8.3.20}$$

the $q \times c$ matrix

$$\mathbf{1}_q \otimes \left(\sum_{i=1}^{r}\sum_{j=1}^{p}\sum_{l=1}^{q} z_{l1}^{(ij)}, \ldots, \sum_{i=1}^{r}\sum_{j=1}^{p}\sum_{l=1}^{q} z_{lm}^{(ij)}, \ldots, \sum_{i=1}^{r}\sum_{j=1}^{p}\sum_{l=1}^{q} z_{lc}^{(ij)}\right) \tag{8.3.21}$$

and the $pq \times c$ matrix

$$\mathbf{U} = \left(\mathbf{U}_1', \ldots, \mathbf{U}_j', \ldots, \mathbf{U}_p'\right)', \tag{8.3.22}$$

where

$$\mathbf{U}_j^{q \times c} = \mathbf{1}_q \otimes \left(\sum_{i=1}^{r}\sum_{j=1}^{p}\sum_{l=1}^{q} z_{l1}^{(ij)}, \ldots, \sum_{i=1}^{r}\sum_{j=1}^{p}\sum_{l=1}^{q} z_{lm}^{(ij)}, \ldots, \sum_{i=1}^{r}\sum_{j=1}^{p}\sum_{l=1}^{q} z_{lc}^{(ij)}\right). \tag{8.3.23}$$

Therefore, from (8.3.15)–(8.3.23), a set of sufficient conditions for (8.3.13) to satisfy is

$$\left.\begin{aligned}
&\sum_{i=1}^{r} z_{lm}^{(ij)} = 0 \quad \forall j = 1, 2, \ldots, p,\ l = 1, 2, \ldots, q,\ m = 1, 2, \ldots, c, \\
\text{and}\quad &\sum_{j=1}^{p}\sum_{l=1}^{q} z_{lm}^{(ij)} = 0 \ \forall i = 1, 2, \ldots, r,\ m = 1, 2, \ldots, c.
\end{aligned}\right\} \tag{8.3.24}$$

It is seen from (8.3.9) that the information matrix for γ under (8.3.24) when $\mathbf{X}_{ij}^{(g)}$'s follow the structure (8.3.3)–(8.3.6), is proportional to $\mathbf{Z}'\mathbf{\Sigma}^{-1}\mathbf{Z}$. Again, by virtue of (8.3.10)

$$\begin{aligned}
\mathbf{Z}'\mathbf{\Sigma}^{-1}\mathbf{Z} &= \tfrac{1}{1-\rho}\left(\sum_{i=1}^{r}\sum_{j=1}^{p} \mathbf{Z}_{ij}'\mathbf{Z}_{ij} - \frac{\rho}{1+(q-1)\rho}\sum_{i=1}^{r}\sum_{j=1}^{p} \mathbf{Z}_{ij}'\mathbf{J}_q\mathbf{Z}_{ij}\right) \\
&\leq \tfrac{1}{1-\rho}\left(\sum_{i=1}^{r}\sum_{j=1}^{p} \mathbf{Z}_{ij}'\mathbf{Z}_{ij}\right)
\end{aligned} \tag{8.3.25}$$

in the sense of Partial Loewner Order (PLO) dominance (cf. Pukelsheim (1993)) since by assumption $\rho > 0$ and $\sum_{i=1}^{r}\sum_{j=1}^{p} \mathbf{Z}'_{ij}\mathbf{J}_q\mathbf{Z}_{ij} = \sum_{i=1}^{r}\sum_{j=1}^{p} \mathbf{Z}'_{ij}\mathbf{1}_q\mathbf{1}'_q\mathbf{Z}_{ij}$ is non-negative definite. Equality holds in (8.3.25) if $\mathbf{Z}'_{ij}\mathbf{1}_q = \mathbf{0}$ $\forall i, j$ or, equivalently

$$\sum_{l=1}^{q} z_{lm}^{(ij)} = 0 \ \forall i = 1, 2, \ldots, r, \ j = 1, 2, \ldots, q, \ m = 1, 2, \ldots, c. \qquad (8.3.26)$$

If, in addition to (8.3.24) and (8.3.26), $\mathbf{Z}_{ij}$ satisfies

$$\sum_{i=1}^{r}\sum_{j=1}^{p}\sum_{l=1}^{q} z_{lm}^{(ij)} z_{lm'}^{(ij)} = 0 \ \forall \ m \neq m = 1, 2, \ldots, c, \qquad (8.3.27)$$

then γ_m's are estimated orthogonally among themselves and orthogonally to the ANOVA effects. Under the above conditions (8.3.24), (8.3.26) and (8.3.27), γ_m can be estimated with the minimum variance $\frac{(1-\rho)\sigma^2}{rpq}$ for each m if $z_{lm}^{ij} = \pm 1 \ \forall \ i, j, l, m$. Hence we get the following theorem given in Dutta et al. (2009).

Theorem 8.3.1 *In the standard split-plot design set-up (8.3.1) the following set of conditions:*

(i) $z_{lm}^{(ij)} = \pm 1 \quad \forall i = 1, 2, \ldots, r, \ j = 1, 2, \ldots, p, \ l = 1, 2, \ldots, q, \ m = 1, 2, \ldots, c$

(ii) $\sum_{l=1}^{q} z_{lm}^{(ij)} = 0 \ \forall i = 1, 2, \ldots, r, \ j = 1, 2, \ldots, p, \ m = 1, 2, \ldots, c$

(iii) $\sum_{i=1}^{r} z_{lm}^{(ij)} = 0 \ \forall j = 1, 2, \ldots, p, \ l = 1, 2, \ldots, q, \ m = 1, 2, \ldots, c$

(iv) $\sum_{i=1}^{r}\sum_{j=1}^{p}\sum_{l=1}^{q} z_{lm}^{(ij)} z_{lm'}^{(ij)} = rpq\delta_{mm'}$ *where* $\delta_{mm'} = 1$ *if* $m = m'$; *=0 if* $m \neq m'$,

is sufficient for the optimum estimation of each of the covariate effects with the minimum possible variance Var $(\widehat{\gamma}_m) = \frac{(1-\rho)\sigma^2}{rpq}$, $m = 1, 2, \ldots, c$.

Note 8.3.1 It must be noted that the conditions laid down above are independent of the actual value of ρ, assumed to be known and positive.

The Optimality Conditions for the Strip-Plot Design Set-Up

In a standard strip-plot design, as the levels of the sub-plot factor B are arranged in strips, the dispersion matrix of the observation vector $\mathbf{Y}$ gets changed though the mean vector remains the same as in (8.3.1). So the linear model (8.3.1) can be adapted by replacing $\mathbf{\Sigma}$ by $\mathbf{\Sigma}^*$, with the elements of $\mathbf{\Sigma}^*$ as

$$\frac{1}{\sigma^2} Cov(y_{ijk}, y_{i'.j'.k'}) = \begin{cases} 1 & \text{if } i = i', \ j = j', \ k = k' \\ \rho_1 & \text{if } i = i', \ j = j', \ k \neq k' \\ \rho_2 & \text{if } i = i', \ j \neq j', \ k = k' \\ 0 & \text{otherwise,} \end{cases} \tag{8.3.28}$$

where y_{ijk}, arranged lexicographically, is the yield of the plot belonging to the kth column-strip and the jth row-strip in the ith block ($i = 1, 2, \ldots, r$; $j = 1, 2, \ldots, p$; $k = 1, 2, \ldots, q$). Therefore, we can write (8.3.28) as

$$\left. \begin{aligned} Disp(\mathbf{Y}) &= \sigma^2 \mathbf{I}_r \otimes \mathbf{\Sigma}^{**} \\ \mathbf{\Sigma}^{**} &= \mathbf{I}_p \otimes \mathbf{\Sigma}_1^* + \mathbf{J}_p \otimes \mathbf{\Sigma}_2^* \\ \mathbf{\Sigma}_1^* &= (1 - \rho_1 - \rho_2)\mathbf{I}_q + \rho_1 \mathbf{J}_q, \quad \mathbf{\Sigma}_2^* = \rho_2 \mathbf{I}_q. \end{aligned} \right\} \tag{8.3.29}$$

Following the same arguments as in split-plot design, here it is also assumed that $\rho_1 > 0$, $\rho_2 > 0$. In a standard strip-plot design, for estimation of the covariate effects orthogonally to the ANOVA effects, we, in analogy to (8.3.11), have from (8.3.1) and (8.3.29)

$$\mathbf{X}_g' \mathbf{\Sigma}^{*-1} \mathbf{Z} = \mathbf{0}, \quad \forall g = 1, 2, 3, 4, \tag{8.3.30}$$

where $\mathbf{X}_g$'s and $\mathbf{Z}$ are defined in (8.3.2). By virtue of (8.3.29), the conditions in (8.3.30) reduce to

$$\sum_{i=1}^{r} \mathbf{X}_i^{(g)'} \mathbf{\Sigma}^{**-1} \mathbf{Z}_{(i)} = \mathbf{0}, \quad \forall g = 1, 2, 3, 4, \tag{8.3.31}$$

where $\mathbf{X}_i^{(g)}$ and $\mathbf{Z}_{(i)}$ are the portions of $\mathbf{X}_g$ and $\mathbf{Z}$ corresponding to the pq observations in the ith block, $i = 1, 2, \ldots, r$.

Again from (8.3.29),

$$\mathbf{\Sigma}_1^{**-1} = \mathbf{I}_p \otimes \mathbf{B}_1 - \mathbf{J}_p \otimes \mathbf{B}_2 \tag{8.3.32}$$

where

$$\left. \begin{aligned} \mathbf{B}_1 &= \mathbf{\Sigma}_1^{*-1} = \frac{1}{1 - \rho_1 - \rho_2} \left(\mathbf{I}_q - \frac{1}{1 + (q-1)\rho_1 - \rho_2} \mathbf{J}_q \right), \\ \mathbf{B}_2 &= (\mathbf{\Sigma}_1^* + p\mathbf{\Sigma}_2^*)^{-1} \mathbf{\Sigma}_2^* \mathbf{\Sigma}_1^{*-1} \\ &= \frac{\rho_2}{(1 - \rho_1 - \rho_2)(1 + \rho_1 + (p-1)\rho_2)} \left(\mathbf{I}_q - \frac{\rho_1(2 + (q-2)\rho_1 + (p-2)\rho_2)}{(1 + (q-1)\rho_1 - \rho_2)(1 + (q-1)\rho_1 + (p-1)\rho_2)} \mathbf{J}_q \right). \end{aligned} \right\} \tag{8.3.33}$$

By virtue of (8.3.2) and (8.3.32), the condition (8.3.31) reduces to

$$\sum_{i=1}^{r} \sum_{j=1}^{p} \mathbf{X}_{ij}^{(g)'} \mathbf{B}_1 \mathbf{Z}_{ij} - \sum_{i=1}^{r} \sum_{j=1}^{p} \mathbf{X}_{ij}^{(g)'} \mathbf{B}_2 (\mathbf{Z}_{i1} + \mathbf{Z}_{i2} + \cdots + \mathbf{Z}_{ip}) = \mathbf{0} \tag{8.3.34}$$

which is satisfied if

$$\sum_{i=1}^{r}\sum_{j=1}^{p}\mathbf{X}_{ij}^{(g)\prime}\mathbf{B}_1\mathbf{Z}_{ij} = \mathbf{0} \qquad (8.3.35)$$

and

$$\sum_{i=1}^{r}\sum_{j=1}^{p}\mathbf{X}_{ij}^{(g)\prime}\mathbf{B}_2(\mathbf{Z}_{i1} + \mathbf{Z}_{i2} + \cdots + \mathbf{Z}_{ip}) = \mathbf{0}. \qquad (8.3.36)$$

Since $\mathbf{B}_1$ is a completely symmetric matrix, it is seen that (8.3.24) is also sufficient for (8.3.35) to hold. Again, using (8.3.33) in (8.3.36), a set of sufficient conditions for (8.3.36) to satisfy is

$$\sum_{i=1}^{r}\sum_{j=1}^{p}\mathbf{X}_{ij}^{(g)\prime}(\mathbf{Z}_{i1} + \mathbf{Z}_{i2} + \cdots + \mathbf{Z}_{ip}) = \mathbf{0}, \qquad (8.3.37)$$

and

$$\sum_{i=1}^{r}\sum_{j=1}^{p}\mathbf{X}_{ij}^{(g)\prime}\mathbf{1}_q \left(\sum_{j=1}^{p}\sum_{l=1}^{q}z_{l1}^{(ij)}, \ldots, \sum_{j=1}^{p}\sum_{l=1}^{q}z_{lm}^{(ij)}, \ldots, \sum_{j=1}^{p}\sum_{l=1}^{q}z_{lc}^{(ij)} \right) = \mathbf{0}. \qquad (8.3.38)$$

It is seen that the condition (8.3.38) holds if (8.3.24) holds. Similarly as before, for $g = 1, 2, 3$ and 4, the left-hand side of (8.3.37) becomes, respectively,
  the $r \times c$ matrix

$$p \left(\sum_{j=1}^{p}\sum_{l=1}^{q}z_{lm}^{(ij)} \right)_{i=1,2,\ldots,r,\ m=1,2,\ldots,c}, \qquad (8.3.39)$$

the $p \times c$ matrix

$$\mathbf{1}_p \otimes \left(\sum_{i=1}^{r}\sum_{j=1}^{p}\sum_{l=1}^{q}z_{l1}^{(ij)}, \ldots, \sum_{i=1}^{r}\sum_{j=1}^{p}\sum_{l=1}^{q}z_{lm}^{(ij)}, \ldots, \sum_{i=1}^{r}\sum_{j=1}^{p}\sum_{l=1}^{q}z_{lc}^{(ij)} \right), \qquad (8.3.40)$$

the $q \times c$ matrix

$$p \left(\sum_{i=1}^{r} \sum_{j=1}^{p} z_{lm}^{(ij)} \right)_{l=1,2,...,q,\ m=1,2,...,c} \tag{8.3.41}$$

and the $pq \times c$ matrix

$$\mathbf{1}_p \otimes \left(\sum_{i=1}^{r} \sum_{j=1}^{p} z_{lm}^{(ij)} \right)_{l=1,2,...,q,\ m=1,2,...,c} . \tag{8.3.42}$$

So (8.3.37) holds whenever (8.3.24) holds. Using $\mathbf{X}_{ij}^{(g)}$'s from (8.3.3) to (8.3.6), and following similar arguments as in a split-plot design, it can be concluded that a set of sufficient conditions to satisfy (8.3.35) and (8.3.36) is

$$\left. \begin{array}{l} \displaystyle\sum_{i=1}^{r} z_{lm}^{(ij)} = 0 \qquad \forall j = 1, 2, \ldots, p,\ l = 1, 2, \ldots, q,\ m = 1, 2, \ldots, c, \\[2mm] \displaystyle\sum_{j=1}^{p} \sum_{l=1}^{q} z_{lm}^{(ij)} = 0 \quad \forall i = 1, 2, \ldots, r,\ m = 1, 2, \ldots, c. \end{array} \right\} \tag{8.3.43}$$

These are the same as the conditions in (8.3.24) for orthogonality in a split-plot design. The z-values satisfying (8.3.43) will ensure estimation of γ orthogonally to the estimates of the ANOVA effects. Under (8.3.43), the information matrix for γ in standard strip-plot design set-up will be proportional to $\mathbf{Z}' \boldsymbol{\Sigma}^{*-1} \mathbf{Z}$. Now from (8.3.32) and (8.3.33)

$$\mathbf{Z}' \boldsymbol{\Sigma}^{*-1} \mathbf{Z} = \sum_{i=1}^{r} \mathbf{Z}_i' \boldsymbol{\Sigma}^{**-1} \mathbf{Z}_i = \sum_{i=1}^{r} \sum_{j=1}^{p} \mathbf{Z}_{ij}' \mathbf{B}_1 \mathbf{Z}_{ij} - \sum_{i=1}^{r} \sum_{j=1}^{p} \mathbf{Z}_{ij}' \mathbf{B}_2 \left(\sum_{j=1}^{p} \mathbf{Z}_{ij} \right). \tag{8.3.44}$$

Here, as the observations in the same row-strip or in the same column-strip, are subject to the influence of the same level of A and the same level of B, respectively, it is expected that $\rho_1 > 0$, $\rho_2 > 0$ (cf. Cochran and Cox 1950) and $\mathbf{B}_2$ is assumed to be a positive definite matrix. Because of this assumption and the structure of $\mathbf{B}_1$, (8.3.44) implies that a design for which

$$\sum_{l=1}^{q} z_{lm}^{(ij)} = 0\ \forall i, j, m;\ \sum_{j=1}^{p} z_{lm}^{(ij)} = 0\ \forall i, l, m \tag{8.3.45}$$

hold, dominates any other design in the sense of PLO. If, in addition,

$$\left.\begin{array}{l} z_{lm}^{(ij)} = \pm 1 \qquad\qquad \forall i, j, l, m \\[2mm] \sum_{i=1}^{r}\sum_{j=1}^{p}\sum_{l=1}^{q} z_{lm}^{(ij)} z_{lm'}^{(ij)} = 0 \ \forall m \neq m' = 1, 2, \ldots, c. \end{array}\right\} \qquad (8.3.46)$$

then

$$\mathbf{Z'\Sigma^{*-1}Z} = \frac{rpq}{1 - \rho_1 - \rho_2}\mathbf{I}_c. \qquad (8.3.47)$$

So from (8.3.43), (8.3.45) and (8.3.46), we get the following theorem which gives a set of sufficient conditions for optimum estimation (in the sense of the minimum variance for the estimator of each γ-component) of the covariate effects in a strip-plot design.

Theorem 8.3.2 *With respect to the linear model (8.3.1) for the standard strip-plot design with variance structure (8.3.29), the following set of conditions:*

(i) $z_{lm}^{(ij)} = \pm 1 \quad \forall i = 1, 2, \ldots, r, \ j = 1, 2, \ldots, p, \ l = 1, 2, \ldots, q, \ m = 1, 2, \ldots, c$

(ii) $\sum_{i=1}^{r} z_{lm}^{(ij)} = 0 \ \ \forall j = 1, 2, \ldots, p, \ l = 1, 2, \ldots, q, \ m = 1, 2, \ldots, c$

(iii) $\sum_{j=1}^{p} z_{lm}^{(ij)} = 0 \ \ \forall i = 1, 2, \ldots, r, \ l = 1, 2, \ldots, q, \ m = 1, 2, \ldots, c$

(iv) $\sum_{l=1}^{q} z_{lm}^{(ij)} = 0 \ \ \forall i = 1, 2, \ldots, r, \ j = 1, 2, \ldots, p, \ m = 1, 2, \ldots, c$

(v) $\sum_{i=1}^{r}\sum_{j=1}^{p}\sum_{l=1}^{q} z_{lm}^{(ij)} z_{lm'}^{(ij)} = rpq\delta_{mm'}$
where $\delta_{mm'} = 1$ if $m = m'$; $=0$ if $m \neq m'$,

are sufficient for the optimum estimation of each of the covariate effects with the minimum variance $Var(\widehat{\gamma}_m) = \frac{(1-\rho_1-\rho_2)\sigma^2}{rpq}, \ \ m = 1, 2, \ldots, c.$

Note 8.3.2 Comparing $Var(\widehat{\gamma}_m)$ in split-plot with that in strip-plot set-up, it is expected that $Var(\widehat{\gamma}_m)$ under strip-plot is less than $Var(\widehat{\gamma}_m)$ under split-plot as ρ is expected to be less than $\rho_1 + \rho_2$. ρ is expected to be equal to ρ_1 if the row-strips are taken to be the strips in split-plot design. The reduction is due to introduction of column strips in strip-plot design.

Note 8.3.3 Condition (iii) of Theorem 8.3.2 for OCDs in strip-plot design is an additional condition with those conditions for OCDs in split-plot design set-up. We can still get an OCD for split-plot design set-up without satisfying this condition. Condition (iii) is called for to meet the condition of orthogonality with respect to row-strip.

8.3.2 Optimum Covariate Designs

We can represent the sufficient conditions of Theorems 8.3.1 and 8.3.2 in terms of a $(3 + c) \times rpq$ rectangular array where the first three rows (forming the first group) contain all possible combinations of the levels of the block (R), the whole-plot factor (A) and the sub-plot factor (B), respectively, arranged lexicographically. The $(3 + i)$th row of the second group which corresponds to the ith row of $\mathbf{Z}'$ have elements ± 1, $i = 1, 2, \ldots, c$. It is easy to verify that if the array satisfies the following conditions, then both Theorems 8.3.1 and 8.3.2 hold true:

(a_1) $\mathbf{Z}'$ is an orthogonal array of strength 2.
(a_2) in any $3 \times rpq$ sub-array containing any two rows from the first group and any one row from the second group every level combinations occur equally often.

Conditions (a_1)–(a_2) imply that the array $(3 + c) \times rpq$ array is obviously an EMOA[rpq, $3+c$, $r \times p \times q \times 2^c$, $(3, 2)$, $(2; 2, 1)$]. Therefore, we get the following theorem.

Theorem 8.3.3 *The existence of an EMOA [rpq, $3+c$, $r \times p \times q \times 2^c$, $(3, 2)$, $(2; 2, 1)$] implies the existence of an OCD for both split- and strip-plot set-ups.*

Below we describe some methods of getting an EMOA [rpq, $3+c$, $r \times p \times q \times 2^c$, $(3, 2)$, $(2; 2, 1)$] which gives an OCD for both split-plot and strip-plot set-ups.

Theorem 8.3.4

(1) If $\mathbf{H}_r$, $\mathbf{H}_p$ and $\mathbf{H}_q$ exist, then an EMOA [rpq, $3+c$, $r \times p \times q \times 2^c$, $(3, 2)$, $(2; 2, 1)$] can be constructed, where $c = (r - 1)(p - 1)(q - 1)$.

(2) If $\mathbf{H}_{2r}$, $\mathbf{H}_p$ and $\mathbf{H}_q$ exist, where r is even, then an EMOA [rpq, $3+c$, $r \times p \times q \times 2^c$, $(3, 2)$, $(2; 2, 1)$] can be constructed, where $c = (r - 1)(p - 1)(q - 1) - (r - 2)(p - 1)$.

(3) If $r \equiv 2 \pmod{4}$, $(r - 1)$ is a prime or a prime power and $\mathbf{H}_p$ and $\mathbf{H}_q$ exist, where r is even, then an EMOA [rpq, $3+c$, $r \times p \times q \times 2^c$, $(3, 2)$, $(2; 2, 1)$] can be constructed, where $c = (r - 1)(p - 1)(q - 1) - (r - 2)$.

Example 8.3.2 Let us take $r = 2$, $p = 2$, $q = 4$. $\mathbf{H}_2$ and $\mathbf{H}_4$ can be, respectively, written as

$$\mathbf{H}_2 = \begin{pmatrix} 1 & -1 \\ 1 & 1 \end{pmatrix} = \begin{pmatrix} \mathbf{H}_2^* \\ \mathbf{1}' \end{pmatrix},$$

$$\mathbf{H}_4 = \begin{pmatrix} 1 & -1 & -1 & 1 \\ 1 & 1 & -1 & -1 \\ 1 & -1 & 1 & -1 \\ 1 & 1 & 1 & 1 \end{pmatrix} = \begin{pmatrix} \mathbf{H}_4^* \\ \mathbf{1}' \end{pmatrix}$$

For $r = 2$, $p = 2$, $q = 4$, $\mathbf{X}_1$, $\mathbf{X}_2$, $\mathbf{X}_3$ and $\mathbf{X}_4$ are written in Example 8.3.1.

The optimum $\mathbf{Z}'$-matrix for split- and strip-plot designs with $r = 2$, $p = 2$, $q = 4$ is given by:

$$\mathbf{Z}' = \mathbf{H}^{3 \times 16} = \mathbf{H}_2^* \otimes \mathbf{H}_2^* \otimes \mathbf{H}_4^*$$

$$= \begin{pmatrix} 1 & -1 & -1 & 1 & -1 & 1 & 1 & -1 & -1 & 1 & 1 & -1 & 1 & -1 & -1 & 1 \\ 1 & 1 & -1 & -1 & -1 & -1 & 1 & 1 & -1 & -1 & 1 & 1 & 1 & 1 & -1 & -1 \\ 1 & -1 & 1 & -1 & -1 & 1 & -1 & 1 & -1 & 1 & -1 & 1 & 1 & -1 & 1 & -1 \end{pmatrix}. \qquad (8.3.48)$$

Let us augment the matrix $\mathbf{Z}'$ with a 3×16 matrix $\mathbf{D}$ whose columns denote the coordinates of the cells of the z-values in lexicographic order. Then $\left(\mathbf{D}', \mathbf{Z}'\right)'$ gives the EMOA$[16, 2, 2 \times 2 \times 4 \times 2^3, (3,2), (2; 2,1)]$ which is as follows:

$$\begin{pmatrix} 1 & 1 & 1 & 1 & 1 & 1 & 1 & 1 & 2 & 2 & 2 & 2 & 2 & 2 & 2 & 2 \\ 1 & 1 & 1 & 1 & 2 & 2 & 2 & 2 & 1 & 1 & 1 & 1 & 2 & 2 & 2 & 2 \\ 1 & 2 & 3 & 4 & 1 & 2 & 3 & 4 & 1 & 2 & 3 & 4 & 1 & 2 & 3 & 4 \\ 1 & -1 & -1 & 1 & -1 & 1 & 1 & -1 & -1 & 1 & 1 & -1 & 1 & -1 & -1 & 1 \\ 1 & 1 & -1 & -1 & -1 & -1 & 1 & 1 & -1 & -1 & 1 & 1 & 1 & 1 & -1 & -1 \\ 1 & -1 & 1 & -1 & -1 & 1 & -1 & 1 & -1 & 1 & -1 & 1 & 1 & -1 & 1 & -1 \end{pmatrix}.$$

Condition (iv) of Theorem 8.3.2 is an additional condition with those conditions for OCDs in split-plot design set-up. We can still get an OCD for split-plot design set-up if we use $\mathbf{H}_2$ instead of instead of $\mathbf{H}_2^*$. Therefore, the optimum $\mathbf{Z}'$-matrix for split-plot design with $r = 2$, $p = 2$, $q = 4$ is given by:

$$\mathbf{Z}' = \mathbf{H}^{6 \times 16} = \mathbf{H}_2^* \otimes \mathbf{H}_2 \otimes \mathbf{H}_4^*$$

$$= \begin{pmatrix} 1 & -1 & -1 & 1 & -1 & 1 & 1 & -1 & -1 & 1 & 1 & -1 & 1 & -1 & -1 & 1 \\ 1 & 1 & -1 & -1 & -1 & -1 & 1 & 1 & -1 & -1 & 1 & 1 & 1 & 1 & -1 & -1 \\ 1 & -1 & 1 & -1 & -1 & 1 & -1 & 1 & -1 & 1 & -1 & 1 & 1 & -1 & 1 & -1 \\ 1 & -1 & -1 & 1 & 1 & -1 & -1 & 1 & -1 & 1 & 1 & -1 & -1 & 1 & 1 & -1 \\ 1 & 1 & -1 & -1 & 1 & 1 & -1 & -1 & -1 & -1 & 1 & 1 & -1 & -1 & 1 & 1 \\ 1 & -1 & 1 & -1 & 1 & -1 & 1 & -1 & -1 & 1 & -1 & 1 & -1 & 1 & -1 & 1 \end{pmatrix}. \qquad (8.3.49)$$

It can be easily be verified that the above $\mathbf{Z}$-matrices in (8.3.48) and (8.3.49) satisfy all the conditions of Theorems 8.3.1 and 8.3.2 respectively.

References

Bose M, Dey A (2009) Optimal crossover designs. World Scientific Publishing Co., Pvt. Ltd., Singapore

Bose RC (1947) Mathematical theory of the symmetrical factorial design. Sankhyā, **8**:107–166

Chakrabarti MC (1962) Mathematics of design and analysis of experiments. Asia Publishing House, New York

Chakravarti IM (1956) Fractional replication in asymmetrical factorial designs and partially balanced arrays. Sankhyā, **17**:143–164

Cheng CS, Wu CF (1980) Balanced repeated measurements designs. Ann Stat 8:1272–1283 (Corrigendum: ibid. 11:349 (1983))

Cochran WG, Cox GM (1950) Experimental designs. Wiley, New York

Das K, Mandal NK, Sinha Bikas K (2003) Optimal experimental designs for models with covariates. J Stat Plan Inference 115:273–285

Dey A, Mukerjee R (1999) Fractional factorial plans. Wiley, New York

Dutta G, Das P, Mandal NK (2009) Optimum covariate designs in split-plot and strip-plot design set-ups. J Appl Stat 36:893–906

Dutta G, Das P (2013) Optimum design for estimation of regression parameters in multi-factor set-up. Commun Stat Theory Methods 42:4431–4443

Dutta G, SahaRay R (2013) Optimal choice of covariates in the set-up of crossover designs. Stat Appl 11(1 and 2):93–109 (Special Issue in Memory of Professor M. N. Das)

Gupta S, Mukerjee R (1989) A calculus for factorial arrangements. Lecture notes in statistics, vol 59. Springer, Berlin

Hedayat AS, Stufken J, Sloane NJA (1999) Orthogonal arrays: theory and applications. Springer, New York

Kshirsagar AM (1983) A course in linear models. Marcel Dekker Inc., New York

Patterson HD (1952) The construction of balanced designs for experiments involving sequences of treatments. Biometrika 39:32–48

Pukelsheim F (1993) Optimal design of experiments. Wiley, New York

Raghavarao D (1971) Constructions and combinatorial problems in design of experiments. Wiley, New York

Rao PSSNVP, Rao SB, Saha GM, Sinha BK (2003) Optimal designs for covariates' models and mixed orthogonal arrays. Electron Notes Discret Math 15:155–158

Sinha BK (2009) A reflection on the choice of covariates in the planning of experimental designs. J Indian Soc Agricul Stat 63:219–225

Stufken, J. (1996) Optimal crossover designs. In: Ghosh S, Rao CR (eds) Design and analysis of experiments. Handbook of statistics 13:63–90, North-Holland, Amsterdam

Williams EJ (1949) Experimental designs balanced for the estimation of residual effects of treatments. Austral. J. Sci. Res. A 2:149–168

Chapter 9
Applications of the Theory of OCDs

9.1 Introduction: Eye-Openers

In this concluding chapter, we propose to discuss at length several examples from standard textbooks. All of these examples deal with ANCOVA models and related analyses of data. We intend to capitalize on our understanding of OCDs in different ANCOVA models as discussed in Chaps. 2–8 and revisit these examples with a view to suggest optimal/highly efficient designs for estimation of the covariate parameter(s). As we will see, for some examples our task is very much routine but for others, it is indeed a highly non-trivial exercise. Most of the material in this chapter is based on Dutta and Sinha (2015).

Example 9.1.1 We started with this example in Chap. 1. It relates to a leprosy study quoted from Snedecor and Cochran's book (1989, p. 377). The point we made is that there is ample scope of improvement in the efficiency of the estimates for the covariates' parameters if we have a 'free' hand in the recruitment of the patients and if a 'pool' is made available to us. Since the basic design is a CRD and there are three 'treatments' under consideration—with ten patients to be recruited under each treatment—an OCD suggests the following scheme of recruitment of the patients in terms of their possession of original pre-treatment score (count of bacilli)—under the supposition that we have a 'free choice' of the patients from a conceivably larger pool. Table 9.1 shows the scheme.

It was further stated that as against the given patients' ad hoc recruitment scheme in Table 1.1 (Chap. 1), the above scheme provides more than 300 % gain in efficiency towards estimation of the covariate parameter. Even with the 'given' pool of 30 patients, a suitable reallocation of the patients across the three treatments, as indicated in Table 1.2 (Chap. 1), would have provided 12 % gain in efficiency against the 'adhoc' allocation in Table 1.1 (Chap. 1). The OCD given in Table 9.1 is based on the theory developed in Chap. 2 with regard to the CRD. Recall the formation of W-matrix with the coded covariate values. In applications, the code -1 (respectively, $+1$) is to be replaced by x_{min} (respectively, x_{max}) which are '3' and '21' in the above example.

© Springer India 2015
P. Das et al., *Optimal Covariate Designs*, DOI 10.1007/978-81-322-2461-7_9

Table 9.1 Recruitment of patients based on pre-treatment score in actual units (patient serial number, covariate value)

1.	Treatment A	(P1, 3), (P2, 3), (P3, 3), (P4, 3), (P5, 3), (P6, 21), (P7, 21), (P8, 21), (P9, 21), (P10, 21)
2.	Treatment D	(P11, 3), (P12, 3), (P13, 3), (P14, 3), (P15, 3), (P16, 21), (P17, 21), (P18, 21), (P19, 21), (P20, 21)
3.	Control F	(P21, 3), (P22, 3), (P23, 3), (P24, 3), (P25, 3), (P26, 21), (P27, 21), (P28, 21), (P29, 21), (P30, 21)

We will now carry out the non-trivial exercise of identifying the design indicated in Table 1.2 (Chap. 1) as obtained through adequate re-allocation of the covariate values of the given pool of 30 patients as in the given design, to be denoted by d_0. For the sake of completeness, we display the allocation of covariate-values over the three treatments as in d_0.

$$
\begin{array}{l}
A: 3, 5, 6, 6, 8, 10, 11, 11, 14, 19 \\
D: 5, 6, 6, 7, 8, 8, 8, 15, 18, 19 \\
F: 7, 9, 11, 12, 12, 12, 13, 16, 16, 21
\end{array} \right\} = d_0, \text{ say.}
$$

It follows that, in terms of the Z-scores ranging in $[-1, 1]$,

$$
\mathbf{I}(\theta) = \begin{pmatrix}
10 & 0 & 0 & -3.0000 \\
0 & 10 & 0 & -2.2222 \\
0 & 0 & 10 & 1.0000 \\
-3.0000 & -2.2222 & 1.0000 & 8.8148
\end{pmatrix}.
$$

Routine computation yields: Information for γ, $I_{d_0}(\gamma) = 7.3210$.

Towards an 'improved' allocation, we arrange the data of pre-treatment scores of all the 30 patients in ascending order: 3, 5, 5, 6, 6, 6, 6, 7, 7, 8, 8, 8, 8, 9, 10, 11, 11, 11, 12, 12, 12, 13, 14, 15, 16, 16, 18, 19, 19, 21.

Now by using the following algorithms, we make an attempt to search for a design for which the information of γ is maximum.

Algorithm 1

Step 1: We conveniently divide ordered Z-scores into three blocks. Block 1 consists of the first nine observations of arranged data, i.e. (3, 5, 5, 6, 6, 6, 6, 7, 7); Block 2 consists of the next 12 observations, i.e. (8, 8, 8, 8, 9, 10, 11, 11, 11, 12, 12, 12); Block 3 consists of the last 9 observations (13, 14, 15, 16, 16, 18, 19, 19, 21).

Step 2: In Block 1 we allocate the first 3 observations i.e. (3, 5, 5) under treatment A, the next 3 observations, i.e. (6, 6, 6) under treatment D and the last 3 observations, i.e. (6, 7, 7) under treatment F.

Step 3: In Block 2 we allocate the first 4 observations, i.e. (8, 8, 8, 8) under treatment D, the next 4 observations, i.e. (9, 10, 11, 11) under treatment F and the last 4 observations, i.e. (11, 12, 12, 12) under treatment A.

Step 4: In Block 3 we allocate the first 3 observations i.e. (13, 14, 15) under treatment F, the next 3 observations, i.e. (16, 16, 18) under treatment A and the last 3 observations, i.e. (19, 19, 21) under treatment D.

Hence we get the following arrangement:

	Block 1	Block 2	Block 3
A	3, 5, 5	11, 12, 12, 12	16, 16, 18
D	6, 6, 6	8, 8, 8, 8	19, 19, 21
F	6, 7, 7	9, 10, 11, 11	13, 14, 15

$= d_1$, say.

The information of γ from $d_1 = I_{d_1}(\gamma) = 8.1852$.

Step 5: Start with d_1. Consider the left block, i.e. Block 1. Permute the rows and generate $3! = 6$ options for this block, while keeping the middle and the right block intact. Work out $I(\gamma)$ for all the 6 options generated from the left block. Identify the best case scenario and hold this intact while passing into the middle block. Here the best design is found to be d_1.

Step 6: For the middle block, i.e. Block 2, we follow a similar step. Here the best design using Step 6 is

	Block 1	Block 2	Block 3
A	3, 5, 5	9, 10, 11, 11	16, 16, 18
D	6, 6, 6	8, 8, 8, 8	19, 19, 21
F	6, 7, 7	11, 12, 12, 12	13, 14, 15

$= d_2$, say.

The information of γ for $d_2 = I_{d_2}(\gamma) = 8.2$.

Step 7: For the last block, i.e. Block 3, we again follow similarly step. Ultimately we get d_2 as the best design.

We now consider other aspects towards improving d_2.

Algorithm 2

Here we consider three allocations:

(I) (ADF—DFA—FAD——ADF—DFA—FAD—ADF——DFA—FAD—ADF)
(II) (ADF—DFA—FAD——ADF—DFA—FAD—ADF——DFA—FAD—DFA)
(III) (ADF—DFA—FAD——ADF—DFA—FAD—ADF——DFA—FAD—FAD)

and the the designs are respectively:

	Block 1	Block 2	Block 3
A	3, 6, 7	8, 10, 11, 12	15, 16, 19
D	5, 6, 7	8, 8, 11, 12	13, 18, 19
F	5, 6, 6	8, 9, 11, 12	14, 16, 21

$= d_{(I)}$, say;

	Block 1	Block 2	Block 3
A	3, 6, 7	8, 10, 11, 12	15, 16, 21
D	5, 6, 7	8, 8, 11, 12	13, 18, 19
F	5, 6, 6	8, 9, 11, 12	14, 16, 19

$= d_{(II)}$, say;

	Block 1	Block 2	Block 3
A	3, 6, 7	8, 10, 11, 12	15, 16, 19
D	5, 6, 7	8, 8, 11, 12	13, 18, 21
F	5, 6, 6	8, 9, 11, 12	14, 16, 19

$= d_{(III)}$, say.

For the above three designs, $I_{d_{(I)}}(\gamma) = 8.2198$, $I_{d_{(II)}}(\gamma) = 8.2148$ and $I_{d_{(III)}}(\gamma) = 8.2148$.

Algorithm 3

We may consider another allocation:

(AFD—FDA—DAF——AFD—FDA—FDA—AFD——DAF—FDA——AFD)

and the corresponding design is

	Block 1	Block 2	Block 3
A	3, 6, 7	8, 10, 11, 12	14, 18, 19
D	5, 6, 6	8, 9, 11, 12	13, 16, 21
F	5, 6, 7	8, 8, 11, 12	15, 16, 19

$= d_3$, say.

Here also $I(\gamma) = 8.2198$.

Heuristic Search:

	G_1	G_2	G_3
A	3, 6, 6	9, 10, 11, 12	14, 18, 19
D	5, 6, 7	8, 8, 11, 12	15, 16, 19
F	5, 6, 7	8, 8, 11, 12	13, 16, 21

$= d_4$, say.

This yields $I_{d_4}(\gamma) = 8.2198$ and d_3 is equivalent to d_4. Further, these are also equivalent to $d_{(I)}$ in the sense of same information.

In the final analysis, we find that there is substantial gain in efficiency (more than 12 %) in the performance of the design $d_{(I)}$ or d_3, as against the original design d_0. This is the design ($d_{(I)}$ or d_3) displayed in Table 1.2 (Chap. 1).

Example 9.1.2 We now elaborate on the second example discussed in Chap. 1. Recall that this refers to an RBD with $b = 5$, $v = 3$. In Chap. 3, we have discussed at length OCDs under RBD set-ups but mostly the 'regular' cases, viz. both b and v being multiples of 4 so that Hadamard matrices exist. Here is a deviation from that and we take this rare opportunity to discuss the example in quite details. Note that we have already provided solutions to two different aspects of the example: (i) For given covariate-values, improved allocation of the available experimental units across the RBD layout; (ii) For a 'free' choice of the covariate values within certain well-defined closed intervals, identification of the experimental units with covariate values of the experimenter's choice. Below we give detailed derivations of the above results. We refer to Tables 1.3, 1.4a, b, 1.5a, b and 1.6a, b in Chap. 1.

Under an RBD ANCOVA model with a single covariate, recall the standard expression for information on γ, viz.

$$
\begin{aligned}
I(\gamma) &= \sum_{i=1}^{5}\sum_{j=1}^{3} z_{ij}^2 - \frac{1}{3}\sum_{i=1}^{5} R_i^2 - \frac{1}{5}\sum_{j=1}^{3} C_j^2 + \frac{G^2}{15} \\
&= \sum_{i=1}^{5}\sum_{j=1}^{3} z_{ij}^2 - \frac{1}{3}\sum_{i=1}^{5} R_i^2 - 5\sum_{j=1}^{3}\left(\bar{z}_{0j} - \bar{z}_{00}\right)^2
\end{aligned}
\tag{9.1.1}
$$

Our aim is to maximize the information of γ given in (9.1.1) by properly allocating the pigs in the two-way RBD layout. This applies to both female and male pigs. Note that towards this, the row totals of the covariate-values should be as close as possible and the same is true of the column totals. We start with the 5×3 table of covariate values for the female pigs and proceed through the following steps:

Step 1: First, we arrange the rows in three sets where the first set consists of the rows where all the covariate values are equal; in the second set, we consider those rows where two of the three covariate values are not equal and the last set consists of the rows where all the covariate values are unequal. The arrangement is shown in Table 9.2.

Step 2: We select the first row of second set (i.e., Pen No. 2) and permute the covariate values keeping the other rows fixed. Next we compute the information of γ_F under each permutation. Then we choose the design in which the information of γ_F will be a maximum. We do the same for the second row of the second set (i.e., Pen No. 4) keeping the other rows of the new design intact. Similarly, we do the same for the third set also (Pen No. 3 and 5).

Step 3: We repeat Step 2 until all C_{Fj}'s are as close as possible to $\frac{G_F}{3}$. Finally, we get the design where the information of γ_F is a maximum with C_{Fj}'s as close as possible to $\frac{G_F}{3}$. We denote it by d_{F1} and $I_{d_{F1}}(\gamma_F) = 81.0667$, where d_{F1} is displayed in Table 9.3.

Table 9.2 Female

Pen	Treatment			Totals
	A	B	C	
1	48	48	48	144
2	32	32	28	92
4	46	46	50	142
3	35	41	33	109
5	32	37	30	99
Totals	193	204	189	586

Table 9.3 d_{F1}

Pen	Treatment			Totals
	A	B	C	
1	48	48	48	144
2	32	28	32	92
4	46	46	50	142
3	41	35	33	109
5	30	37	32	99
Totals	197	194	195	586

Table 9.4 An alternate arrangement

Treatment			Totals
A	B	C	
30	48	*	78
32	46	*	78
41	37	*	78
46	35	*	81
48	28	*	76

Table 9.5 d_{F2}

	Treatment			Totals
	A	B	C	
	30	48	48	126
	32	46	33	111
	41	37	32	110
	46	35	32	113
	48	28	50	126
Totals	197	194	195	586

Step 4: We arrange the initial weights under treatment A in ascending order and the initial weights under treatment B in descending order. The arrangement is shown in Table 9.4.

Since the sum of the two entries in each of 5 rows are 78, 78, 78, 81, 76, we fill the entries under treatment C as 48, 33, 32, 32, 50. Then we get the design d_{F2} and here $I_{d_{F2}}(\gamma_F) = 782.4$. We display the design d_{F2} in Table 9.5.

For another option, we arrange the initial weights under treatment A in ascending order and the initial weights under treatment C in descending order. The arrangement is shown in Table 9.6.

Since the sum of the two entries in each of 5 rows are 80, 80, 74, 78, 80, we fill the entries under treatment B as 37, 35, 48, 46, 28. Then we get the design d_{F3} displayed in Table 9.7 and here $I_{d_{F3}}(\gamma_F) = 817.0667$.

Table 9.6 A second alternative

Treatment			Totals
A	B	C	
30	*	50	80
32	*	48	80
41	*	33	74
46	*	32	78
48	*	32	80

Table 9.7 d_{F3}

Treatment			Totals
A	B	C	
30	37	50	117
32	35	48	115
41	48	33	122
46	46	32	124
48	28	32	108
Totals 197	194	195	586

Table 9.8 A third alternative

Pen	Treatment			Totals
	A	B	C	
1	*	28	50	78
2	*	35	48	83
4	*	37	33	70
3	*	46	32	78
5	*	48	32	80

Table 9.9 d_{F4}

Treatment			Totals
A	B	C	
41	28	50	119
30	35	48	113
48	37	33	118
46	46	32	124
32	48	32	112
Totals 197	194	195	586

Lastly, we arrange the initial weights under treatment B in ascending order and the initial weights under treatment C in descending order. The arrangement is shown in Table 9.8.

Since the sum of the two entries in each of 5 rows are 78, 83, 70, 78, 80, we fill the entries under treatment A as 41, 30, 48, 46, 32. Then we get the design d_{F4} shown in Table 9.9 and here $I_{d_{F4}}(\gamma_G) = 838.4$.

Table 9.10 d_{F5}

	Treatment			Totals
	A	B	C	
	41	28	48	117
	30	35	50	115
	48	37	33	118
	46	46	32	124
	32	48	32	112
Totals	197	194	195	586

Table 9.11 d_{F6}

	Treatment			Totals
	A	B	C	
	41	28	48	117
	30	37	50	117
	48	35	33	116
	46	46	32	124
	32	48	32	112
Totals	197	194	195	586

Table 9.12 d_{F7}

	Treatment			Totals
	A	B	C	
	46	28	48	122
	30	37	50	117
	48	35	33	116
	41	46	32	119
	32	48	32	112
Totals	197	194	195	586

Now we start with d_{F4} and proceed with Step 1 and Step 2. Then we observe that d_{F4} is a better design. Next we can improve over d_{F4} by interchanging the first element and the second element under Treatment C and denote the design by d_{F5} shown in Table 9.10. Here $I_{d_{F5}}(\gamma_F) = 843.7333$.

Again we can improve d_{F5} by interchanging the second element and the third element under treatment B and we denote the design by d_{F6} shown in Table 9.11. Here $I_{d_{F6}}(\gamma_F) = 845.0667$.

We can further improve d_{F6} by interchanging the first element and the fourth element under Treatment A and denote the design by d_{F7} shown in Table 9.12. Here $I_{d_{F7}}(\gamma_F) = 851.7333$.

Lastly, we improve d_{F7} by interchanging the third element and the fourth element under Treatment C and denote the design by d_{F8} shown in Table 9.13. Here $I_{d_{F8}}(\gamma_F) = 853.7333$.

Table 9.13 d_{F8}

	Treatment			Totals
	A	B	C	
	46	28	48	122
	30	37	50	117
	48	35	32	115
	41	46	32	119
	32	48	33	113
Totals	197	194	195	586

Table 9.14 d_{F9}

	Treatment			Totals
	A	B	C	
	46	28	48	122
	30	37	50	117
	46	35	32	113
	41	48	32	121
	32	48	33	113
Totals	195	196	195	586

Table 9.15 d_{M1}

Pen	Treatment			Totals
	A	B	C	
5	43	40	40	125
1	38	39	48	110
2	37	38	35	129
3	41	46	42	130
4	48	42	40	123
Totals	207	205	205	617

Now we construct design d_{F9} shown in Table 9.14 by interchanging the third element under Treatment A and the fourth element under Treatment B of d_{F8}. Here $I_{d_{F9}}(\gamma_G) = 846.5333$ which is less than $I_{d_{F8}}(\gamma_F)$ even though column sums are more or less equal. We stop here and recommend the design d_{F8} for use.

As is indicated in the above, the designs d_{F6} to d_{F9} are displayed in Tables 9.11, 9.12, 9.13 and 9.14 respectively. Similarly, for increasing the information of γ_M, we follow the Steps 1, 2, 3 and get the design denoted by d_{M1} and shown in Table 9.15.

Here $I_{d_{M1}}(\gamma_H) = 119.4667$.

In the same way as mentioned above for female pigs, to get three designs we follow Step 4 and find d_{M2} shown in Table 9.16 with $I_{d_{M2}}(\gamma_M) = 195.4667$, d_{M3} shown in Table 9.17 with $I_{d_{M3}}(\gamma_H) = 202.1333$ and d_{M4} displayed in Table 9.18 with $I_{d_{M4}}(\gamma_M) = 193.4667$. Therefore, d_{M3} is a better design. Now we start with d_{M3} and

Table 9.16 d_{M2}

	Treatment			Totals
	A	B	C	
	37	46	40	123
	38	42	48	128
	41	40	42	123
	43	39	40	122
	48	38	35	121
Totals	207	205	205	617

Table 9.17 d_{M3}

	Treatment			Totals
	A	B	C	
	37	38	48	123
	38	46	42	126
	41	42	40	123
	43	40	40	123
	48	39	35	122
Totals	207	205	205	617

Table 9.18 d_{M4}

	Treatment			Totals
	A	B	C	
	37	38	48	123
	43	39	42	124
	48	40	40	128
	38	42	40	120
	41	46	35	122
Totals	207	205	205	617

we improve it further and denote the design by d_{M5} with $I_{d_{M5}}(\gamma_M) = 202.5333$ using Steps 1, 2, 3. This design d_{M5} is displayed in Table 9.19. Again we improve d_{M5} by interchanging the second element and the fourth element under Treatment C and denote it by d_{M6} which is displayed in Table 9.20. Here $I_{d_{M6}}(\gamma_M) = 203.8667$. Next we improve d_{M6} by interchanging the third element and the fourth element under treatment C and denote it by d_{M7}. Here $I_{d_{M7}}(\gamma_M) = 205.2$. Lastly there is another possibility to improve d_{M7} by interchanging the 4th element and the 5th element under Treatment B. We denote it by d_{M8}. Here $I_{d_{M8}}(\gamma_M) = 204.5333$.

We stop here and accept the design d_{M7} for the use of male pigs. It is a routine task to compute the percent gain. The designs d_{M7} and d_{M8} are displayed in Tables 9.21 and 9.22 respectively.

Now we consider the 'hypothetical' situation for female data wherein the row totals are equal or nearly equal to each other and also the column totals are equal or nearly equal to each other. For female data, that would amount to the row sums being

Table 9.19 d_{M5}

	Treatment			Totals
	A	B	C	
	37	38	48	123
	38	46	42	126
	40	42	41	123
	43	40	40	123
	48	39	35	122
Totals	206	205	206	617

Table 9.20 d_{M6}

	Treatment			Totals
	A	B	C	
	37	38	48	123
	38	46	40	124
	40	42	41	124
	43	40	42	124
	48	39	35	122
Totals	206	205	206	617

Table 9.21 d_{M7}

	Treatment			Totals
	A	B	C	
	37	38	48	123
	38	46	40	124
	40	42	42	124
	43	39	41	123
	48	40	35	123
Totals	206	205	206	617

Table 9.22 d_{M8}

	Treatment			Totals
	A	B	C	
	37	38	48	123
	38	46	40	124
	40	42	42	124
	43	40	41	124
	48	39	35	122
Totals	206	205	206	617

117, 117, 117, 117 and 118 and column sums being 195, 195, 196 since the total is fixed at $G_F = 586$. In this hypothetical situation $I(\gamma_F) = 870.5333$. Therefore, the relative efficiency of $d_{F8} = 98.0702\,\%$ whereas the relative efficiency of γ_F under design d_0 is $6.6473\,\%$. In other words, relative gain in efficiency by use of d_{F8} as against d_0 is more than 1300 %. Similarly, we consider the hypothetical situation for male data where the row sums are meant to be 123, 123, 123, 124, 124 and column sums are also meant to be 206, 206, 205. Here $I(\gamma_M) = 205.2$ which is equal to d_{M7}. Therefore, the relative efficiency of $d_{M7} = 100\%$ whereas the relative efficiency of γ_M under design d_0 is 56.6602 %. In other words, relative gain in efficiency by the use of d_{M7} as against d_0 is more than 75 %. Hence we can improve the information of covariate parameter by properly allocating the covariate values in the rows and columns separately.

Recall the expression for $I(\gamma)$ given in (9.1.1) above. It is readily seen that

$$I(\gamma) \le \sum_{i=1}^{5} \sum_{j=1}^{3} z_{ij}^2 - \frac{1}{3} \sum_{i=1}^{5} R_i^2 \qquad (9.1.2)$$

equality holds if $\bar{z}_{0j} = \bar{z}_{00}$ for all j, where $\bar{z}_{0j} = \frac{C_j}{5}$ and $\bar{z}_{00} = \frac{G}{15}$ and since we fix z_{ij}

at ± 1, $I(\gamma) \le \sum_{i=1}^{5} \sum_{j=1}^{3} z_{ij}^2 - \frac{1}{3} \sum_{i=1}^{5} R_i^2 \le 15 - \frac{5}{3} = \frac{40}{3} = 13.33$.

Again,

$$I(\gamma) \le \sum_{i=1}^{5} \sum_{j=1}^{3} z_{ij}^2 - \frac{1}{5} \sum_{j=1}^{3} C_j^2 = I_2(\gamma) \text{ (say)} \qquad (9.1.3)$$

equality holds if $\bar{z}_{i0} = \bar{z}_{00}$ for all i, where $\bar{z}_{i0} = \frac{R_i}{3}$ and $\bar{z}_{00} = \frac{G}{15}$ and $I_2(\gamma) \le 15 - \frac{3}{5} = \frac{72}{5} = 14.4$. Therefore, combining the two inequalities, we deduce that $I(\gamma) \le 13.33$. We display a design in Table 9.23 for which $I(\gamma)$ attains the bound 13.33.

The above exercise suggests that if we have such flexibility to choose the initial weights for males and females separately, then our suggestion is the design shown in Table 9.23 in terms of z-values and this applies to both males and females. For female and male pigs, the arrangement of weights for OCD are already shown in Table 1.6a, b in Chap. 1 in terms of original weights.

Table 9.23 Design where $I_1(\gamma)$ is attained

	A	B	C	Totals
	−1	1	1	1
	1	−1	1	1
	1	1	−1	1
	−1	1	1	1
	1	−1	−1	−1
Totals	1	1	1	$3 = G$

9.2 Other Application Areas

In this section, we undertake four different types of examples of application of OCD.

Example 9.2.1 We consider the observations and the design (see Goos and Jones (2011), p. 79) in Table 9.24.

The response variable is peel strength, which measures the amount of force required to open the package. Raw material from three suppliers (S_1, S_2 and S_3) is used in the sealing process. We consider temperature (Z_1), pressure (Z_2) and speed (Z_3) on the peel strength as covariates. These three covariates are controllable. The range and unit of each covariate is given in Table 9.25.

Note that the authors present a very general ANCOVA model in the book and discuss some aspects of D-optimal designs. Instead, we will consider the simplest model:

Table 9.24 Data for the robustness experiment

Run number	Temperature	Pressure	Speed	Supplier	Peel strength
1	211.5	2.2	32	1	4.36
2	193.0	2.7	32	1	5.20
3	230.0	3.2	32	1	4.75
4	230.0	2.2	41	1	5.73
5	193.0	3.2	41	1	4.49
6	193.0	2.2	50	1	6.38
7	230.0	2.7	50	1	5.59
8	211.5	3.2	50	2	5.40
9	193.0	2.2	32	2	5.78
10	230.0	2.7	32	2	4.80
11	193.0	3.2	32	2	4.93
12	211.5	2.7	41	2	5.96
13	211.5	2.7	41	2	6.55
14	230.0	2.2	50	2	6.92
15	193.0	2.7	50	2	6.18
16	230.0	3.2	50	2	6.55
17	230.0	2.2	32	3	5.44
18	193.0	2.7	32	3	4.57
19	211.5	3.2	32	3	4.48
20	193.0	2.2	41	3	4.78
21	211.5	2.7	41	3	5.03
22	230.0	3.2	41	3	3.98
23	211.5	2.2	50	3	4.73
24	193.0	3.2	50	3	4.70

Table 9.25 Range and unit of covariates

Covariate	Range	Unit
Temperature	193–230	°C
Pressure	2.2–3.2	bar
Speed	32–50	cpm

$$E(y_{ij}) = s_i + \sum_{l=1}^{3} \gamma_l z_{ij}^{(l)},$$

where y_{ij} is the jth observation corresponding to ith supplier and s_i is the effect due to ith supplier and γ_l is the lth covariate effect, $z_{ij}^{(l)}$ is the lth covariate value corresponding the (i, j)th observation and $|z_{ij}^{(l)}| \leq 1$ for all i, j, l. Further, we will take up the problem of estimating the three covariate parameters most efficiently by selecting the design. With the replication numbers each equal to 8, it would have been a trivial exercise in a CRD model with three covariates. Vide Chap. 2. However, with $(7, 8, 9)$ as the replication numbers, standard theory breaks down and we run into what has been termed as 'non-regular case'. As usual, the covariates are all coded to lie inside the closed interval $[-1, 1]$. From Theorem 2.3.1 in Chap. 2, for fixed $\{r_i\}$, i.e. $r_1 = 7$, $r_2 = 9$, $r_3 = 8$,

$$det(\mathbf{I}(\boldsymbol{\theta})) \leq (7 \times 9 \times 8)\,(a + 2b)\,(a - b)^2$$

where $a = 24 - \delta$, $b = \delta = \frac{1}{7} + \frac{1}{9}$ and r_i is the number of times the ith supplier replicated.

Therefore, $det(\mathbf{I}(\boldsymbol{\theta})) \leq 7 \times 9 \times 8 \times 13385.21$. But it is very difficult to construct **Z**-matrix where $det(\mathbf{I}(\boldsymbol{\theta}))$ attains the upper bound. Here we construct the design (say d^*) whose $det(\mathbf{I}(\boldsymbol{\theta}))$ is very close to the upper bound and the **Z**-matrix is written as:

$$\mathbf{Z}' = \left(\mathbf{Z}'(S_1), \mathbf{Z}'(S_2), \mathbf{Z}'(S_3)\right)$$

where

$$\mathbf{Z}'(S_1) = \begin{pmatrix} 1 & -1 & 1 & -1 & -1 & 1 & -1 \\ 1 & -1 & -1 & 1 & -1 & -1 & 1 \\ -1 & 1 & -1 & 1 & -1 & 1 & -1 \end{pmatrix};$$

$$\mathbf{Z}'(S_2) = \begin{pmatrix} 1 & -1 & 1 & -1 & 1 & -1 & 1 & -1 & 1 \\ 1 & -1 & -1 & 1 & 1 & -1 & -1 & 1 & 1 \\ -1 & 1 & -1 & 1 & 1 & -1 & 1 & -1 & 1 \end{pmatrix};$$

$$\mathbf{Z}'(S_3) = \begin{pmatrix} 1 & -1 & 1 & -1 & 1 & -1 & 1 & -1 \\ 1 & -1 & -1 & 1 & 1 & -1 & -1 & 1 \\ -1 & 1 & -1 & 1 & 1 & -1 & 1 & -1 \end{pmatrix}.$$

The information matrix for $\theta = (s_1, s_2, s_3, \gamma_1, \gamma_2, \gamma_3)'$ is given below:

$$\mathbf{I}_{d^*}(\theta) = \begin{pmatrix} 7 & 0 & 0 & -1 & -1 & -1 \\ 0 & 9 & 0 & 1 & 1 & 1 \\ 0 & 0 & 8 & 0 & 0 & 0 \\ -1 & 1 & 0 & 24 & 0 & 0 \\ -1 & 1 & 0 & 0 & 24 & 0 \\ -1 & 1 & 0 & 0 & 0 & 24 \end{pmatrix}$$

whence the information matrix for γ is

$$\mathbf{I}_{d^*}(\gamma) = \begin{pmatrix} 23.746 & -0.254 & -0.254 \\ -0.254 & 23.746 & -0.254 \\ -0.254 & -0.254 & 23.746 \end{pmatrix}.$$

Here $det(\mathbf{I}_{d^*}(\theta)) = 7 \times 9 \times 8 \times 13385.14$ and the relative D-efficiency $= \frac{13385.14}{13385.21} \times 100 = 99.9995\,\%$. It is interesting to note that the three parts of the design d^* are derived essentially from $\mathbf{H}_8$.

Example 9.2.2 Here we consider the data given in van Belle et al. (2004) and reproduced below in Tables 9.26 and 9.27. It relates to 'exercise' data for healthy active males (44) and females (43). There are four covariates, viz. Heart Rate, Age, Height and Weight. The response variable is the VO_2 Max, measured in a suitable unit.

As usual, we introduce coded covariates Z_1, Z_2, Z_3, Z_4, each in the range $[-1, 1]$. We reproduce the above tables only in terms of the four covariates in coded forms, skipping the responses (vide Tables 9.28 and 9.29).

It is a routine task to carry out data analysis using a 4-variate linear regression model for each data set involving the four covariates. We skip that part. Instead, we ask a non-trivial problem. If an experimenter is to design the survey to accommodate 10 males and 10 females out of the above 'pool' and to maximize information-content for the joint estimation of the covariate parameters, what would have been our recommendation? Again, if we had a 'larger' pool of healthy males and females with specified Z_{min} and Z_{max} for each covariate, would our recommendation be far better off? We propose to address both the problems below. We assume that the covariate parameters are the same for both males and females for each of the characteristics. In a way, we set $\gamma_{1,M} = \gamma_{1,F}$ and so on.

For the first problem, we may use the following selection methods.

Method 1: Use of Heart Rate Data We start with the above data set and arrange the data set separately for males and females both where the values of heart rate Z_1 are in the ascending order. Then select 10 persons (5 males and 5 females) from the smallest values of the Z_1-scores and 10 persons (5 males and 5 females) from the largest values of the Z_1-scores and denote this design by d_1. The data on covariates of the selected persons are shown in Table 9.30.

Table 9.26 Exercise data for healthy active males

Sl. no.	(VO$_2$ max)/Duration	Heart rate	Age	Height	Weight
1	0.0588	192	46	165	57
2	0.0627	190	25	193	95
3	0.0694	190	25	187	82
4	0.0670	174	31	191	84
5	0.0670	194	30	171	67
6	0.0662	168	36	177	78
7	0.0579	185	29	174	70
8	0.0618	200	27	185	76
9	0.0670	164	56	180	78
10	0.0652	175	47	180	80
11	0.0604	175	46	180	81
12	0.0609	162	55	180	79
13	0.0656	190	50	161	66
14	0.0688	175	52	174	76
15	0.0603	164	46	173	84
16	0.0547	156	60	169	69
17	0.0655	174	49	178	78
18	0.0545	166	54	181	101
19	0.0615	184	57	179	74
20	0.0695	160	50	170	66
21	0.0621	186	41	175	75
22	0.0713	175	58	173	79
23	0.0615	175	55	160	79
24	0.0579	175	46	164	65
25	0.0632	174	47	180	81
26	0.0500	174	56	183	100
27	0.0695	168	82	183	82
28	0.0549	164	48	181	77
29	0.0490	146	68	166	65
30	0.0606	156	54	177	80
31	0.0566	180	56	179	82
32	0.0638	164	50	182	87
33	0.0628	166	48	174	72
34	0.0626	184	56	176	75
35	0.0619	186	45	179	75
36	0.0619	174	45	179	79
37	0.0717	188	43	179	73
38	0.0413	180	54	180	75
39	0.0642	168	55	172	71
40	0.0702	174	41	187	84
41	0.0712	166	44	185	81
42	0.0753	174	41	186	83
43	0.0672	180	50	175	78
44	0.0746	182	42	176	85

Data source van Belle et al. (2004), p. 294

Table 9.27 Exercise data for healthy active females

Sl. no.	(VO$_2$ max)/Duration	Heart rate	Age	Height	Weight
1	0.0577	184	23	177	83
2	0.0611	183	21	163	52
3	0.0655	200	21	174	61
4	0.0583	170	42	160	50
5	0.0485	188	34	170	68
6	0.0398	190	43	171	68
7	0.0527	190	30	172	63
8	0.0552	180	49	157	53
9	0.0615	184	30	178	63
10	0.0566	162	57	161	63
11	0.0578	188	58	159	54
12	0.0597	170	51	162	55
13	0.0649	184	32	165	57
14	0.0594	175	42	170	53
15	0.0564	180	51	158	47
16	0.0590	200	46	161	60
17	0.0558	190	37	173	56
18	0.0488	170	50	161	62
19	0.0614	158	65	165	58
20	0.0552	186	40	154	69
21	0.0511	166	52	166	67
22	0.0495	170	40	160	58
23	0.0662	188	52	162	64
24	0.0480	190	47	161	72
25	0.0556	194	43	164	56
26	0.0540	190	48	176	82
27	0.0509	190	43	165	61
28	0.0520	188	45	166	62
29	0.0497	184	52	167	62
30	0.0560	170	52	168	62
31	0.0661	168	56	162	66
32	0.0634	175	56	159	56
33	0.0560	156	51	161	61
34	0.0525	184	44	154	56
35	0.0544	180	56	167	79
36	0.0555	184	40	165	56
37	0.0654	156	53	157	52
38	0.0508	194	52	161	65
39	0.0564	190	40	178	64
40	0.0587	188	55	162	61
41	0.0647	164	39	166	59
42	0.0575	185	57	168	68
43	0.0542	178	46	156	53

Data source van Belle et al. (2004), p. 341

Table 9.28   Coded data for males

Sl. no.	Heart rate	Age	Height	Weight
1	0.7037	−0.2632	−0.6970	−1.0000
2	0.6296	−1.0000	1.0000	0.7273
3	0.6296	−1.0000	0.6364	0.1364
4	0.0370	−0.7895	0.8788	0.2273
5	0.7778	−0.8246	−0.3333	−0.5455
6	−0.1852	−0.6140	0.0303	−0.0455
7	0.4444	−0.8596	−0.1515	−0.4091
8	1.0000	−0.9298	0.5152	−0.1364
9	−0.3333	0.0877	0.2121	−0.0455
10	0.0741	−0.2281	0.2121	0.0455
11	0.0741	−0.2632	0.2121	0.0909
12	−0.4074	0.0526	0.2121	0.0000
13	0.6296	−0.1228	−0.9394	−0.5909
14	0.0741	−0.0526	−0.1515	−0.1364
15	−0.3333	−0.2632	−0.2121	0.2273
16	−0.6296	0.2281	−0.4545	−0.4545
17	0.0370	−0.1579	0.0909	−0.0455
18	−0.2593	0.0175	0.2727	1.0000
19	0.4074	0.1228	0.1515	−0.2273
20	−0.4815	−0.1228	−0.3939	−0.5909
21	0.4815	−0.4386	−0.0909	−0.1818
22	0.0741	0.1579	−0.2121	0.0000
23	0.0741	0.0526	−1.0000	0.0000
24	0.0741	−0.2632	−0.7576	−0.6364
25	0.0370	−0.2281	0.2121	0.0909
26	0.0370	0.0877	0.3939	0.9545
27	−0.1852	1.0000	0.3939	0.1364
28	−0.3333	−0.1930	0.2727	−0.0909
29	−1.0000	0.5088	−0.6364	−0.6364
30	−0.6296	0.0175	0.0303	0.0455
31	0.2593	0.0877	0.1515	0.1364
32	−0.3333	−0.1228	0.3333	0.3636
33	−0.2593	−0.1930	−0.1515	−0.3182
34	0.4074	0.0877	−0.0303	−0.1818
35	0.4815	−0.2982	0.1515	−0.1818
36	0.0370	−0.2982	0.1515	0.0000
37	0.5556	−0.3684	0.1515	−0.2727
38	0.2593	0.0175	0.2121	−0.1818
39	−0.1852	0.0526	−0.2727	−0.3636
40	0.0370	−0.4386	0.6364	0.2273
41	−0.2593	−0.3333	0.5152	0.0909
42	0.0370	−0.4386	0.5758	0.1818
43	0.2593	−0.1228	−0.0909	−0.0455
44	0.3333	−0.4035	−0.0303	0.2727

Table 9.29 Coded data for females

Sl. no.	Heart rate	Age	Height	Weight
1	0.2727	−0.9091	0.9167	1.0000
2	0.2273	−1.0000	−0.2500	−0.7222
3	1.0000	−1.0000	0.6667	−0.2222
4	−0.3636	−0.0455	−0.5000	−0.8333
5	0.4545	−0.4091	0.3333	0.1667
6	0.5455	0.0000	0.4167	0.1667
7	0.5455	−0.5909	0.5000	−0.1111
8	0.0909	0.2727	−0.7500	−0.6667
9	0.2727	−0.5909	1.0000	−0.1111
10	−0.7273	0.6364	−0.4167	−0.1111
11	0.4545	0.6818	−0.5833	−0.6111
12	−0.3636	0.3636	−0.3333	−0.5556
13	0.2727	−0.5000	−0.0833	−0.4444
14	−0.1364	−0.0455	0.3333	−0.6667
15	0.0909	0.3636	−0.6667	−1.0000
16	1.0000	0.1364	−0.4167	−0.2778
17	0.5455	−0.2727	0.5833	−0.5000
18	−0.3636	0.3182	−0.4167	−0.1667
19	−0.9091	1.0000	−0.0833	−0.3889
20	0.3636	−0.1364	−1.0000	0.2222
21	−0.5455	0.4091	0.0000	0.1111
22	−0.3636	−0.1364	−0.5000	−0.3889
23	0.4545	0.4091	−0.3333	−0.0556
24	0.5455	0.1818	−0.4167	0.3889
25	0.7273	0.0000	−0.1667	−0.5000
26	0.5455	0.2273	0.8333	0.9444
27	0.5455	0.0000	−0.0833	−0.2222
28	0.4545	0.0909	0.0000	−0.1667
29	0.2727	0.4091	0.0833	−0.1667
30	−0.3636	0.4091	0.1667	−0.1667
31	−0.4545	0.5909	−0.3333	0.0556
32	−0.1364	0.5909	−0.5833	−0.5000
33	−1.0000	0.3636	−0.4167	−0.2222
34	0.2727	0.0455	−1.0000	−0.5000
35	0.0909	0.5909	0.0833	0.7778
36	0.2727	−0.1364	−0.0833	−0.5000
37	−1.0000	0.4545	−0.7500	−0.7222
38	0.7273	0.4091	−0.4167	0.0000
39	0.5455	−0.1364	1.0000	−0.0556
40	0.4545	0.5455	−0.3333	−0.2222
41	−0.6364	−0.1818	0.0000	−0.3333
42	0.3182	0.6364	0.1667	0.1667
43	0.0000	0.1364	−0.8333	−0.6667

Table 9.30 Choice of males and females based on heart rate data

Sl. no.	Heart rate	Age	Height	Weight	Sex	
1	−1.0000	0.5088	−0.6364	−0.6364	M	
2	−0.6296	0.2281	−0.4545	−0.4545	M	
3	−0.6296	0.0175	0.0303	0.0455	M	
4	−0.4815	−0.1228	−0.3939	−0.5909	M	
5	−0.4074	0.0526	0.2121	0.0000	M	
6	0.6296	−1.0000	1.0000	0.7273	M	
7	0.6296	−1.0000	0.6364	0.1364	M	
8	0.7037	−0.2632	−0.6970	−1.0000	M	
9	0.7778	−0.8246	−0.3333	−0.5455	M	
10	1.0000	−0.9298	0.5152	−0.1364	M	$= d_1$, say
1	−1.0000	0.3636	−0.4167	−0.2222	F	
2	−1.0000	0.4545	−0.7500	−0.7222	F	
3	−0.9091	1.0000	−0.0833	−0.3889	F	
4	−0.7273	0.6364	−0.4167	−0.1111	F	
5	−0.6364	−0.1818	0.0000	−0.3333	F	
6	0.5455	0.0000	0.4167	0.1667	F	
7	0.7273	0.0000	−0.1667	−0.5000	F	
8	0.7273	0.4091	−0.4167	0.0000	F	
9	1.0000	−1.0000	0.6667	−0.2222	F	
10	1.0000	0.1364	−0.4167	−0.2778	F	

The information matrix of γ under d_1 is

$$\mathbf{I}_{d_1}(\gamma) = \mathbf{I}_{Md_1}(\gamma) + \mathbf{I}_{Fd_1}(\gamma) = \begin{pmatrix} 5.0485 & -3.4438 & 1.8097 & 0.6033 \\ -3.4438 & 2.8317 & -2.0649 & -1.1962 \\ 1.8097 & -2.0649 & 3.0785 & 2.4386 \\ 0.6033 & -1.1962 & 2.4386 & 2.2241 \end{pmatrix}$$

$$+ \begin{pmatrix} 7.1085 & -2.5908 & 1.5553 & 0.7469 \\ -2.5908 & 2.6322 & -1.4470 & -0.1490 \\ 1.5553 & -1.4470 & 1.6592 & 0.4199 \\ 0.7469 & -0.1490 & 0.4199 & 0.5682 \end{pmatrix} = \begin{pmatrix} 12.1570 & -6.0346 & 3.3650 & 1.3502 \\ -6.0346 & 5.4639 & -3.5119 & -1.3452 \\ 3.3650 & -3.5119 & 4.7377 & 2.8585 \\ 1.3502 & -1.3452 & 2.8585 & 2.7923 \end{pmatrix},$$

$det(\mathbf{I}_{d_1}) = 61.6096, 20^{-4} \times det(\mathbf{I}_{d_1}) = 0.0004.$

Method 2: **Use of Age Data** Similarly as in Selection 1, we select 20 persons (10 males and 10 females) by arranging the available reported information on age Z_2 in ascending order. The selected data set is shown in Table 9.31.

Table 9.31 Choice of males and females based on reported data on age

Sl. no.	Heart rate	Age	Height	Weight	Sex	
1	0.6296	−1.0000	1.0000	0.7273	M	
2	0.6296	−1.0000	0.6364	0.1364	M	
3	1.0000	−0.9298	0.5152	−0.1364	M	
4	0.4444	−0.8596	−0.1515	−0.4091	M	
5	0.7778	−0.8246	−0.3333	−0.5455	M	
6	0.4074	0.1228	0.1515	−0.2273	M	
7	0.0741	0.1579	−0.2121	0.0000	M	
8	−0.6296	0.2281	−0.4545	−0.4545	M	
9	−1.0000	0.5088	−0.6364	−0.6364	M	
10	−0.1852	1.0000	0.3939	0.1364	M	$= d_2$, say
1	0.2273	−1.0000	−0.2500	−0.7222	F	
2	1.0000	−1.0000	0.6667	−0.2222	F	
3	0.2727	−0.9091	0.9167	1.0000	F	
4	0.5455	−0.5909	0.5000	−0.1111	F	
5	0.2727	−0.5909	1.0000	−0.1111	F	
6	−0.4545	0.5909	−0.3333	0.0556	F	
7	0.3182	0.6364	0.1667	0.1667	F	
8	−0.7273	0.6364	−0.4167	−0.1111	F	
9	0.4545	0.6818	−0.5833	−0.6111	F	
10	−0.9091	1.0000	−0.0833	−0.3889	F	

The information matrix of γ under d_2 is

$$\mathbf{I}_{d_2}(\gamma) = \mathbf{I}_{Md_2}(\gamma) + \mathbf{I}_{Fd_2}(\gamma) = \begin{pmatrix} 3.7360 & -3.4305 & 1.9192 & 0.9087 \\ -3.4305 & 4.9602 & -1.5228 & -0.6202 \\ 1.9192 & -1.5228 & 2.5565 & 1.7466 \\ 0.9087 & -0.6202 & 1.7466 & 1.5144 \end{pmatrix}$$

$$+ \begin{pmatrix} 3.2678 & -3.0323 & 1.5652 & 0.0854 \\ -3.0323 & 6.1191 & -2.8871 & -0.6283 \\ 1.5652 & -2.8871 & 3.0063 & 1.3940 \\ 0.0854 & -0.6283 & 1.3940 & 2.0522 \end{pmatrix} = \begin{pmatrix} 7.0038 & -6.4628 & 3.4844 & 0.9941 \\ -6.4628 & 11.0793 & -4.4099 & -1.2485 \\ 3.4844 & -4.4099 & 5.5628 & 3.1406 \\ 0.9941 & -1.2485 & 3.1406 & 3.5666 \end{pmatrix},$$

$det(\mathbf{I}_{d_2}) = 196.608, 20^{-4} \times det(\mathbf{I}_{d_2}) = 0.0012.$

Method 3: **Use of Height Data** Similarly for height Z_3, we show the arranged data in Table 9.32.

Table 9.32 Choice of males and females based on height data

Sl. no.	Heart rate	Age	Height	Weight	Sex	
1	0.0741	0.0526	−1.0000	0.0000	M	
2	0.6296	−0.1228	−0.9394	−0.5909	M	
3	0.0741	−0.2632	−0.7576	−0.6364	M	
4	0.7037	−0.2632	−0.6970	−1.0000	M	
5	0.0370	−0.4386	0.5758	0.1818	M	
6	−1.0000	0.5088	−0.6364	−0.6364	M	
7	0.6296	−1.0000	0.6364	0.1364	M	
8	0.0370	−0.4386	0.6364	0.2273	M	
9	0.0370	−0.7895	0.8788	0.2273	M	
10	0.6296	−1.0000	1.0000	0.7273	M	$= d_3$, say
1	0.3636	0.1364	−1.0000	0.2222	F	
2	0.2727	0.0455	−1.0000	−0.5000	F	
3	0.0000	0.1364	−0.8333	−0.6667	F	
4	0.0909	0.2727	−0.7500	−0.6667	F	
5	−1.0000	0.4545	−0.7500	−0.7222	F	
6	1.0000	−1.0000	0.6667	−0.2222	F	
7	0.5455	0.2273	0.8333	0.9444	F	
8	0.2727	−0.9091	0.9167	1.0000	F	
9	0.2727	−0.5909	1.0000	−0.1111	F	
10	0.5455	−0.1364	1.0000	−0.0556	F	

The information matrix of γ under d_3 is

$$
\mathbf{I}_{d_3}(\gamma) = \mathbf{I}_{Md_3}(\gamma) + \mathbf{I}_{Fd_3}(\gamma) =
\begin{pmatrix}
2.3566 & -1.4126 & 0.5880 & 0.3333 \\
-1.4126 & 2.0137 & -2.8538 & -1.5551 \\
0.5880 & -2.8538 & 6.2519 & 3.3611 \\
0.3333 & -1.5551 & 3.3611 & 2.6572
\end{pmatrix}
$$

$$
+
\begin{pmatrix}
2.4000 & -1.4396 & 2.2153 & 1.2949 \\
-1.4396 & 2.2983 & -2.5925 & -1.1803 \\
2.2153 & -2.5925 & 7.7979 & 3.2704 \\
1.2949 & -1.1803 & 3.2704 & 3.6061
\end{pmatrix}
=
\begin{pmatrix}
4.7566 & -2.8522 & 2.8033 & 1.6282 \\
-2.8522 & 4.3120 & -5.4463 & -2.7354 \\
2.8033 & -5.4463 & 14.0498 & 6.6315 \\
1.6282 & -2.7354 & 6.6315 & 6.2633
\end{pmatrix}
$$

$det(\mathbf{I}_{d_3}) = 267.4351$, $20^{-4} \times det(\mathbf{I}_{d_3}) = 0.0017$.

Method 4: Use of Weight Data Lastly for weight Z_4, we show the arranged data in Table 9.33.

Table 9.33 Choice of males and females based on reported weight data

Sl. no.	Heart rate	Age	Height	Weight	Sex	
1	0.7037	−0.2632	−0.6970	−1.0000	M	
2	0.0741	−0.2632	−0.7576	−0.6364	M	
3	0.6296	−0.1228	−0.9394	−0.5909	M	
4	−0.4815	−0.1228	−0.3939	−0.5909	M	
5	−1.0000	0.5088	−0.6364	−0.6364	M	
6	0.3333	−0.4035	−0.0303	0.2727	M	
7	−0.3333	−0.1228	0.3333	0.3636	M	
8	0.6296	−1.0000	1.0000	0.7273	M	
9	0.0370	0.0877	0.3939	0.9545	M	
10	−0.2593	0.0175	0.2727	1.0000	M	$= d_4$, say
1	0.0909	0.3636	−0.6667	−1.0000	F	
2	−0.3636	−0.0455	−0.5000	−0.8333	F	
3	0.2273	−1.0000	−0.2500	−0.7222	F	
4	−1.0000	0.4545	−0.7500	−0.7222	F	
5	0.0000	0.1364	−0.8333	−0.6667	F	
6	0.3636	−0.1364	−1.0000	0.2222	F	
7	0.5455	0.1818	−0.4167	0.3889	F	
8	0.0909	0.5909	0.0833	0.7778	F	
9	0.5455	0.2273	0.8333	0.9444	F	
10	0.2727	−0.9091	0.9167	1.0000	F	

The information matrix of γ under d_4 is

$$\mathbf{I}_{d_4}(\gamma) = \mathbf{I}_{Md_4}(\gamma) + \mathbf{I}_{Fd_4}(\gamma) = \begin{pmatrix} 2.8050 & -1.4000 & 0.1887 & 0.0062 \\ -1.4000 & 1.3298 & -1.0116 & -0.5517 \\ 0.1887 & -1.0116 & 3.6323 & 3.8410 \\ 0.0062 & -0.5517 & 3.8410 & 5.1531 \end{pmatrix}$$

$$+ \begin{pmatrix} 1.9424 & -0.6423 & 1.1352 & 1.9689 \\ -0.6423 & 2.6365 & -0.9936 & -0.2254 \\ 1.1352 & -0.9936 & 4.0618 & 3.5874 \\ 1.9689 & -0.2254 & 3.5874 & 5.8422 \end{pmatrix} = \begin{pmatrix} 4.7474 & -2.0423 & 1.3239 & 1.9751 \\ -2.0423 & 3.9663 & -2.0052 & -0.7771 \\ 1.3239 & -2.0052 & 7.6941 & 7.4284 \\ 1.9751 & -0.7771 & 7.4284 & 10.9953 \end{pmatrix}$$

$det(\mathbf{I}_{d_4}) = 292.4578, 20^{-4} \times det(\mathbf{I}_{d_4}) = 0.00183.$

Selection Method Based on Principal Component

Selection of 10 Males

A PCA is concerned with explaining the variance–covariance structure of a set of responses through a few linear combinations of these responses. In this study, only two eigenvalues were larger (14.6743 and 11.3163, respectively); so two components

Table 9.34 Details for PCA for Male Data

	Comp.1	Comp.2	Comp.3	Comp.4
Standard deviation	0.5594	0.4865	0.2595	0.2102
Proportion of variance	0.4733	0.3580	0.1018	0.0668
Cumulative proportion	0.4733	0.8313	0.9332	1.0000

were extracted, based on Kaiser principle (cf. Kaiser 1960). The first component (PC_{M1}) accounts for about 47.33 % and the second component (PC_{M2}) accounts for about 35.80 % of the total variance in the data set. Therefore the first two components account for 83.13 % of the variance (vide Table 9.34).

The eigenvalues of $\mathbf{Z}'_M \mathbf{Z}_M$ are $14.6743 = \lambda_{M1}$, $11.3163 = \lambda_{M2}$, $3.4597 = \lambda_{M3}$, $1.9446 = \lambda_{M4}$; where $\mathbf{Z}_M^{44 \times 4} = (z_{Mj}^{(l)})$ is design matrix for covariate parameters for male data and the corresponding eigenvectors $(\boldsymbol{\xi}_{M1}, \boldsymbol{\xi}_{M2}, \boldsymbol{\xi}_{M3}, \boldsymbol{\xi}_{M4})$ are:

$$
\begin{pmatrix}
\boldsymbol{\xi}_{M1} & \boldsymbol{\xi}_{M2} & \boldsymbol{\xi}_{M3} & \boldsymbol{\xi}_{M4} \\
-0.4520 & 0.5225 & 0.7182 & -0.0828 \\
0.5955 & -0.3808 & 0.6107 & -0.3569 \\
-0.5887 & -0.4745 & -0.0998 & -0.6468 \\
-0.3074 & -0.5973 & 0.3182 & 0.6689
\end{pmatrix}
$$

Therefore

$$PC_{M1} = -0.4520 Z_{M1} + 0.5955 Z_{M2} - 0.5887 Z_{M3} - 0.3074 Z_{M4}$$
$$PC_{M2} = 0.5225 Z_{M1} - 0.3808 Z_{M2} - 0.4745 Z_{M3} - 0.5973 Z_{M4}$$

Now we take a convex combination of PC_{M1} and PC_{M2} and get a new score and denote it by $P_M = \frac{\lambda_{M1}}{\lambda_{M1}+\lambda_{M2}} PC_{M1} + \frac{\lambda_{M2}}{\lambda_{M1}+\lambda_{M2}} PC_{M2} = -0.0277 Z_{M1} + 0.1704 Z_{M2}$ $- 0.5390 Z_{M3} - 0.4336 Z_{M4}$. Now we arrange P_M score in ascending order and select 10 males where 5 are from the top and 5 are from the bottom. The data of the selected males are shown in Table 9.35.

Selection of 10 Females

A similar PCA is carried out for female data. In this study, three eigenvalues were larger (21.3051, 10.9512 and 7.2187, respectively); so three components were extracted, based on Kaiser principle. The first component (PC_{F1}) accounts for about 51.29 %, the second component (PC_{F2}) accounts for about 22.92 % and the third component (PC_{F3}) accounts for about 17.65 % of the total variance in the data set. Therefore, the first three components account for 91.86 % of the variance (vide Table 9.36).

The eigenvalues of $\mathbf{Z}'_F \mathbf{Z}_F$ are $21.3051 = \lambda_{F1}$, $10.9512 = \lambda_{F2}$, $7.2187 = \lambda_{F3}$, $3.8364 = \lambda_{F4}$; where $\mathbf{Z}_F^{44 \times 4} = (z_{Fj}^{(l)})$ is design matrix for covariate parameters for

Table 9.35 Male data

Heart rate	Age	Height	Weight	
0.6296	−1.0000	1.0000	0.7273	
0.6296	−1.0000	0.6364	0.1364	
0.0370	−0.7895	0.8788	0.2273	
−0.2593	0.0175	0.2727	1.0000	
0.0370	0.0877	0.3939	0.9545	$= d_{M5}$, say
0.7037	−0.2632	−0.6970	−1.0000	
0.6296	−0.1228	−0.9394	−0.5909	
0.0741	0.0526	−1.0000	0.0000	
0.0741	−0.2632	−0.7576	−0.6364	
−1.0000	0.5088	−0.6364	−0.6364	

Table 9.36 Details of PCA for Female Data

	Comp.1	Comp.2	Comp.3	Comp.4
Standard deviation	0.6910	0.4618	0.4054	0.2753
Proportion of variance	0.5129	0.2292	0.1765	0.0814
Cumulative proportion	0.5129	0.7421	0.9186	1.0000

female data and the corresponding eigenvectors $\left(\xi_{F1}, \xi_{F2}, \xi_{F3}, \xi_{F4}\right)$ are:

$$\begin{pmatrix} \xi_{F1} & \xi_{F2} & \xi_{F3} & \xi_{F4} \\ 0.4400 & 0.7258 & -0.5234 & -0.0753 \\ -0.4615 & -0.2532 & -0.6623 & -0.5333 \\ 0.6504 & -0.3185 & 0.2000 & -0.6599 \\ 0.4128 & -0.5547 & -0.4975 & 0.5239 \end{pmatrix}$$

Therefore

$$PC_{F1} = 0.4400Z_{F1} - 0.4615Z_{F2} + 0.6504Z_{F3} + 0.4128Z_{F4}$$
$$PC_{F2} = 0.7258Z_{F1} - 0.2532Z_{F2} - 0.3185Z_{F3} - 0.5547Z_{F4}$$
$$PC_{F3} = -0.5234Z_{F1} - 0.6623Z_{F2} + 0.2000Z_{F3} - 0.4975Z_{F4}$$

Now we take convex combination of PC_{F1}, PC_{F2} and PC_{F3} and get a new score and denote it by $P_F = \frac{\lambda_{F1}}{\lambda_{F1}+\lambda_{F2}+\lambda_{F3}}PC_{F1} + \frac{\lambda_{F2}}{\lambda_{F1}+\lambda_{F2}+\lambda_{F3}}PC_{F2} + \frac{\lambda_{F3}}{\lambda_{F1}+\lambda_{F2}+\lambda_{F3}}$ $PC_{F3} = 0.3431Z_{F1} - 0.4404Z_{F2} + 0.2992Z_{F3} - 0.0220Z_{F4}$. Now we arrange P_F score in ascending order and select 10 females where 5 are from top and 5 are from bottom. The data of the selected females are shown in Table 9.37.

Table 9.37 Female data

Heart rate	Age	Height	Weight	
−0.7273	0.6364	−0.4167	−0.1111	
−0.9091	1.0000	−0.0833	−0.3889	
−0.4545	0.5909	−0.3333	0.0556	
−1.0000	0.3636	−0.4167	−0.2222	
−1.0000	0.4545	−0.7500	−0.7222	$= d_{F5}$, say
0.2727	−0.9091	0.9167	1.0000	
1.0000	−1.0000	0.6667	−0.2222	
0.5455	−0.5909	0.5000	−0.1111	
0.2727	−0.5909	1.0000	−0.1111	
0.5455	−0.1364	1.0000	−0.0556	

Table 9.38 Selection of 10 males based on total scoring

Heart rate	Age	Height	Weight	
0.7037	−0.2632	−0.6970	−1.0000	
−0.6296	0.2281	−0.4545	−0.4545	
−0.4815	−0.1228	−0.3939	−0.5909	
0.0741	−0.2632	−0.7576	−0.6364	
−1.0000	0.5088	−0.6364	−0.6364	$= d_{M6}$, say
0.6296	−1.0000	1.0000	0.7273	
−0.2593	0.0175	0.2727	1.0000	
0.0370	0.0877	0.3939	0.9545	
−0.1852	1.0000	0.3939	0.1364	
0.2593	0.0877	0.1515	0.1364	

Therefore, $d_5 = \begin{pmatrix} d_{M5} \\ d_{F5} \end{pmatrix}$.

Therefore $det(\mathbf{I}_{d_5}(\gamma)) = det(\mathbf{I}_{Md_5}(\gamma) + \mathbf{I}_{Fd_5}(\gamma)) = 255.4019$ and $20^{-4} \times det(\mathbf{I}_{d_5}(\gamma)) = 0.00160$.

Selection Method Based on Total Scoring

We may also adopt one more ad hoc method of selection. This time we select males or females by use of total Z-scores from all the covariates. For males, we base on the $Z_M = Z_{M1} + Z_{M2} + Z_{M3} + Z_{M4}$ values. We arrange Z_M in ascending order and select 10 males where 5 are from the top and 5 are from the bottom. The data of the selected males are shown in Table 9.38.

Likewise, we select 10 females based on $Z_F = Z_{F1} + Z_{F2} + Z_{F3} + Z_{F4}$ values. We arrange Z_F in ascending order and select 10 females where 5 are from the top and 5 are from the bottom. The data of the selected females are shown in Table 9.39.

Table 9.39 Selection of 10 females based on total scoring

Heart rate	Age	Height	Weight	
0.2273	−1.0000	−0.2500	−0.7222	
−0.3636	−0.0455	−0.5000	−0.8333	
−0.3636	−0.1364	−0.5000	−0.3889	
−1.0000	0.4545	−0.7500	−0.7222	
0.0000	0.1364	−0.8333	−0.6667	$= d_{F6}$, say
0.2727	−0.9091	0.9167	1.0000	
0.5455	0.2273	0.8333	0.9444	
0.0909	0.5909	0.0833	0.7778	
0.5455	−0.1364	1.0000	−0.0556	
0.3182	0.6364	0.1667	0.1667	

Therefore, $d_6 = \begin{pmatrix} d_{M6} \\ d_{F6} \end{pmatrix}$. Consequently, $det(\mathbf{I}_{d_6}(\gamma)) = det(\mathbf{I}_{Md_6}(\gamma) + \mathbf{I}_{Fd_6}(\gamma)) = 294.0333$ and $20^{-4} \times det(\mathbf{I}_{d_6}(\gamma)) = 0.00184$.

At the end, we conclude that $d_1 \prec d_2 \prec d_5 \prec d_3 \prec d_4 \prec d_6$, since $20^{-4} \times det(\mathbf{I}_{d_1}(\gamma)) = 0.0004$, $20^{-4} \times det(\mathbf{I}_{d_2}(\gamma)) = 0.0012$, $20^{-4} \times det(\mathbf{I}_{d_3}(\gamma)) = 0.0017$, $20^{-4} \times det(\mathbf{I}_{d_4}(\gamma)) = 0.00183$, $20^{-4} \times det(\mathbf{I}_{d_5}(\gamma)) = 0.00160$, $20^{-4} \times det(\mathbf{I}_{d_6}(\gamma)) = 0.00184$ and $87^{-4} \times det(\mathbf{I}_d(\gamma)) = 0.0006$, where d is a design based on whole data set.

It thus transpires that the criterion of selection (d_6) based on total scoring of the participating males or females fares much better than the rest for efficient estimation of the covariate parameters. It may be noted that implicitly we have exploited our understanding of OCDs in carrying out this exercise.

Example 9.2.3 The data in Table 9.40 are from a study to evaluate the effect of roll gap and variety of wheat on the amount of flour produced (percent of total wheat ground) during a run of a pilot flour mill (cf. Milliken and Johnson (2001), p. 406). Three batches of wheat from each of three varieties were used in the study with enough raw material for all where a single batch of wheat was used for all four roll gap setting. The experiment was conducted on three different days.

For each variety type, there are three batches of wheat with fixed but varying moisture contents and the batches have been assigned across the 3 days of the experiment as shown in Table 9.40.

Our purpose is to revisit this data set and examine the possibility of improving the design for increased information on the covariate parameter. We will examine the possibility of re-allocation of the batches of wheats across different days of the experiment. Towards this, we start with a model description.

Model for 3 way layout (Day × Variety × Roll Gap):

$$y_{ijl} = \mu + \beta_i + \tau_j + \delta_l + z_{ij}\gamma + e_{ijl}; \quad i = 1, 2, \ldots, r; \ j = 1, 2, \ldots, p; \ l = 1, 2, \ldots, q;$$
(9.2.1)

Table 9.40 Data for amount of flour milled (percent) during the first break of a flour mill operation from three varieties of wheat using four roll gaps

Day	Variety	Moist	Roll gap			
			0.02in	0.04in	0.06in	0.08in
1	A	12.4	18.3	14.6	12.2	9.0
1	B	12.8	18.8	14.9	11.7	8.3
1	C	12.1	18.7	15.5	12.7	9.2
2	A	14.4	19.1	15.4	12.4	8.7
2	B	12.4	17.5	14.4	11.3	8.3
2	C	13.2	18.9	15.4	12.5	8.2
3	A	13.1	18.2	15.0	12.1	8.4
3	B	14.0	20.4	16.2	12.9	9.1
3	C	13.4	19.7	16.9	13.4	9.5

Source Milliken and Johnson (2001). Analysis of Messy Data: Volume III: Analysis of covariance, Chapman and Hall/CRC

where μ is general effect, β_i is effect due to ith day, τ_j is effect due to jth variety, δ_l is effect due to lth roll gap, γ is the covariate effect and z_{ij} is the covariate value corresponding to ith day and jth variety. Without loss of generality, we assume $|z_{ij}| \leq 1$ for all i, j. Note that z_{ij}'s are to be computed by using the standard transformation of the original covariate values x_{ij}'s and it is given by $z_{ij} = \frac{2(x_{ij}-x_{min})}{x_{max}-x_{min}} - 1$. These x_{min} and x_{max} are to be based on the entire collection of x_{ij} values. In (9.2.1), e_{ijl}s are random errors with

$$V(e_{ijl}) = \sigma^2 \qquad \forall i, j, l$$
$$Cov(e_{ijl}, e_{ijl'}) = \rho\sigma^2 \ \forall i, j, l \neq l'$$
$$Cov(e_{ijl}, e_{ij'l}) = 0 \qquad \forall i, j \neq j', l$$
$$Cov(e_{ijl}, e_{i'jl}) = 0 \qquad \forall i \neq i', j, l$$

Therefore, $Disp(\mathbf{y}_{ij}) = \sigma^2((1-\rho)\mathbf{I}_q + \rho\mathbf{J}_q)$, $\forall i, j$ where $\mathbf{I}_q$ is an identity matrix of order q and J_q is a matrix of order q with all elements unity.

Since for each day × variety combination, there are four correlated observations, observational contrasts would provide information on the contrasts involving effects of roll gaps, eliminating the effects of all other parameters, including the covariate parameter, viz. the effect of moisture content. The average of the four observations, on the other hand, will provide information on all the three components, viz. day effect, variety effect and covariate effect, as in a standard RBD set-up.

We will start from there and proceed to examine the possibilities of rearrangement of the moist batches of wheat for extracting maximum information on the covariate effect.

In a general RBD set-up involving q correlated observations, we would proceed as follows:

$$\mathbf{L} = \begin{pmatrix} \frac{1}{\sqrt{q}} & \frac{1}{\sqrt{q}} & \frac{1}{\sqrt{q}} & \frac{1}{\sqrt{q}} & \cdots & \frac{1}{\sqrt{q}} & \frac{1}{\sqrt{q}} \\ \frac{1}{\sqrt{2}} & -\frac{1}{\sqrt{2}} & 0 & 0 & \cdots & 0 & 0 \\ \frac{1}{\sqrt{6}} & \frac{1}{\sqrt{6}} & -\frac{2}{\sqrt{q}} & 0 & \cdots & 0 & 0 \\ \vdots & \vdots & \vdots & \vdots & \vdots & \vdots & \vdots \\ \frac{1}{\sqrt{q(q-1)}} & \frac{1}{\sqrt{q(q-1)}} & \frac{1}{\sqrt{q(q-1)}} & \frac{1}{\sqrt{q(q-1)}} & \cdots & \frac{1}{\sqrt{q(q-1)}} & -\frac{q-1}{\sqrt{q(q-1)}} \end{pmatrix}$$

such that $\mathbf{LL}' = \mathbf{I}_q$.

Then

$$Disp(\mathbf{Ly}_{ij}) = \sigma^2 \mathbf{L}((1-\rho)\mathbf{I}_q + \rho\mathbf{J}_q)\mathbf{L}' = \sigma^2 \begin{pmatrix} 1+(q-1)\rho & 0 & 0 & \cdots & 0 \\ 0 & 1-\rho & 0 & \cdots & 0 \\ 0 & 0 & 1-\rho & \cdots & 0 \\ \vdots & \vdots & \vdots & \vdots & \vdots \\ 0 & 0 & 0 & \cdots & 1-\rho \end{pmatrix}$$

Now we take

$$\mathbf{M} = \begin{pmatrix} \frac{1}{\sqrt{1+(q-1)\rho}} & 0 & 0 & \cdots & 0 \\ 0 & \frac{1}{\sqrt{1-\rho}} & 0 & \cdots & 0 \\ 0 & 0 & \frac{1}{\sqrt{1-\rho}} & \cdots & 0 \\ \vdots & \vdots & \vdots & \vdots & \vdots \\ 0 & 0 & 0 & \cdots & \frac{1}{\sqrt{1-\rho}} \end{pmatrix}$$

Then $Disp(\mathbf{MLy}_{ij}) = \sigma^2 \mathbf{I}_q$. Let $\mathbf{MLy}_{ij} = \mathbf{y}^*_{ij}$. Now our Model (9.2.1) can be written as:

$$\frac{1}{\sqrt{q(1+(q-1)\rho)}} \sum_{l=1}^{q} y_{ijl} = \frac{1}{\sqrt{q(1+(q-1)\rho)}} \left(q(\mu + \beta_i + \tau_j + z_{ij}\gamma) + \sum_{l=1}^{q}\delta_l + \sum_{l=1}^{q} e_{ijl} \right)$$

$$\frac{1}{\sqrt{2(1-\rho)}}(y_{ij1} - y_{ij2}) = \frac{1}{\sqrt{2(1-\rho)}}(\delta_1 - \delta_2) + \frac{1}{\sqrt{2(1-\rho)}}(e_{ij1} - e_{ij2})$$

$$\frac{1}{\sqrt{6(1-\rho)}}(y_{ij1} + y_{ij2} - 2y_{ij3}) = \frac{1}{\sqrt{6(1-\rho)}}(\delta_1 + \delta_2 - 2\delta_3) + \frac{1}{\sqrt{6(1-\rho)}}(e_{ij1} + e_{ij2} - 2e_{ij3})$$

$$\vdots$$

$$\frac{1}{\sqrt{q(q-1)(1-\rho)}}(y_{ij1} + y_{ij2} + \cdots + y_{ijq-1} - (q-1)y_{ijq})$$

$$= \frac{1}{\sqrt{q(q-1)(1-\rho)}}((\delta_1 + \delta_2 + \cdots + \delta_{q-1} - (q-1)\delta_q) + (e_{ij1} + e_{ij2} + \cdots + e_{ijq-1} - (q-1)e_{ijq}))$$

$$\tag{9.2.2}$$

As explained above, the only source of information on the covariate parameter γ is the set of rp subtotals or means on Day $\times$ Treatment with the roll gap observations averaged out. Hence the model is:

$$\bar{y}_{ij0} = \mu^* + \beta_i + \tau_j + z_{ij}\gamma + \bar{e}_{ij0} \tag{9.2.3}$$

where $\bar{y}_{ij0} = \frac{1}{q}\sum_{l=1}^{q} y_{ijl}$, $\bar{e}_{ij0} = \frac{1}{q}\sum_{l=1}^{q} e_{ijl}$, $\mu^* = \mu + \frac{1}{q}\sum_{l=1}^{q} \delta_l$ with $V(\bar{y}_{ij0}) = $
$\frac{(1+(q-1)\rho)}{q}\sigma^2$ and $\bar{y}_{ij0}$'s are uncorrelated. This is simply covariate model for RBD with single covariate. The information of γ is

$$I(\gamma) = \sum_{i=1}^{r}\sum_{j=1}^{p} z_{ij}^2 - \frac{1}{p}\sum_{i=1}^{r} R_i^2 - \frac{1}{r}\sum_{j=1}^{p} C_j^2 + \frac{G^2}{rp}, \tag{9.2.4}$$

where $R_i = \sum_{j=1}^{p} z_{ij}$, $i = 1, 2, \ldots, r$ and $C_j = \sum_{i=1}^{r} z_{ij}$, $j = 1, 2, \ldots, p$, and
$G = \sum_{i=1}^{r}\sum_{j=1}^{p} z_{ij} = \sum_{i=1}^{r} R_i = \sum_{j=1}^{p} C_j$. Now we transform the moisture values of
Table 9.40 into z-values and the coded covariate values are shown in Table 9.41.
Information of $\theta = (\mu, \beta', \tau', \gamma)$ is

$$I_{d_1}(\theta) = \begin{pmatrix} 9 & 3 & 3 & 3 & 3 & 3 & 3 & -1.261 \\ 3 & 3 & 0 & 0 & 1 & 1 & 1 & -2.130 \\ 3 & 0 & 3 & 0 & 1 & 1 & 1 & 0.217 \\ 3 & 0 & 0 & 3 & 1 & 1 & 1 & 0.652 \\ 3 & 1 & 1 & 1 & 3 & 0 & 0 & 0.13 \\ 3 & 1 & 1 & 1 & 0 & 3 & 0 & -0.478 \\ 3 & 1 & 1 & 1 & 0 & 0 & 3 & -0.913 \\ -1.261 & -2.13 & 0.217 & 0.652 & 0.13 & -0.478 & -0.913 & 3.707 \end{pmatrix}$$

$I_{d_1}(\gamma) = 1.8533$.

Table 9.41 Coded z-values	Day	Variety	Moist	
	1	A	−0.7391	
	1	B	−0.3913	
	1	C	−1.0000	
	2	A	1.0000	$= d_1$, say
	2	B	−0.7391	
	2	C	−0.0435	
	3	A	−0.1304	
	3	B	0.6522	
	3	C	0.1304	

Now we consider the design given in Table 9.42.

Information of $\theta = (\mu, \beta', \tau', \gamma)$ is

$$I_{d_2}\theta) = \begin{pmatrix} 9 & 3 & 3 & 3 & 3 & 3 & 3 & -3 \\ 3 & 3 & 0 & 0 & 1 & 1 & 1 & -1 \\ 3 & 0 & 3 & 0 & 1 & 1 & 1 & -1 \\ 3 & 0 & 0 & 3 & 1 & 1 & 1 & -1 \\ 3 & 1 & 1 & 1 & 3 & 0 & 0 & -1 \\ 3 & 1 & 1 & 1 & 0 & 3 & 0 & -1 \\ 3 & 1 & 1 & 1 & 0 & 0 & 3 & -1 \\ -3 & -1 & -1 & -1 & -1 & -1 & -1 & 9 \end{pmatrix}$$

$I_{d_2}(\gamma) = 8$. The D-efficiency of d_1 with respect to d_2 is 23.1674 %. Hence if we use d_2 instead of d_1, then we improve the information of γ a lot. The moisture values of d_2 are given in Table 9.43.

Here we conclude that the design which is optimum under uncorrelated model is also optimum under correlated model.

Table 9.42 Coded z-values

Day	Variety	Moist	
1	A	−1	
1	B	1	
1	C	−1	
2	A	−1	$= d_2$, say
2	B	−1	
2	C	1	
3	A	1	
3	B	−1	
3	C	−1	

Table 9.43 Uncoded z-values

Day	Variety	Moist	
1	A	12.1	
1	B	14.4	
1	C	12.1	
2	A	12.1	$= d_2$, say
2	B	12.1	
2	C	14.4	
3	A	14.4	
3	B	12.1	
3	C	12.1	

Remark 9.2.1 The reader may note that the allocation design d_2 relates to a hypothetical scenario under the assumption that there are indeed batches of wheat varieties available with the stipulated moisture contents coded ± 1's. The uncoded values are shown in Table 9.43. Naturally, d_2 does not address the reality which is governed by the given values of the moisture contents of nine bundles of the wheat. Therefore, we turn back to the original collection of z-values as in d_1 and try to suggest improvements over the given allocation.

Now we want to improve d_1 by reallocating the batches of wheat of each type across the 3 days so that SSDays $(=\frac{1}{p}\sum_{i=1}^{r} R_i^2 - \frac{G^2}{rp})$ is the least i.e. R_i's are as equal as possible. So we have three treatments A, B, C and each has three bundles of input material with naturally given day specific moisture contents. Consider Treatment A so that we have $3! = 6$ ways of distributing the bundles across the days. Likewise $3! = 6$ for those of B and $3! = 6$ for those of C and all are independent. So in effect given the experimental material and no further input, there are $6 \times 6 \times 6 = 216$ ways of distribution of the bundles across 3 days and the given design is just one of these 216 possible allocations. Now we rewrite d_1 as in Table 9.44 and follow certain steps for reallocation of coded z-values in each column.

Step 1: We select the first column and permute the covariate values keeping the other columns fixed. Next we compute the information of γ under each permutation. Then we choose the design in which the information of γ will be a maximum. We denote the design by d_3 (Table 9.45).

The information of γ is $I_{d_3}(\gamma) = 3.1942$.

Step 2: Now replace the second column by $(-0.3913, 0.6522, -0.7391)$ and keep the third column intact, and then permute the first column in all possible ways. Then we get the design d_4 which improves the information of γ and is equal to 3.2849 (Table 9.46).

Step 3: Now replace the second column by $(-0.7391, -0.3913, 0.6522)$ and keep the third column intact, and then permute the first column in all possible ways. But

Table 9.44 d_1

Day	A	B	C	Total
1	−0.7391	−0.3913	−1.0000	−2.1304
2	1.0000	−0.7391	−0.0435	0.2174
3	−0.1304	0.6522	0.1304	0.6522

Table 9.45 d_3

Day	A	B	C	Total
1	1.0000	−0.3913	−1.0000	−0.3913
2	−0.1304	−0.7391	−0.0435	−0.913
3	−0.7391	0.6522	0.1304	0.0435

Table 9.46 d_4

Day	A	B	C	Total
1	1.0000	−0.3913	−1.0000	−0.3913
2	−0.7391	0.6522	−0.0435	−0.1304
3	−0.1304	−0.7391	0.1304	−0.7391

Table 9.47 d_5

Day	A	B	C	Total
1	1.0000	−0.7391	−1.0000	−0.7391
2	−0.7391	0.6522	−0.0435	−0.1304
3	−0.1304	−0.3913	0.1304	−0.3913

Table 9.48 d_6

Day	A	B	C	Total
1	1.0000	−0.3913	−1.0000	−0.3913
2	−0.1304	−0.7391	0.1304	−0.7391
3	−0.7391	0.6522	−0.0435	−0.1304

Table 9.49 d_7

Day	A	B	C	Total
1	1.0000	−0.7391	−1.0000	−0.7391
2	−0.1304	−0.3913	0.1304	−0.3913
3	−0.7391	0.6522	−0.0435	−0.1304

we cannot improve the information of γ than d_4 (since here we get the design where the information of γ is 3.2345 and the design is best among the designs obtained from all six permutations of the first column).

Step 4: Replace the second column by (−0.7391, 0.6522, −0.3913) and keep the third column intact, and then permute the first column in all possible ways. But we cannot improve the information of γ than d_4 (since here we get the design where the information of γ is 3.2849 and the design is best among the designs obtained from all six permutations of the first column). We denote this design by d_5 (Table 9.47).

Step 5: Replace the second column by (0.6522, −0.3913, −0.7391) and keep the third column intact, and then permute the first column in all possible ways. But we cannot improve the information of γ than d_4.

Now replace the third column of d_1 by (−1.000, 0.1304, −0.0435) and by following all the steps mentioned above we get the designs d_6 and d_7 where the information of γ is the same as d_4 (Tables 9.48 and 9.49).

Table 9.50 d_8

Day	A	B	C	Total
1	−0.7391	0.6522	−0.0435	−0.1304
2	1.0000	−0.3913	−1.0000	−0.3913
3	−0.1304	−0.7391	0.1304	−0.1304

Table 9.51 d_9

Day	A	B	C	Total
1	−0.7391	0.6522	−0.0435	−0.1304
2	1.0000	−0.7391	−1.0000	−0.7391
3	−0.1304	−0.3913	0.1304	−0.3913

Table 9.52 d_{10}

Day	A	B	C	Total
1	−0.7391	0.6522	−0.0435	−0.1304
2	−0.1304	−0.3913	0.1304	−0.3913
3	1.0000	−0.7391	−1.0000	−0.7391

Table 9.53 d_{11}

Day	A	B	C	Total
1	−0.7391	0.6522	−0.0435	−0.1304
2	−0.1304	−0.7391	0.1304	−0.7391
3	1.0000	−0.3913	−1.0000	−0.3913

Now replace the third column of d_1 by $(-0.0435, -1.000, 0.1304)$ and by follow-ing all the steps mentioned above we get the designs d_8, d_9 where the information of γ is the same as d_4 (Tables 9.50 and 9.51).

Now replace the third column of d_1 by $(-0.0435, 0.1304, -1.000)$ and by follow-ing all the steps mentioned above we get the designs d_{10}, d_{11} where the information of γ is same as d_4 (Tables 9.52 and 9.53).

Now replace the third column of d_1 by $(0.1304, -1.000, -0.0435)$ and by follow-ing all the steps mentioned above we get the designs d_{12}, d_{13} where the information of γ is same as d_4 (Tables 9.54 and 9.55).

Now replace the third column of d_1 by $(0.1304, -0.0435, -1.000)$ and by follow-ing all the steps mentioned above we get the designs d_{14}, d_{15} where the information of γ is same as d_4 (Tables 9.56 and 9.57).

Therefore, d_4 is the best design when the covariate values are given and the relative efficiency of d_1 with respect to d_4 is 56.42 %. Therefore, we are in a position to improve the information of γ in the design d_1 almost twice. But if we have such

Table 9.54 d_{12}

Day	A	B	C	Total
1	−0.1304	−0.3913	0.1304	−0.3913
2	1.0000	−0.7391	−1.0000	−0.7391
3	−0.7391	0.6522	−0.0435	−0.1304

Table 9.55 d_{13}

Day	A	B	C	Total
1	−0.1304	−0.7391	0.1304	−0.7391
2	1.0000	−0.3913	−1.0000	−0.3913
3	−0.7391	0.6522	−0.0435	−0.1304

Table 9.56 d_{14}

Day	A	B	C	Total
1	−0.1304	−0.3913	0.1304	−0.3913
2	−0.7391	0.6522	−0.0435	−0.1304
3	1.0000	−0.7391	−1.0000	−0.7391

Table 9.57 d_{15}

Day	A	B	C	Total
1	−0.1304	−0.7391	0.1304	−0.7391
2	−0.7391	0.6522	−0.0435	−0.1304
3	1.0000	−0.3913	−1.0000	−0.3913

flexibility to choose covariate values then d_2 is the best design. Here we note that we continued our search till all R_i's are same or almost the same.

Example 9.2.4 The data in Table 9.58 are from an experiment on the effects of two drugs on Mental Activity (MA). The mental activity score is the sum of the scores on seven items in a questionnaire given to each of 24 volunteer subjects. The treatments are Morphine, Heroin and a Placebo (an inert substance) given in subcutaneous injections. On different occasions, each subject received each drug in turn. The mental activity is measured before taking the drug (Z) and at $\frac{1}{2}$, 2, 3 and 4 h after. The response data (Y) in Table 9.58 are those at 2 h after. As a common precaution in these experiments, eight subjects took Morphine first, eight took Heroin first and eight took the Placebo first, and similarly on the second and third occasions. These data show no apparent effect of the order in which drugs were given, and the order is ignored here.

Since each subject gets three treatments in three different time points in some pre-determined order, we have a flexibility to allocate subjects to the three treatments at

Table 9.58 Mental Activity scores before (Z) and 2 h after (Y) a drug

Subject	Morphine (M)		Heroin (H)		Placebo (P)	
	Z	Y	Z	Y	Z	Y
1	7	4	0	2	0	7
2	2	2	4	0	2	1
3	14	14	14	13	14	10
4	14	0	10	0	5	10
5	1	2	4	0	5	6
6	2	0	5	0	4	2
7	5	6	6	1	8	7
8	6	0	6	2	6	5
9	5	1	4	0	6	6
10	6	6	10	0	8	6
11	7	5	7	2	6	3
12	1	3	4	1	3	8
13	0	0	1	0	1	0
14	8	10	9	1	10	11
15	8	0	4	13	10	10
16	0	0	0	0	0	0
17	11	1	11	0	10	8
18	6	2	6	4	6	6
19	7	9	0	0	8	7
20	5	0	6	1	5	1
21	4	2	11	5	10	8
22	7	7	7	7	6	5
23	0	2	0	0	0	1
24	12	12	12	0	11	5
Total	138		141		144	

the first time point only. Thereafter, we repeat the allocation of the subjects of Time point 1 for the next two time points. We analyse the data for each of the three time points and employ CRD model for respective time points separately.

Although RBD analysis was carried out in the book, it is more appropriate to treat this as an exercise in repeated CRD analysis and we may assume that the covariate effect is the same across all the three time points. The point to be noted is that the experimental subjects may be classified into three groups based on their covariate values at Time Point 1 only. Subsequently, there is no scope to alter their classifications to other groups.

We now proceed to objectively look into the given data on the covariates.

Towards this, we rearrange Z-values in Table 9.59a–c corresponding to time point 1, time point 2 and time point 3, respectively, and denote it by d_0.

Now we consider the CRD model for single covariate for each Time point.

Table 9.59 Arrangement of the Z-values of the 24 subjects in the three different time points

(a) Time point 1

	M	H	P
	7	4	10
	2	10	6
	14	7	8
	14	4	5
	1	1	10
	2	9	6
	5	4	0
	6	0	11
Totals	51	39	56

(b) Time point 2

	H	P	M
	0	6	11
	4	8	6
	14	6	7
	10	3	5
	4	1	4
	5	10	7
	6	10	0
	6	0	12
Totals	49	44	52

(c) Time point 3

	P	M	H
	0	5	11
	2	6	6
	14	7	0
	5	1	6
	5	0	11
	4	8	7
	8	8	0
	6	0	12
Totals	44	35	53

For Table 9.59a–c, routine computations for a CRD yield, $I_{d_0}(\gamma_1) = 364.75$, $I_{d_0}(\gamma_2) = 330.875$ and $I_{d_0}(\gamma_3) = 365.75$. We assume that $\gamma_1 = \gamma_2 = \gamma_3 = \gamma$ (say), i.e. there is no effect of time on the mental activity within the subject. Therefore $I_{d_0}(\gamma) = I_{d_0}(\gamma_1) + I_{d_0}(\gamma_2) + I_{d_0}(\gamma_3) = 1061.375$.

It is evident that we have some flexibilities in suggesting an OCD for a CRD model based on the covariate-values at Time Point 1 and, naturally, we should utilize that provision at least to maximize $I(\gamma_1)$. We have considered all possible permutations of Z-values at Time point 1 (TP 1) and by computer search we have found

$I(\gamma_1) = 383.75$, which is the maximum among all and we get 162 OCDs for Time point 1 where $I(\gamma_1) = 383.75$. These constitute one 'Equivalence Group' for the subjects. This exercise is very special. We recommend one member of this family in the absence of any other information and that presumably is the end of our task. What is the impact of our choice on subsequent TPs 2 and 3? For various choices of the family members, we obtain a range of $I(\gamma)$ values combining all the three TPs. We should look at the entire spectrum of $I(\gamma)$ and see if there is wide variation among them. One acceptable criterion may be $I(\gamma)_{min}$ is within 2 % of $I(\gamma)_{max}$. If that has been the case in this example as assessed by the covariate values at TP 2 and TP 3, we need not worry about our specific recommendation. Otherwise, there is a scope of introspection and dig into 'possible relations' among the covariate values for each subject over the TPs.

Since the size of EG is very large, therefore we have considered one member of this group and then found out the impact of our choice on subsequent Time Points 2 and 3.

We fix the same allocation of the subjects, shown in Table 9.60, for Time point 2 and Time point 3. In Table 9.61 we show the allocation of Z-values under each treatment for Time point 1, Time point 2 and Time point 3. We denote the design by d_1.

If we look at Table 9.60, then there is still chance to improve d_1, i.e. if we permute the allocation of the subjects where same covariate values appear in the rows of Table 9.60, then there is a possibility to improve the value of $I_2(\gamma)$ and $I_3(\gamma)$. Here the possible rows are R2, R3, R4, R5, R7, R8. We have considered all $2! \times 3! \times 2! \times 2! \times 3! \times 2! = 576$ permutations. By computer search, we have found $I(\gamma)_{max} = 1102.875$ and $I(\gamma)_{min} = 1018.625$ and the improvement of the design with $I(\gamma)_{max}$ (=1102.875) over the design with $I(\gamma)_{min}$ (=1018.625) is 8.27 % ($= \frac{1102.875 - 1018.625}{1018.625} \times 100\,\%$) only. In Table 9.62, we show the design with maximum $I(\gamma)$ and denote the design by d_2.

Now we consider the design with $I_{min}(\gamma) = 1018.625$ and denote it by d_3. This design is shown in Table 9.63.

Table 9.60 Allocation of subjects based on OCD for Time Point 1 [TP1]

	M	H	P	Allocation of subjects		
				M	H	P
	0	0	1	16	23	5
	2	2	1	2	6	13
	4	4	4	9	12	15
	6	5	5	8	7	20
	6	6	7	18	22	1
	7	8	9	11	19	14
	10	10	10	10	17	21
	14	14	11	3	4	24
Totals	49	49	48			

Table 9.61 Arrangement of the Z-values of the 24 subjects for Time point 1, Time point 2 and Time point 3

(a) Time point 1

	M	H	P
	0	0	1
	2	2	1
	4	4	4
	6	5	5
	6	6	7
	7	8	9
	10	10	10
	14	14	11
Totals	49	49	48

(b) Time point 2

	H	P	M
	0	0	4
	4	5	1
	6	3	10
	6	6	5
	6	7	0
	6	7	10
	8	11	4
	14	10	12
Totals	50	49	46

(c) Time point 3

	P	M	H
	0	0	5
	2	4	0
	5	1	8
	6	8	6
	6	7	0
	7	0	8
	6	11	11
	14	5	12
Totals	46	36	50

Here $I_{d_2}(\gamma_2) = 333.875$, $I_{d_2}(\gamma_3) = 385.25$, $I_{d_3}(\gamma_2) = 313.625$ and $I_{d_3}(\gamma_3) = 321.25$. It follows that the relative efficiencies of d_2 and d_3 with respect to d_0 are 103.91 % and 95.97 %. Therefore, though we start with optimal design for Time point 1, yet there might be a chance of getting less efficient design (viz., d_3) than d_0 at the end of Time point 3. Fortunately, still d_3 has a very high relative efficiency as against d_0. So our decision is not too bad to look for an optimal design with reference to Time point 1 and follow it through.

Table 9.62 Design d_2 with maximum I (γ)

	Time Point 1			Time Point 2			Time Point 3			Allocation		
	M	H	P	M	H	P	M	H	P	M	H	P
	0	0	1	0	0	4	0	0	5	16	23	5
	2	2	1	4	5	1	2	4	0	2	6	13
	4	4	4	6	3	10	5	1	8	9	12	15
	6	5	5	6	6	5	6	8	6	8	7	20
	6	6	7	6	7	0	6	7	0	18	22	1
	7	8	9	6	7	10	7	0	8	11	19	14
	10	10	10	11	4	8	11	11	6	17	21	10
	14	14	11	10	14	12	5	14	12	4	3	24
Totals	49	49	48	49	46	50	42	45	45			

Table 9.63 Design d_3 with minimum I (γ)

	Time Point 1			Time Point 2			Time Point 3			Allocation		
	M	H	P	M	H	P	M	H	P	M	H	P
	0	0	1	0	0	4	0	0	5	16	23	5
	2	2	1	5	4	1	4	2	0	6	2	13
	4	4	4	10	3	6	8	1	5	15	12	9
	6	5	5	6	5	6	6	6	8	8	20	7
	6	6	7	7	6	0	7	6	0	22	18	1
	7	8	9	6	7	10	7	0	8	11	19	14
	10	10	10	11	8	4	11	6	11	17	10	21
	14	14	11	14	10	12	14	5	12	3	4	24
Totals	49	49	48	59	43	43	57	26	49			

References

Dutta G, Sinha BK (2015) Search for optimal covariates' designs in some real-life experiments (Unpublished Manuscript)

Goos P, Jones B (2011) Optimal design of experiments: a case study approach. Wiley, Chichester

Kaiser HF (1960) The application of electronic computers to factor analysis. Educ Psychol Meas 20:141–151

Milliken GA, Johnson DE (2001) Analysis of messy data, volume III: analysis of covariance. Chapman & Hall/CRC, New York

Snedecor GW, Cochran WG (1989) Statistical methods, 8th edn. Iowa State University Press, Ames

van Belle G, Fisher LD, Heagerty PJ, Lumley T (2004) Biostatistics: a methodology for the health sciences, 2nd edn. Wiley-Interscience; Wiley, New Jersey

Author Index

© Springer India 2015
P. Das et al., *Optimal Covariate Designs*, DOI 10.1007/978-81-322-2461-7

Subject Index

A

Allocation, 1, 2, 4, 5, 8, 13, 14, 16, 18, 177–
180, 203, 208, 212, 214, 216
ANCOVA models, 1–3, 6, 177, 181, 189
ANOVA models, 1, 6
Association scheme, 66, 67, 77, 79, 127, 128

B

Balanced crossover design, 131, 135, 136,
144–146
Balanced incomplete block design (BIBD),
1, 6, 9, 41–47, 50, 51, 56, 58, 65,
67–70, 72, 73, 77, 85, 89, 90, 92, 95,
97, 100, 102, 104, 105, 113, 119–123,
125, 126, 129
Balanced treatment incomplete block design
(BTIBD), 1, 9, 113, 123–125, 127–
129
Binary proper equireplicated block design
(BPEBD), 1, 9, 42, 56, 89, 90, 92,
100–102, 104, 105
Block designs, 6, 8, 28, 56, 65, 89, 90, 114
Bose's method of difference, 9, 41–43, 45,
56

C

Catalogue of designs, 9, 30, 65, 80, 89, 105
Conference matrices, 35
Covariate models, 1
Covariate parameters, 1, 2, 6, 8, 13, 27, 28,
41, 43, 57, 113, 152, 163, 177, 188,
190, 191, 200, 203, 205

Covariates, 1, 3, 4, 6–10, 13–18, 26–28, 41,
44, 45, 51, 54, 65, 94, 95, 113, 114,
116, 131, 132, 135, 138, 141, 147,
150, 152, 153, 155, 156, 159, 160,
163, 177, 189–191, 202, 212
CRD-Non-regular case, 13, 16, 18, 24, 190
regular case, 13, 16, 180
Crossover designs, 9, 131, 132, 135
Cyclic association scheme, 66
Cyclic designs, 9, 56, 66, 94–98, 105, 110

D

Design of experiments, 1
D-optimal design, 6, 16, 18, 24, 25, 41, 56,
62, 189

E

Experimental units, 1, 5–8, 13, 27, 113, 131,
132, 180, 181
Extended and mixed orthogonal arrays
(EMOA), 147, 149–156, 158–163,
174, 175

G

Generator design, 113, 116, 119, 122, 123,
129
Globally optimal designs, 15
Group divisible association scheme, 66
Group divisible design (GDD), 1, 9, 65–67,
80, 89, 106, 113, 129

© Springer India 2015
P. Das et al., *Optimal Covariate Designs*, DOI 10.1007/978-81-322-2461-7

Printed in the United States
By Bookmasters

Hence, by summing $\frac{w_c}{w_g}$ rows of G, chosen at random, we get a random codeword with Hamming weight about w_c. Actually, due to some overlapping ones, the resulting weight could result smaller than w_c. In this case, some other row can be added, or some row replaced, or another combination of rows can be tested, in order to approach w_c. Moreover, as we will see in the following, using a random codeword with Hamming weight slightly smaller than w_c is not a problem in the proposed system. Based on the above considerations, the number of codewords with weight close to w_c which can be easily selected at random from an LDGM code having G with rows of weight w_g, with $w_g|w_c$, can be roughly estimated as

$$A_{w_c} \approx \binom{k}{\frac{w_c}{w_g}}. \tag{4}$$

3   System Description

In this section we describe the main steps of the proposed digital signature system.

3.1   Key Generation

The first part of the private key for the proposed system is formed by the $r \times n$ parity-check matrix H of an LDGM code $C(n, k)$, having length n and dimension k ($r = n - k$). The matrix H is in systematic form, with an identity block in the rightmost part. The private key also includes two other non-singular matrices: an $r \times r$ transformation matrix Q and an $n \times n$ scrambling matrix S (both defined below). The public key is then obtained as $H' = Q^{-1} \cdot H \cdot S^{-1}$.

The matrix S is a sparse non-singular matrix, with average row and column weight $m_S \ll n$. The matrix Q, instead, is a *weight controlling* transformation matrix as defined in [5]. For this kind of matrices, when s is a suitably chosen sparse vector, the vector $s' = Q \cdot s$ has a small Hamming weight, which is only a few times greater than that of s. As shown in [5], a matrix Q with such a feature can be obtained as the sum of an $r \times r$ low-rank dense matrix R and a sparse matrix T, chosen in such a way that $Q = R + T$ is non singular. In order to design R, we start from two $z \times r$ matrices, a and b, with $z < r$ properly chosen (see below), and define R as:

$$R = a^T \cdot b. \tag{5}$$

This way, R has rank $\leq z$. The matrix T is then chosen as a sparse matrix with row and column weight m_T, such that $Q = R + T$ is full rank.

It can be easily verified that, if the vector s is selected in such a way that $b \cdot s = 0_{z \times 1}$, where $0_{z \times 1}$ is the $z \times 1$ all-zero vector, then $R \cdot s = 0_{r \times 1}$ and $s' = Q \cdot s = T \cdot s$. Hence, the Hamming weight of s' is, at most, equal to m_T times that of s, and Q actually has the weight controlling feature we desire.

As we will see in Section 4.3, although it is relatively simple for an attacker to obtain the space defined by the matrix b, and its dual space, this does not help to mount a key recovery attack. Hence, the matrix b, which is only a small part of Q, can even be made public.

When a QC code is used as the private code, H is formed by $r_0 \times n_0$ circulant matrices of size $p \times p$, and it is desirable to preserve this QC structure also for H', in such a way as to exploit its benefits in terms of key size. For this purpose, both Q and S must be formed by circulant blocks with the same size as those forming H. Concerning the matrix S, it is obtained in QC form (S_{QC}) by simply choosing at random a block of $n_0 \times n_0$ sparse or null circulant matrices such that the overall row and column weight is m_S.

Concerning the matrix Q, instead, a solution to obtain it in QC form is to define R as follows:

$$R_{QC} = \left(a_{r_0}^T \cdot b_{r_0}\right) \otimes 1_{p \times p}, \tag{6}$$

where a_{r_0} and b_{r_0} are two $z \times r_0$ binary matrices, $1_{p \times p}$ is the all-one $p \times p$ matrix and $\otimes$ denotes the Kronecker product. Then, T_{QC} is chosen in the form of $n_0 \times n_0$ sparse circulant blocks with overall row and column weight m_T and Q_{QC} is obtained as $R_{QC} + T_{QC}$. This way, if H is in QC form, $H' = Q_{QC}^{-1} \cdot H \cdot S_{QC}^{-1}$ is in QC form as well. In the QC case, the condition we impose on s, that is, $b \cdot s = 0_{z \times 1}$ becomes $(b_{r_0} \otimes 1_{1 \times p}) \cdot s = 0_{z \times 1}$.

Such a condition, both in the generic and in the QC case, is equivalent to a set of z parity-check constraints for a code with length r and redundancy z. Hence, if we fix b such that this code has minimum distance d, then a vector s with weight $w < d$ cannot satisfy such condition, and Q loses its weight controlling feature on s. This is useful to reinforce the system against some vulnerabilities, and justifies the form used for the matrix Q.

Apart from the private and public key pair, the system needs two functions which are made public as well: a hash function $\mathcal{H}$ and a function $\mathcal{F}_\Theta$ that converts the output vector of $\mathcal{H}$ into a sparse r-bit vector s of weight $w \ll r$. The output of $\mathcal{F}_\Theta$ depends on the parameter Θ, which is associated to the message to be signed and made public by the signer. An example of implementation of $\mathcal{F}_\Theta$ is provided in the next section.

3.2   Signature Generation

In order to get a unique digital signature from some document M, the signer computes the digest $h = \mathcal{H}(M)$ and then finds Θ_M such that $s = \mathcal{F}_{\Theta_M}(h)$ verifies $b \cdot s = 0_{z \times 1}$. Since s has weight w, $s' = Q \cdot s$ has weight $\leq m_T w$. Concerning the implementation of the function $\mathcal{F}_\Theta(h)$, an example is as follows. Given a message digest $h = \mathcal{H}(M)$ of length x bits, similarly to what is done in the CFS scheme, it is appended with the y-bit value l of a counter, thus obtaining $[h|l]$. The value of $[h|l]$ is then mapped uniquely into one of the $\binom{r}{w}$ r-bit vectors of weight w, hence it must be $\binom{r}{w} \geq 2^{x+y}$. The counter is initially set to zero by the signer, and then progressively increased. This way, a different r-bit vector is obtained each time, until one orthogonal to b is found, for $l = \bar{l}$. This step requires the signer to test 2^z different values of the counter, on average. With this implementation of $\mathcal{F}_\Theta(h)$, we have $\Theta_M = \bar{l}$, and different signatures correspond to different vectors s, unless a hash collision occurs.

After having obtained s, the signer has to find a vector e of weight $\leq m_T w$ which corresponds to the private syndrome $s' = Q \cdot s$ through C. Since H is in systematic form, it can be written as $H = [X|I_r]$, where X is an $r \times k$ matrix and I_r is the $r \times r$ identity matrix. Hence, the private syndrome s' can be obtained from the error vector $e = [0_{1 \times k}|s'^T]$. So, in this case, finding e simply translates into a vector transposition and some zero padding.

The signer finally selects a random codeword $c \in C$ with small Hamming weight (w_c), and computes the public signature of M as $e' = (e + c) \cdot S^T$. If the choice of the codeword c is completely random and independent of the document to be signed, the signature obtained for a given document changes each time it is signed, and the system becomes vulnerable to attacks exploiting many signatures of the same document. This can be simply avoided by choosing the codeword c as a deterministic function of the document M and, hence, of the public syndrome s. For example, s or, equivalently, $[h|\bar{l}]$ can be used as the initial state of the pseudo-random integer generator through which the signer extracts the indexes of the rows of G that are summed to obtain c. This way, the same codeword is always obtained for the same public syndrome.

To explain the role of the codeword c, let us suppose for a moment that the system does not include any random codeword, that is equivalent to fix $c = 0_{1 \times n}, \forall M$. In this case, we could write $e' = W(s)$, where W is a linear bijective map from the set of public syndromes to the set of valid signatures. This can be easily verified, since it is simple to check that $W(s_1 + s_2) = W(s_1) + W(s_2)$. So, an attacker who wants to forge a signature for the public syndrome s_x could simply express s_x as a linear combination of previously intercepted public syndromes, $s_x = s_{i_1} + s_{i_2} + \ldots s_{i_N}$, and forge a valid signature by linearly combining their corresponding signatures: $e'_x = e'_{i_1} + e'_{i_2} + \ldots e'_{i_N}$.

As mentioned, to prevent this risk, the signer adds a random codeword c, with weight $w_c \ll n$, to the error vector e, before multiplication by S^T. This way, the map W becomes an affine map which depends on the random codeword c, and it no longer has the set of valid signatures as its image. In fact, we can denote this new map as $W_c(s)$, such that $e'_1 = W_{c_1}(s_1)$ and $e'_2 = W_{c_2}(s_2)$, where c_1 and c_2 are two randomly selected codewords of the private code with weight w_c. If we linearly combine the signatures, we obtain $e'_f = e'_1 + e'_2 = W_{c_1}(s_1) + W_{c_2}(s_2) = W_{c_1 + c_2}(s_1 + s_2)$. The vector $c_1 + c_2$ is still a valid codeword of the secret code, but it has weight $> w_c$ with a very high probability.

3.3 Signature Verification

After receiving the message M, its signature e' and the associated parameter Θ_M, the verifier first checks that the weight of e' is $\leq (m_T w + w_c) m_S$. If this condition is not satisfied, the signature is discarded. Then the verifier computes $\hat{s} = \mathcal{F}_{\Theta_M}(\mathcal{H}(M))$ and checks that $\hat{s}$ has weight w, otherwise the signature is discarded. If the previous checks have been positive, the verifier then computes $H' \cdot e'^T = Q^{-1} \cdot H \cdot S^{-1} \cdot S \cdot (e^T + c^T) = Q^{-1} \cdot H \cdot (e^T + c^T) = Q^{-1} \cdot H \cdot e^T = Q^{-1} \cdot s' = s$. If $s = \hat{s}$, the signature is accepted; otherwise, it is discarded.

3.4   Number of Different Signatures

An important parameter for any digital signature scheme is the total number of different signatures. In our case, a different signature corresponds to a different r-bit vector s, having weight w. Only vectors s satisfying the z constraints imposed by b are acceptable, so the maximum number of different signatures is:

$$N_s \approx \frac{\binom{r}{w}}{2^z}. \tag{7}$$

4   Possible Vulnerabilities

For a security assessment of the proposed system, it would be desirable to find possible security reductions to some well known hard problems, and then to evaluate the complexity of practical attacks aimed at solving such problems. This activity is still at the beginning, and work is in progress in this direction. Hence, in this paper we only provide a sketch of some possible vulnerabilities we have already devised, which permit to obtain a first rough estimate of the security level of the system. Completing the security assessment will allow to improve the security level estimation, and possibly to find more effective choices of the system parameters.

From the definition of the proposed system, it follows that the published signature e' associated to a document M is always a sparse vector, with Hamming weight $\leq (m_T w + w_c) m_S$. Since e' is an error vector corresponding to the public syndrome s through the public code parity-check matrix H', having a low Hamming weight ensures that e' is difficult to find, starting from s and H'. This is achieved by using the weight controlling matrix Q and the sparse matrix S. If this was not the case, and e' was a dense vector, it would be easy to forge signatures, since a dense vector corresponding to s through H' is easy to find.

Based on these considerations, one could think that choosing both Q and S as sparse as possible would be a good solution. Let us suppose that they are two permutation matrices, P_1 and P_2. In this case, the public matrix would be $H' = P_1^T \cdot H \cdot P_2^T$, and both s' and e' would be sparse, thus avoiding easy forgeries. Actually, a first reason for avoiding to use permutation matrices is that, when masked only through permutations, the security of H decreases. In fact, using a doubly permuted version of H may still allow to perform decoding through the public code. However, neglecting for a moment this fact, we find that, in this case, e' would have weight $\leq w + w_c$. If e and c have disjoint supports, which is very likely true, since we deal with sparse vectors, w non-zero bits in e' would correspond to a reordered version of the non-zero bits in s. So, apart from the effect of the random codeword, we would simply have a disposition of the non-zero bits in s within e', according to a fixed pattern. This pattern could be discovered by an attacker who observes a sufficiently large number of signatures, so that the effect of the random codeword could be eliminated. In fact, by computing the intersection of the supports of many vectors s and their corresponding vectors e', the support of e' could be decomposed and the reordering of each bit disclosed.

Based on these considerations, we can conclude that the density of e' must be carefully chosen between two opposite needs:

- being sufficiently low to avoid forgeries;
- being sufficiently high to avoid support decompositions.

4.1 Forgery Attacks

In order to forge signatures, an attacker could search for an $n \times r$ right-inverse matrix H'_r of H'. Then, he could compute $f = (H'_r \cdot s)^T$, which is a forged signature. It is easy to find a right-inverse matrix able to forge dense signatures. In fact, provided that $H' \cdot H'^T$ is invertible, $H'_r = H'^T \cdot (H' \cdot H'^T)^{-1}$ is a right-inverse matrix of H'. The matrix H' is dense and the same occurs, in general, for $(H' \cdot H'^T)^{-1}$; so, H'_r is dense as well. It follows that, when multiplied by s, H'_r produces a dense vector, thus allowing to forge dense signatures. By using sparse signatures, with weight $\leq (m_T w + w_c) m_S$, the proposed system is robust against this kind of forged signatures.

However, the right-inverse matrix is not unique. So, the attacker could search for an alternative, possibly sparse, right-inverse matrix. In fact, given an $n \times n$ matrix Z such that $H' \cdot Z \cdot H'^T$ is invertible, $H''_r = Z \cdot H'^T \cdot (H' \cdot Z \cdot H'^T)^{-1}$ is another valid right-inverse matrix of H'. We notice that $H''_r \neq Z \cdot H'_r$. When H' contains an invertible $r \times r$ square block, there is also another simple way to find a right-inverse. It is obtained by inverting such block, putting its inverse at the same position (in a transposed matrix) in which it is found within H', and padding the remaining rows with zeros.

In any case, there is no simple way to find a right-inverse matrix that is also sparse, which is the aim of an attacker. Actually, for the matrix sizes considered here, the number of possible choices of Z is always huge. Moreover, there is no guarantee that any of them produces a sparse right-inverse. Searching for an $r \times r$ invertible block within H' and inverting it would also produce unsatisfactory results, since the overall density of H'^{-1} is reduced, but the inverse of the square block is still dense. So, the attacker would be able to forge signatures with a number of symbols 1 on the order of $r/2$, that is still too large for the system considered here. In fact, in the system examples we provide in Section 5, we always consider public signatures with weight on the order of $r/3$ or less.

A further chance is to exploit Stern's algorithm [29] (or other approaches for searching low weight codewords) to find a sparse representation of the column space of H'_r. If this succeeds, it would result in a sparse matrix $H_S = H'_r \cdot B$, for some $r \times r$ transformation matrix B. However, in this case, H_S would not be a right-inverse of H'.

For these reasons, approaches based on right-inverses seem to be infeasible for an attacker. An alternative attack strategy could be based on decoding. In fact, an attacker could try syndrome decoding of s through H', hoping to find a sparse vector f. He would have the advantage of searching for one out of many possible vectors, since he is not looking for a correctable error vector. Several algorithms could be exploited for solving such problem [8, 10, 11, 20, 27]. These algorithms are commonly used to search for low weight vectors with a null syndrome, but, with a small modification, they can also be used to find vectors corresponding to a given (non-zero) syndrome. In addition, their complexity decreases when an attacker has access to a high number of decoding instances, and wishes to solve only one of them [28], which is the case for the proposed system. We will discuss the complexity issue in Section 5.

4.2 Support Decomposition Attacks

Concerning support decomposition, let us suppose that e and c have disjoint supports. In this case, the overall effect of the proposed scheme on the public syndrome s can be seen as the expansion of an $r \times 1$ vector s of weight w into a subset of the support of the $1 \times n$ vector e', having weight $\leq m_T m_S w$, in which each symbol 1 in s corresponds, at most, to $m = m_T m_S$ symbols 1 in e'.

An attacker could try to find the w sets of m (or less) symbols 1 within the support of e' in order to compute valid signatures. In this case, he will work as if the random codeword was absent, that is, $c = 0_{1 \times n}$. Thus, even after succeeding, he would be able to forge signatures that are sparser than the authentic ones. In any case, this seems a rather dangerous situation, so we should aim at designing the system in such a way as to avoid its occurrence.

To reach his target, the attacker must collect a sufficiently large number L of pairs (s, e'). Then, he can intersect the supports (that is, compute the bit-wise AND) of all the s vectors. This way, he obtains a vector s_L that may have a small weight $w_L \geq 1$. If this succeeds, the attacker analyzes the vectors e', and selects the $m w_L$ set bit positions that appear more frequently. If these bit positions are actually those corresponding to the w_L bits set in s_L, then the attacker has discovered the relationship between them. An estimate of the probability of success of this attack can be obtained through combinatorial arguments.

An even more efficient strategy could be to exploit information set decoding to remove the effect of the random codeword. In fact, an attacker knows that $e' = (e + c) \cdot S^T = e'' + c''$, with c'' such that $H'c'' = 0$. Hence, e'' can be considered as an error vector with weight $\leq m_T m_S w$ affecting the codeword c'' of the public code. So the attacker could consider a random subset of k coordinates of the public signature e' and assume that no errors occurred on these coordinates. In this case, he can easily recover c'' and, hence, remove the effect of the random codeword c. The probability that there are no errors in the chosen k coordinates is $\binom{n - m_T m_S w}{k} / \binom{n}{k}$, and its inverse provides a rough estimate of the work factor of this attack.

4.3 Key Recovery Attacks

An attacker could aim to mount a key recovery attack, that is, to obtain the private code. A potential vulnerability in this sense comes from the use of LDGM codes. As we have seen in Section 2, LDGM codes offer the advantage of having a predictable (and sufficiently high) number of codewords with a moderately low weight w_c, and of making their random selection very easy for the signer.

On the other hand, when the private code is an LDGM code, the public code admits a generator matrix in the form $G'_I = G \cdot S^T$, which is still rather sparse. So, the public code contains low weight codewords, coinciding with the rows of G'_I, which have weight approximately equal to $w_g \cdot m_S$. Since G'_I has k rows, and summing any two of them gives higher weight codewords with a very high probability, we can consider that the multiplicity of these words in the public code is k. They could be searched by using again Stern's algorithm [29] and its improved versions [8, 10, 11, 20, 27], in such a way as to recover G'_I. After that, G'_I could be separated into G and S^T by exploiting their sparsity. In Section 5 we discuss how to estimate the work factor of this attack.

Another possible vulnerability comes from the fact that the matrix b is public. Even if b was not public, an attacker could obtain the vector space generated by b, as well as its dual space, by observing $O(r)$ public syndromes s, since $b \cdot s = 0_{z \times 1}$. Hence, we must suppose that an attacker knows an $r \times r$ matrix V such that $R \cdot V = 0 \Rightarrow Q \cdot V = T \cdot V$. The attacker also knows that $H' = Q^{-1} \cdot H \cdot S^{-1}$ and that the public code admits any non-singular generator matrix in the form $G'_X = X \cdot G \cdot S^T$, which becomes $G'_Q = Q \cdot G \cdot S^T$ for $X = Q$. Obviously, G'_I is the sparsest among them, and it can be attacked by searching for low weight codewords in the public code, as we have already observed. Instead, knowing V is useless to reduce the complexity of attacking either H' or one of the possible G'_X, hence it cannot be exploited by an attacker to perform a key recovery attack.

4.4 Other Attacks

As for any other hash-and-sign scheme, classical collision birthday attacks represent a threat for the proposed system. Since the system admits up to N_s different signatures, it is sufficient to collect $\approx \sqrt{N_s}$ different signatures to have a high probability of finding a collision [19]. Hence, the security level reached by the system cannot exceed $\sqrt{N_s}$.

However, N_s can be made sufficiently high by increasing the value of w. The definition of the proposed system allows to choose its parameters in such a way as to guarantee this fact, as we will see in Section 5. This is possible since the choice of w is not constrained by the row weight of the private generator matrix. In fact, in the proposed scheme we do not actually need a private code of minimum distance greater than $2w$, because we rely on a decoding procedure which uniquely associates to a syndrome of a given weight an error vector with the same weight, though such an error vector is not necessarily unique.

Finally, it is advisable to consider the most dangerous attacks against the CFS scheme. It was successfully attacked by exploiting syndrome decoding based on the generalized birthday algorithm [14], even if the proposed attacking algorithm was not the optimal generalization of the birthday algorithm [22]. If we do not take into account some further improvement due to the QC structure of the public key, these algorithms provide a huge work factor for the proposed system parameters, since they try to solve the decoding problem for a random code. Just to give an idea, we obtain a work factor of more than 2^{200} binary operations even for the smallest key sizes we consider. However, there are some strategies that can be implemented to improve the efficiency of the attack on structured matrices, like those of dyadic codes [25]. This improvement could be extended to QC codes as well, but the attack work factor, for the cryptosystems analyzed in [25], is lowered by (at most) 2^{10} binary operations, starting from a maximum value of 2^{344}. Hence, it is very unlikely that this strategy can endanger the signature scheme we propose.

5   System Examples

By using the preliminary security assessment reported in Section 4, we can find some possible choices of the system parameters aimed at reaching fixed security levels. For this purpose, we have considered all the vulnerabilities described in Section 4, and we have estimated the work factor needed to mount a successful attack exploiting each of them.

We have used the implementation proposed in [27] for estimating the work factor of low weight codeword searches. Actually, [27] does not contain the most up-to-date and efficient implementation of information set decoding. In fact, some improvements have appeared in the literature concerning algorithms for decoding binary random codes (as [20], [8]). These papers, however, aim at finding algorithms which are asymptotically faster, by minimizing their asymptotic complexity exponents. Instead, for computing the work factor of attacks based on decoding, we need actual operation counts, which are not reflected in these recent works. Also "ball collision decoding" [10] achieves significant improvements asymptotically, but they become negligible for finite code lengths and not too high security levels. For these reasons, we prefer to resort to [27], which provides a detailed analysis of the algorithm, together with a precise operation count for given code parameters. On the other hand, attacks against the proposed system which exploit decoding, i.e., trying to recover the rows of G or to forge valid signatures through decoding algorithms, are far away from providing the smallest work factors, and, hence, to determine the security level. For the choices of the system parameters we suggest, the smallest work factor is always achieved by attacks aiming at decomposing the signature support, which hence coincides with the security level. For the instances proposed in this section, the work factor of attacks based on decoding is on the order of 2^{2SL}, where SL is the claimed security level. Hence, even considering some reduction in the work factor of decoding attacks would not change the security level of the considered instances of the system. This situation does not change even if we consider the improvement coming from the "decoding one out of many" approach [28]. In fact, as shown in [28], even if an attacker has access to an unlimited number of decoding instances, the attack complexity is raised by a power slightly larger than 2/3.

Concerning support decomposition attacks, a rough estimation of their complexity has instead been obtained through simple combinatorial arguments, which are not reported here for the sake of brevity.

A more detailed analysis of the attacks work factor is out of scope of this paper, and will be addressed in future works, together with a more complete

Table 1. System parameters for some security levels (SL), with $d = 2$ and $w_L = 2$

SL (bits)	n	k	p	w	w_g	w_c	z	m_T	m_S	A_{w_c}	N_s	S_k (KiB)
80	9800	4900	50	18	20	160	2	1	9	$2^{82.76}$	$2^{166.10}$	117
120	24960	10000	80	23	25	325	2	1	14	$2^{140.19}$	$2^{242.51}$	570
160	46000	16000	100	29	31	465	2	1	20	$2^{169.23}$	$2^{326.49}$	1685

security assessment. This will also permit to refine the choice of the system parameters, in such a way as to find the best trade-off between the security level and the key size.

Table 1 provides three choices of the system parameters which are aimed at achieving 80-bit, 120-bit and 160-bit security, respectively. All these instances of the system use QC-LDGM codes with different values of p, also reported in the table, and consider the case in which the matrix Q is such that $d = w_L = 2$. Actually, achieving minimum distance equal to 2 (or more) is very easy: it is sufficient to choose $z > 1$ and to guarantee that the matrix b does not contain all-zero columns. For each instance of the system, the value of the key size S_k is also shown, expressed in kibibytes (1 KiB = $1024 \cdot 8$ bits).

In the original version of the CFS system, to achieve an attack time and memory complexity greater than 2^{80}, we need to use a Goppa code with length $n = 2^{21}$ and redundancy $r = 21 \cdot 10 = 210$ [14]. This gives a key size on the order of $4.4 \cdot 10^8$ bits = 52.5 MiB. By using the parallel CFS proposed in [15], the same work factor can be reached by using keys with size ranging between $1.05 \cdot 10^7$ and $1.7 \cdot 10^8$ bits, that is, between 1.25 MiB and 20 MiB. As we notice from Table 1, the proposed system requires a public key of only 117 KiB to achieve 80-bit security. Hence, it is able to achieve a dramatic reduction in the public key size compared to the CFS scheme, even when using a parallel implementation of the latter.

Another advantage of the proposed solution compared to the CFS scheme is that it exploits a straightforward decoding procedure for the secret code. On the other hand, 2^z attempts are needed, on average, to find an s vector such that $b \cdot s = 0_{z \times 1}$. However, such a check is very simple to perform, especially for very small values of z, like those considered here.

6   Conclusion

In this paper, we have addressed the problem of achieving efficient code-based digital signatures with small public keys.

We have proposed a solution that, starting from the CFS schemes, exploits LDGM codes and sparse syndromes to achieve good security levels with compact keys. The proposed system also has the advantage of using a straightforward decoding procedure, which reduces to a transposition and a concatenation with an all-zero vector. This is considerably faster than classical decoding algorithms used for common families of codes.

The proposed scheme allows to use a wide range of choices of the code parameters. In particular, the low code rates we adopt avoid some problems of the classical CFS scheme, due to the use of codes with high rate and small correction capability.

On the other hand, using sparse vectors may expose the system to new attacks. We have provided a sketch of possible vulnerabilities affecting this system, together with a preliminary evaluation of its security level. Work is in progress to achieve more precise work factor estimates for the most dangerous attacks.

References

1. Baldi, M., Chiaraluce, F.: Cryptanalysis of a new instance of McEliece cryptosystem based on QC-LDPC codes. In: Proc. IEEE International Symposium on Information Theory (ISIT 2007), Nice, France, pp. 2591–2595 (June 2007)
2. Baldi, M., Chiaraluce, F., Garello, R., Mininni, F.: Quasi-cyclic low-density parity-check codes in the McEliece cryptosystem. In: Proc. IEEE International Conference on Communications (ICC 2007), Glasgow, Scotland, pp. 951–956 (June 2007)
3. Baldi, M., Bodrato, M., Chiaraluce, F.: A new analysis of the McEliece cryptosystem based on QC-LDPC codes. In: Ostrovsky, R., De Prisco, R., Visconti, I. (eds.) SCN 2008. LNCS, vol. 5229, pp. 246–262. Springer, Heidelberg (2008)
4. Baldi, M., Bambozzi, F., Chiaraluce, F.: On a Family of Circulant Matrices for Quasi-Cyclic Low-Density Generator Matrix Codes. IEEE Trans. on Information Theory 57(9), 6052–6067 (2011)
5. Baldi, M., Bianchi, M., Chiaraluce, F., Rosenthal, J., Schipani, D.: Enhanced public key security for the McEliece cryptosystem (2011), http://arxiv.org/abs/1108.2462
6. M. Baldi, M. Bianchi, and F. Chiaraluce. "Security and complexity of the McEliece cryptosystem based on QC-LDPC codes. IET Information Security (in press), http://arxiv.org/abs/1109.5827
7. Baldi, M., Bianchi, M., Chiaraluce, F.: Optimization of the parity-check matrix density in QC-LDPC code-based McEliece cryptosystems. To be presented at the IEEE International Conference on Communications (ICC 2013) - Workshop on Information Security over Noisy and Lossy Communication Systems, Budapest, Hungary (June 2013)
8. Becker, A., Joux, A., May, A., Meurer, A.: Decoding random binary linear codes in $2^{n/20}$: How $1 + 1 = 0$ improves information set decoding. In: Pointcheval, D., Johansson, T. (eds.) EUROCRYPT 2012. LNCS, vol. 7237, pp. 520–536. Springer, Heidelberg (2012)
9. Bernstein, D.J., Lange, T., Peters, C.: Attacking and defending the mcEliece cryptosystem. In: Buchmann, J., Ding, J. (eds.) PQCrypto 2008. LNCS, vol. 5299, pp. 31–46. Springer, Heidelberg (2008)
10. Bernstein, D.J., Lange, T., Peters, C.: Smaller decoding exponents: ball-collision decoding. In: Rogaway, P. (ed.) CRYPTO 2011. LNCS, vol. 6841, pp. 743–760. Springer, Heidelberg (2011)
11. Chabaud, F., Stern, J.: The cryptographic security of the syndrome decoding problem for rank distance codes. In: Kim, K.-c., Matsumoto, T. (eds.) ASIACRYPT 1996. LNCS, vol. 1163, pp. 368–381. Springer, Heidelberg (1996)

12. Cheng, J.F., McEliece, R.J.: Some high-rate near capacity codecs for the Gaussian channel. In: Proc. 34th Allerton Conference on Communications, Control and Computing, Allerton, IL (October 1996)

13. Courtois, N.T., Finiasz, M., Sendrier, N.: How to achieve a McEliece-based digital signature scheme. In: Boyd, C. (ed.) ASIACRYPT 2001. LNCS, vol. 2248, pp. 157–174. Springer, Heidelberg (2001)

14. Finiasz, M., Sendrier, N.: Security bounds for the design of code-based cryptosystems. In: Matsui, M. (ed.) ASIACRYPT 2009. LNCS, vol. 5912, pp. 88–105. Springer, Heidelberg (2009)

15. Finiasz, M.: Parallel-CFS strengthening the CFS McEliece-based signature scheme. In: Proc. PQCrypto, Darmstadt, Germany, pp. 61–72, May 25-28 (2010)

16. Garcia-Frias, J., Zhong, W.: Approaching Shannon performance by iterative decoding of linear codes with low-density generator matrix. IEEE Commun. Lett. 7(6), 266–268 (2003)

17. González-López, M., Vázquez-Araújo, F.J., Castedo, L., Garcia-Frias, J.: Serially-concatenated low-density generator matrix (SCLDGM) codes for transmission over AWGN and Rayleigh fading channels. IEEE Trans. Wireless Commun. 6(8), 2753–2758 (2007)

18. Kabatianskii, G., Krouk, E., Smeets, B.: A digital signature scheme based on random error correcting codes. In: Darnell, M.J. (ed.) Cryptography and Coding 1997. LNCS, vol. 1355, pp. 161–167. Springer, Heidelberg (1997)

19. Lim, C.H., Lee, P.J.: On the length of hash-values for digital signature schemes. In: Proc. CISC 1995, Seoul, Korea, November 1995, pp. 29–31 (1995)

20. May, A., Meurer, A., Thomae, E.: Decoding random linear codes in $\tilde{O}(2^{0.054n})$. In: Lee, D.H., Wang, X. (eds.) ASIACRYPT 2011. LNCS, vol. 7073, pp. 107–124. Springer, Heidelberg (2011)

21. McEliece, R.J.: A public-key cryptosystem based on algebraic coding theory. DSN Progress Report, pp. 114–116 (1978)

22. Minder, L., Sinclair, A.: The extended k-tree algorithm. Journal of Cryptology 25(2), 349–382 (2012)

23. Misoczki, R., Tillich, J.-P., Sendrier, N., Barreto, P.S.L.M.: MDPC-McEliece: New McEliece variants from moderate density parity-check codes (2012), http://eprint.iacr.org/2012/409

24. Monico, C., Rosenthal, J., Shokrollahi, A.: Using low density parity check codes in the McEliece cryptosystem. In: Proc. IEEE International Symposium on Information Theory (ISIT 2000), Sorrento, Italy, p. 215 (June 2000)

25. Niebuhr, R., Cayrel, P.-L., Buchmann, J.: Improving the efficiency of Generalized Birthday Attacks against certain structured cryptosystems. In: Proc. WCC 2011, Paris, France, April 11-15 (2011)

26. Otmani, A., Tillich, J.-P.: An efficient attack on all concrete KKS proposals. In: Yang, B.-Y. (ed.) PQCrypto 2011. LNCS, vol. 7071, pp. 98–116. Springer, Heidelberg (2011)

27. Peters, C.: Information-set decoding for linear codes over F_q. In: Sendrier, N. (ed.) PQCrypto 2010. LNCS, vol. 6061, pp. 81–94. Springer, Heidelberg (2010)

28. Sendrier, N.: Decoding one out of many. In: Yang, B.-Y. (ed.) PQCrypto 2011. LNCS, vol. 7071, pp. 51–67. Springer, Heidelberg (2011)

29. Stern, J.: A method for finding codewords of small weight. In: Wolfmann, J., Cohen, G. (eds.) Coding Theory and Applications 1988. LNCS, vol. 388, pp. 106–113. Springer, Heidelberg (1989)

Quantum Algorithms
for the Subset-Sum Problem

Daniel J. Bernstein[1,2], Stacey Jeffery[3], Tanja Lange[2], and Alexander Meurer[4]

[1] Department of Computer Science
University of Illinois at Chicago
Chicago, IL 60607–7045, USA
djb@cr.yp.to
[2] Department of Mathematics and Computer Science
Technische Universiteit Eindhoven
P.O. Box 513, 5600 MB Eindhoven, The Netherlands
tanja@hyperelliptic.org
[3] Institute for Quantum Computing, University of Waterloo, Canada
sjeffery@uwaterloo.ca
[4] Ruhr-University Bochum, Horst Görtz Institute for IT-Security
alexander.meurer@rub.de

Abstract. This paper introduces a subset-sum algorithm with heuristic asymptotic cost exponent below 0.25. The new algorithm combines the 2010 Howgrave-Graham–Joux subset-sum algorithm with a new streamlined data structure for quantum walks on Johnson graphs.

Keywords: subset sum, quantum search, quantum walks, radix trees, decoding, SVP, CVP.

1 Introduction

The subset-sum problem is the problem of deciding, given integers $x_1, x_2, \ldots, x_n$ and s, whether there exists a subset I of $\{1, 2, \ldots, n\}$ such that $\sum_{i \in I} x_i = s$; i.e., whether there exists a subsequence of $x_1, x_2, \ldots, x_n$ with sum s. Being able to solve this decision problem implies being able to find such a subset if one exists: for $n > 1$ one recursively tries $x_1, x_2, \ldots, x_{n-1}$ with sum s or, if that fails, sum $s - x_n$.

The reader should imagine, as a typical "hard" case, that $x_1, x_2, \ldots, x_n$ are independent uniform random integers in $\{0, 1, \ldots, 2^n\}$, and that s is chosen as a uniform random integer between $(n/2 - \sqrt{n})2^{n-1}$ and $(n/2 + \sqrt{n})2^{n-1}$. The number of subsets $I \subseteq \{1, 2, \ldots, n\}$ with $\sum_{i \in I} x_i = s$ then has a noticeable chance of being larger than 0 but is quite unlikely to be much larger, say larger than n.

This work was supported by the National Science Foundation under grant 1018836, by the Netherlands Organisation for Scientific Research (NWO) under grant 639.073.005, by NSERC Strategic Project Grant FREQUENCY, by the US ARO, and by the European Commission under Contract ICT-2007-216676 ECRYPT II. Permanent ID of this document: 797dcbe9c3389410ae4d5ae6dba33ab7. Date: 2013.04.06.

P. Gaborit (Ed.): PQCrypto 2013, LNCS 7932, pp. 16–33, 2013.
© Springer-Verlag Berlin Heidelberg 2013

The subset-sum problem is, historically, one of the first problems to be proven NP-complete. A polynomial-time non-quantum algorithm for the subset-sum problem would violate the standard P $\neq$ NP conjecture; a polynomial-time quantum algorithm for the subset-sum problem would violate the standard NP $\not\subseteq$ BQP conjecture. There is, however, a very large gap between polynomial time and the time needed for a naive search through all 2^n subsets of $\{1, 2, \ldots, n\}$. The standard NP $\not\subseteq$ BQP conjecture does not rule out faster exponential-time algorithms, or even subexponential-time algorithms, or even algorithms that take polynomial time for *most* inputs. This paper studies faster exponential-time algorithms.

Variations. Often one is interested only in sums of fixed weight, or of limited weight. We are now given integers $x_1, x_2, \ldots, x_n$, s, and w; the problem is to decide whether there is a subset I of $\{1, 2, \ldots, n\}$ such that $\sum_{i \in I} x_i = s$ and $\#I = w$. In the special case $s = 0$ with $w \neq 0$, such a subset I immediately produces a short nonzero vector in the lattice L of vectors $v \in \mathbf{Z}^n$ satisfying $\sum_i x_i v_i = 0$: specifically, the characteristic function of I is a vector of length $\sqrt{w}$ in L. In many applications this is the shortest nonzero vector in L; in some applications this vector can be found by standard SVP algorithms.

For $s \neq 0$ one can instead compute a vector $r \in \mathbf{R}^n$ satisfying $\sum_i x_i r_i = s$, and then observe that subtracting the characteristic function of I from r produces an element of L. In many applications this is the vector in L closest to r; in some applications this vector can be found by standard CVP algorithms.

A variant of the same problem is the central algorithmic problem in coding theory. The input now consists of vectors $x_1, x_2, \ldots, x_n$, a vector s, and an integer w; these vectors all have the same length and have entries in the field $\mathbf{F}_2$ of integers modulo 2. The problem, as above, is to decide whether there is a subset I of $\{1, 2, \ldots, n\}$ such that $\sum_{i \in I} x_i = s$ and $\#I = w$.

We do not mean to suggest that these problems are identical. However, the algorithmic techniques used to attack subset-sum problems are among the central algorithmic techniques used to attack lattice problems and decoding problems. For example, the best attack known against code-based cryptography, at least asymptotically, is a very recent decoding algorithm by Becker, Joux, May, and Meurer [4], improving upon a decoding algorithm by May, Meurer, and Thomae [24]; the algorithm of [4] is an adaptation of a subset-sum algorithm by Becker, Coron, and Joux [3], improving analogously upon a subset-sum algorithm by Howgrave-Graham and Joux [17].

There is also a line of work on building cryptographic systems whose security is more directly tied to the subset-sum problem. For example, Lyubashevsky, Palacio, and Segev in [22] propose a public-key encryption system and prove that being able to break it implies being able to solve modular subset-sum problems of the following type: find a random subset $I \subseteq \{1, 2, \ldots, n\}$ given random $x_1, x_2, \ldots, x_n$ modulo M and given $\sum_{i \in I} x_i$ modulo M, where M is roughly $(10n \log n)^n$. They claim in [22, Section 1] that "there are currently no known quantum algorithms that perform better than classical ones on the subset sum problem".

Table 1.1. Heuristic asymptotic performance of various subset-sum algorithms. An algorithm using $2^{(e+o(1))n}$ operations is listed as "exponent" e.

Exponent	Quantum	Split	Algorithm
1	No	1	Brute force
0.5	Yes	1	Quantum search; §2
0.5	No	1/2	Left-right split; §2
0.5	No	1/4	Left-right split with a modulus; §4
0.375	Yes	1/4	Quantum search with a modulus; §4
0.337...	No	1/16	Moduli + representations; §5
0.333...	Yes	1/3	Quantum left-right split; §2
0.333...	Yes	1/2	Quantum walk; §3
0.3	Yes	1/4	Quantum walk with a modulus; §4
0.291...	No	1/16	Moduli + representations + overlap; [3]
0.241...	Yes	1/16	**New**; quantum walk + moduli + representations; §5

Contents of This Paper. We introduce the first subset-sum algorithm that beats $2^{n/4}$. Specifically, we introduce a quantum algorithm that, under reasonable assumptions, uses at most $2^{(0.241...+o(1))n}$ qubit operations to solve a subset-sum problem. This algorithm combines quantum walks with the central "representations" idea of [17]. Table 1.1 compares this exponent 0.241... to the exponents of other algorithms.

One can reasonably speculate that analogous quantum speedups can also be applied to the algorithms of [24] and [4]. However, establishing this will require considerable extra work, similar to the extra work of [24] and [4] compared to [17] and [3] respectively.

Cost Metric and Other Conventions. This paper follows the tradition of measuring algorithm cost as the number of bit operations or, more generally, qubit operations. In particular, random access to an array of size $2^{O(n)}$ is assumed to cost only $n^{O(1)}$, even if the array index is a quantum superposition.

We systematically suppress cost factors polynomial in n; our concern is with asymptotic exponents such as the 0.241... in $2^{(0.241...+o(1))n}$. We also assume that the inputs $x_1, x_2, \ldots, x_n, s$ have $n^{O(1)}$ bits. These conventions mean, for example, that reading the entire input $x_1, x_2, \ldots, x_n, s$ costs only 1.

Almost all of the algorithms discussed here split size-n sets into parts, either 2 or 3 or 4 or 16 parts, as indicated by the "Split" column in Table 1.1. Any reasonably balanced split is adequate, but to simplify the algorithm statements we assume that n is a multiple of 2 or 3 or 4 or 16 respectively.

The algorithms in this paper are designed to work well for random inputs, particularly in the "hard" case that $x_1, x_2, \ldots, x_n, s$ each have about n bits. Our analyses — like the analyses of state-of-the-art algorithms for integer factorization, discrete logarithms, and many other problems of cryptographic interest — are heuristic. We do not claim that the algorithms work for *all* inputs, and we do not claim that what we call the "hard" case is the worst case. Even for random inputs we do not claim that our analyses are proven, but we do speculate that they are provable by an adaptation of the proof ideas stated in [17, eprint version].

Acknowledgments. This work was initiated during the Post-Quantum Cryptography and Quantum Algorithms workshop at the Lorentz Center in November 2012. We acknowledge helpful discussions with Andris Ambainis, Frédéric Magniez, Nicolas Sendrier, and Jean-Pierre Tillich.

2   Searches

Define Σ as the function that maps $I \subseteq \{1, 2, \ldots, n\}$ to $\sum_{i \in I} x_i$. Recall that we assume that $x_1, x_2, \ldots, x_n, s$ have $n^{O(1)}$ bits, and that we suppress polynomial cost factors; evaluating Σ therefore has cost 1.

The subset-sum problem is the problem of deciding whether there exists I with $\Sigma(I) = s$, i.e., whether the function $\Sigma - s$ has a root. A classical search for a root of $\Sigma - s$ uses 2^n evaluations of $\Sigma - s$, for a total cost of 2^n. Of course, the search can finish much sooner if it finds a root (one expects only 2^{n-1} evaluations on average if there is 1 root, and fewer if there are more roots); but as discussed in Section 1 we focus on "hard" cases where there are not many roots, and then the cost is 2^n (again, suppressing polynomial factors) with overwhelming probability.

This section reviews two standard ways to speed up this brute-force search. The first way is Grover's quantum search algorithm. The second way is decomposing $\Sigma(I)$ as $\Sigma(I_1) + \Sigma(I_2)$, where $I_1 = I \cap \{1, 2, \ldots, n/2\}$ and $I_2 = I \cap \{n/2 + 1, \ldots, n\}$; this split was introduced by Horowitz and Sahni in [16].

Review: The Performance of Quantum Search. Consider any computable function f with a b-bit input and a unique root. Grover's algorithm [15] finds the root (with negligible failure chance) using approximately $2^{b/2}$ quantum evaluations of f and a small amount of overhead.

More generally, consider any computable function f with a b-bit input and $r > 0$ roots. Boyer, Brassard, Høyer, and Tapp in [6] introduced a generalization of Grover's algorithm (almost exactly the same as Grover's original algorithm but stopping after a particular r-dependent number of iterations) that finds a root (again with negligible failure chance) using approximately $(2^b/r)^{1/2}$ quantum evaluations of f and a small amount of overhead. One can easily achieve the same result by using Grover's original algorithm sensibly (as mentioned in [15]): choose a fast but sufficiently random map from $b - \lceil \lg r \rceil$ bits to b bits; the composition of this map with f has a good chance of having a unique root; apply Grover's algorithm to this composition; repeat several times so that the failure chance becomes negligible.

Even more generally, consider any computable function f with a b-bit input. A more general algorithm in [6] finds a root using approximately $(2^b/r)^{1/2}$ quantum evaluations of f and a small amount of overhead, where r is the number of roots. If no root exists then the algorithm says so after approximately $2^{b/2}$ quantum evaluations of f. As above, the algorithm can fail (or take longer than expected), but only with negligible probability; and, as above, the same result can also be obtained from Grover's algorithm.

As a trivial application, take $b = n$ and $f = \Sigma - s$: finding a root of $\Sigma - s$ costs $2^{n/2}$. Some implementation details of this quantum subset-sum algorithm appeared in [9] in 2009. We emphasize, however, that the same operation count is achieved by well-known non-quantum algorithms, and is solidly beaten by recent non-quantum algorithms.

Review: Left-Right Split. Define $L_1 = \{(\Sigma(I_1), I_1) : I_1 \subseteq \{1, 2, \ldots, n/2\}\}$ and $L_2 = \{(s - \Sigma(I_2), I_2) : I_2 \subseteq \{n/2 + 1, n/2 + 2, \ldots, n\}\}$. Note that each of these sets has size just $2^{n/2}$.

Compute L_1 by enumerating sets I_1. Store the elements of L_1 in a table, and sort the table by its first coordinate. Compute L_2 by enumerating sets I_2. For each $(s - \Sigma(I_2), I_2) \in L_2$, look for $s - \Sigma(I_2)$ by binary search in the sorted table. If there is a collision $\Sigma(I_1) = s - \Sigma(I_2)$, print out $I_1 \cup I_2$ as a root of $\Sigma - s$ and stop. If there are no collisions, print "there is no subset-sum solution" and stop.

This algorithm costs $2^{n/2}$. It uses $2^{n/2}$ memory, and one can object that random access to memory is expensive, but we emphasize that this paper follows the tradition of simply counting operations. There are several standard variants of this algorithm: for example, one can sort L_1 and L_2 together, or one can store the elements of L_1 in a hash table.

Quantum Left-Right Split. Redefine L_1 as $\{(\Sigma(I_1), I_1) : I_1 \subseteq \{1, 2, \ldots, n/3\}\}$; note that $n/2$ has changed to $n/3$. Compute and sort L_1 as above; this costs $2^{n/3}$.

Consider the function f that maps a subset $I_2 \subseteq \{n/3 + 1, n/3 + 2, \ldots, n\}$ to 0 if $s - \Sigma(I_2)$ is a first coordinate in L_1, otherwise to 1. Binary search in the sorted L_1 table costs only 1, so computing f costs only 1.

Now use quantum search to find a root of f, i.e., a subset $I_2 \subseteq \{n/3 + 1, \ldots, n\}$ such that $s - \Sigma(I_2)$ is a first coordinate in L_1. There are $2n/3$ bits of input to f, so the quantum search costs $2^{n/3}$.

Finally, find an I_1 such that $s - \Sigma(I_2) = \Sigma(I_1)$, and print $I_1 \cup I_2$. Like the previous algorithm, this algorithm finds a root of $\Sigma - s$ if one exists; any root I of $\Sigma - s$ can be expressed as $I_1 \cup I_2$.

Note that, with the original split of $\{1, \ldots, n\}$ into left and right halves, quantum search would not have reduced cost (modulo polynomial factors). Generalizing the original algorithm to allow an unbalanced split, and in particular a split into $n/3$ and $2n/3$, is pointless without quantum computers but essential for the quantum optimization. The split into $n/3$ and $2n/3$ imitates the approach used by Brassard, Høyer, and Tapp in [8] to search for hash-function collisions.

This algorithm uses $2^{n/3}$ memory, and as before one can object that random access to memory is expensive, especially when memory locations are quantum superpositions. See [5] for an extended discussion of the analogous objection to [8]. We again emphasize that this paper follows the tradition of simply counting operations; we do not claim that improved operation counts imply improvements in other cost models.

3   Walks

This section summarizes Ambainis's unique-collision-finding algorithm [2] (from the edge-walk perspective of [23]); introduces a new way to streamline Ambainis's algorithm; and applies the streamlined algorithm to the subset-sum context, obtaining cost $2^{n/3}$ in a different way from Section 2. This section's subset-sum algorithm uses collision finding as a black box, but the faster algorithms in Section 5 do not.

Review: Quantum Walks for Finding Unique Collisions. Consider any computable function f with b-bit inputs such that there is a unique pair of colliding inputs, i.e., exactly one pair (x, y) of b-bit strings such that $f(x) = f(y)$. The problem tackled in [2] is to find this pair (x, y).

The algorithm has a positive integer parameter $r < 2^b$, normally chosen on the scale of $2^{2b/3}$. At each step the algorithm is in a superposition of states of the form $(S, f(S), T, f(T))$. Here S and T are sets of b-bit strings such that $\#S = r$, $\#T = r$, and $\#(S \cap T) = r - 1$; i.e., S and T are adjacent vertices in the "Johnson graph" of r-subsets of the set of b-bit strings, where edges are between sets that differ in exactly one element. The notation $f(S)$ means $\{f(x) : x \in S\}$.

The algorithm begins in a uniform superposition of states; setting up this superposition uses $O(r)$ quantum evaluations of f. The algorithm then performs a "quantum walk" that alternates two types of steps: diffusing each state to a new choice of T while keeping S fixed, and diffusing each state to a new choice of S while keeping T fixed. Only one element of T changes when S is fixed (and vice versa), so each step uses only $O(1)$ quantum evaluations of f.

Periodically (e.g., after every $2\lceil\sqrt{r}\rceil$ steps) the algorithm negates the amplitude of every state in which S contains a colliding pair, i.e., in which $\#f(S) < r$. Because $f(S)$ has already been computed, checking whether $\#f(S) < r$ does not involve any evaluations of f. One can object that this check is nevertheless extremely expensive; this objection is discussed in the "data structures" subsection below.

Ambainis's analysis shows that after roughly $2^b/\sqrt{r}$ steps the algorithm has high probability of being in a state in which S contains a colliding pair. Observing this state and then sorting the pairs $(f(x), x)$ for $x \in S$ reveals the colliding pair. Overall the algorithm uses only $O(2^{2b/3})$ evaluations of f.

As in the case of Grover's algorithm, this algorithm is easily generalized to the case that there are p pairs of colliding inputs, and to the case that p is not known in advance. The algorithm is also easily generalized to functions of S more complicated than "contains a colliding pair".

Data Structures. The most obvious way to represent a set of b-bit strings is as a sorted array. The large overlap between S and T suggests storing the union $S \cup T$, together with a pointer to the element not in S and a pointer to the element not in T; similar comments apply to the multisets $f(S)$ and $f(T)$. Keeping a running tally of $\#f(S)$ allows easily checking whether $\#f(S) < r$.

To decide whether a b-bit string x is suitable as a new element of T, one must check whether $x \in S$. Actually, what the diffusion steps need is not merely knowing whether $x \in S$, but also knowing the number of elements of S smaller than x. ("Smaller" need not be defined lexicographically; the real objective is to compute a bijective map from b-bit strings to $\{1, 2, 3, \ldots, 2^b\}$ that maps S to $\{1, 2, 3, \ldots, r\}$.) The obvious sorted-array data structure allows these questions to be efficiently answered by binary search.

The big problem with this data structure is that inserting the new string into T requires, in the worst case, moving the other r elements of the array. This cost-r operation is performed at every step of the quantum walk, and dominates the total cost of the algorithm (unless evaluating f is very slow).

There is an extensive literature on classical data structures that support these operations much more efficiently. However, adapting a data structure to the quantum context raises three questions:

- Is the data-structure performance a random variable? Many data structures in the literature are randomized and provide good *average-case* performance but not good *worst-case* performance. The standard conversion of an algorithm to a quantum circuit requires first expressing the algorithm as a classical combinatorial circuit; the size of this circuit reflects the *worst-case* performance of the original algorithm.
- Does the performance of the data structure depend on S? For example, a standard hash table provides good performance for *most* sets S but not for *all* sets S.
- Is the data structure history-dependent? For most data structures, the representation of a set S depends on the order in which elements were added to and removed from the set. This spoils the analysis of the quantum walk through sets S, and presumably spoils the actual performance of the walk.

The first problem is usually minor: one can simply stop each algorithm after a constant time, where the constant is chosen so that the chance of an incorrect answer is negligible. The second problem can usually be converted into the first problem by some extra randomization: for example, one can choose a random hash function from a suitable family (as suggested by Wegman and Carter in [33]), or encrypt the b-bit strings before storing them. But the third problem is much more serious: it rules out balanced trees, red-black trees, most types of hash tables, etc.

Ambainis handles these issues in [2, Section 6] with an ad-hoc "combination of a hash table and a skip list", requiring several pages of analysis. We point out a much simpler solution: storing S etc. in a *radix tree*. Presumably this also saves time, although the speedup is not visible at the level of detail of our analysis.

The simplest type of radix tree is a binary tree in which the left subtree stores $\{x : (0, x) \in S\}$ and the right subtree stores $\{x : (1, x) \in S\}$; subtrees storing empty sets are omitted. To check whether $x \in S$ one starts from the root of the tree and follows pointers according to the bits of x in turn; the worst-case number of operations is proportional to the number of bits in x. Counting the number of elements of S smaller than x is just as easy if each tree node is augmented by a count of the number of elements below that node. A tree storing $f(S)$, with each leaf node accompanied by its multiplicity, allows an efficient running tally of the number $\#f(S)$ of distinct elements of $f(S)$, and in particular quickly checking whether $\#f(S) < r$.

Randomizing the memory layout of the nodes for the radix tree for S (inductively, by placing each new node at a uniform random free position) provides history-independence for the classical data structure: each possible representation of S has equal probability to appear. Similarly, creating a uniform superposition over all possible memory layouts of the nodes produces a unique quantum data structure representing S.

Subset-Sum Solutions via Collisions. It is straightforward to recast the subset-sum problem as a collision-finding problem.

Consider the function f that maps $(1, I_1)$ to $\Sigma(I_1)$ for $I_1 \subseteq \{1, 2, \dots, n/2\}$, and maps $(2, I_2)$ to $s - \Sigma(I_2)$ for $I_2 \subseteq \{n/2 + 1, n/2 + 2, \dots, n\}$. Use the algorithm described above to find a collision in f. There are only $n/2 + 1$ bits of input to f, so the cost of this algorithm is only $2^{n/3}$.

In the "hard" cases of interest in this paper, there are not likely to be many collisions among inputs $(1, I_1)$, and there are not likely to be many collisions among inputs $(2, I_2)$, so the collision found has a good chance of having the form $\Sigma(I_1) = s - \Sigma(I_2)$, i.e., $\Sigma(I_1 \cup I_2) = s$. One can, alternatively, tweak the algorithm to ignore collisions among $(1, I_1)$ and collisions among $(2, I_2)$ even if such collisions exist.

4 Moduli

This section discusses the use of a "modulus" to partition the spaces being searched in Section 2. The traditional view is that this is merely a method to reduce memory consumption (which we do not measure in this paper); but moduli are also an essential building block for the faster algorithms of Section 5. This section reviews the traditional left-right split with a modulus, and then states a quantum algorithm with a modulus, as a warmup for the faster quantum algorithm of Section 5.

Schroeppel and Shamir in [29] introduced an algorithm with essentially the same reduction of memory consumption, but that algorithm does not use moduli and does not seem helpful in Section 5. Three decades later, Howgrave-Graham and Joux in [17, eprint version, Section 3.1] described the left-right split with a modulus as a "useful practical variant" of the Schroeppel–Shamir algorithm; Becker, Coron, and Joux stated in [3] that this was a "simpler but heuristic variant of Schroeppel–Shamir". A more general algorithm (with one minor restriction, namely a prime choice of modulus) had already been stated a few years earlier by Elsenhans and Jahnel; see [10, Section 4.2.1] and [11, page 2]. There are many earlier papers that used moduli to partition input spaces without stating the idea in enough generality to cover subset sums; we have not attempted to comprehensively trace the history of the idea.

Review: Left-Right Split with a Modulus. Choose a positive integer $M \approx 2^{n/4}$, and choose $t \in \{0, 1, 2, \ldots, M-1\}$. Compute

$$L_1 = \{(\Sigma(I_1), I_1) : I_1 \subseteq \{1, 2, \ldots, n/2\}, \quad \Sigma(I_1) \equiv t \pmod{M}\}.$$

The problem of finding all I_1 here is a size-$n/2$ subset-sum problem modulo M. This problem can, in turn, be solved as a small number of size-$n/2$ subset-sum problems without moduli, namely searching for subsets of $x_1 \bmod M, x_2 \bmod M, \ldots, x_{n/2} \bmod M$ having sum t or $t + M$ or $\ldots$ or $t + (n/2 - 1)M$. Note, however, that it is important for this size-$n/2$ subroutine to find *all* solutions rather than just *one* solution.

A reasonable choice of subroutine here is the original left-right-split algorithm (without a modulus). This subroutine costs $2^{n/4}$ for a problem of size $n/2$, and is trivially adapted to find all solutions. For this adaptation one must add the number of solutions to the cost, but in this context one expects only about $2^{n/2}/M \approx 2^{n/4}$ subsets I_1 to satisfy $\Sigma(I_1) \equiv t \pmod{M}$, for a total cost of $2^{n/4}$. One can also tweak this subroutine to work directly with sums modulo M, rather than separately handling $t, t + M$, etc.

Similarly compute

$$L_2 = \{(s - \Sigma(I_2), I_2) : I_2 \subseteq \{n/2 + 1, \ldots, n\}, \quad \Sigma(I_2) \equiv s - t \pmod{M}\}.$$

Store L_1 in a sorted table, and for each $(s - \Sigma(I_2), I_2) \in L_2$ check whether $s - \Sigma(I_2)$ appears in the table. If there is a collision $\Sigma(I_1) = s - \Sigma(I_2)$, print $I_1 \cup I_2$ as a root of $\Sigma - s$ and stop. Otherwise try another value of t, repeating until all choices of $t \in \{0, 1, 2, \ldots, M-1\}$ are exhausted.

One expects each choice of t to cost $2^{n/4}$, as discussed above. There are $M \approx 2^{n/4}$ choices of t, for a total cost of $2^{n/2}$. If there is a subset-sum solution then it will be found for some choice of t.

Quantum Search with a Modulus. The algorithm above is a classical search for a root of the function that maps t to 0 if there is a collision $\Sigma(I_1) = s - \Sigma(I_2)$ satisfying $\Sigma(I_1) \equiv t \pmod{M}$ (and therefore also satisfying $\Sigma(I_2) \equiv s - t \pmod{M}$). One way to take advantage of quantum computers here is to instead search for t by Grover's algorithm, which finds the root with only $\sqrt{M} \approx 2^{n/8}$ quantum evaluations of the same function, for a total cost of $2^{n/8}2^{n/4} = 2^{3n/8}$.

Quantum Walks with a Modulus. A different way to take advantage of quantum computers is as follows.

Recall that the collision-finding algorithm of Section 3 walks through adjacent pairs of size-r sets S, searching for sets that contain collisions under a function f. Each set S is stored in a radix tree, as is the multiset $f(S)$. Each radix tree is augmented to record at each node the number of distinct elements below that node, allowing fast evaluation of the number of elements of S smaller than a specified input and fast evaluation of whether S contains a collision.

One can design this collision-finding algorithm in four steps:

- Start with a simple classical collision-finding algorithm that computes $f(S)$ where S is the set of all 2^b b-bit strings.

- Generalize to a lower-probability algorithm that computes $f(S)$ where S is a set of only r strings and that checks whether S contains the collision.
- Build a data structure that expresses the entire computation of the lower-probability algorithm. Observe that this data structure allows efficiently moving from S to an adjacent set: a single element of S has only a small impact on the computation.
- Apply a quantum walk, walking through adjacent pairs of size-r sets S while maintaining this data structure for each S. This takes $O(\sqrt{r}/\sqrt{p})$ steps where p is the success probability of the previous algorithm, plus cost r to set up the data structure in the first place.

We now imitate the same four-step approach, starting from the classical left-right split with a modulus and ending with a new quantum subset-sum algorithm. The following description assumes that the correct value of t is already known; the overhead of searching for t is discussed after the algorithm.

First step: Classical algorithm. Recall that the subroutine to find all I_1 with $\Sigma(I_1) \equiv t$ computes $\Sigma(I_{11}) \bmod M$ for all $I_{11} \subseteq \{1, 2, \ldots, n/4\}$; computes $t - \Sigma(I_{12}) \bmod M$ for all $I_{12} \subseteq \{n/4+1, n/4+2, \ldots, n/2\}$; and finds collisions between $\Sigma(I_{11}) \bmod M$ and $t - \Sigma(I_{12}) \bmod M$. Similarly, the subroutine to find all I_2 finds collisions between $\Sigma(I_{21}) \bmod M$ for $I_{21} \subseteq \{n/2+1, n/2+2, \ldots, 3n/4\}$ and $s - t - \Sigma(I_{22}) \bmod M$ for $I_{22} \subseteq \{3n/4+1, 3n/4+2, \ldots, n\}$. The high-level algorithm finishes by finding collisions between $\Sigma(I_1)$ and $s - \Sigma(I_2)$.

Second step: Generalize to a lower-probability computation by restricting the sets that contain collisions. Specifically, instead of enumerating all subsets I_{11}, take a random collection S_{11} containing exactly r such subsets; here $r \leq 2^{n/4}$ is an algorithm parameter optimized below. Similarly take a random collection S_{12} of exactly r subsets I_{12}. Find collisions between $\Sigma(I_{11}) \bmod M$ and $t - \Sigma(I_{12}) \bmod M$; one expects about r^2/M collisions, producing r^2/M sets $I_1 = I_{11} \cup I_{12}$ satisfying $\Sigma(I_1) \equiv t \pmod{M}$. Similarly take random size-r sets S_{21} and S_{22} consisting of, respectively, subsets I_{21} and I_{22}; find collisions between $\Sigma(I_{21}) \bmod M$ and $s - t - \Sigma(I_{22}) \bmod M$, obtaining about r^2/M sets I_2 satisfying $\Sigma(I_2) \equiv s - t \pmod{M}$. Finally check for collisions between $\Sigma(I_1)$ and $s - \Sigma(I_2)$. One can visualize the construction of $I = I_1 \cup I_2$ as a three-level binary tree with I as the root, I_1 and I_2 as its left and right children, and $I_{11}, I_{12}, I_{21}, I_{22}$ as the leaves.

Recall that we are assuming that the correct value of t is known, i.e., that the desired subset-sum solution is expressible as $I_1 \cup I_2$ with $\Sigma(I_1) \equiv t \pmod{M}$ and $\Sigma(I_2) \equiv s - t \pmod{M}$. Then S_{11} has probability $r/2^{n/4}$ of containing the set $I_{11} = I_1 \cap \{1, 2, \ldots, n/4\}$. Similar comments apply to S_{12}, S_{21}, and S_{22}, for an overall success probability of $(r/2^{n/4})^4$.

The optimal choice of r is discussed later, and is far below the classical extreme $2^{n/4}$. This drop is analogous to the drop in list sizes from a simple left-right split (list size $2^{n/2}$) to the quantum-walk subset-sum algorithm of Section 3 (list size $2^{n/3}$). This drop has an increasingly large impact on subsequent levels of the tree: the number of sets $I_{11}, I_{12}, I_{21}, I_{22}$ is reduced by a factor of $2^{n/4}/r$, and the number of sets I_1, I_2 is reduced by a factor of $(2^{n/4}/r)^2$.

An interesting consequence of this drop is that one can reduce M without creating bottlenecks at subsequent levels of the tree. Specifically, taking $M \approx r$ means that one expects about r sets I_1 and about r sets I_2.

Third step: Data structure. This lower-probability computation is captured by a data structure that contains the following sets in augmented radix trees:

- The size-r set S_{11} of subsets $I_{11} \subseteq \{1, 2, \ldots, n/4\}$.
- The set $\{(\Sigma(I_{11}) \bmod M, I_{11}) : I_{11} \in S_{11}\}$.
- The size-r set S_{12} of subsets $I_{12} \subseteq \{n/4 + 1, n/4 + 2, \ldots, n/2\}$.
- The set $\{(t - \Sigma(I_{12}) \bmod M, I_{12}) : I_{12} \in S_{12}\}$.
- The set S_1 of $I_{11} \cup I_{12}$ for all pairs $(I_{11}, I_{12}) \in S_{11} \times S_{12}$ such that $\Sigma(I_{11}) \equiv t - \Sigma(I_{12}) \pmod{M}$, subject to the limit discussed below.
- The size-r set S_{21} of subsets $I_{21} \subseteq \{n/2 + 1, n/2 + 2, \ldots, 3n/4\}$.
- The set $\{(\Sigma(I_{21}) \bmod M, I_{21}) : I_{21} \in S_{21}\}$.
- The size-r set S_{22} of subsets $I_{22} \subseteq \{3n/4 + 1, 3n/4 + 2, \ldots, n\}$.
- The set $\{(s - t - \Sigma(I_{22}) \bmod M, I_{22}) : I_{22} \in S_{22}\}$.
- The set S_2 of $I_{21} \cup I_{22}$ for all pairs $(I_{21}, I_{22}) \in S_{21} \times S_{22}$ such that $\Sigma(I_{21}) \equiv s - t - \Sigma(I_{22}) \pmod{M}$.
- The set $\{(\Sigma(I_1), I_1) : I_1 \in S_1\}$.
- The set $\{(s - \Sigma(I_2), I_2) : I_2 \in S_2\}$.
- The set S of $I_1 \cup I_2$ for all pairs $(I_1, I_2) \in S_1 \times S_2$ such that $\Sigma(I_1) = s - \Sigma(I_2)$, subject to the limit discussed below.

Note that this data structure supports, e.g., fast removal of an element I_{11} from S_{11} followed by fast insertion of a replacement element I'_{11}. Checking for $\Sigma(I_{11}) \bmod M$ in the stored set $\{(t - \Sigma(I_{12}) \bmod M, I_{12})\}$ efficiently shows which elements have to be removed from S_1, and then a similar check shows which elements have to be removed from S.

Each element I_{11} of S_{11} affects very few elements of S_1; on average one expects "very few" to be $r/M \approx 1$. To control the time taken by each step of the algorithm we put a polynomial limit on the number of elements of S_1 involving any particular I_{11}. If this limit is reached then (to ensure history-independence) we use a random selection of elements, but this limit has negligible chance of affecting the algorithm output. Similar comments apply to I_{12}, I_{21}, and I_{22}.

Fourth step: Walk through adjacent pairs of 4-tuples $(S_{11}, S_{12}, S_{21}, S_{22})$ of size-r sets, maintaining the data structure above and searching for tuples for which the final set S is nonempty. Amplifying the $(r/2^{n/4})^4$ success probability mentioned above to a high probability requires a quantum walk consisting of $O(\sqrt{r}(2^{n/4}/r)^2)$ steps. Setting up the data structure in the first place costs $O(r)$.

For r on the scale of $2^{0.2n}$ these costs are balanced at $2^{0.2n}$; but recall that this assumes that t is already known. A classical search for t means repeating this algorithm $M \approx r \approx 2^{0.2n}$ times, for a total cost of $2^{0.4n}$. We do better by using amplitude amplification [7], repeating the quantum walk only $2^{0.1n}$ times, for a total cost of $2^{0.3n}$. We do not describe amplitude amplification in detail; this subset-sum algorithm is superseded by the approach of Section 5.

5   Representations

"Representations" are a technique to improve the "left-right split with a modulus" algorithm of Section 4. Howgrave-Graham and Joux introduced this technique in [17] and obtained a subset-sum algorithm that costs just $2^{(0.337...+o(1))n}$. Beware that [17] incorrectly claimed a cost of $2^{(0.311...+o(1))n}$; the underlying flaw in the analysis was corrected in [3] with credit to May and Meurer.

This section reviews the Howgrave-Graham–Joux algorithm, and then presents a new quantum subset-sum algorithm with cost only $2^{(0.241...+o(1))n}$. The new quantum algorithm requires the quantum-walk machinery discussed in Section 3.

We simplify the algorithm statements in this section by considering only half-weight sets I; i.e., we search only for sets I with $\#I = n/2$ and $\Sigma(I) = s$. We comment, however, that straightforward generalizations of these algorithms, still within the same cost bound, handle smaller known weights (adjusting the set sizes shown below, at the expense of some complications in notation); also handle larger known weights (replacing I and s with their complements); and handle arbitrary unknown weights (trying all $n+1$ possible weights).

Review: The Basic Idea of Representations. Recall that the original left-right split partitions I as $I_1 \cup I_2$ where $I_1 \subseteq \{1,\ldots,n/2\}$ and $I_2 \subseteq \{n/2+1,\ldots,n\}$. The main idea of representations is to partition I in a different, ambiguous way as $I_1 \cup I_2$ with $I_1, I_2 \subseteq \{1,2,\ldots,n\}$ and $\#I_1 = \#I_2 = n/4$. Note that there are $\binom{n/2}{n/4} \approx 2^{n/2}$ such partitions. The key observation is that finding only one out of these exponentially many *representations* (I_1, I_2) of I is sufficient to solve the subset-sum problem.

Recall also the idea of moduli: pick $t \in \{0,1,2,\ldots,M-1\}$ and hope that $\Sigma(I_1) \equiv t \pmod{M}$. In Section 4, there was only one choice of I_1, so one expects each choice of t to work with probability only about $1/M$, forcing a search through choices of t. In this section, there are $\approx 2^{n/2}$ choices of I_1, so one expects a single choice of t to work with high probability for M as large as $2^{n/2}$.

These observations motivate the following strategy. Pick a modulus $M \approx 2^{n/2}$ and choose a random target value $t \in \{0,1,\ldots,M-1\}$. Compute

$$L_1 = \{(\Sigma(I_1), I_1) : I_1 \subseteq \{1,\ldots,n\},\ \#I_1 = n/4,\ \Sigma(I_1) \equiv t \pmod{M}\}$$

and

$$L_2 = \{(s - \Sigma(I_2), I_2) : I_2 \subseteq \{1,\ldots,n\},\ \#I_2 = n/4,\ \Sigma(I_2) \equiv s - t \pmod{M}\}.$$

If there is a collision $\Sigma(I_1) = s - \Sigma(I_2)$ satisfying $I_1 \cap I_2 = \{\}$, print $I_1 \cup I_2$ and stop. If there are no such collisions, repeat with another choice of t. One expects a negligible failure probability after a polynomial number of repetitions.

(We point out that if $t \equiv s - t \pmod{M}$ then computing L_1 immediately produces L_2. One can arrange for this by choosing a random odd M and taking t to be half of s modulo M; one can also choose a random even M if s is even. If other speed constraints prevent M from being chosen randomly then one can still try these special values of t first. Similar comments apply to the next level of the Howgrave-Graham–Joux algorithm described below. Of course, the resulting speedup is not visible at the level of detail of our analysis.)

One expects $\#L_1 \approx \binom{n}{n/4}/2^{n/2} \approx 2^{0.311\ldots n}$: there are $\binom{n}{n/4}$ sets $I_1 \subseteq \{1,\ldots,n\}$ with $\#I_1 = n/4$, and one expects each I_1 to satisfy $\Sigma(I_1) \equiv t \pmod{M}$ with probability $1/M \approx 1/2^{n/2}$. (The calculation of $0.311\ldots$ relies on the standard approximation $\binom{n}{\alpha n} \approx 2^{H(\alpha)n}$, where $H(\alpha) = -\alpha \log_2 \alpha - (1-\alpha)\log_2(1-\alpha)$.) The same comment applies to L_2. One also expects the number of collisions between L_1 and L_2 to be exponentially large, about $\#L_1\#L_2/2^{n/2} \approx 2^{0.122\ldots n}$, since each $\Sigma(I_1)$ is already known to match each $s - \Sigma(I_2)$ modulo M; but the only collisions satisfying $I_1 \cap I_2 = \{\}$ are collisions arising from subset-sum solutions.

The remaining task is to compute L_1 and L_2 in the first place. Howgrave-Graham and Joux solve these two weight-$n/4$ modular subset-sum problems by first applying another level of representations (using a smaller modulus that divides M), obtaining four weight-$n/8$ modular subset-sum problems; they solve each of those problems with a weight-$n/16$ left-right split. The details appear below.

Review: The Complete Howgrave-Graham–Joux Algorithm. Choose a positive integer $M_1 \approx 2^{n/4}$. Choose a positive integer $M \approx 2^{n/2}$ divisible by M_1. Choose randomly $s_1 \in \{0,1,\ldots,M-1\}$ and define $s_2 = s - s_1$. Choose randomly $s_{11} \in \{0,1,\ldots,M_1-1\}$ and define $s_{12} = s_1 - s_{11}$. Choose randomly $s_{21} \in \{0,1,\ldots,M_1-1\}$ and define $s_{22} = s_2 - s_{21}$. Also choose random subsets $R_{111}, R_{121}, R_{211}, R_{221}$ of $\{1,2,\ldots,n\}$, each of size $n/2$, and define R_{ij2} as the complement of R_{ij1}.

The following algorithm searches for a weight-$n/2$ subset-sum solution I decomposed as follows: $I = I_1 \cup I_2$ with $\#I_i = n/4$ and $\Sigma(I_i) \equiv s_i \pmod{M}$; furthermore $I_i = I_{i1} \cup I_{i2}$ with $\#I_{ij} = n/8$ and $\Sigma(I_{ij}) \equiv s_{ij} \pmod{M_1}$; furthermore $I_{ij} = I_{ij1} \cup I_{ij2}$ with $\#I_{ijk} = n/16$ and $I_{ijk} \subseteq R_{ijk}$. These constraints are shown as a tree in Figure 5.1. One expects a weight-$n/2$ subset-sum solution to decompose in this way with high probability (inverse polynomial in n), as discussed later, and if it does decompose in this way then it is in fact found by this algorithm.

Start with, for each $(i,j,k) \in \{1,2\} \times \{1,2\} \times \{1,2\}$, the set S_{ijk} of all subsets $I_{ijk} \subseteq R_{ijk}$ with $\#I_{ijk} = n/16$. Compute the sets

$$L_{ij1} = \{(\Sigma(I_{ij1}) \bmod M_1, I_{ij1}) : I_{ij1} \in S_{ij1}\},$$
$$L_{ij2} = \{(s_{ij} - \Sigma(I_{ij2}) \bmod M_1, I_{ij2}) : I_{ij2} \in S_{ij2}\}.$$

Merge L_{ij1} and L_{ij2} to obtain the set S_{ij} of $I_{ij1} \cup I_{ij2}$ for all pairs $(I_{ij1}, I_{ij2}) \in S_{ij1} \times S_{ij2}$ such that $\Sigma(I_{ij1}) \equiv s_{ij} - \Sigma(I_{ij2}) \pmod{M_1}$. Note that each $I_{ij} \in S_{ij}$ has $\Sigma(I_{ij}) \equiv s_{ij} \pmod{M_1}$ and $\#I_{ij} = n/8$. Next compute the sets

$$L_{i1} = \{(\Sigma(I_{i1}) \bmod M, I_{i1}) : I_{i1} \in S_{i1}\},$$
$$L_{i2} = \{(s_i - \Sigma(I_{i2}) \bmod M, I_{i2}) : I_{i2} \in S_{i2}\}.$$

Merge L_{i1} and L_{i2} to obtain the set S_i of $I_{i1} \cup I_{i2}$ for all pairs $(I_{i1}, I_{i2}) \in S_{i1} \times S_{i2}$ such that $\Sigma(I_{i1}) \equiv s_i - \Sigma(I_{i2}) \pmod{M}$ and $I_{i1} \cap I_{i2} = \{\}$. Note that each $I_i \in S_i$

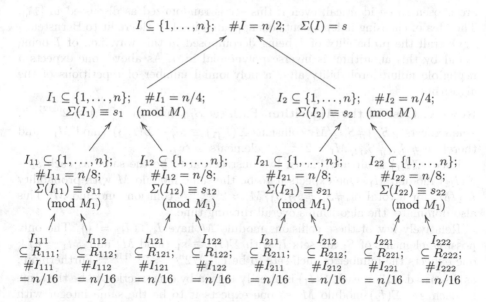

Fig. 5.1. Decomposition of a weight-$n/2$ subset-sum solution $I \subseteq \{1, 2, \ldots, n\}$

has $\Sigma(I_i) \equiv s_i \pmod{M}$ and $\#I_i = n/4$. Next compute the sets

$$L_1 = \{(\Sigma(I_1), I_1) : I_1 \in S_1\},$$
$$L_2 = \{(s - \Sigma(I_2), I_2) : I_2 \in S_2\}.$$

Merge L_1 and L_2 to obtain the set S of $I_1 \cup I_2$ for all pairs $(I_1, I_2) \in S_1 \times S_2$ such that $\Sigma(I_1) = s - \Sigma(I_2)$ and $I_1 \cap I_2 = \{\}$. Note that each $I \in S$ has $\Sigma(I) = s$ and $\#I = n/2$. If S is nonempty, print its elements and stop.

Review: Success Probability of the Algorithm. Consider any weight-$n/2$ subset $I \subseteq \{1, \ldots, n\}$ with $\Sigma(I) = s$. There are $\binom{n/2}{n/4} \approx 2^{n/2} \approx M$ ways to partition I as $I_1 \cup I_2$ with $\#I_1 = n/4$ and $\#I_2 = n/4$, and as discussed earlier one expects that with high probability at least one of these ways will satisfy $\Sigma(I_1) \equiv s_1 \pmod{M}$, implying $\Sigma(I_2) \equiv s_2 \pmod{M}$.

Similarly, there are $\binom{n/4}{n/8} \approx 2^{n/4} \approx M_1$ ways to partition I_1 as $I_{11} \cup I_{12}$ with $\#I_{11} = n/8$ and $\#I_{12} = n/8$. One expects that with high probability at least one of these ways will satisfy $\Sigma(I_{11}) \equiv s_{11} \pmod{M_1}$, implying $\Sigma(I_{12}) \equiv s_{12} \pmod{M_1}$ (since M_1 divides M and $s_{11} + s_{12} = s_1$). Analogous comments apply to I_2, I_{21}, I_{22}.

A uniform random subset of a set of size $n/8$ has size exactly $n/16$ with probability $\Theta(1/\sqrt{n})$, so with probability $\Theta(1/n^2)$ each of the four sets $I_{ij1} = I_{ij} \cap R_{ij1}$ has size exactly $n/16$, implying that each of the four complementary sets $I_{ij2} = I_{ij} \cap R_{ij2}$ also has size exactly $n/16$. (Experiments indicate that the probability is somewhat worse, although still inverse polynomial in n, if all R_{ij1}

are chosen to be identical, even if this set is randomized as discussed in [17]. The idea of choosing independent sets appeared in [4] with credit to Bernstein.)

Overall the probability of I being decomposed in this way, i.e., of I being found by this algorithm, is inverse polynomial in n. As above, one expects a negligible failure probability after a polynomial number of repetitions of the algorithm.

Review: Cost of the Algorithm. Each set S_{ijk} has size $\binom{n/2}{n/16} \approx 2^{0.271...n}$. One expects $\#S_{ij1}\#S_{ij2}/M_1$ collisions $\Sigma(I_{ij1}) \equiv s_{ij} - \Sigma(I_{ij2}) \pmod{M_1}$, and therefore $\#S_{ij1}\#S_{ij2}/M_1 \approx 2^{0.293...n}$ elements in S_{ij}.

Each $\Sigma(I_{i1})$ is already known by construction to be the same as each $s_i - \Sigma(I_{i2})$ modulo M_1. One expects it to be the same modulo M with probability M_1/M, for a total of $\#S_{i1}\#S_{i2}M_1/M \approx 2^{0.337...n}$ collisions modulo M. This also dominates the algorithm's overall running time.

Relatively few of these collisions modulo M have $I_{i1} \cap I_{i2} = \{\}$. The only possible elements of S_1 are sets I_1 with $\Sigma(I_1) \equiv s_1 \pmod{M}$ and $\#I_1 = n/4$; one expects the number of such sets to be $\binom{n}{n/4}/2^{n/2} \approx 2^{0.311...n}$. Furthermore, as discussed earlier, each $\Sigma(I_1)$ is already known by construction to be the same as each $s - \Sigma(I_2)$ modulo M, so one expects it to be the same integer with probability about $M/2^n$, for a total of $\#S_1\#S_2M/2^n \approx 2^{0.122...n}$ collisions.

New: Quantum Walks with Moduli and Representations. We now combine quantum walks with the idea of representations. The reader is assumed to be familiar with the simpler quantum algorithm of Section 4.

We introduce a parameter $r \leq \binom{n/2}{n/16}$ into the Howgrave-Graham–Joux algorithm, and take each S_{ijk} as a random collection of exactly r weight-$n/16$ subsets I_{ijk} of R_{ijk}. The extreme case $r = \binom{n/2}{n/16}$ is the same as the original Howgrave-Graham–Joux algorithm: in this case S_{ijk} is the set of all weight-$n/16$ subsets of R_{ijk}. For smaller r this generalized algorithm has lower success probability, as discussed below, but is also faster. The resulting computation is captured by the following 29 sets, which we store in augmented radix trees:

- For each i, j, k, a set S_{ijk} consisting of exactly r weight-$n/16$ subsets of R_{ijk}.
- For each i, j, the set $L_{ij1} = \{(\Sigma(I_{ij1}) \bmod M_1, I_{ij1}) : I_{ij1} \in S_{ij1}\}$.
- For each i, j, the set $L_{ij2} = \{(s_{ij} - \Sigma(I_{ij2}) \bmod M_1, I_{ij2}) : I_{ij2} \in S_{ij2}\}$.
- For each i, j, the set S_{ij} of $I_{ij1} \cup I_{ij2}$ for all pairs $(I_{ij1}, I_{ij2}) \in S_{ij1} \times S_{ij2}$ such that $\Sigma(I_{ij1}) \equiv s_{ij} - \Sigma(I_{ij2}) \pmod{M_1}$, subject to the limit discussed below.
- For each i, the set $L_{i1} = \{(\Sigma(I_{i1}) \bmod M, I_{i1}) : I_{i1} \in S_{i1}\}$.
- For each i, the set $L_{i2} = \{(s_i - \Sigma(I_{i2}) \bmod M, I_{i2}) : I_{i2} \in S_{i2}\}$.
- For each i, the set S_i of $I_{i1} \cup I_{i2}$ for all pairs $(I_{i1}, I_{i2}) \in S_{i1} \times S_{i2}$ such that $\Sigma(I_{i1}) \equiv s_i - \Sigma(I_{i2}) \pmod{M}$ and $I_{i1} \cap I_{i2} = \{\}$, subject to the limit discussed below.
- The set $L_1 = \{(\Sigma(I_1), I_1) : I_1 \in S_1\}$.
- The set $L_2 = \{(s - \Sigma(I_2), I_2) : I_2 \in S_2\}$.
- The set S of $I_1 \cup I_2$ for all pairs $(I_1, I_2) \in S_1 \times S_2$ such that $\Sigma(I_1) = s - \Sigma(I_2)$ and $I_1 \cap I_2 = \{\}$, subject to the limit discussed below.

Like the data structure in the quantum walk of Section 4, this data structure supports, e.g., fast removal of an element I_{111} from S_{111} followed by fast insertion of a replacement element I'_{111}.

The optimal choice of r is discussed later; it is far below the starting list size $\binom{n/2}{n/16} \approx 2^{0.272n}$ used by Howgrave-Graham and Joux, and is even below $2^{n/4}$. One expects the number of collisions modulo M_1 to be $r^2/2^{n/4}$, which is smaller than r, and the list sizes on subsequent levels to be even smaller. Consequently this quantum algorithm ends up being bottlenecked at the bottom level of the tree, while the original algorithm is bottlenecked at a higher level.

Furthermore, one expects each element I_{ijk} of S_{ijk} to affect, on average, approximately $r/2^{n/4}$ elements of the set of collisions modulo M, and therefore to affect 0 elements of S_{ij} in almost all cases, 1 element of S_{ij} with exponentially small probability, 2 elements of S_{ij} with far smaller probability, etc. As in Section 4, to control the time taken by each step of the algorithm we put a polynomial limit on the number of elements of S_{ij} involving any particular I_{ijk}, a polynomial limit on the number of elements of S_i involving any particular I_{ij}, and a polynomial limit on the number of elements of S involving any particular I_i. A constant limit of 100 seems ample, and there is no obvious problem with a limit of 1.

Compared to the original Howgrave-Graham–Joux algorithm, the success probability of the generalized algorithm drops by a factor of $(r/\binom{n/2}{n/16}))^8$, since each target I_{ijk} has chance only $r/\binom{n/2}{n/16}$ of being in S_{ijk}. Amplifying this to a high probability requires a quantum walk consisting of $O(\sqrt{r}(\binom{n/2}{n/16}/r)^4)$ steps. Setting up the data structure in the first place costs $O(r)$. For r on the scale of $\binom{n/2}{n/16}^{4/4.5}$ these costs are balanced at $\binom{n/2}{n/16}^{4/4.5}$, i.e., at $2^{(0.241...+o(1))n}$.

References

[1] — (no editor): 20th annual symposium on foundations of computer science. IEEE Computer Society, New York (1979). MR 82a:68004. See [32]

[2] Ambainis, A.: Quantum walk algorithm for element distinctness. SIAM Journal on Computing 37, 210–239 (2007). http://arxiv.org/abs/quant-ph/0311001. Citations in this document: §3, §3, §3

[3] Becker, A., Coron, J.-S., Joux, A.: Improved generic algorithms for hard knapsacks. In: Eurocrypt 2011 [27] (2011). http://eprint.iacr.org/2011/474. Citations in this document: §1, §1, §1, §4, §5

[4] Becker, A., Joux, A., May, A., Meurer, A.: Decoding random binary linear codes in $2^{n/20}$: how $1 + 1 = 0$ improves information set decoding. In: Eurocrypt 2012 [28] (2012). http://eprint.iacr.org/2012/026. Citations in this document: §1, §1, §1, §1, §5

[5] Bernstein, D.J.: Cost analysis of hash collisions: Will quantum computers make SHARCS obsolete? In: Workshop Record of SHARCS'09: Special-purpose Hardware for Attacking Cryptographic Systems (2009). http://cr.yp.to/papers.html#collisioncost. Citations in this document: §2

[6] Boyer, M., Brassard, G., Høyer, P., Tapp, A.: Tight bounds on quantum searching. Fortschritte Der Physik 46, 493–505 (1998). http://arxiv.org/abs/quant-ph/9605034v1. Citations in this document: §2, §2

[7] Brassard, G., Høyer, P., Mosca, M., Tapp, A.: Quantum amplitude amplification and estimation. In: [20], pp. 53–74 (2002). http://arxiv.org/abs/quant-ph/0005055. Citations in this document: §4

[8] Brassard, G., Høyer, G., Tapp, A.: Quantum cryptanalysis of hash and claw-free functions. In: LATIN'98 [21], pp. 163–169 (1998). MR 99g:94013. Citations in this document: §2, §2

[9] Chang, W.-L., Ren, T.-T., Feng, M., Lu, L.C., Lin, K.W., Guo, M.: Quantum algorithms of the subset-sum problem on a quantum computer. International Conference on Information Engineering 2, 54–57 (2009). Citations in this document: §2

[10] Elsenhans, A.-S., Jahnel, J.: The diophantine equation $x^4 + 2y^4 = z^4 + 4w^4$. Mathematics of Computation 75, 935–940 (2006). http://www.uni-math.gwdg.de/jahnel/linkstopaperse.html. Citations in this document: §4

[11] Elsenhans, A.-S., Jahnel, J.: The Diophantine equation $x^4 + 2y^4 = z^4 + 4w^4$— a number of improvements (2006). http://www.uni-math.gwdg.de/jahnel/linkstopaperse.html. Citations in this document: §4

[12] Gilbert, H. (ed.): Advances in cryptology — EUROCRYPT 2010, 29th annual international conference on the theory and applications of cryptographic techniques, French Riviera, May 30–June 3, 2010, proceedings. LNCS, vol. 6110. Springer (2010). See [17]

[13] Goldwasser, S. (ed.): 35th annual IEEE symposium on the foundations of computer science. Proceedings of the IEEE symposium held in Santa Fe, NM, November 20–22, 1994. IEEE (1994). ISBN 0-8186-6580-7. MR 98h:68008. See [30]

[14] Grover, L.K.: A fast quantum mechanical algorithm for database search. In: [26], pp. 212–219 (1996). MR 1427516. http://arxiv.org/abs/quant-ph/9605043

[15] Grover, L.K.: Quantum mechanics helps in searching for a needle in a haystack. Physical Review Letters 79, 325–328 (1997). http://arxiv.org/abs/quant-ph/9706033. Citations in this document: §2, §2

[16] Horowitz, E., Sahni, S.: Computing partitions with applications to the knapsack problem. Journal of the ACM 21, 277–292 (1974). Citations in this document: §2

[17] Howgrave-Graham, N., Joux, A.: New generic algorithms for hard knapsacks. In: Eurocrypt 2010 [12] (2010). http://eprint.iacr.org/2010/189. Citations in this document: §1, §1, §1, §1, §4, §5, §5, §5

[18] Johnson, D.S., Feige, U. (eds.): Proceedings of the 39th annual ACM symposium on the theory of computing, San Diego, California, USA, June 11–13, 2007. Association for Computing Machinery (2007). ISBN 978-1-59593-631-8. See [23]

[19] Lee, D.H., Wang, X. (eds.): Advances in cryptology — ASIACRYPT 2011, 17th international conference on the theory and application of cryptology and information security, Seoul, South Korea, December 4–8, 2011, proceedings. LNCS, vol. 7073. Springer (2011). ISBN 978-3-642-25384-3. See [24]

[20] Lomonaco Jr., S.J., Brandt, H.E. (eds.): Quantum computation and information. Papers from the AMS Special Session held in Washington, DC, January 19–21, 2000. Contemporary Mathematics, vol. 305. American Mathematical Society (2002). ISBN 0-8218-2140-7. MR 2003g:81006. See [7]

[21] Lucchesi, C.L., Moura, A.V. (eds.): LATIN'98: theoretical informatics. Proceedings of the 3rd Latin American symposium held in Campinas, April 20–24, 1998. LNCS, vol. 1380. Springer (1998). ISBN 3-540-64275-7. MR 99d:68007. See [8]

[22] Lyubashevsky, V., Palacio, A., Segev, G.: Public-key cryptographic primitives provably as secure as subset sum. In: TCC 2010 [25], pp. 382–400 (2010). http://eprint.iacr.org/2009/576. Citations in this document: §1, §1

[23] Magniez, F., Nayak, A., Roland, J., Santha, M.: Search via quantum walk. In: STOC 2007 [18], pp. 575–584 (2007). http://arxiv.org/abs/quant-ph/0608026. Citations in this document: §3

[24] May, A., Meurer, A., Thomae, E.: Decoding random linear codes in $\tilde{O}(2^{0.054n})$. In: Asiacrypt 2011 [19] (2011). http://www.cits.rub.de/imperia/md/content/may/paper/decoding.pdf. Citations in this document: §1, §1, §1

[25] Micciancio, D. (ed.): Theory of cryptography, 7th theory of cryptography conference, TCC 2010, Zurich, Switzerland, February 9–11, 2010, proceedings. LNCS, vol. 5978. Springer (2010). ISBN 978-3-642-11798-5. See [22]

[26] Miller, G.L. (ed.): Proceedings of the twenty-eighth annual ACM symposium on the theory of computing, Philadelphia, PA, May 22–24, 1996. Association for Computing Machinery (1996). ISBN 0-89791-785-5. MR 97g:68005. See [14]

[27] Paterson, K.G. (ed.): Advances in cryptology — EUROCRYPT 2011, 30th annual international conference on the theory and applications of cryptographic techniques, Tallinn, Estonia, May 15–19, 2011, proceedings. LNCS, vol. 6632. Springer (2011). ISBN 978-3-642-20464-7. See [3]

[28] Pointcheval, D., Johansson, T. (eds.): Advances in cryptology — EUROCRYPT 2012 — 31st annual international conference on the theory and applications of cryptographic techniques, Cambridge, UK, April 15–19, 2012, proceedings. LNCS, vol. 7237. Springer (2012). ISBN 978-3-642-29010-7. See [4]

[29] Schroeppel, R., Shamir, A.: A $T = O(2^{n/2})$, $S = O(2^{n/4})$ algorithm for certain NP-complete problems. SIAM Journal on Computing 10, 456–464 (1981). Citations in this document: §4

[30] Shor, P.W.: Algorithms for quantum computation: discrete logarithms and factoring. In: [13], pp. 124–134 (1994); see also newer version [31]. MR 1489242

[31] Shor, P.W.: Polynomial-time algorithms for prime factorization and discrete logarithms on a quantum computer. SIAM Journal on Computing 26, 1484–1509 (1997); see also older version [30]. MR 98i:11108. http://arxiv.org/abs/quant-ph/9508027

[32] Wegman, M.N., Lawrence Carter, J.: New classes and applications of hash functions. In: [1], pp. 175–182 (1979); see also newer version [33]

[33] Wegman, M.N., Lawrence Carter, J.: New hash functions and their use in authentication and set equality. Journal of Computer and System Sciences 22, 265–279 (1981); see also older version [32]. ISSN 0022-0000. MR 82i:68017. Citations in this document: §3

Improved Lattice-Based Threshold Ring Signature Scheme

Slim Bettaieb and Julien Schrek

XLIM-DMI, Université de Limoges,
123, av. Albert Thomas
87060 Limoges Cedex, France
{slim.bettaieb,julien.schrek}@xlim.fr

Abstract. We present in this paper an improvement of the lattice-based threshold ring signature proposed by Cayrel, Lindner, Rückert and Silva (CLRS) [LATINCRYPT '10]. We generalize the same identification scheme CLRS to obtain a more efficient threshold ring signature. The security of our scheme relies on standard lattice problems. The improvement is a significant reduction of the size of the signature. Our result is a t-out-of-N threshold ring signature which can be seen as t different ring signatures instead of N for the other schemes. We describe the ring signature induced by the particular case of only one signer. To the best of our knowledge, the resulted signatures are the most efficient lattice-based ring signature and threshold signature.

Keywords: Threshold ring signatures, lattices.

1 Introduction

Lattices were first used in cryptography with the LLL algorithm [1] in order to cryptanalyse some number theory primitives [2]. In 1996, the NTRU cryptosystem proposed an original idea to base cryptography on lattices assumptions.

In 1996, Ajtai [3] showed how to use the standard lattice problem GapSVP in order to build cryptographic schemes. More specifically he proved that the worst-case hardness of GapSVP is reduced to the average-case hardness of SIS problem. Lattice-based cryptography became a good alternative to number theory cryptography in regard to the quantum computer assumption. Nowadays several lattice-based cryptosystems show good results with strong security proofs.

The notion of group signature was first formalised by Chaum and van Heyst in 1991 [4]. A group signature scheme allows a user in a group to sign anonymously a message on behalf of the group. The group is administered by a group manager, who can accept to add users to the group and has the ability to determine the real identity of the signer when needed. Motivated by the following situation: a high-ranking official in the government wants to leak anonymously an important information to a journalist. Rivest, Shamir, and Tauman [5] introduced the concept of ring signature and proposed a scheme based on RSA.

P. Gaborit (Ed.): PQCrypto 2013, LNCS 7932, pp. 34–51, 2013.

As opposed to group signatures, ring signatures have no group manager and no anonymity revocation system.

The concept of ring signature was extended by Bresson, Stern and Szydlo to threshold ring signatures [6]. In this setting, t-out-of-N users interact together in order to produce a signature without giving any information of the set of signers which produced the signature. Since their introduction, several threshold ring signature schemes were proposed [7–9]. Those constructions have a particularity in common: a complexity in $\mathcal{O}(Nt)$.

Aguilar, Cayrel and Gaborit [10] proposed a new threshold ring signature in 2008 of size in $\mathcal{O}(N)$. It is a code-based signature considered as a very efficient scheme due to its complexity.

In 2010, Cayrel, Lindner, Rückert and Silva [11] presented a lattice based version of the scheme in [10]. This version generalised an identification scheme given in [12], which is more efficient than Stern's identification protocol used by Aguilar et al. [10]. The security of the new scheme is based on the shortest independent vectors problem SIVP.

Our Contributions. In this work, we present an improvement of the lattice-based threshold ring signature scheme given in [11]. The improvement is due to a different way to achieve anonymity. Our scheme looks like t different digital signature. The difference is an extra challenge-answer part to deal with anonymity. The first part has the same size as t digital signatures (we use seeds to represent random permutations and other masks). The other part which deals with anonymity is in complexity $O(Nt)$ which is more important than $O(N)$ in [11] but implies smaller signatures when Nt is smaller than the size of t digital signatures. We also detail a ring signature in the particular case $t = 1$, which has the smallest size among others lattice-based ring signatures [13, 14, 11]. A table of comparison can be found in section 6.

Organization. In Section 2, we give some definitions and notions related to lattices and cryptography as well as a description of the identification scheme from [12]. In section 3, we present and detail our ring signature scheme. In section 4, we construct the lattice-based threshold ring signature scheme. The proofs of security for the scheme are detailed in section 5. In section 6, we give some concrete instantiations of our schemes and a comparison with the schemes from [11].

2 Preliminaries

This section is split into three parts. In the first one we give some basic definitions about lattices and cryptography. The second part details formal security definitions about ring signatures. In the third part we describe the identification scheme CLRS presented in [12].

Notations. We use bold upper-case letters to denote matrices and bold lower-case letters to denote vectors. We denote the Euclidean norm of the vector $\mathbf{v}$

by $\|\mathbf{v}\|$. For q an integer, $\mathbb{Z}_q$ denotes the group of integers modulo q. For a set E, we use the notation $w \xleftarrow{\$} E$ to mean that w is chosen randomly at uniform from E. Let $wh(\mathbf{v})$ denote the Hamming weight of $\mathbf{v}$ (the number of non-zero elements in $\mathbf{v}$) and for any integer m we denote by S_m the set of permutation of $\{1, \ldots, m\}$. Let $N \in \mathbb{N}$, we denote by δ_j the vector in $\{0,1\}^N$ with 1 in the j-th position and 0 elsewhere.

2.1 Lattices in Cryptography

We recall the definition of lattices, the small integer solution problem SIS, the inhomogeneous small integer solution problem ISIS, the standard hard lattice problem SIVP and a lattice-based commitment scheme.

Definition 1. *Let $B = \{\mathbf{b}_1, \ldots, \mathbf{b}_n\}$ be set of n linearly independent vectors in $\mathbb{R}^m$. The lattice generated by B is the set*

$$\mathcal{L}(B) = \left\{ \sum_{i=1}^n x_i \mathbf{b}_i \mid x_i \in \mathbb{Z} \right\}$$

of all integer combination of vectors in B. We denote by $\lambda_1(\mathcal{L}(B))$ the shortest vector of the lattice $\mathcal{L}(B)$. For $i \in \{1, \ldots, n\}$, we denote by $\lambda_i(\mathcal{L})$ the successive minima which is the smallest values $\lambda_i(\mathcal{L})$ such that the sphere of radius $\lambda_i(\mathcal{L})$ of center the origin contains at least i linearly independent lattice vectors.

Definition 2 (SIS$_{q,n,m,\alpha}$ problem). *Given a random matrix $\mathbf{A} \xleftarrow{\$} \mathbb{Z}_q^{n \times m}$ find a non zero vector $\mathbf{v} \in \mathbb{Z}^m$ such that $\mathbf{A}\mathbf{v}^T = 0$ and $\|\mathbf{v}\| \leq \alpha$.*

Definition 3 (ISIS$_{q,n,m,\alpha}$ problem). *Given a random matrix $\mathbf{A} \xleftarrow{\$} \mathbb{Z}_q^{n \times m}$ a vector $\mathbf{v} \in \mathbb{Z}_q^n$ and a real α find a vector $\mathbf{s} \in \mathbb{Z}^m$ such that $\mathbf{A}\mathbf{s}^T = \mathbf{v} \bmod q$ and $\|\mathbf{s}\| \leq \alpha$.*

Definition 4 (SIVP$_\gamma$ problem). *Given an n dimensional lattice $\mathcal{L}$, find n linearly independent lattice vectors of length at most $\gamma \cdot \lambda_n(\mathcal{L})$.*

In [15], Gentry *et al.* proved that the worst-case hardness of SIVP is reduced to the average-case hardness of SIS or ISIS problem. Their result is in the following theorem.

Theorem 1 ([15]). *Let $m, \alpha = poly(n)$ and for any prime $q \geq \alpha \cdot \omega(\sqrt{n \log n})$. There is a probabilistic polynomial time reduction from solving SIVP$_\gamma$ for $\gamma = \alpha \cdot \tilde{O}(\sqrt{n})$ in the worst case to solving SIS$_{q,n,m,\alpha}$ and ISIS$_{q,n,m,\alpha}$ on average with non negligible probability.*

In [16], Kawachi, Tanaka and Xagawa introduced a lattice-based commitment scheme COM. The proposed scheme satisfies the essential security properties, namely, the statistically-hiding and computationally-binding properties. Let COM(μ, ρ) be a commitment scheme.

- COM is said to be a statistically hiding scheme if for any messages μ_1, μ_2, any attacker cannot distinguish between $\mathsf{COM}(\mu_1, \rho_1)$ and $\mathsf{COM}(\mu_2, \rho_2)$.
- The computational binding property ensure that no polynomial time attacker can change the committed message to another one.

2.2 Ring Signature

In this subsection we review some definitions and properties about ring signatures that will be used in the following sections.

Ring Signature. We use the same definition as in [17]. A ring signature is a digital signature where the signer is not known, however his membership of a particular set can be verified.

Definition 5 (Ring Signature Scheme). *A ring signature scheme consists of the three following algorithms:*

- R.KeyGen : *A probabilistic polynomial time algorithm that takes as input a security parameter and outputs a key pair formed by a public key PK and a secret key SK.*
- R.Sign : *A probabilistic polynomial time algorithm that takes as input a set of public keys $PK_1, \ldots, PK_N$, a message μ, and a signing key. The output is a ring signature of the message μ.*
- R.Verify : *A deterministic algorithm that takes as input a ring signature, a message μ and the list of public keys of the users of the ring. The output of this algorithm is* accept *if the ring signature is valid or* reject *otherwise.*

Threshold Ring Signature. In 2002, Bresson, Stern and Szydlo introduced the concept of threshold ring signatures [6]. In such a scheme, a set of t users wants to collaborate to produce a signature while preserving the anonymity of the participating signers. We give a formal definition of t-out-of-N threshold ring signature scheme. We denote the set of users by $U' = \{1, \ldots, N\}$ and by S' the set of signers with $S' \subset U'$. Each user i in U' has a public/secret key pair (PK_i, SK_i).

Definition 6 (Threshold Ring Signature Scheme). *A threshold ring signature scheme consists of the three following algorithms:*

- T.KeyGen : *outputs a public key PK and a secret key SK.*
- T.Sign(μ, U', S') : *an interactive protocol between t users that take on input a set of public keys correponding to users in U' a set of t secret keys corresponding to users in S' and a message μ. The output is a t-out-of-N-threshold ring signature σ on μ.*
- T.Verify(μ, σ, t, U') : *deterministic algorithm that takes as input a value t, a set of public keys corresponding to users in U', a message-signature pair (μ, σ), and outputs* accept *or* reject.

A t-out-of-N threshold ring signature scheme is said to be secure if it is source hiding and unforgeable.

Anonymity. The source hiding definition described in [17] and [11] does not fit with threshold ring signature schemes obtained using the Fiat-Shamir transform. Indeed, the one-way functions used during the commitments do not allow to build two identical signatures for different set of signers. Moreover, this remains almost impossible even if two sets of signers have the same secret keys. We introduce a new definition generalised from the anonymity property for ring signatures see Definition 4 in [18]. With this definition we introduce the notion of indistinguishability.

Definition 7 (Indistinguishable source hiding). *Given a t-out-of-N threshold ring signature and a probabilistic polynomial time adversary $\mathcal{A}$, consider the following game*

1. *For $i = 1$ to N, generate (PK_i, SK_i). Give to $\mathcal{A}$ the set of public keys $P = \{PK_1, \ldots, PK_N\}$. The adversary $\mathcal{A}$ is also given access to a signing oracle $\mathsf{OT.Sign}(\cdot, \cdot, \cdot)$, that returns $\sigma = \mathsf{T.Sign}(\mu, P, S)$ on input (μ, P, S) with S a set signers.*
2. *$\mathcal{A}$ outputs a message μ, distinct sets $\{i_{1,0}, \ldots, i_{t,0}\}$, $\{i_{1,1}, \ldots, i_{t,1}\}$ and a ring P for which $PK_{i_{l,j}} \in P$ for $l \in \{0, 1\}$ and $j \in \{1, \ldots, t\}$. Adversary $\mathcal{A}$ is given $\{SK_1, \ldots, SK_N\} \setminus \{SK_{i_{1,0}}, \ldots, SK_{i_{t,0}}\}$. Furthermore, a random bit b is chosen and $\mathcal{A}$ is given $\sigma \leftarrow \mathsf{T.Sign}(\mu, P, \{SK_{i_{1,b}}, \ldots, SK_{i_{t,b}}\})$.*
3. *The adversary outputs a bit b', and succeeds if $b' = b$.*

Unforgeability. We give a formal definition of existential unforgeability under chosen message attacks in the setting of t-out-of-N threshold ring signature. The definition is described using a game between an existential forger $\mathcal{F}$ and a challenger $\mathcal{C}$, and it is similar to one used in [17] and [11].

Definition 8 (Existensial Unforgeability). *A threshold ring signature scheme is existentially unforgeable under a chosen message attack if for any probabilistic polynomial time $\mathcal{F}$, the probability that $\mathcal{F}$ succeeds in the following game is negligible:*

1. *The challenger $\mathcal{C}$ generates key pairs $\{PK_i, SK_i\}_{i=1}^{N}$, and gives the set of public keys $P = \{PK_i\}_{i=1}^{N}$ to $\mathcal{F}$.*
2. *$\mathcal{F}$ is given access to a signing oracle as in Definition 7.*
3. *$\mathcal{F}$ is also given access to a key exposure oracle $\mathsf{OExp}(\cdot)$, that returns a secret key SK_i on input i.*
4. *$\mathcal{F}$ outputs a t-out-of-N threshold ring signature $\sigma^\star$ for a new message $\mu^\star$.*

The adversary $\mathcal{F}$ succeeds if the verification $\mathsf{T.Verify}(\mu^\star, \sigma^\star, t, P) = 1$, $\mu^\star$ has not been asked by $\mathcal{F}$ in a signing query in step 2 of the game and the number of key exposure queries is strictly less than t.

2.3   CLRS Identification Scheme

CLRS is an identification scheme proposed by Cayrel, Lindner, Rückert and Silva [12]. It is a lattice-based version of a generalisation of the Stern code-based identification scheme presented in [19].

Input: n, m, q

$\quad \mathbf{x} \xleftarrow{\$} \{0,1\}^m$, with $wh(\mathbf{x}) = m/2$

$\quad \mathbf{A} \xleftarrow{\$} \mathbb{Z}_q^{n \times m}$

$\quad \mathbf{y} \leftarrow \mathbf{A}\mathbf{x}^T \bmod q$

$\quad \mathsf{COM} \xleftarrow{\$} \mathcal{F}$, a suitable family of commitment functions.

Output: $(SK, PK) = (\mathbf{x}, (\mathbf{y}, \mathbf{A}, \mathsf{COM}))$

Fig. 1. Key generation algorithm

P chooses $\sigma \xleftarrow{\$} S_m$, $\mathbf{u} \xleftarrow{\$} \mathbb{Z}_q^m$, $\mathbf{r}_0 \xleftarrow{\$} \{0,1\}^n$ and $\mathbf{r}_1 \xleftarrow{\$} \{0,1\}^n$.

P computes $c_0 \leftarrow \mathsf{COM}(\sigma \| \mathbf{A}\mathbf{u} \| \mathbf{r}_0)$ and $c_1 \leftarrow \mathsf{COM}(\sigma(\mathbf{u}) \| \sigma(\mathbf{x}) \| \mathbf{r}_1)$.

1. [**first commitment**] P sends c_0 and c_1 to V.
2. [**first challenge**] V sends $\alpha \xleftarrow{\$} \mathbb{Z}_q$ to P.
3. [**second commitment**] P sets $\beta = \sigma(\mathbf{u} + \alpha\mathbf{x})$ and sends β to V.
4. [**second challenge**] V sends $b \xleftarrow{\$} \{0,1\}$, to P.
5. [**final answer**]
   If $b = 0$ then
   $\qquad P$ sends σ and $\mathbf{r}_0$ to V.
   If $b = 1$ then
   $\qquad P$ sends $\sigma(\mathbf{x})$ and r_1 to V.

Verification:
If $b = 0$ then V checks if

$$c_0 \overset{?}{=} \mathsf{COM}(\sigma \| \mathbf{A}\sigma^{-1}(\beta)^T - \alpha\mathbf{y} \| \mathbf{r}_0)$$

If $b = 1$ then V checks, $wh(\sigma(\mathbf{x})) \overset{?}{=} m/2$ and

$$c_1 \overset{?}{=} \mathsf{COM}(\beta - \alpha\sigma(\mathbf{x}) \| \sigma(\mathbf{x}) \| \mathbf{r}_1)$$

Fig. 2. CLRS Identification protocol

The Stern protocol was introduced in 1993, it was the first efficient code-based identification scheme. It has the specific property to be zero-knowledge, indeed nobody can give any information about the involved secret from the transcripts of the interactions between a prover and a verifier. The zero-knowledge property allows to obtain a digital signature scheme by applying the Fiat-Shamir transform. Stern's scheme uses challenges and answers to establish the authentication of the user. There is only one challenge by round, we call it a three-pass protocol.

A generalisation of that scheme was presented in [20] by using two challenges for each round. Hence, the digital signatures obtained from this five-pass protocol have a smaller size. The CLRS identification scheme is a lattice-based version of this protocol.

We use a variation of the CLRS identification scheme in our construction, as well as it was done in [11].

The Scheme. In Figure 2, we have an identification between a prover P and a verifier V. The prover's secret key is a vector $\mathbf{x}$ of Hamming weight $m/2$. The public key $\mathbf{y}$ is related to $\mathbf{x}$ by the equation $\mathbf{y} = \mathbf{A}\mathbf{x}^T \bmod q$.

High Level Description. The protocol is made with challenges and answers. The verifier asks some challenges to the prover who needs to answer correctly. The challenges focus on the secret value $\mathbf{x}$. The vector $\mathbf{x}$ has an Hamming weight of $m/2$ and verify $\mathbf{A}\mathbf{x}^T = \mathbf{y}$. We can see that two different properties are required to identify the secret. It is a fundamental observation to understand the scheme. Indeed, the verifer will either ask to verify the Hamming weight of $\mathbf{x}$ or verify the relation $\mathbf{A}\mathbf{x}^T = \mathbf{y}$.

The masks used to protect the secret properties are from two different kinds. A permutation σ allows to mask which of the vectors of some given Hamming weight is used. The random vector $\mathbf{u}$ allows to mask which one of the vector verifying $\mathbf{A}\mathbf{x} = \mathbf{y}$ is used. After masking the secret twice (with each mask), the prover can unmask the masked secret with the permutation σ or the additional random vector to prove either that the secret is a vector of given Hamming weight or that it verify $\mathbf{A}\mathbf{x} = \mathbf{y}$.

We can identify 5 steps in the scheme: first commitment step, first challenge, second commitment, second challenge and final answer. The first commitment aims to set the random permutation and the random vector. The first challenge α is here to prevent replay attacks. The second commitment β is the secret masked two times. The second challenge b consists on asking to reveal either the weight of the secret ($b = 1$) or a way to compute $\mathbf{A}\mathbf{x}^T$ ($b = 0$).

This protocol is a probabilistic protocol in the fact that it is possible to anticipate the challenges and respond correctly without knowing the secret key. The soundness is close to $1/2$. Therefore, to reduce the cheating probability, the protocol has to be repeated many times.

2.4 Fiat-Shamir Transform

To build signature schemes from canonical identification schemes, one can use the Fiat-Shamir transform.

The challenges of the scheme are built with a random oracle H instead of an honest verifier. They are generated randomly at uniform from the message and commitments. In this way, a malicious prover cannot anticipate challenges one at a time and needs to anticipate them at once. The cheating probability remains the same as in the authentication protocol. Pointcheval and Stern [21] proved that an unforgeable signature scheme can be obtained when applying the Fiat-Shamir transform.

Recently Cayrel et al [22] proved that the Fiat-Shamir transform can be extended to fit with the $(2n + 1)$-pass identification schemes (n is an integer in our case $n = 2$). We briefly review this transformation. In order to transform a five-pass identification scheme to a signature scheme, the signer proceeds as follows. He uses two random oracle $\mathcal{O}_1$ and $\mathcal{O}_2$ to generate the challenges C_1 and C_2 given by the verifier in the identification scheme. Let μ be a message, the signature σ of μ is formed by R_1, C_1, R_2, C_2 and RSP, where R_1, R_2 and RSP are the values calculated by the prover in the identification scheme.

3 Ring Signature

A ring signature is a digital signature where the signer is not known, however his membership of a particular set can be verified. In the following we explain the idea described in [10] and [11], then we show the differences with our scheme. In this section we consider U as a set of N users.

3.1 Description of CLRS Ring Signature Scheme

The ring signature scheme in [11] is obtained by applying the Fiat-Shamir transform to a ring identification scheme. The idea is to construct, for each user i, a secret key $\mathbf{x}_i$ with the same property $\mathbf{A}\mathbf{x}_i^T = 0$. Hence, we cannot distinguish two users from their public keys. The problem is to construct several such secret keys. In fact, statistically, there is only one vector $\mathbf{v}$ of small enough weight w such that $\mathbf{A}\mathbf{v}^T = 0$. The construction consists on changing the matrix $\mathbf{A}$ for each user. Now they are identified by their public matrix $\mathbf{A}_i$ instead of their public value $\mathbf{A}_i\mathbf{x}_i^T$. To generate a ring identification, N identifications are done, one for each user. The real signer uses his secret key $\mathbf{x}_i$ for only one identification and the null vector for the others. The anonymity is obtained with a permutation of the users' commitments.

3.2 Our Ring Signature Scheme

This scheme uses the same key generation algorithm as the CLRS scheme presented in section 2.3. On the other hand, a matrix $\mathbf{M}$ is added to the public values. The matrix $\mathbf{M}$ is composed of all public elements $\mathbf{y}_1, \ldots, \mathbf{y}_N$ with $\mathbf{A}\mathbf{x}_i^T = \mathbf{y}_i$ for $i \in \{1, \ldots, N\}$.

The ring signature scheme can be obtained from the ring identification scheme in Figure 3 by applying the Fiat-Shamir transform. This scheme is very similar to the CLRS identification scheme described in section 2.3. As it is detailed in Figure 3, a new commitment β' and two masks $\mathbf{u}'$ and Σ were added as well as a secret value δ_j. Those new elements were designed to guarantee the anonymity of the real signer. The idea is to use δ_j as the secret identifying the real signer and to mask it in the same way as the secret key $\mathbf{x}_i$.

In this scheme, the $\mathbf{y}_j$ which defines the signer and which is used to compute the first commitment is masked in β'. Indeed, we have $\mathbf{y}_j = \mathbf{M}\delta_j^T$ with δ_j masked by Σ and $\mathbf{u}'$. The first commitment can be computed in the same way for each signer as far as $\mathbf{A}(\mathbf{x}_i + \mathbf{u})^T - \mathbf{M}(\delta_i + \mathbf{u}')^T$ is equal for each signer i. We use the same construction to mask $\mathbf{x}_i$ and δ_i because both values are identified by their weight and their image by a particular matrix ($\mathbf{A}$ or $\mathbf{M}$).

The fact that $wh(\Sigma(\delta_j)) = 1$ with Σ a permutation guarantees that one user in the ring U' is the real signer.

3.3 Features of the Scheme

This scheme, without β', u' and Σ, is the same as the identification scheme CLRS. We do not need to send N identifications to obtain anonymity like it was done in previous ring signature schemes. Therefore, a significant reduction of signature size is obtained for reasonable values of N and t.

An other notable property is the use of a unique random matrix $\mathbf{A}$ instead of N public matrices $\mathbf{A}_i$ in [11]. A consequence is a reduction of the size of the public keys.

4 Threshold Ring Signature

The threshold ring signature is a generalisation of the ring signature. The threshold ring signature is made by many signers instead of only one for the ring signature.

The authors of [6] and [10] claim that a threshold ring signature is not a repetition of several ring signatures. The reason is that we have to prove that at least t signer has been involved in the process. For example, a signer alone should not be able to sign twice in the same signature and produce a valid signature.

In our case, the threshold ring signature is more like t different ring signatures. That is why the signature size is close to t signatures instead of N for other schemes [10, 11].

4.1 From Ring Signature to Threshold Ring Signature

In this subsection we explain how to prove that t different users can make a threshold ring signature with our scheme. The threshold ring signature scheme is obtained by applying the Fiat-Shamir transform on the threshold ring identification scheme given in Figure 4. The idea is very simple, each δ_i represents a different user, which can be identified with the matrix $\mathbf{M}$. We only have to show that the δ_i, used in the identification, are different. We prove that point by verifying that the $\Sigma(\delta_i)$ are different because if $\Sigma(\delta_i) \neq \Sigma(\delta_j)$ then $\delta_i \neq \delta_j$, with Σ a permutation. From security perspective, Σ is known by all the signers, and this does not affect the unforgeability because Σ only masks δ_i which defines the identity of the signer. Moreover, no information about $\mathbf{x}_i$ is revealed. The

Let $U' = \{users\}$. Let $\mathbf{M}$ the matrix of all public elements $\mathbf{y}_1, \ldots, \mathbf{y}_N$ of users in U'.

$$\mathbf{M} = \begin{pmatrix} | & | & \cdots & | \\ \mathbf{y}_1 & \mathbf{y}_2 & \cdots & \mathbf{y}_N \\ | & | & \cdots & | \end{pmatrix}, \text{ with } \mathbf{y}_i = \mathbf{A}\mathbf{x}_i^T \text{ for all } i \in \{1, \ldots, N\}.$$

The user S with index j in $\{1, \ldots, N\}$, do the following:

He sets $\delta_j \in \{0,1\}^N$ with 1 in the j-th position and 0 elsewhere. He chooses a random permutation Σ of $\{1, \ldots, N\}$, $\mathbf{u}' \overset{\$}{\leftarrow} \mathbb{Z}_q^N, \sigma \overset{\$}{\leftarrow} S_m, \mathbf{u} \overset{\$}{\leftarrow} \mathbb{Z}_q^m, \mathbf{r}_0 \overset{\$}{\leftarrow} \{0,1\}^n$ and $\mathbf{r}_1 \overset{\$}{\leftarrow} \{0,1\}^n$.

He computes

$$c_0 \leftarrow \mathsf{COM}(\sigma\|\Sigma\|\mathbf{A}\mathbf{u}^T - \mathbf{M}\mathbf{u}'^T\|\mathbf{r}_0) \text{ and } c_1 \leftarrow \mathsf{COM}(\sigma(\mathbf{u})\|\Sigma(\mathbf{u}')\|\sigma(\mathbf{x}_j)\|\Sigma(\delta_j)\|\mathbf{r}_1).$$

1. [**first commitment**] S sends c_0 and c_1 to the verifier V.

2. [**first challenge**] V sends $\alpha \overset{\$}{\leftarrow} \mathbb{Z}_q$ to S.

3. [**second commitment**] S sends β and β' to V, with

$$\beta = \sigma(\mathbf{u} + \alpha\mathbf{x}_j) \text{ and } \beta' = \Sigma(\mathbf{u}' + \alpha\delta_j).$$

4. [**second challenge**] V sends $b \overset{\$}{\leftarrow} \{0,1\}$ to S.

5. [**final answer**]
   If $b = 0$ then
     S sends $\phi = \Sigma, \psi = \sigma$, and $\mathbf{a} = \mathbf{r}_0$ to V.
   If $b = 1$ then
     S sends $\chi = \Sigma(\delta_j), \mathbf{d} = \sigma(\mathbf{x}_j)$ and $\mathbf{e} = \mathbf{r}_1$ to V.

Verification :

If $b = 0$ then V checks if
$$c_0 \overset{?}{=} \mathsf{COM}(\psi\|\phi\|\mathbf{A}\psi^{-1}(\beta)^T - \mathbf{M}\phi^{-1}(\beta')^T\|\mathbf{a})$$

If $b = 1$ then V checks if

$$c_1 \overset{?}{=} \mathsf{COM}((\beta - \alpha\mathbf{d})\|(\beta' - \alpha\chi)\|\mathbf{d}\|\chi\|\mathbf{e})$$

$$\mathbf{d} \overset{?}{\in} \{0,1\}^m, \ wh(\mathbf{d}) \overset{?}{=} m/2 \text{ and } wh(\chi) \overset{?}{=} 1.$$

Fig. 3. Ring identification scheme

anonymity property remains unchanged because signers are supposed to know each other when they produce the signature.

During the verification, when $b = 0$, we have just to verify that each $\Sigma(\delta_i)$ is different from the others. Thus, this proves that t different signers have cooperated to generate the threshold ring signature.

Let $\mathbf{M}$ the matrix of all public elements $\mathbf{y}_1, \ldots, \mathbf{y}_N$ of users in U'.

$\mathbf{M} = \begin{pmatrix} | & | & \cdots & | \\ \mathbf{y}_1 & \mathbf{y}_2 & \cdots & \mathbf{y}_N \\ | & | & \cdots & | \end{pmatrix}$, with $\mathbf{y}_i = \mathbf{A}\mathbf{x}_i^T$ for all $i \in \{1, \ldots, N\}$.

$\delta_i \in \{0,1\}^N$ the vector with 1 on the i-th position and 0 elsewhere else.
$U' = \{$users$\}$ and $S' = \{$signers$\}$, with $S' \subset U'$, $|S'| = t$ and $|U'| = N$.
L the leader, with $L \in S'$ and V the verifier.

1. **[first commitment]**

   L construct at random Σ a permutation of $\{1, \ldots, N\}$.

   for i from 1 to t do
       L choose the i-th signer S_j
       S_j receives Σ from L.
       S_j constructs $\sigma_i \xleftarrow{\$} S_m$, $\mathbf{u}_i \xleftarrow{\$} \mathbb{Z}_q^m$, $\mathbf{u}_i' \xleftarrow{\$} \mathbb{Z}_q^N$, $\mathbf{r}_{0,i} \xleftarrow{\$} \{0,1\}^n$ and $\mathbf{r}_{1,i} \xleftarrow{\$} \{0,1\}^n$
       S_j computes : $c_{0,i} \leftarrow \mathsf{COM}(\sigma_i \| \Sigma \| \mathbf{A}\mathbf{u}_i^T - \mathbf{M}\mathbf{u}_i'^T \| \mathbf{r}_{0,i})$
                    $c_{1,i} \leftarrow \mathsf{COM}(\sigma_i(\mathbf{u}_i) \| \Sigma(\mathbf{u}_i') \| \sigma_i(\mathbf{x}_j) \| \Sigma(\delta_j) \| \mathbf{r}_{1,i})$
       S_j sends $c_{0,i}$ and $c_{1,i}$ to L.
   end for

   L computes $\mathbf{r}_0 \xleftarrow{\$} \{0,1\}^n$ and $\mathbf{r}_1 \xleftarrow{\$} \{0,1\}^n$.
   L computes the master commitments :
   $C_0 = \mathsf{COM}(c_{0,1} \| \ldots \| c_{0,t} \| \mathbf{r}_0)$ and $C_1 = \mathsf{COM}(c_{1,1} \| \ldots \| c_{1,t} \| \mathbf{r}_1)$

   L sends C_0 and C_1 to V.

2. **[first challenge]** V sends α, such as $\alpha \xleftarrow{\$} \mathbb{Z}_q$.

3. **[second commitment]**

   We denote $\bar{\mathbf{x}}_i$ (respectively $\bar{\delta}_i$) the $\mathbf{x}_j$ (respectively δ_j) corresponding to the i-th signer S_j.
   for i from 1 to t do
       L choose the i-th signer S_j
       S_j computes $\beta_i \leftarrow \sigma_i(\mathbf{u}_i + \alpha\bar{\mathbf{x}}_i)$
       S_j computes $\beta_i' \leftarrow \Sigma(\mathbf{u}_i' + \alpha\bar{\delta}_i)$
       S_j sends β_i, β_i' to L.
   end for

   L sends $\mathbf{v}_1 = \beta_1, \ldots, \mathbf{v}_t = \beta_t$ and $\mathbf{w}_1 = \beta_1', \ldots, \mathbf{w}_t = \beta_t'$ to V.

4. **[second challenge]** V sends $b = 0$ or $b = 1$ to L.

5. **[final answer]**

   if $b = 0$ then
       for i from 1 to t do
           L choose the i-th signer S_j
           S_j sends $\sigma_i, \mathbf{r}_{0,i}$ to L
       end for
   if $b = 1$ then
       for i from 1 to t do
           L choose the i-th signer S_j
           S_j sends $\mathbf{u}_i, \mathbf{u}_i'$ and $\mathbf{r}_{1,i}$ to L
       end for

   if $b = 0$ then
       L sends $\phi = \Sigma$, $\psi_1 = \sigma_1, \ldots, \psi_t = \sigma_t$, $\mathbf{a}_1 = \mathbf{r}_{0,1}, \ldots, \mathbf{a}_t = \mathbf{r}_{0,t}$ and $\rho = \mathbf{r}_0$ to V.
   if $b = 1$ then
       L sends $\chi_1 = \Sigma(\bar{\delta}_1), \ldots \chi_t = \Sigma(\bar{\delta}_t)$, $\mathbf{d}_1 = \sigma_1(\bar{\mathbf{x}}_1), \ldots, \mathbf{d}_t = \sigma_t(\bar{\mathbf{x}}_t)$, $\mathbf{e}_1 = \mathbf{r}_{1,1}, \ldots, \mathbf{e}_t = \mathbf{r}_{1,t}$
       and $\varrho = \mathbf{r}_1$ to V.

Fig. 4. Threshold ring identification scheme

if $b = 0$ then
  for i from 1 to t
    Set $c_i = \mathsf{COM}(\psi_i\|\phi\|\mathbf{A}\psi_i^{-1}(\mathbf{v}_i)^T - \mathbf{M}\phi^{-1}(\mathbf{w}_i)^T\|\mathbf{a}_i)$
  end for
  V checks that $C_0 \overset{?}{=} \mathsf{COM}(c_1\|\ldots\|c_t\|\rho)$
if $b = 1$ then
  for i from 1 to t
    Set $c_i = \mathsf{COM}(\mathbf{v}_i - \alpha\mathbf{d}_i\|\mathbf{w}_i - \alpha\chi_i\|\mathbf{d}_i\|\chi_i\|\mathbf{e}_i)$

    V checks that $wh(\mathbf{d}_i) = m/2$ and $\mathbf{d}_i \in \{0,1\}^m$

    V checks that $wh(\chi_i) = 1$
  end for
  V checks that $\sum_{i=1}^{t} wh(\chi_i) = t$ and $\chi_i \in \{0,1\}^N$
  V checks that $C_1 \overset{?}{=} \mathsf{COM}(c_1\|\ldots\|c_t\|\varrho)$

Fig. 5. Verification algorithm

4.2   Lattice-Based Threshold Identification Scheme

In this subsection we present our threshold identification scheme. The scheme is in Figure 4, the verification algorithm is given in Figure 5.

5   Security Analysis

In this section we prove the security of our scheme. A threshold ring signature scheme must satisfy two properties. The source hiding, which ensures the anonymity of the users and the existential unforgeability, which is common to every digital signature.

5.1   Source Hiding

Lemma 1. *For any transcript τ of the threshold identification scheme in Figure 4 performed by a set S of t signers we have that:*

For any set S' of t signers there exists a transcript σ' performed by S' such that the differences between τ and τ' are in the commitment values.

Proof. We will prove the lemma for one round of the threshold identification scheme. Let τ be the transcript of the threshold identification, it is easy to see that τ consists of $((C_0, C_1), \alpha, (\beta_i, \beta_i'; i \in \{1,\ldots,t\}), b, RSP)$ where RSP is the response sent by the leader L to the verifier V.

Let $(\mathbf{x}_i, \delta_i; i \in \{1,\ldots,t\})$, be the secrets used by the signers in S and $(\mathbf{x}_i', \delta_i'; i \in \{1,\ldots,t\})$, be the secrets used by the signers in S'. We will show how to choose $(\mathbf{u}_i, \mathbf{u}_i', \sigma_i, \Sigma)$ such that the transcript does not change except C_0 and C_1 when the secret keys $(\mathbf{x}_i, \delta_i; i \in \{1,\ldots,t\})$ are replaced by $(\mathbf{x}_i', \delta_i'; i \in \{1,\ldots,t\})$.

If $b = 0$, we set $U_i = \mathbf{u}_i + \alpha \mathbf{x}_i - \alpha \mathbf{x}'_i$ and $U'_i = \mathbf{u}'_i + \alpha \delta_i - \alpha \delta'_i$. Therefore, when we replace $(\mathbf{x}_i, \mathbf{u}_i, \mathbf{u}'_i, \delta_i; i \in \{1, \ldots, t\})$ by $(\mathbf{x}'_i, U_i, U'_i, \delta'_i; i \in \{1, \ldots, t\})$, we obtain that β_i and β'_i keep the same values. Since $\mathbf{u}_i$ and $\mathbf{u}'_i$ do not appears in RSP, then the only thing that change in the transcript are the values C_0 and C_1.

If $b = 1$, we choose π_i and Π such that $\pi_i(\mathbf{x}'_i) = \mathbf{x}_i$ and $\Pi(\delta'_i) = \delta_i$ and we set $U_i = \pi^{-1}(\mathbf{u}_i)$ and $U'_i = \Pi^{-1}(\mathbf{u}'_i)$. Therefore, when we replace $(\sigma_i, \mathbf{u}_i, \Sigma, \mathbf{u}'_i)$ by $(\sigma_i \circ \pi_i, U_i, \Sigma \circ \Pi, U'_i)$, we obtain that β_i and β'_i keep the same values as well as in RSP. Then, the only thing that change in the transcript are the values C_0 and C_1.

Applying this construction for all the rounds finishes the proof.

Lemma 2. *Let n, q be the parameters of the commitment scheme* COM *and rd the number of rounds of the threshold identification scheme. For any message μ and any threshold signature σ generated by a set S of t signers, we have : For any set S' of t signers there exists a signature σ' with probability*

$$p = 1 - \left(1 - \frac{1}{(2q)^{rd}}\right)^{2^{n \times rd}}$$

performed by S' such that the difference between σ and σ' are in the commitment values.

Proof. Using the Fiat-Shamir transform we can see the threshold signature as $(R_1, CH_1, R_2, CH_2, RSP)$ with R_1 is the concatenation of C_0 and C_1 of all the rounds, R_2 is the concatenation of all β_i and β'_i of all the rounds and RSP is the concatenation of final answers of all the rounds. The values of CH_1 and CH_2 are computed by two random oracle $\mathcal{O}_1$ and $\mathcal{O}_2$ such that $CH_1 = \mathcal{O}_1(\mu, R_1)$ with μ the message and $CH_2 = \mathcal{O}_2(\mu, CH_1, R_2)$. The Lemma 1 states the existence of another transcript equal to the signature except the values of C_0 and C_1. Those commitments are computed using the commitment scheme COM with random values $\mathbf{r}_0$ and $\mathbf{r}_1$. For each round of the transcript we can take an other $\mathbf{r}_0$ or $\mathbf{r}_1$ and we get another transcript equal to the signature except the values of C_0 and C_1. To finish the proof we compute the probability that one of those transcripts corresponds to a signature of μ. Most of the time the transcript is not a signature because the challenges are different than $\mathcal{O}_1(\mu, R_1)$ and $\mathcal{O}_2(\mu, CH_1, R_2)$. The probability that the transcript corresponds to a signature of μ can be computed by a Bernoulli distribution with parameters $p_B = \frac{1}{(2q)^{rd}}$ and $n_B = 2^{n \times rd}$. The parameter p_B corresponds to the probability to obtain a particular challenge and n_B correspond to the number of transcript. Then, the probability to obtain a signature is

$$1 - \left(1 - \frac{1}{(2q)^{rd}}\right)^{2^{n \times rd}}.$$

Theorem 2. *If there is a probabilistic polynomial time attacker $\mathcal{A}$ that can win the game of source hiding, then $\mathcal{A}$ can break the statistically hiding property of the commitment scheme* COM.

Proof. Let S_0 and S_1 be two sets of t signers. In the game described in Definition 7, the attacker is given as a challenge a threshold ring signature generated by S_b (with $b \in \{0, 1\}$) and he has to guess with non negligible probability b' such that $b' = b$.

We consider that the attacker has access to the set of all the secret keys and chooses two sets of t signers S_0 and S_1. Then, he chooses a message μ and requests a challenge. After receiving the threshold ring signature σ of μ signed under S_b, the attacker has to guess the set which generated it.

By lemma 2, we obtain that there exists a signature σ' generated by $S_{\bar{b}}$ with probability p such that the difference between σ and σ' are in the commitments calculated by COM. In the parameters of the commitment scheme COM, q is polynomial on n, therefore the probability p is not negligible. If the attacker $\mathcal{A}$ wins the source hiding game, then $\mathcal{A}$ can distinguish between two outputs of the commitment scheme COM with non negligible probability.

5.2   Unforgeability

In this subsection we prove that if a forger can win the game in Definition 8, then a polynomial time algorithm can be built to solve a standard lattice problem. The proof of unforgeability is given in Theorem 3. We start by giving the following lemma which is used in the proof of unforgeability.

Lemma 3. *If there exist a probabilistic polynomial time algorithm $\mathcal{A}$ that is able to produce a t-out-of-N threshold signature scheme with probability greater than $\left(\frac{q+2}{2q}\right)^{rd}$, than either he can produce t values which can be used as secret keys or he can find a collision in the commitment scheme COM.*

Proof. If $\mathcal{A}$ produces a t-out-of-N threshold signature with probability greater than $\left(\frac{q+2}{2q}\right)^{rd}$, then he succeeds, in the corresponding threshold identification scheme, in at least one round with probability greater than $\frac{q+2}{2q}$. In each round there are $2q$ possible challenges, q possibilities for the first challenge step and 2 possibilities for the second challenge step. By the pigeon-hole principle we deduce that $\mathcal{A}$ can answer correctly with a particular commitment for two different challenges α, β in the first challenge step and any possible challenge in the second challenge step. Therefore, he can build the following transcript:

$$((C_0, C_1), \alpha, (\mathbf{v}_i, \mathbf{w}_i), 0, (\phi, \psi_i, \mathbf{a}_i, \rho)), \quad i \in \{1, \dots, t\}$$

$$((C_0, C_1), \alpha, (\mathbf{v}_i, \mathbf{w}_i), 1, (\chi_i, \mathbf{d}_i, \mathbf{e}_i, \varrho)), \quad i \in \{1, \dots, t\}$$

$$((C_0, C_1), \beta, (\mathbf{v}'_i, \mathbf{w}'_i), 0, (\phi', \psi'_i, \mathbf{a}'_i, \rho')), \quad i \in \{1, \dots, t\}$$

$$((C_0, C_1), \beta, (\mathbf{v}'_i, \mathbf{w}'_i), 1, (\chi_i, \mathbf{d}_i, \mathbf{e}_i, \varrho)), \quad i \in \{1, \dots, t\}$$

such that all those transcript succeeds in the verification protocol for that round. We have that either $\mathcal{A}$ can find a collision in COM or we obtain the following equations from the verification protocol.

1. $\psi_i = \psi'_i$, $\phi = \phi'$, $\mathbf{A}\psi_i^{-1}(\mathbf{v}_i)^T - \mathbf{M}\phi^{-1}(\mathbf{w}_i)^T = \mathbf{A}\psi_i'^{-1}(\mathbf{v}'_i)^T - \mathbf{M}\phi'^{-1}(\mathbf{w}'_i)^T$,
   $\mathbf{a}_i = \mathbf{a}'_i$ for $i \in \{1, \ldots, t\}$.
2. $\mathbf{v}_i - \alpha\mathbf{d}_i = \mathbf{v}'_i - \beta\mathbf{d}'_i$, $\mathbf{w}_i - \alpha\chi_i = \mathbf{w}'_i - \beta\chi'_i$, $\mathbf{d}_i = \mathbf{d}'_i$, $\chi_i = \chi'_i$, $\mathbf{e}_i = \mathbf{e}'_i$ for
   $i \in \{1, \ldots, t\}$.
3. $wh(\mathbf{d}_i) = wh(\mathbf{d}'_i) = m/2$, $\mathbf{d}_i \in \{0,1\}^m$, $\mathbf{d}'_i \in \{0,1\}^m$ for $i \in \{1, \ldots, t\}$.
4. $\sum_{i=1}^t wh(\chi_i) = t$, $\sum_{i=1}^t wh(\chi'_i) = t$, $\chi_i \in \{0,1\}^N$, $\chi'_i \in \{0,1\}^N$ for $i \in$
   $\{1, \ldots, t\}$.

From the equations in 2, we have that $\mathbf{v}'_i = \mathbf{v}_i - \alpha\mathbf{d}_i + \beta\mathbf{d}'_i$ and $\mathbf{w}'_i = \mathbf{w}_i - \alpha\chi_i + \beta\chi'_i$. When we replace $\mathbf{v}'_i$ and $\mathbf{w}'_i$ in the third equation in 1, we obtain:

$$\mathbf{A}\psi_i^{-1}(\mathbf{v}_i)^T - \mathbf{M}\phi^{-1}(\mathbf{w}_i)^T = \mathbf{A}\psi_i^{-1}(\mathbf{v}_i - \alpha\mathbf{d}_i + \beta\mathbf{d}_i)^T - \mathbf{M}\phi^{-1}(\mathbf{w}_i - \alpha\chi_i + \beta\chi_i)^T$$

$$0 = (\beta - \alpha)(\mathbf{A}\psi_i^{-1}(\mathbf{d}_i)^T - \mathbf{M}\phi^{-1}(\chi_i)^T)$$

Since $\alpha \neq \beta$ we have that $\mathbf{A}\psi_i^{-1}(\mathbf{d}_i)^T = \mathbf{M}\phi^{-1}(\chi_i)^T$. From the equation in 4, we have that $\mathbf{M}\phi^{-1}(\chi_i)^T$ correspond to t different public keys and $\psi_i^{-1}(\mathbf{d}_i)$ can be used to simulate t different secret keys.   □

In the following theorem we prove the unforgeability of the threshold signature scheme.

Theorem 3 (Unforgeability). *If a forger wins the game in Definition 8 with probability p' in polynomial time, then the forger can solve an $ISIS_{q,n,m,\sqrt{m}}$ instance in polynomial time with probability $p'p\frac{1}{N^2}$ or find a collision in the commitment scheme* COM.

Proof. The challenger $\mathcal{C}$ is given an ISIS instance $(\mathbf{A}, \mathbf{y})$. Then, $\mathcal{C}$ chooses $k \in \{1 \ldots N\}$ and sets $\mathbf{x}_k := 0$, $\mathbf{y}_k := \mathbf{y}$. $\mathcal{C}$ generates $N - 1$ keys $(\mathbf{x}_i, \mathbf{y}_i)$ with $i \in \{1 \ldots N\}$ and $i \neq k$.

We start the game in Definition 8 with the forger $\mathcal{F}$ and the pairs $(\mathbf{x}_i, \mathbf{y}_i)$ with $i \in \{1 \ldots N\}$, as key pairs. We notice that only the key pair $(\mathbf{x}_k, \mathbf{y}_k)$ is not generated as a valid key pair and since $\mathbf{x}_k$ is not revealed, the set of key pairs $(\mathbf{x}_i, \mathbf{y}_i)$ is indistinguishable from a valid set of key pairs.

The challenger $\mathcal{C}$ simulates the signing oracle OT.Sign. When $\mathcal{F}$ requests the signing oracle, he send a query to the challenger $\mathcal{C}$ involving the public keys $\mathbf{y}_i$, $i \in I$ and $I \subseteq N$. If k is not in the set I, the challenger $\mathcal{C}$ is able to produce the corresponding signature. Otherwise the challenger produces a signature σ by replacing $\mathbf{x}_k$ by $\mathbf{x}_j$ with $j \notin I$. By Lemma 2, with probability p there exist a signature σ' such that σ' could be produced using the secret key corresponding to $\mathbf{y}_i$.

The challenger $\mathcal{C}$ also simulates the oracle OExp. If the forger $\mathcal{F}$ asks for the value of $\mathbf{x}_k$, the game ends without a forged signature. Since the forger do not ask for all the secret keys, the probability that he ask for $\mathbf{x}_k$ is less than $\frac{N-1}{N}$. Otherwise, the forger $\mathcal{F}$ wins the game and outputs a valid signature in polynomial time with probability p'.

According to Lemma 3, $\mathcal{F}$ can either simulate t different secret keys or find a collision on the commitment scheme COM. If he succeeds to simulate t secret keys, at least one of them, $\mathbf{x}_l$, was not a query for the key exposure oracle. The vector $\mathbf{x}_l$ is such that $wh(\mathbf{x}_l) = m/2$ and $\mathbf{Ax}_l^T = \mathbf{y}_l$. Since k was chosen randomly at uniform in $\{1, \dots, N\}$, then the probability that k is equal to l is $1/N$. If l is equal to k, then $\mathbf{x}_l$ is a solution of the $\mathrm{ISIS}_{q,n,m,\sqrt{m}}$ instance $(\mathbf{A}, \mathbf{y})$.

With the interaction of the challenger and the forger, we can build a probabilistic polynomial time algorithm that can solve an $\mathrm{ISIS}_{q,n,m,\sqrt{m}}$ instance with probability greater than $pp'\frac{1}{N^2}$. □

It was shown in [16] that the security of the commitment scheme COM is based on the average hardness of $\mathrm{ISIS}_{q,n,m,\sqrt{m}}$. By Theorem 1 we have that there exist average-case/worst-case reduction from SIVP_γ to $\mathrm{ISIS}_{q,n,m,\alpha}$ with an approximation factor $\gamma = \alpha \cdot \tilde{O}(\sqrt{n})$. Therefore, we have that for a prime $q = \tilde{O}(n)$, $m = O(n \log q)$ and $\alpha = \sqrt{m}$, the unforageability of our threshold ring signature scheme is based on $\mathrm{SIVP}_{\tilde{O}(n)}$.

6 Parameters

In this section we compare our threshold ring signature and our ring signature to the CLRS ring and threshold ring signature in [11]. The comparison is only made with this scheme because others schemes in [13, 14] don't give instantiation parameters.

6.1 Parameters Assumption

We use the same parameters used in [11] (subsection 5.1) to compare the performance of the schemes. The parameters are $n = 64, m = 2048, q = 257$ and the length of the commitment of COM is 224 bits for bit-security equal to 111. To compute the size of the signature, we use a seed to represent each random permutation σ_i and Σ. In fact, the signer only has to send a seed from which the verifier can obtain the desired element and thus reduce the communication cost. The vector $\sigma_i(\mathbf{x}_i)$ is a vector in $\{0,1\}^m$, its length is so m-bits instead of $\log_2 q \times m$ bits.

6.2 Ring Signature

In the following table we can see the significant reduction obtained with our ring signature scheme. Our scheme stay reasonable even for huge size of ring. The number of member of the ring is given by N.

Table 1. Comparison of lattice-based ring signature schemes in Mbytes

N	100	1000	5000	10000	100000
CLRS ring	24.43	244.24	1221.21	2442.42	24424.20
Scheme in Figure 3	0.26	0.37	0.84	1.43	12.05

6.3   Threshold Ring Signature

In the following table we can see different signature sizes for some values of t and N.

Table 2. Comparison of lattice-based threshold ring signature schemes in Mbytes

N	100	100	100	100	100	100	200	200	200	1000	1000	1000
t	2	10	30	50	70	100	2	10	50	2	10	50
CLRS threshold ring	24.43	24.43	24.43	24.43	24.43	24.43	48.85	48.85	48.85	244.24	244.24	244.24
Scheme in Figure 4	0.52	2.56	7.68	12.80	17.92	25.60	0.54	2.68	13.39	0.73	3.63	18.11

We see that our threshold scheme has a size close to t ring signatures and do not depend so much on the parameter N for the given parameters.

Acknowledgement. We deeply thank one of the anonymous reviewer of the conference which comments and suggestions were very helpful.

References

1. Lenstra, A.K., Lenstra, H.W., Lovász, L.: Factoring polynomials with rational coefficients. Mathematische Annalen 261(4), 515–534 (1982)
2. Coppersmith, D.: Finding small solutions to small degree polynomials. In: Silverman, J.H. (ed.) CaLC 2001. LNCS, vol. 2146, pp. 20–31. Springer, Heidelberg (2001)
3. Ajtai, M.: Generating hard instances of lattice problems. In: Proceedings of the Twenty-eighth Annual ACM Symposium on Theory of Computing., pp. 99–108. ACM (1996)
4. Chaum, D., van Heyst, E.: Group signatures. In: Davies, D.W. (ed.) EUROCRYPT 1991. LNCS, vol. 547, pp. 257–265. Springer, Heidelberg (1991)
5. Rivest, R.L., Shamir, A., Tauman, Y.: How to leak a secret. In: Boyd, C. (ed.) ASIACRYPT 2001. LNCS, vol. 2248, pp. 552–565. Springer, Heidelberg (2001)
6. Bresson, E., Stern, J., Szydlo, M.: Threshold ring signatures and applications to ad-hoc groups. In: Yung, M. (ed.) CRYPTO 2002. LNCS, vol. 2442, pp. 465–480. Springer, Heidelberg (2002)
7. Liu, J.K., Wei, V.K., Wong, D.S.: A separable threshold ring signature scheme. In: Lim, J.-I., Lee, D.-H. (eds.) ICISC 2003. LNCS, vol. 2971, pp. 12–26. Springer, Heidelberg (2004)
8. Dallot, L., Vergnaud, D.: Provably secure code-based threshold ring signatures. In: Parker, M.G. (ed.) Cryptography and Coding 2009. LNCS, vol. 5921, pp. 222–235. Springer, Heidelberg (2009)
9. Zheng, D., Li, X., Chen, K.: Code-based ring signature scheme. IJ Network Security 5(2), 154–157 (2007)
10. Aguilar Melchor, C., Cayrel, P.-L., Gaborit, P.: A new efficient threshold ring signature scheme based on coding theory. In: Buchmann, J., Ding, J. (eds.) PQCrypto 2008. LNCS, vol. 5299, pp. 1–16. Springer, Heidelberg (2008)

11. Cayrel, P.-L., Lindner, R., Rückert, M., Silva, R.: A lattice-based threshold ring signature scheme. In: Abdalla, M., Barreto, P.S.L.M. (eds.) LATINCRYPT 2010. LNCS, vol. 6212, pp. 255–272. Springer, Heidelberg (2010)
12. Cayrel, P.-L., Lindner, R., Rückert, M., Silva, R.: Improved zero-knowledge identification with lattices. In: Heng, S.-H., Kurosawa, K. (eds.) ProvSec 2010. LNCS, vol. 6402, pp. 1–17. Springer, Heidelberg (2010)
13. Wang, J., Sun, B.: Ring signature schemes from lattice basis delegation. In: Qing, S., Susilo, W., Wang, G., Liu, D. (eds.) ICICS 2011. LNCS, vol. 7043, pp. 15–28. Springer, Heidelberg (2011)
14. Brakerski, Z., Kalai, Y.: A framework for efficient signatures, ring signatures and identity based encryption in the standard model. Technical report, Cryptology ePrint Archive, Report 2010/086 (2010)
15. Gentry, C., Peikert, C., Vaikuntanathan, V.: Trapdoors for hard lattices and new cryptographic constructions. In: Proceedings of the 40th Annual ACM Symposium on Theory of Computing, pp. 197–206. ACM (2008)
16. Kawachi, A., Tanaka, K., Xagawa, K.: Concurrently secure identification schemes based on the worst-case hardness of lattice problems. In: Pieprzyk, J. (ed.) ASIACRYPT 2008. LNCS, vol. 5350, pp. 372–389. Springer, Heidelberg (2008)
17. Aguilar Melchor, C., Cayrel, P., Gaborit, P., Laguillaumie, F.: A new efficient threshold ring signature scheme based on coding theory. IEEE Transactions on Information Theory 57(7), 4833–4842 (2011)
18. Bender, A., Katz, J., Morselli, R.: Ring signatures: Stronger definitions, and constructions without random oracles. In: Halevi, S., Rabin, T. (eds.) TCC 2006. LNCS, vol. 3876, pp. 60–79. Springer, Heidelberg (2006)
19. Stern, J.: A new identification scheme based on syndrome decoding. In: Stinson, D.R. (ed.) CRYPTO 1993. LNCS, vol. 773, pp. 13–21. Springer, Heidelberg (1994)
20. Cayrel, P.L., Veron, P.: Improved code-based identification scheme. arXiv preprint arXiv:1001.3017 (2010)
21. Pointcheval, D., Stern, J.: Security proofs for signature schemes. In: Maurer, U.M. (ed.) EUROCRYPT 1996. LNCS, vol. 1070, pp. 387–398. Springer, Heidelberg (1996)
22. El Yousfi Alaoui, S.M., Dagdelen, Ö., Véron, P., Galindo, D., Cayrel, P.-L.: Extended security arguments for signature schemes. In: Mitrokotsa, A., Vaudenay, S. (eds.) AFRICACRYPT 2012. LNCS, vol. 7374, pp. 19–34. Springer, Heidelberg (2012)

Degree of Regularity for HFEv and HFEv-

Jintai Ding[1,2,*] and Bo-Yin Yang[3,**]

[1] University of Cincinnati, Cincinnati OH, USA
[2] Chongqing University, China,
jintai.ding@gmail.com
[3] Academia Sinica, Taipei, Taiwan,
by@crypto.tw

Abstract. In this paper, we first prove an explicit formula which bounds the degree of regularity of the family of HFEv ("HFE with vinegar") and HFEv- ("HFE with vinegar and minus") multivariate public key cryptosystems over a finite field of size q. The degree of regularity of the polynomial system derived from an HFEv- system is less than or equal to

$$\frac{(q-1)(r+v+a-1)}{2} + 2 \text{ if } q \text{ is even and } r+a \text{ is odd,}$$

$$\frac{(q-1)(r+v+a)}{2} + 2 \text{ otherwise,}$$

where the parameters v, D, q, and a are parameters of the cryptosystem denoting respectively the number of vinegar variables, the degree of the HFE polynomial, the base field size, and the number of removed equations, and r is the "rank" paramter which in the general case is determined by D and q as $r = \lfloor \log_q(D-1) \rfloor + 1$. In particular, setting $a = 0$ gives us the case of HFEv where the degree of regularity is bound by

$$\frac{(q-1)(r+v-1)}{2} + 2 \text{ if } q \text{ is even and } r \text{ is odd,}$$

$$\frac{(q-1)(r+v)}{2} + 2 \text{ otherwise.}$$

This formula provides the first solid theoretical estimate of the complexity of algebraic cryptanalysis of the HFEv- signature scheme, and as a corollary bounds on the complexity of a direct attack against the QUARTZ digital signature scheme. Based on some experimental evidence, we evaluate the complexity of solving QUARTZ directly using F_4/F_5 or similar Gröbner methods to be around 2^{92}.

Keywords: Degree of regularity, HFE, HFEv, HFEv-.

* Partially sponsored by National Science Foundation of China, Grant #60973131 and U1135004, and the Charles Phelps Taft Research Center.
** Partially sponsored by Taiwan's National Science Council project #100-2628-E-001-004-MY3 and 101-2915-I-001-019.

P. Gaborit (Ed.): PQCrypto 2013, LNCS 7932, pp. 52–66, 2013.
© Springer-Verlag Berlin Heidelberg 2013

1 Introduction

1.1 Questions

HFE (Hidden Field Equations) and its derivatives form one of the best known families of multivariate quadratic public-key cryptosystems. It was invented by Patarin as a modification of the Matsumoto-Imai cryptosystem C^* in 1997.

Shor's algorithm from 1994 and its extensions [30,34] will break RSA and ECC when large quantum computers became available. In this context, multivariate PKCs and in particular HFE [28] had been viewed as a possible candidate to replace RSA. Although it was shown by Faugère and Joux [19] that the basic form can be cryptanalyzed by a direct algebraic attack, simple HFE variations had already been designed to guard against known attacks. The best known of these is probably QUARTZ, a very conservatively designed HFE variant over $\mathbb{F}_2$ using both the "Vinegar" and "Minus" modifications [29]. QUARTZ (and all HFEv variants) have never been credibly cryptanalyzed.

We want to give a solid theoretical bound on the degree of regularity of HFEv and associated systems, such as QUARTZ, and thereby obtain a good estimate on the complexity of attacking HFEv, HFEv-, and ipHFE cryptosystems using Gröbner Bases.

1.2 Answers

One usually solves $p_1(x_1, \ldots, x_n) = p_2(x_1, \ldots, x_n) = \cdots = p_m(x_1, \ldots, x_n) = 0$ over $\mathbb{F}_q$ using Gröbner basis algorithms such as F_4/F_5. The critical parameter which determines the complexity is known as "the degree of regularity", which is the maximum degree of monomials that appear in the computation. If we denote by $(p_i)^h$ the homogeneous leading part of p_i, the degree of regularity of the system is the first degree at which we find non-trivial relations among the $(p_i)^h$'s, or if we set as the graded ring $B := \mathbb{F}_q[x_1, \ldots, x_n]/\langle x_1^q, \ldots, x_n^q \rangle$ and B_d its degree-d slice, we may state a definition as follows for the case of degree-2 equations (generalizable to higher/mixed degrees):

Definition 1.1. *For homogeneous quadratic polynomials* $(\lambda_1, \ldots, \lambda_m) \in B_2^m$, *let* $\psi_d : B_d^m \to B_{d+2}$ *be the map defined as* $\psi(b_1, \ldots, b_m) = \sum_{i=1}^m b_i \lambda_i$. *Then* $R_d(\lambda_1, \ldots, \lambda_m) := \ker \psi_d$ *defines the subspace of relations* $\sum_{i=1}^m b_i \lambda_i = 0$. *Further let* $T_d(\lambda_1, \ldots, \lambda_n)$ *be the subspace of trivial relations generated by elements*

$$\{b(\lambda_i e_j - \lambda_j e_i) | 1 \leq i < j \leq m, b \in B_{d-2}\}, \text{ and}$$

$$\{b(\lambda_i^{q-1} - 1)e_i | 1 \leq i \leq m, b \in B_{d-2(q-1)}\}.$$

Here e_i *means the i-th unit vector consisting of all zeros except one 1 at the i-th position. The <u>degree of regularity</u> of a homogeneous quadratic set is then*

$$D_{reg}(\lambda_1, \ldots, \lambda_m) := \min\{d | R_{d-2}(\lambda_1, \ldots, \lambda_m)/T_{d-2}(\lambda_1, \ldots, \lambda_m) \neq \{0\}\},$$

and $D_{reg}(p_1, \ldots, p_m) := D_{reg}((p_1)^h, \ldots, (p_m)^h)$ *for polynomials in general.*

We find an upper bound to D_{reg} for HFEv and HFEv-, which like in earlier studies depends on the size of the base field q, the rank of the HFE polynomial r, the number of removed equations a (if "minus" is used), and additionally the number of vinegar variables v, but in general on not the number of variables n:

$$D_{\text{reg}} \leq \frac{(q-1)(r+v+a-1)}{2} + 2, \quad \text{if } q \text{ is even and } r+a \text{ is odd,}$$

$$D_{\text{reg}} \leq \frac{(q-1)(r+v+a)}{2} + 2, \quad \text{otherwise.}$$

For small numbers we evaluated D_{reg} of random tests for HFEv and HFEv- using MAGMA and in each case the bound is relatively tight (see Section 4.1) which lends credence to predictions using our bound above for the Gröbner bases complexity.

As an example, substituting the actual parameters of QUARTZ we get $D_{\text{reg}} \leq 9$. Assuming that it is indeed 9, we can compute the number of bit-operations required to break it as $\approx 2^{92}$ (see Section 4.1), so QUARTZ should be reasonably secure for now.

This also shows that the break of an instance of internally perturbed HFE in [18], which is very much related to HFEv, is likely a case of overly aggressive parameters rather than of systematic problems.

1.3 Related Work

The C^* cryptosystem can be seen as a simple case of an HFE cryptosystem, and [14] noted that Patarin's linearization attack [27] was equivalent to the degree of regularity of C^* being three (in line with the formula in that paper).

The Square cryptosystem [7] is a C^* system with rank 1 and an odd base field. [14] proves a *lower* bound on its degree of regularity, showing a direct algebraic attack with Gröbner basis to be infeasible. However, such a result does not mean that the system is secure, because Square is actually broken by a different attack.

[9] was the first to claim to "break" HFE (cryptanalyze in significantly under design security), and [10] the earliest to mention HFEv- and HFE- specifically. But neither was followed up with a concrete implementation, and all interest was attracted to the news of Faugère's actually breaking HFE Challenge 1 [19].

[21] started to investigate algebraically the degree of regularity of HFE, but [17] seems to be the first rigorous study of the subject, which is continued by [14, 15].

2 Background

In the standard formulation of a multivariate public-key cryptosystem over a finite field $\mathbb{F}$, the public-key $P : \mathbb{F}^n \mapsto \mathbb{F}^m = T \circ Q \circ S$ is a composition of two invertible affine maps $S : \mathbb{F}^n \mapsto \mathbb{F}^n$ and $T : \mathbb{F}^m \mapsto \mathbb{F}^m$, and a quadratic map (possibly with some parameters) $Q : \mathbb{F}^n \mapsto \mathbb{F}^m$ which is easily invertible when all parameters are given. The maps S and T are part of the secret key, and properties of the central map Q determines most of the properties of the cryptosystem.

2.1 The HFEv, ipHFE and HFEv- Cryptosystems

Let $\mathbb{F} \cong \mathbb{F}_q$ be a finite field of order q and $\mathbb{K}$ a degree-n extension of $\mathbb{F}$, with a "canonical" isomorphism ϕ identifying $\mathbb{K}$ with the vector space $\mathbb{F}^n$. That is, $\mathbb{F}^n \xrightarrow{\phi} \mathbb{K}$, $\mathbb{K} \xrightarrow{\phi^{-1}} \mathbb{F}^n$. Any function or map F from $\mathbb{K}$ to $\mathbb{K}$ can be expressed *uniquely* as a polynomial function with coefficients in $\mathbb{K}$ and degree less than q^n, namely

$$F(X) = \sum_{i=0}^{q^n-1} a_i X^i, \quad a_i \in \mathbb{K}.$$

Denote by $\deg_{\mathbb{K}}(F)$ the degree of $F(X)$ for any map F. Using ϕ, we can build a new map $F' : \mathbb{F}^n \to \mathbb{F}^n$

$$P(x_1, .., x_n) = (p_1(x_1, .., x_n), \ldots, p_n(x_1, .., x_n)) = \phi^{-1} \circ F \circ \phi(x_1, .., x_n),$$

which is essentially F but viewed from the perspective of $\mathbb{F}^n$. We can identify F and F' unless there is a chance of confusion.

An $\mathbb{F}$-degree-2 or $\mathbb{F}$-quadratic function from $\mathbb{K}$ to $\mathbb{K}$ can in this framework be seen to be a polynomial all of whose monomials have exponent $q^i + q^j$ or q^i or 0 for some i and j. The general form of this $\mathbb{F}$-quadratic function is $Q(X) = \sum_{i,j=0}^{n-1} a_{ij} X^{q^i+q^j} + \sum_{i=0}^{n-1} b_i X^{q^i} + c.$, the *extended Dembowski-Ostrom polynomial map*. Such a $Q(X)$ with a fixed low $\mathbb{K}$-degree is used to build the HFE multivariate public key cryptosystems, as in the following

$$Q(X) = \sum_{i,j=0}^{q^i+q^j \le D, j \le i} a_{ij} X^{q^i+q^j} + \sum_{i=0}^{q^i \le D} b_i X^{q^i} + c;$$

Note that the coefficients are values in $\mathbb{K}$, and all coefficients $a_{ii} = 0$ if $q = 2$, since those are covered by the b-part of the coefficients.

For a recent overview of multivariate cryptosystems, including all the common modifiers such as "minus", "internal perturbation", and "vinegar" see [16]. It gives this formulation of HFEv, which uses the vinegar modification [23], built from this polynomial:

$$Q(X, \bar{X}) := \sum_{i,j} a_{ij} X^{q^i+q^j} + \sum_{i,j} b_{ij} X^{q^i} \bar{X}^{q^j} + \sum_{i,j} \alpha_{ij} \bar{X}^{q^i+q^j} + \sum_i b_i X^{q^i} + \sum_i \beta'_i \bar{X}^{q^i} + c$$

(1)

where the auxiliary variable $\bar{X}$ occupies only a subspace of small rank v in $\mathbb{K} \cong \mathbb{F}^n$. The function Q is quadratic in the components of X and $\bar{X}$, and so is $P = T \circ Q \circ S$ for affine bijections T and S in $\mathbb{F}^n$ and $\mathbb{F}^{n+v}$. We hope that P is hard to invert to the adversary, while the legitimate user, with the knowledge of

(S, T) can compute X by substituting a random $\bar{X}$, then solving for X via root-finding algorithms such as Berlekamp (or Cantor-Zassenhaus, if $q \neq 2$). To limit the effort of Berlekamp, we restrict the maximum degree D of the polynomial. QUARTZ has the parameter set $(q, n, D, v, a) = (2, 103, 129, 4, 3)$.

We note that to verify in QUARTZ, one invokes the public map multiple times, but the ability to defeat QUARTZ still principally rests on inverting an HFEv- public map.

In an HFEv- cryptosystem, the public key P becomes P^-, that is, it is released minus the last a equations. Again we hope that inverting P^- is intractible without the trapdoor. The legitimate user can invert P^- simply by appending a random numbers from $\mathbb{F}_q$ to to the ciphertext or signature before inverting P.

Another closely related scheme to HFEv is ipHFE (internally perturbed HFE). Suppose in Eq. 1, $\bar{X}$ is not a free variable, but is instead the image of ℓ, a map from $\mathbb{F}^n$ onto $\mathbb{F}^v$. So the central map is really $Q'(X) := Q(X, \ell(X))$. To invert Q', the legitimate user would *guess* the values at positions in $V = \ell(X)$, solve for X, and then check whether $V = \ell(X)$. So the inversion process becomes less efficient in the sense that it takes in the worst case q^v tries to get one answer. From this description, we can see that ipHFE is the same as HFEv with the prefix modification (i.e., one or more limbs of the plaintext in a multivariate scheme becomes pre-determined).

2.2   Conventional Wisdom about HFE Security

There is no "proof of security" for any variant of HFE or any of the usual multivariate PKC proposals that reduce to a difficult computational problem commonly used for cryptography. However, similarly the security of NTRU depends on the hardness of lattice problems, but does not reduce to them. There are lattice-based systems which reduce to hard lattice problems, but these are much less efficient than NTRU. Analogously, there are multivariate PKCs that are "provably secure" in the sense that a break of such a PKC would imply an advance in the solution to an MQ-related computational problem [22,32,33], which happen to be much less efficient. Hence we take the approach that only careful study of cryptanalytic techniques can determine the security of a cryptosystem.

It is unfortunate, then, that HFE Challenge 1 was proposed when we understood the algebra behind it much less. It is even more unfortunate that some of the proposed HFE variants were overly aggressive and were promptly broken [3, 4, 18] just like many other multivariate schemes, because the public perception became biased against the HFE family.

HFE variants also gained a further reputation for being flimsy, more specifically poly-time-solvable [17, 21] with further mathematical studies. In particular [21] sketched a way to bound the degree of regularity for HFE when $q = 2$, using an approach to lift the problem back to the extension $\mathbb{K}$, an idea first suggested by Kipnis-Shamir [24]. They managed to describe a connection of the degree of regularity of the HFE system to the degree of regularity of a lifted system over the big field. Heuristic asymptotic bounds were found when $q = 2$

leading to the conclusion that if $D = O(n)$ the complexity of Gröbner basis solvers for the corresponding HFE systems is quasi-polynomial.

In some ways, this reputation is actually somewhat unfair, since simple HFE variations such as QUARTZ have resisted known attacks for a long time, and it is actually known in various contexts how the degree of operations in an algebraic attack varies (cf. [14]). We hope to achieve a more realistic evaluation of the security of HFE-related schemes. In particular we hope that better understanding of the degree of regularity under algebraic attcks can establish some HFE variants as fundamentally sound cryptosystems which had previously been proposed with overly aggressive parameters, rather than fundamentally broken systems (like $C^{*}-$).

2.3 Algebraic Cryptanalysis

Aside from cases in which brute-force enumeration [5] seems to the best *practical* way to solve systems, almost all of today's algebraic algorithms to solve

$$p_1(x_1, \ldots, x_n) = p_2(x_1, \ldots, x_n) = \cdots = p_m(x_1, \ldots, x_n) = 0$$

over $\mathbb{F}_q$ go back to Buchberger's algorithm for computing Gröbner bases [6]. Lazard proposed the following critical simplification (later reinvented as the XL Method): multiply the equations with monomials to form a collection of relations up to a some degree d. Linearize (i.e., treat each individual monomial as a variable), and use well-studied matrix algorithms over $\mathbb{F}_q$ on the resulting matrix (the *extended Macaulay matrix*) [11, 25].

The Degree of Regularity. The critical concept in the complexity analysis of algebraic polynomial solving algorithms is the concept of *degree of regularity*. As given in Definition 1.1, the degree of regularity of the polynomial system is the lowest degree where we find a *non-trivial* degree drop. Conventional wisdom has it that in general this is the degree at which F_4/F_5 and similar algorithms usually terminate. Therefore D_{reg} is used to characterize the complexity of the algorithms.

We first note that almost all modern Gröbner Bases methods improve on XL as follows: suppose we fix a degree d and multiply each p_i with all monomials of degree $d - \deg p_i$ to create a large collection of relations of degree d. Order the monomials and linearize these equations to obtain the Macaulay matrix $\text{Mac}^{(d)}(p_1, \ldots, p_m)$. Try to eliminate the highest degree monomials from $\text{Mac}^{(d)}(p_1, \ldots, p_m)$ to create relations of degree $d - 1$ or lower.

After we find such polynomials with degree drop, we multiply them by individual variables, and we obtain equations of at most degree d, which are effectively elimination remnants from higher-degree relations. If necessary, we can repeat this process many times until we can solve for all the variables. This describes MutantXL or XL2 [13, 36] which will terminate at the same degree as F_4/F_5 [36]. Any superiority of the latter comes from having fewer redundant equations being generated or going through the elimination.

In Definition 1.1, we can see that the subspace T_d of trivial syzygies represents a "known-to-be-useless" degree drop in the following sense: Let $p_i = c^{(i)} + \sum_k b_k^{(i)} x_k + \sum_{k \le \ell} a_{k\ell}^{(i)} x_k x_\ell$. For a polynomial p, let $(p)^h$ represent the homogeneous highest degree part of the polynomial p, and (p) a corresponding row in a Macaulay-type matrix. Clearly $(p_j)^h (p_i)^h - (p_i)^h (p_j)^h = 0$ is a trivial syzygy, which is equivalent to the combination of degree-4 rows $\left(\sum_{k\ell} a_{k\ell}^{(i)} (x_k x_\ell p_j) \right) - \left(\sum_{k\ell} a_{k\ell}^{(j)} (x_k x_\ell p_i) \right)$ being of degree-3 (or fewer). Equally clearly this "degree-drop" will not give us anything useful since

$$\left(c^{(i)}(p_j) + \sum_k b_k^{(i)}(x_k p_j) + \sum_{k\ell} a_{k\ell}^{(i)}(x_k x_\ell p_j) \right)$$
$$= \left(c^{(j)}(p_i) + \sum_k b_k^{(j)}(x_k p_i) + \sum_{k\ell} a_{k\ell}^{(j)}(x_k x_\ell p_i) \right),$$

given that both give $(p_i p_j)$. Thus we just "found" a linear combination of polynomials we already have at degree 3. So a trivial or *principal* syzygy between the top-degree parts $(p_i)^h$ leads to a *trivial* degree drop useless for generating new equations. We must verify that a degree-drop is non-trivial before we can claim that we have reached the degree of regularity.

Issue of Terminology. There is some confusion about the term "the degree of regularity". The rank of Macaulay matrices at a given degree can be derived as the coefficients of certain generating functions, with the heuristic assumption that there are no non-trivial syzygies. A system where this holds for all degrees is called *regular*. However this can be the case only for underdetermined systems over characteristic zero fields. Otherwise at a sufficiently high degree the generating function eventually has a non-positive coefficient, and regularity becomes impossible. Systems for which the rank of the Macaulay matrices follows the heuristic for as long as possible are called "semi-regular" [12]. Definition 1.1 follows [17] in that the degree of regularity is defined as "the first appearance of non-trivial degree fall", i.e., where the system ceases to behave as semi-regular.

The heuristic formulas that have since long been known to hold for the degree of regularity of most random systems (including asymptotics) are given by Bardet et al [1,2,37]. However, this formula does not hold for most systems with structure.

Conventional wisdom also accepts that when $m/n = h + o(1)$ where h is a constant not far removed from 1, solving m "generic" or "random" equations in n variables is exponential in time and space in n. We can do a tiny bit better. That is, for sufficiently large h we may solve the system faster than just guessing variables first (cf. e.g. [8, 35]), but it is still exponential time and space in m (and/or n).

Invariance of Degree of Regularity. The degree of regularity is invariant under invertible linear transformation in both the domain and the codomain.

So if $P = T \circ Q \circ S$ is the public map of a multivariate PKC with the central map Q with both S and T invertible affine transformations, then the degree of regularity in solving X from $P(X) = Y$ depends only on Q, and can be written $D_{\mathrm{reg}}(Q)$.

3 Main Results

To recap, suppose we wish to solve an HFEv system with $\mathbb{K} \cong \mathbb{F}^n$, where $\mathbb{F} = \mathbb{F}_q$, with degree D and v vinegar variables. We would have then $n + v$ variables and n equations. However, MutantXL or F_4/F_5 algorithms deal with determined or overdetermined equations. The standard way to get around this problem is to guess some v variables and bring it down to a system with n variables. As noted earlier, we have now an ipHFE instance. We try to analyze the direct attack as in [14, 15, 17]. First, let us present our main results.

Theorem 3.1. *Let r be the rank of the HFE polynomial and v the number of vinegar variables. We may bound the degree of regularity of HFEv as follows:*

$$D_{\mathrm{reg}} \leq \frac{(q-1)(r+v-1)}{2} + 2, \quad \text{if } q \text{ is even and } r \text{ is odd}, \tag{2}$$

$$D_{\mathrm{reg}} \leq \frac{(q-1)(r+v)}{2} + 2, \quad \text{otherwise}. \tag{3}$$

This result is sufficient to bound the complexity of a direct algebraic attack against HFEv. If we assume that the direct algebraic attack is the best attack on HFEv systems, this would be the most important bound required to evaluate the security of odd-field HFEv and derivatives.

However, QUARTZ is an instance of HFEv-, not just HFEv. We recall that HFEv-, of which QUARTZ is a special case is derived from HFEv by removing a few public key polynomials. We normally have $n+v$ variables and $n-a$ equations. To solve a HFEv- case, we again first guess v-values. Then we have n variables and $n-a$ equations. As we mentioned, this is essentially an ipHFE system. Now we need to bound the degree of regularity of a direct algebraic attack on HFEv on such a system.

Theorem 3.2. *Let r be the rank of the HFE polynomial, v the number of vinegar variables, and a the number of "minus" equations, then we may bound the degree of regularity as follows:*

$$D_{\mathrm{reg}} \leq \frac{(q-1)(r+a+v-1)}{2} + 2, \quad \text{if } q \text{ is even and } r+a \text{ is odd},$$

$$D_{\mathrm{reg}} \leq \frac{(q-1)(r+a+v)}{2} + 2, \quad \text{otherwise}.$$

We will now show how our main results is proved.

To prove Equation (3) in Theorem 3.1, we must use a result that link the degree of regularity on a big-field multivariate to the rank of the central map.

Proposition 3.3. *[14, Theorem 4.1] For central maps Q that corresponds to quadratic maps, we have*

$$D_{reg}(Q) \leq \frac{(q-1)\operatorname{Rank}(Q)}{2} + 2.$$

We now need to show that the rank of an HFEv central polynomial with v vinegar variables is no higher than that of the original HFE polynomial plus v. First, we rewrite the HFEv polynomial so that it is more easily handled.

Proposition 3.4. *The associated polynomial when solving an HFEv or an ipHFE system over the big field $\mathbb{K}$ can be written as:*

$$\bar{P}(X) = \sum_{i=0}^{q^i < D} \left(\left(\sum_{j=0}^{q^i + q^j \leq D, j \leq i} a_{ij} X^{q^i + q^j} \right) + \left(\sum_{l=0}^{v-1} a'_{il} X^{q^i} \bar{X}_l \right) \right)$$

$$+ \sum_{i=0}^{v-1} \sum_{j=i}^{v-1} a''_{ij} \bar{X}_i \bar{X}_j + \sum_{i=0}^{q^i \leq D} b_i X^{q^i} + \sum_{i=0}^{v-1} u_i \bar{X}_i + c, \qquad (4)$$

where $\bar{X}_i := \operatorname{Tr}(\alpha_i X)$ for suitably chosen α_i. The map Tr is the trace function, which is also given by $\operatorname{Tr}(X) := \sum_{j=0}^{n-1} (X)^{q^j}$.,

Proof. We note that Tr is a nontrivial linear map of $\mathbb{F}^n \to \mathbb{F}$. For some representation of $\mathbb{F}^n \cong \mathbb{K}$, we can write it as a projection into the first component. With a suitably chosen α_i, we can make the first component of $\alpha_i X$ any nontrivial linear map of the components of X. So we can express each of the components of $\bar{X} = \ell(X)$ in Eq. 1 as $\operatorname{Tr}(\alpha_i X)$ for some α_i.

So Theorem 3.1 can be proved if we can show that:

Proposition 3.5. *The rank of the quadratic form associated with the polynomial $\bar{P}$ above, written $R(\bar{P})$, is bounded by:*

$$R(\bar{P}) \leq R(P) + v.$$

To obtain this we need this result about quadratic forms:

Proposition 3.6. *[26, Chapter 6] The rank of a quadratic form F is less than or equal to the minimum number of linear forms one needs to express F as a quadratic function in them. That is, if one can write F as a quadratic function of linear forms $\ell_1, \dots, \ell_r$, then $\operatorname{Rank} F \leq r$.*

Definition 3.7. *Let F be a quadratic form over a field k, and $F(X,Y) := X^t F Y$ be the bilinear (symmetric) form associated with F over the field k^n. Let*

$$N_F = \{X \in k^n | F(X,Y) = 0, \quad \text{for any } Y \in k^n\}.$$

N_F as linear subspace is called the radical for the bilinear form F.

Note that for any F of rank r, we can write F in the linear forms $\ell_1, \dots, \ell_r$, and any X with $\ell_1(X) = \dots = \ell_r(X) = 0$ is in N_F. So by using the following observation, we see that the dimension of N_F is $n - r$.

Proposition 3.8. *Let z_ℓ, $\ell = 0, \ldots, v-1$, be linear functions from $\mathbb{F}^n$ to $\mathbb{F}$, i.e., $z_\ell : (x_1, \ldots, x_n) \mapsto \sum \beta_i^{(\ell)} x_i$. Then the dimension of the intersection of kernels $K(z_i) := \{X \in \mathbb{F}^n \mid z_i(X) = 0\}$ is bounded by*

$$\dim \left(\bigcap_i^{v-1} K(z_i) \right) \geq n - v.$$

Proposition 3.9. *Under the conditions and notation of Definition 3.7 and Proposition 3.8,*

$$\dim(N_F \bigcap K(Z)) \geq n - r - v.$$

The last proposition follows from 3.7 and 3.8, basically by inclusion-exclusion.

Proposition 3.10. *Let $F(x_0, ..x_{n-1})$ be a quadratic form (or polynomial) whose rank is r. Here each variable x_i can additionally be considered as a linear map or function from $\mathbb{F}^n$ to $\mathbb{F}$. In this manner it would be viewed the i-th component map $x_i(u_0, \ldots, u_{n-1}) = u_i$, for $(u_0, \ldots, u_{n-1}) \in \mathbb{F}^n$. Let $A : \mathbb{F}^n \to \mathbb{F}^n$ be an invertible linear transformation (with A^{-1} its inverse), such that*

$$F(A(x_0), \ldots, A(x_{n-1}))) = \sum_{i=0}^{r} \sum_{j=i}^{r} a_{ij} x_i x_j,$$

where $A(x_i)$ is the function from $\mathbb{F}^n$ to $\mathbb{F}$ derived from $x_i \circ A$. Let

$$\bar{F}(x) = F(x_0, ..x_{n-1}) + \sum_{i=0}^{r} (\sum_{\ell=0}^{v-1} a'_{i\ell} A^{-1}(x_i) z_\ell),$$

where each z_ℓ is a linear function from $\mathbb{F}^n$ to $\mathbb{F}$, i.e., $z_\ell : (x_0, \ldots, x_{n-1}) \mapsto \sum \beta_i^{(\ell)} x_i$. Then

$$\mathrm{Rank}(\bar{F}) \leq \mathrm{Rank}(F) + v.$$

This follows from Proposition 3.9.

Now we further note that the process of fixing v variables to get a determined system corresponds to introducing v linear relations of the form

$$\sum_i a_i X^{q^i} + \sum_{j=1}^{v} b_j \overline{X}_j = 0.$$

From this we can substitute for each of the $\overline{X}_j$, a linear combination of the X^{q^i} (which is itself linear in X), which shows that the quadratic form $\overline{P}$ of HFEv or ipHFE can be expressed using v extra linear forms than the Dembowski-Ostrom polynomial map P (that is, without the v forms $\overline{X}_i$). Then since the rank of a quadratic form is bounded by the number of linear forms used to express it, we have Proposition 3.5, and Equation (3) then follows.

We note that the above line of reasoning is good only for odd q because in binary fields the rank of the associated matrix to a symmetric form is always even, creating various off-by-one errors in the above process, we may go through steps akin to that in [14] to patch those off-by-one problems (to be included in a full journal version), and account for the binary field cases in Theorem 3.1.

A note on HFE over tower fields. An HFE-derivative cryptosystem built over $\mathbb{F}_{q^k}$ is also one over $\mathbb{F}_q$. So we can (for example) attack an HFE-type instance built over $\mathbb{F}_{16}$ by solve it as a system over $\mathbb{F}_2$. However, in this situation the rank parameter r would usually be $\lfloor \log_{16}(D-1) \rfloor + 1$, not $\lfloor \log_2(D-1) \rfloor + 1$. The reason is that the central Dembowski-Ostrom polynomial, and therefore the rank r, is an entity in the big field and does not vary according to our viewpoint.

Proving Theorem 3.2 Again let us examine only odd characteristic cases for now. From the definition of HFEv-, it may be viewed as (HFE-)v. I.e., just as a central map of HFEv is one of HFE *plus a quadratic function with the extra variables in the form of a vector in an unknown subspace of dimension v (the "vinegar subspace"),* in exactly the fashion, HFEv- is HFEv plus a quadratic function with extra unknowns in that same vinegar subspace.

Put another way, let $\hat{P}^-$ be the public key of an HFEv- instance which is derived from the corresponding public key of an HFEv instance:

$$\hat{P}^-(x_1, \ldots, x_n) = (\hat{p}_1(x_1, \ldots, x_n), \ldots, \hat{p}_{n-a}(x_1, \ldots, x_n), 0, ..0).$$

We can then depict $\hat{P}^-$ as the vinegar form of an HFE- instance with central map Q^-. Q^- is a quadratic map, and can hence expressible as an extended Dembowski-Ostrom map. In other words, Q^- is also the central map of an HFE instance.

Now, according to [15, Section 4, Proposition 1] we have $\text{Rank}(Q) \le \text{Rank}(Q) + a$, where a is the number of "minus" equations. This holds because all the arguments there depend only on rank and not on exponents in the formulas.

Finally, we use Proposition 3.10 with $\hat{P}^-$ as the central map of an HFEv instance. We conclude that $\text{Rank}(P^-) < \text{Rank}(Q) + r + a$ which leads to the odd-q half of Theorem 3.2.

4   Testing, Implication and Discussion

Having given a bound for the degree of regularity for HFEv- (and ipHFE) systems, we give some experimental results and discuss what this means for QUARTZ.

4.1   Tests and Results

We ran MAGMA-2.7.12 on random systems for each parameter $n \le 13$, $r \le 4$, $a, v \le 2$, and $q \le 5$, on a workstation (with 2x Opteron 6212 and 32GB of RAM) to find D_{reg} on 4–20 randomly generated HFEv and HFEv- systems, and for $q = 2$ further for 1 random system each up to $n = 29$. We added $x_i^q - x_i$ for each i as part of the system of equations, so as to trigger field-specific optimizations that MAGMA might have for $q = 2$. In each case, D_{reg} proves to be the *smaller* of *either* the minimum of the bound in the formula above *or*, if we use $[u]S$ to mean the coefficient of the term u in a corresponding series expansion of S:

$$\min\left\{ d : [x^d]\left(\left(\frac{1-x^q}{1-x}\right)^n \left(\frac{1-x^2}{1-x^{2q}}\right)^m \right) < 0 \right\},$$

for m equations and n variables in $\mathbb{F}_q$. The cryptic expression above denotes the smallest d such that the coefficient of x^d in the Maclaurin expansion of $\left(\frac{1-x^q}{1-x}\right)^n \left(\frac{1-x^2}{1-x^{2q}}\right)^m$ becomes negative. It is actually the usual heuristic expression for D_{reg} for random systems, such as those found in [1] (for $q = 2$ only).

The numbers may seem too small to be conclusive, but for 13 variables and equations over $\mathbb{F}_7$ or 14 variables and equations over $\mathbb{F}_5$ MAGMA is already running out of memory, and these results lend credence to predictions using our bound for the Gröbner Bases complexity for HFEv and HFEv- systems. We can now try to justify the predictions for QUARTZ given in Section 1.

4.2 Implications for QUARTZ

We have obtained a bound on the degree of regularity of 9 for QUARTZ (which has $q = 2$, $n = 103$, $r = 7$, $a = 3$, $v = 4$), which represents a big drop already compared to degree 13 for a random system of that size (cf. formula above). However, if the bound is reasonably tight, the number of columns (monomials) involved in the elimination should be roughly the number of top-level monomials, which are $T := \binom{n}{D_{\text{reg}}} = \binom{100}{9} \gtrsim 2^{40}$ in total. A dense-matrix elimination would require 2^{80} bits of storage which is clearly not feasible.

Let us assume an extremely optimistic scenario for the attacker, such that a putative sparse-matrix-enabled F_4/F_5 attack is possible. Since each row has $\tau = \binom{100}{2} \geq 2^{12}$ terms, we will require about 2^{52} bits of memory. This is very large still, but not impossible in the mid-term future. We further use the number of bit-operations in the most time-consuming Wiedemann or Block Wiedemann type elimination methods as the estimate of the attack complexity, then we get the evaluation of the complexity given in Section 1:

$$C_{\mathrm{F}_4/\mathrm{F}_5} \geq 3\tau T^2 \gtrsim 3 \cdot 2^{12} \cdot (2^{40})^2 \geq 2^{92}.$$

Note: This evaluation above is highly optimistic in that it makes the implicit assumption that there is no penalty for accessing large memory. This may be very wrong in two ways:

– There is a very perceptible cost penalty in assembling a large amount of RAM which is either accessible on one machine or is networked using high speed interconnect to every other machine.
– Accessing a large amount of memory is slower; most server motherboards takes a speed penalty when using the maximum number of memory modules, and accessing memory on other machines of course incurs terrible latency.

What this might mean practically is that *it might be more advantageous to attack QUARTZ by brute-force [5]*, which imposes no communication requirements (i.e., networking and memory bandwidth and latencies) and is embarassingly parallelizable (hence perfectly scalable).

Final Remark: In some of the cases previously studied, we can *prove* tightness of the bounds. Clearly more of this type of work is needed, where theoretical bounds for attacks are given, just like the studies of theoretical bounds on differential probabilities in AES.

References

1. Bardet, M., Faugère, J.-C., Salvy, B.: On the complexity of Gröbner basis computation of semi-regular overdetermined algebraic equations. In: Proceedings of the International Conference on Polynomial System Solving, pp. 71–74 (2004); Previously INRIA report RR-5049
2. Bardet, M., Faugère, J.-C., Salvy, B., Yang, B.-Y.: Asymptotic expansion of the degree of regularity for semi-regular systems of equations. In: Gianni, P. (ed.) MEGA 2005 Sardinia, Italy (2005)
3. Bettale, L., Faugère, J.-C., Perret, L.: Cryptanalysis of multivariate and odd-characteristic HFE variants. In: Catalano, D., Fazio, N., Gennaro, R., Nicolosi, A. (eds.) PKC 2011. LNCS, vol. 6571, pp. 441–458. Springer, Heidelberg (2011)
4. Billet, O., Macario-Rat, G.: Cryptanalysis of the square cryptosystems. In: Matsui, M. (ed.) ASIACRYPT 2009. LNCS, vol. 5912, pp. 451–468. Springer, Heidelberg (2009)
5. Bouillaguet, C., Chen, H.-C., Cheng, C.-M., Chou, T., Niederhagen, R., Shamir, A., Yang, B.-Y.: Fast exhaustive search for polynomial systems in $\mathbb{F}_2$. In: Mangard, S., Standaert, F.-X. (eds.) CHES 2010. LNCS, vol. 6225, pp. 203–218. Springer, Heidelberg (2010)
6. Buchberger, B.: Ein Algorithmus zum Auffinden der Basiselemente des Restklassenringes nach einem nulldimensionalen Polynomideal. PhD thesis, Innsbruck (1965)
7. Clough, C., Baena, J., Ding, J., Yang, B.-Y., Chen, M.-S.: Square, a new multivariate encryption scheme. In: Fischlin, M. (ed.) CT-RSA 2009. LNCS, vol. 5473, pp. 252–264. Springer, Heidelberg (2009)
8. Courtois, N., Goubin, L., Meier, W., Tacier, J.-D.: Solving underdefined systems of multivariate quadratic equations. In: Naccache, D., Paillier, P. (eds.) PKC 2002. LNCS, vol. 2274, pp. 211–227. Springer, Heidelberg (2002)
9. Courtois, N.T.: The security of hidden field equations (HFE). In: Naccache, D. (ed.) CT-RSA 2001. LNCS, vol. 2020, pp. 266–281. Springer, Heidelberg (2001)
10. Courtois, N.T., Daum, M., Felke, P.: On the security of HFE, HFEv- and Quartz. In: Desmedt, Y.G. (ed.) PKC 2003. LNCS, vol. 2567, pp. 337–350. Springer, Heidelberg (2002)
11. Courtois, N.T., Klimov, A., Patarin, J., Shamir, A.: Efficient algorithms for solving overdefined systems of multivariate polynomial equations. In: Preneel, B. (ed.) EUROCRYPT 2000. LNCS, vol. 1807, pp. 392–407. Springer, Heidelberg (2000), http://www.minrank.org/xlfull.pdf
12. Diem, C.: The XL-algorithm and a conjecture from commutative algebra. In: Lee, P.J. (ed.) ASIACRYPT 2004. LNCS, vol. 3329, pp. 323–337. Springer, Heidelberg (2004)
13. Ding, J., Buchmann, J., Mohamed, M.S.E., Mohamed, W.S.A.E., Weinmann, R.-P.: Mutant XL. Talk at the First International Conference on Symbolic Computation and Cryptography (SCC 2008), Beijing (2008)
14. Ding, J., Hodges, T.J.: Inverting HFE systems is quasi-polynomial for all fields. In: Rogaway [31], pp. 724–742
15. Ding, J., Kleinjung, T.: Degree of regularity for HFE−. Cryptology ePrint Archive, Report 2011/570 (2011), http://eprint.iacr.org/

16. Ding, J., Yang, B.-Y.: Multivariate public key cryptography. In: Bernstein, D.J., Buchmann, J., Dahmen, E. (eds.) Post Quantum Cryptography, 1st edn., pp. 193–241. Springer, Berlin (2008) ISBN 3-540-88701-6

17. Dubois, V., Gama, N.: The degree of regularity of HFE systems. In: Abe, M. (ed.) ASIACRYPT 2010. LNCS, vol. 6477, pp. 557–576. Springer, Heidelberg (2010)

18. Dubois, V., Granboulan, L., Stern, J.: Cryptanalysis of HFE with internal perturbation. In: Okamoto, T., Wang, X. (eds.) PKC 2007. LNCS, vol. 4450, pp. 249–265. Springer, Heidelberg (2007)

19. Faugère, J.-C., Joux, A.: Algebraic Cryptanalysis of Hidden Field Equation (HFE) Cryptosystems Using Gröbner Bases. In: Boneh, D. (ed.) CRYPTO 2003. LNCS, vol. 2729, pp. 44–60. Springer, Heidelberg (2003)

20. Fischlin, M., Buchmann, J., Manulis, M. (eds.): PKC 2012. LNCS, vol. 7293. Springer, Heidelberg (2012)

21. Granboulan, L., Joux, A., Stern, J.: Inverting HFE is quasipolynomial. In: Dwork, C. (ed.) CRYPTO 2006. LNCS, vol. 4117, pp. 345–356. Springer, Heidelberg (2006)

22. Huang, Y.-J., Liu, F.-H., Yang, B.-Y.: Public-key cryptography from new multivariate quadratic assumptions. In: Fischlin et al. [20], pp. 190–205

23. Kipnis, A., Patarin, J., Goubin, L.: Unbalanced oil and vinegar signature schemes. In: Stern, J. (ed.) EUROCRYPT 1999. LNCS, vol. 1592, pp. 206–222. Springer, Heidelberg (1999)

24. Kipnis, A., Shamir, A.: Cryptanalysis of the HFE public key cryptosystem by relinearization. In: Wiener, M. (ed.) CRYPTO 1999. LNCS, vol. 1666, pp. 19–30. Springer, Heidelberg (1999), http://www.minrank.org/hfesubreg.ps

25. Lazard, D.: Gröbner-bases, Gaussian elimination and resolution of systems of algebraic equations. In: van Hulzen, J.A. (ed.) ISSAC 1983 and EUROCAL 1983. LNCS, vol. 162, pp. 146–156. Springer, Heidelberg (1983)

26. Lidl, R., Niederreiter, H.: Finite Fields, 2nd edn. Encyclopedia of Mathematics and its Application, vol. 20. Cambridge University Press (2003)

27. Patarin, J.: Cryptanalysis of the Matsumoto and Imai Public Key Scheme of Eurocrypt '88. In: Coppersmith, D. (ed.) CRYPTO 1995. LNCS, vol. 963, pp. 248–261. Springer, Heidelberg (1995)

28. Patarin, J.: Hidden Fields Equations (HFE) and Isomorphisms of Polynomials (IP): Two New Families of Asymmetric Algorithms. In: Maurer, U.M. (ed.) EUROCRYPT 1996. LNCS, vol. 1070, pp. 33–48. Springer, Heidelberg (1996), http://www.minrank.org/hfe.pdf

29. Patarin, J., Courtois, N.T., Goubin, L.: QUARTZ, 128-bit long digital signatures. In: Naccache, D. (ed.) CT-RSA 2001. LNCS, vol. 2020, pp. 282–288. Springer, Heidelberg (2001)

30. Proos, J., Zalka, C.: Shor's discrete logarithm quantum algorithm for elliptic curves. Quantum Information & Computation 3(4), 317–344 (2003)

31. Rogaway, P. (ed.): Advances in Cryptology – CRYPTO 2011. LNCS, vol. 6841. Springer, Heidelberg (2011)

32. Sakumoto, K.: Public-key identification schemes based on multivariate cubic polynomials. In: Fischlin et al. [20], pp. 172–189

33. Sakumoto, K., Shirai, T., Hiwatari, H.: Public-key identification schemes based on multivariate quadratic polynomials. In: Rogaway [31], pp. 706–723

34. Shor, P.W.: Polynomial-time algorithms for prime factorization and discrete logarithms on a quantum computer. SIAM Journal on Computing 26(5), 1484–1509 (1997)

35. Thomae, E., Wolf, C.: Solving underdetermined systems of multivariate quadratic equations revisited. In: Fischlin et al. [20], pp. 156–171.

36. Yang, B.-Y., Chen, J.-M.: All in the XL family: Theory and practice. In: Park, C.-s., Chee, S. (eds.) ICISC 2004. LNCS, vol. 3506, pp. 67–86. Springer, Heidelberg (2005)

37. Yang, B.-Y., Chen, J.-M.: Theoretical analysis of XL over small fields. In: Wang, H., Pieprzyk, J., Varadharajan, V. (eds.) ACISP 2004. LNCS, vol. 3108, pp. 277–288. Springer, Heidelberg (2004)

Software Speed Records
for Lattice-Based Signatures

Tim Güneysu[1], Tobias Oder[1], Thomas Pöppelmann[1], and Peter Schwabe[2,*]

[1] Horst Görtz Institute for IT-Security, Ruhr-University Bochum, Germany
[2] Digital Security Group, Radboud University Nijmegen, The Netherlands

Abstract. Novel public-key cryptosystems beyond RSA and ECC are urgently required to ensure long-term security in the era of quantum computing. The most critical issue on the construction of such cryptosystems is to achieve security *and* practicability at the same time. Recently, lattice-based constructions were proposed that combine both properties, such as the lattice-based digital signature scheme presented at CHES 2012. In this work, we present a first highly-optimized SIMD-based software implementation of that signature scheme targeting Intel's Sandy Bridge and Ivy Bridge microarchitectures. This software computes a signature in only 634988 cycles on average on an Intel Core i5-3210M (Ivy Bridge) processor. Signature verification takes only 45036 cycles. This performance is achieved with full protection against timing attacks.

Keywords: Post-quantum cryptography, lattice-based cryptography, cryptographic signatures, software implementation, AVX, SIMD.

1 Introduction

Besides breakthroughs in classical cryptanalysis the potential advent of quantum computers is a serious threat to the established discrete-logarithm problem (DLP) and factoring-based public-key encryption and signature schemes, such as RSA, DSA and elliptic-curve cryptography. Especially when long-term security is required, all DLP or factoring-based schemes are somewhat risky to use. The natural consequence is the need for more diversification and investigation of potential alternative cryptographic systems that resist attacks by quantum computers. Unfortunately, it is challenging to design secure *post-quantum* signature schemes that are efficient in terms of speed and key sizes. Those which are known to be very efficient, such as the lattice-based NTRU-sign [15] have been shown to be easily broken [19]. Multivariate quadratic (MQ) signatures, e.g., Unbalanced Oil and Vinegar (UOV), are fast and compact, but their public keys are huge with around 80 kB and thus less suitable on embedded systems – even with optimizations the keys are still too large (around 8 Kb) [20].

* This work was supported by the National Institute of Standards and Technology under Grant 60NANB10D004. Permanent ID of this document: ead67aa537a6de60813845a45505c313. Date: March 28, 2013

P. Gaborit (Ed.): PQCrypto 2013, LNCS 7932, pp. 67–82, 2013.
© Springer-Verlag Berlin Heidelberg 2013

The introduction of special ring-based (ideal) lattices and their theoretical analysis (see, e.g., [18]) provides a new class of signature and encryption schemes with a good balance between key size, signature size, and speed. The speed advantage of ideal lattices over standard lattice constructions usually stems from the applicability of the Number Theoretic Transform (NTT), which allows operations in quasi-linear runtime of $\mathcal{O}(n \log n)$ instead of quadratic complexity. In particular, two implementations of promising lattice-based constructions for encryption [12] and digital signatures [14] were recently presented and demonstrate that such constructions can be efficient in reconfigurable hardware. However, as the proof-of-concept implementation in [12] is based on the generic NTL library [22], it remains still somewhat unclear how these promising schemes perform on high-performance processors that include modern SIMD multimedia extensions such as SSE and AVX.

Contribution. The main contribution of this work is the first optimized software implementation of the lattice-based signature scheme proposed in [14]. It is an aggressively optimized variant of the scheme originally proposed by Lyubashevsky [17] without Gaussian sampling. We use security parameters $p = 8383489, n = 512, k = 2^{14}$ that are assumed to provide an equivalent of about 80 bits of security against attacks by quantum computers and 100 bits of security against classical computers. With these parameters, public keys need only 1536 bytes, private keys need 256 bytes and signatures need 1184 bytes. On one core of an Intel Core i5-3210M processor (Ivy Bridge microarchitecture) running at 2.5 GHz, our software can compute more than 3900 signatures per second or verify more than 55000 signatures per second. To maximize reusability of our results we put the software into the public domain[1]. We will additionally submit our software to the eBACS benchmarking project [4] for public benchmarking.

Outline. In Section 2 we first provide background information on the implemented signature scheme. Our implementation and optimization techniques are described in Section 3 and evaluated and compared to previous work in Section 4. We conclude with future work in Section 5.

2   Signature Scheme Background

In this section we briefly revisit the lattice-based signature scheme implemented in this work. For more detailed information as well as security proofs, please refer to [14,17].

2.1   Notation

In this section we briefly recall the notation from [14]. We use a similar notation and denote by $\mathcal{R}^{p^n}$ the polynomial ring $\mathbb{Z}[x]_p \langle x^n + 1 \rangle$ with integer coefficients in the range $[-\frac{p-1}{2}, \frac{p-1}{2}]$ where n is a power of two. The prime p must satisfy the

---

[1] The software is available at http://cryptojedi.org/crypto/#lattisigns

congruence relation $p \equiv 1 \pmod{2n}$ to allow us to use the quasi-linear-runtime NTT-based multiplication. For any positive integer k, we denote by $\mathcal{R}_k^{p^n}$ the set of polynomials in $\mathcal{R}^{p^n}$ with coefficients in the range $[-k, k]$. The expression $a \xleftarrow{\$} D$ denotes the uniformly random sampling of a polynomial a from the set D.

Algorithm 1. KEY GENERATION ALGORITHM $\mathrm{GEN}(p, n)$

    Input: Parameters p, n
    Output: $(t)_{pk}$, $(s_1, s_2)_{sk}$
 1 $s_1, s_2 \xleftarrow{\$} \mathcal{R}_1^{p^n}$;
 2 $t \leftarrow a s_1 + s_2$;

2.2 Definition

According to the description in [14] we have chosen a to be a randomly generated global constant. For the key generation described in Algorithm 1 we therefore basically perform sampling of random values from the domains $\mathcal{R}_1^{p^n}$ followed by a polynomial multiplication with the global constant and an addition. The private key sk consists of the values s_1, s_2 while t is the public key pk. Algorithm 2 signs a message m specified by the user. In step 1 two polynomials y_1, y_2 are chosen uniformly at random with coefficients in the range $[-k, k]$. In step 2 a hash function is applied on the higher-order bits of $ay_1 + y_2$ which outputs a polynomial c by interpreting the first 160-bit of the hash output as a sparse polynomial. In step 3 and 4, y_1 and y_2 are used to mask the private key by computing z_1 and z_2. The algorithm only continues if z_1 and z_2 are in the range $[-(k-32), k-32]$ and restarts otherwise. The polynomial z_2 is then compressed into z_2' in step 7 by Compress. This compression is part of the aggressive size reduction of the signature $\sigma = (z_1, z_2', c)$ since only some portions of z_2 are necessary to maintain the security of the scheme. For the implemented parameter set Compress has a chance of failure of less than two percent which results in the restart of the whole signing process.

    The verification algorithm VER as described in Algorithm 3 first ensures that all coefficients of z_1, z_2' are in the range $[-(k-32), k-32]$ and rejects the input otherwise by returning $b = 0$ to indicate an invalid signature. In the next step, $az_1 + z_2' - tc$ is computed, transformed into the higher-order bits and then hashed. If the polynomial c from the signature and the output of the hash match, the signature is valid and the algorithm outputs $b = 1$ to indicate its success.

    In Algorithm 4 the transformation of a polynomial into a higher-order representation is described. This algorithm exploits the fact that every polynomial $Y \in \mathcal{R}^{p^n}$ can be written as

$$Y = Y^{(1)}(2(k - 32) + 1) + Y^{(0)}$$

where $Y^{(0)} \in \mathcal{R}_{k-32}^{p^n}$ and thus every coefficient of $Y^{(0)}$ is in the range $[-(k - 32), k - 32]$. Due to this bijectional relationship, every polynomial Y can be also written as the tuple $(Y^{(1)}, Y^{(0)})$.

Algorithm 2. SIGNING ALGORITHM $\text{SIGN}(s_1, s_2, m)$

Input: $s_1, s_2 \in \mathcal{R}_1^{p^n}$, message $m \in \{0,1\}^*$
Output: $z_1, z_2' \in \mathcal{R}_{k-32}^{p^n}$, $c \in \{0,1\}^{160}$

1 $y_1, y_2 \xleftarrow{\$} \mathcal{R}_k^{p^n}$;
2 $c \leftarrow \text{H}(\text{Transform}(ay_1 + y_2), m)$;
3 $z_1 \leftarrow s_1 c + y_1$;
4 $z_2 \leftarrow s_2 c + y_2$;
5 **if** z_1 *or* $z_2 \notin \mathcal{R}_{k-32}^{p^n}$ **then**
6 $\quad$ go to step 1;

7 $z_2' \leftarrow \text{Compress}(ay_1 + y_2 - z_2, z_2, p, k-32)$;
8 **if** $z_2' = \perp$ **then**
9 $\quad$ go to step 1;

Algorithm 3. VERIFICATION ALGORITHM $\text{VER}(z_1, z_2', c, t, m)$

Input: $z_1, z_2' \in \mathcal{R}_{k-32}^{p^n}$, $t \in \mathcal{R}^{p^n}$, $c \in \{0,1\}^{160}$, message $m \in \{0,1\}^*$
Output: b

1 **if** z_1 *or* $z_2' \notin \mathcal{R}_{k-32}^{p^n}$ **then**
2 $\quad$ $b \leftarrow 0$;

3 **else**
4 $\quad$ **if** $c = H(\text{Transform}(az_1 + z_2' - tc), m)$ **then**
5 $\quad\quad$ $b \leftarrow 1$;

6 $\quad$ **else**
7 $\quad\quad$ $b \leftarrow 0$;

Algorithm 5 describes the compression algorithm Compress which takes a polynomial y, a polynomial z with small coefficients and the security parameter k as well as p as input. It is designed to return a polynomial z' that is compacted but still maintains the equality between the higher-order bits of $y + z$ and $y + z'$ so that $(y + z)^{(1)} = (y + z')^{(1)}$. In particular, the parameters of the scheme are chosen in a way that the if-condition specified in step 3 is true only for rare cases. This is important since only values assigned to $z'[i]$ in step 6 to step 13 can be efficiently encoded.

The hash function H maps an arbitrary-length input $\{1, 0\}^*$ to a 512-coefficient polynomial with 32 coefficients in $\{-1, 1\}$ and all other coefficients zero. The whole process of generating this string and its transformation into a polynomial with the above described character is shown in Algorithm 6. In step 1 the message is concatenated with a binary representation of the polynomial x generated by the algorithm BinRep. It takes a polynomial $x \in \mathcal{R}^{p^n}$ as input and outputs a (somehow standardized) binary representation of this polynomial. The 160-bit hash value is processed by partitioning it into 32 blocks of 5 side-by-side bits (beginning with the lowest ones) that each correspond to a particular region in the

---

Algorithm   4.   Higher-Order   Transformation   Algorithm
Transform(y, k)

Input: $y \in \mathcal{R}^{p^n}$, k
Output: $y^{(1)}$
1 **for** $i=0$ **to** $n-1$ **do**
2 $\quad\big\lfloor\; y^{(0)}[i] \leftarrow y[i] \mod (2(k - 32) + 1);$
3 $\quad\;\; y^{(1)}[i] \leftarrow \frac{y[i] - y^{(0)}[i]}{2(k-32)+1};$
4 **return** $y^{(1)}$;

---

polynomial c. These bits are $r_4 r_3 r_2 r_1 r_0$ where $(r_3 r_2 r_1 r_0)_2$ represents the position in the region interpreted as a 4-bit unsigned integer and the bit r_4 determines if the value of the coefficient is -1 or 1.

2.3   Parameters and Security

Parameters that offer a reasonable security margin of approximately 100 bits of comparable classical symmetric security are $n = 512$, $p = 8383489$, and $k = 2^{14}$ This parameter set is the primary target of this work. For some intuition on how these parameters were selected, how the security level has been computed, for a second parameter set and a security proof in the random-oracle model we refer again to [14].

In general, the security of the signature scheme is based on the Decisional Compact Knapsack (DCK$_{p,n}$) problem and the hardness of finding a preimage in the hash function. For solving the DCK problem one has to distinguish between uniform samples from $\mathcal{R}^{p^n} \times \mathcal{R}^{p^n}$ and samples from the distribution $(a, as_1 + s_2)$ with a being chosen uniformly at random from $\mathcal{R}^{p^n}$ and s_1, s_2 being chosen uniformly at random from $\mathcal{R}_1^{p^n}$. In comparison to the Ring-LWE problem [18], where s_1, s_2 are chosen from a Gaussian distribution of a certain range, this just leads to s_1, s_2 with coefficients being either ± 1 or zero. Therefore, the DCK problem is an "aggressive" variant of the LWE problem but is not affected by the Arora-Ge algorithm as only one sample is given for the DCK problem and not the required polynomially-many [1]. Note also that extraction of the private key from the public key requires to solve the search variant of the DCK problem. In [14] the hardness of breaking the signature scheme for the implemented parameter set is computed based on the root Hermite factor of 1.0066 and stated to provide roughly 100 bits of security. Finding a preimage in the hash function has classical time complexity of 2^l but is lowered to $2^{l/2}$ by Grover's quantum algorithm [13]. As we use an output bit length of $l = 160$ from the hash function the implemented scheme achieves a security level of roughly 80 bits of security against attacks by a quantum computer.

Algorithm 5. COMPRESSION ALGORITHM COMPRESS(y, z, p, k)

Input: $y \in \mathcal{R}_k^{p^n}$, $z \in \mathcal{R}_{k-32}^{p^n}$, p, k

Output: $z' \in \mathcal{R}_k^{p^n}$

1 $uncompressed \leftarrow 0$;
2 **for** $i=0$ **to** $n-1$ **do**
3   **if** $|y[i]| > \frac{p-1}{2} - k$ **then**
4        $z'[i] \leftarrow z[i]$;
5        $uncompressed \leftarrow uncompressed + 1$;
6   **else**
7   write $y[i] = y[i]^{(1)}(2k+1) + y[i]^{(0)}$ where $-k \leq y[i]^{(0)} \leq k$
8   **if** $y[i]^{(0)} + z[i] > k$ **then**
9           $z[i]' \leftarrow k$;
10   **else if** $y[i]^{(0)} + z[i] < -k$ **then**
11          $z[i]' \leftarrow -k$;
12   **else**
13          $z[i]' \leftarrow 0$;

14 **if** $uncompressed \leq \frac{6kn}{p}$ **then**
15   **return** z' ;
16 **else**
17   **return** $\perp$;

---

3 Software Optimization

In this section we show our approach to high-level optimization of algorithms and low-level optimization to make best use of the target micro-architecture.

3.1 High-Level Optimization

In the following we present high-level ideas to speed-up the polynomial multiplication, runtime behavior as well as randomness generation.

Polynomial Multiplication. In order to achieve quasi-linear speed in $\mathcal{O}(n \log n)$ when performing the essential polynomial-multiplication operation we use the Fast Fourier Transform (FFT) or more specifically the Number Theoretic Transform (NTT) [21]. The advantages offered by the NTT have recently been shown by a hard- and software implementation of an ideal lattice-based public key cryptosystem [12]. The NTT is defined in a finite field or ring for a given primitive n-th root of unity ω. The generic forward $\mathrm{NTT}_\omega(a)$ of a sequence $\{a_0, .., a_{n-1}\}$ to $\{A_0, \ldots, A_{n-1}\}$ with elements in $\mathbb{Z}_p$ and length n is defined as $A_i = \sum_{j=0}^{n-1} a_j \omega^{ij} \bmod p$, $i = 0, 1, \ldots, n-1$ with the inverse $\mathrm{NTT}_\omega^{-1}(A)$ just using ω^{-1} instead of ω.

Algorithm 6. HASH FUNCTION INVOCATION $H(x, m)$

Input: Polynomial $x \in \mathcal{R}^{p^n}$, message $m \in \{0,1\}^*$, hash function
$\quad\quad \tilde{H}(\{0,1\}^*) \to \{0,1\}^{160}$
Output: $c \in \mathcal{R}_1^{p^n}$ with at most 32 coefficients being -1 or 1
1 $r \leftarrow \tilde{H}(m\|\mathsf{BinRep}(x))$;
2 **for** $i=0$ **to** $n-1$ **do**
3 $\quad\quad c[i] = 0$;
4 **for** $i=0$ **to** 31 **do**
5 $\quad\quad pos \leftarrow 8 \cdot r_{5i+3} + 4 \cdot r_{5i+2} + 2 \cdot r_{5i+1} + r_{5i}$;
6 $\quad\quad$ **if** $r_{5i+4} = 0$ **then**
7 $\quad\quad\quad\quad c[i \cdot 16 + pos] \leftarrow -1$;
8 $\quad\quad$ **else**
9 $\quad\quad\quad\quad c[i \cdot 16 + pos] \leftarrow 1$;

For lattice-based cryptography it is also convenient that most schemes are defined in $\mathbb{Z}_p[\mathbf{x}]/\langle x^n + 1\rangle$ and require reduction modulo $x^n + 1$. As a consequence, let ω be a primitive n-th root of unity in $\mathbb{Z}_p$ and $\psi^2 = \omega$. Then when $a = (a_0, \ldots a_{n-1})$ and $b = (b_0, \ldots b_{n-1})$ are vectors of length n with elements in $\mathbb{Z}_p$ let $d = (d_0, \ldots d_{n-1})$ be the negative wrapped convolution of a and b (thus $d = a * b \bmod x^n + 1$). Let $\bar{a}, \bar{b}$ and $\bar{d}$ be defined as $(a_0, \psi a_1, \ldots, \psi^{n-1}a_{n-1})$, $(b_0, \psi b_1, \ldots, \psi^{n-1}b_{n-1})$, and $(d_0, \psi d_1, \ldots, \psi^{n-1}d_{n-1})$. It then holds that $\bar{d} = NTT_w^{-1}(NTT_w(\bar{a}) \circ NTT_w(\bar{b}))$ [24], where $\circ$ means componentwise multiplication. This avoids the doubling of the input length of the NTT and also gives us a modular reduction by $x^n + 1$ for free. If parameters are chosen such that n is a power of two and that $p \equiv 1 \bmod 2n$, the NTT exists and the negative wrapped convolution can be implemented efficiently.

In order to achieve high NTT performance, we precompute all constants $\omega^i, \omega^{-i}, \psi^i$ as well as $n^{-1} \cdot \psi^i$ for $i \in 0 \ldots n-1$. The multiplication by n^{-1}, which is necessary in the NTT^{-1} step, is directly performed as we just multiply by $n^{-1} \cdot \psi^{-i}$.

Storing Parameters in NTT Representation. The polynomial a is used as input to the key-generation algorithm and can be chosen as a global constant. By setting $\tilde{a} = \mathrm{NTT}(a)$ and storing $\tilde{a}$ we just need to perform $\mathrm{NTT}^{-1}(\tilde{a} \circ \mathrm{NTT}(y_1))$, which consists of one forward transform, one point multiplication and one backward transform. This is implemented in the `poly_mul_a` function and is superior to the general-purpose NTT multiplication, which requires three transforms.

Random Polynomials. During signature generation we need to generate two polynomials with random coefficients uniformly distributed in $[-k, k]$. To obtain these polynomials, we first generate $4 \cdot (n + 16) = 2112$ random bytes using the Salsa20 stream cipher [2] and a seed from the Linux kernel random-number generator /dev/urandom. We interpret these bytes as an array of $n+16$ unsigned 32-bit integers. To convert one such a 32-bit integer r to a polynomial coefficient c in $[-k, k]$ we first check whether $r \geq (2k+1) \cdot \lfloor 2^{32}/(2k+1) \rfloor$. If it is, we discard

this integer and move to the next integer in the array. Otherwise we compute
$c = (r \mod (2k + 1)) - k$.

The probability that an integer is discarded is $(2^{32} \mod (2k + 1))/2^{32}$. For
our parameters we have $(2^{32} \mod (2k + 1)) = 4$. The probability to discard a
randomly chosen 32-bit integer is thus $4/2^{32} = 2^{-30}$. The 16 additional elements
in our array (corresponding to one block of Salsa20) make it extremely unlikely
that we do not sample enough random elements to set all coefficients of the
polynomial. In this highly unlikely case we simply sample another 2112 bytes of
randomness.

During key generation we use the same approach to generate polynomials
with coefficients in $\{-1, 0, 1\}$. The difference is that we sample bytes instead of
32-bit integers. We again sample one additional block of Salsa20 output, now
corresponding to 64 additional elements. A byte is discarded only if its value is
255, the chance to discard a random byte is thus 2^{-8}.

3.2   Low-Level Optimization

The performance of the signature scheme is largely determined by a small set
of operations on polynomials with $n = 512$ coefficients over $\mathbb{Z}_p$ where p is a 23-
bit prime. This section first describes how we represent polynomials and what
implementation techniques we use to accelerate operations on these polynomials.

Representation of Polynomials. We represent each 512-coefficient polynomial as an array of 512 double-precision floating-point values. Each such array
is aligned on a 32-byte boundary, meaning that the address in memory is divisible by 32. This representation has the advantage that we can use the single-
instruction multiple-data (SIMD) instructions of the AVX instruction-set extension in modern Intel and AMD CPUs. These instructions operate on vectors of
4 double-precision floats in 256-bit-wide, so called ymm vector registers. These
registers and the corresponding AVX instructions can be found, for example, in
the Intel Sandy Bridge, Intel Ivy Bridge, and AMD Bulldozer processors. The
following performance analysis focuses on Ivy Bridge processors; Section 4 also
reports benchmarks from a Sandy Bridge processor.

Both Sandy Bridge and Ivy Bridge processors can perform one AVX double-
precision-vector multiplication and one addition every cycle. This corresponds
to 4 multiplications (vmulpd instruction) and 4 additions (vaddpd instruction)
of polynomial coefficients each cycle. However, arithmetic cost is not the main
bottleneck in our software as loads and stores are often necessary because only
64 polynomial coefficients fit into the 16 available ymm registers. The performance
of loads and stores is more complex to determine than arithmetic throughput.
In principle, the processor can perform two loads and one store every two cycles.
However, this maximal throughput can be reduced by bank conflicts. For details
see [10, Section 8.13].

Modular Reduction of Coefficients. To perform a modular reduction of a
coefficient x, we first compute $c = x \cdot \overline{p^{-1}}$, then round c, then multiply c by p
and then subtract c from x. The first step uses a precomputed double-precision

approximation $\overline{p^{-1}}$ of the inverse of p. When reducing all coefficients of a polynomial, the multiplications and the subtraction are performed on four coefficients in parallel with the vmulpd and vsubpd AVX instructions, respectively. The rounding is also done on four coefficients in parallel using the vroundpd instruction. Note that depending on the rounding mode we can obtain the reduced value of x in different intervals. If we perform a truncation we obtain x in $[0, p-1]$, if we round to the nearest integer we obtain x in $[-((p-1)/2), (p-1)/2]$. We only need rounding to the nearest integer (vroundpd with rounding-mode constant 0x08). Both representations are required at different stages of the computation; vroundpd supports choosing the rounding mode.

Lazy Reduction. The prime p has 23 bits. A double-precision floating-point value has a 53-bit mantissa and one sign bit. Even the product of two coefficients does not use the whole available precision, so we do not have to perform modular reduction after each addition, subtraction or even multiplication. We can thus make use of the technique known as *lazy reduction*, i.e., of performing reduction modulo p only when necessary.

Optimizing the NTT. The most speed-critical operation for signing is polynomial multiplication and we can thus use the NTT transformation as described above. We start from a standard fast iterative algorithm (see, e.g., [9]) for computing the FFT/NTT and adapt it to the target architecture. The transformation of a polynomial f with coefficients $f_0, \ldots, f_{511}$ to or from NTT representation consist of an initial permutation of the coefficients followed by $\log_2 n = 9$ levels of operations on coefficients. On level 0, pick up f_0 and f_1, multiply f_1 with a constant (a power of ω), add the result to f_0 to obtain the new value of f_0 and subtract the result from f_0 to obtain the new value of f_1. Then pick up f_2 and f_3 and perform the same operations to find the new values for f_2 and f_3 and so on. The following levels work in a similar way except that the distance of pairs of elements that are processed together is different: on level i process elements that are 2^i positions apart. For example, on level 2 pick up and transform f_0 and f_4, then f_1 and f_5 etc. On level 0 we can omit the multiplication by a constant, because the constant is 1.

The obvious bottleneck in this computation are additions (and subtractions): Each level performs 256 additions and 256 subtractions accounting for a total of $9 \cdot 512 = 4608$ additions requiring at least 1152 cycles. In fact the lower bound of cycles is much higher, because after each multiplication by a constant we need to reduce the coefficients modulo p. This takes one vroundpd instruction and one subtraction. The vroundpd instruction is processed in the same port as additions and subtractions, we thus get a lower bound of $(9 \cdot 512 + 8 \cdot 512)/4 = 2176$ cycles. To get close to this lower bound, we need to make sure that all the additions can be efficiently processed in AVX instructions by minimizing overhead from memory access, multiplications or vector-shuffle instructions.

Starting from level 2, the structure of the algorithm is very friendly for 4-way vector processing: For example, we can load (f_0, f_1, f_2, f_3) into one vector register, load (f_4, f_5, f_6, f_7) in another vector register, load the required constants (c_0, c_1, c_2, c_3) into a third vector register and then use one vector multiplication, one vector addition and one vector subtraction to obtain $(f_0+c_0f_4, f_1+c_1f_5, f_2+c_2f_6, f_3+c_3f_7)$ and $(f_0-c_0f_4, f_1-c_1f_5, f_2-c_2f_6, f_3-c_3f_7)$. However, on levels 0 and 1 the transformations are not that straightforwardly done in vector registers. On level 0 we do the following: Load f_0, f_1, f_2, f_3 into one register; perform vector multiplication of this register with $(1, -1, 1, -1)$ and store the result in another register; perform a vhaddpd instruction of these two registers which results exactly in $(f_0 + v_1, f_0 - f_1, f_2 + f_3, f_2 - f_3)$. On level 1 we do the following: Load f_0, f_1, f_2, f_3; multiply with a vector of constants, reduce the result modulo p; use the vperm2f128 instruction with constant argument 0x01 to obtain $c_2f_2, c_3f_3, c_0f_0, c_1f_1$ in another register and perform vector register multiplication of this register by $(1, 1, -1, -1)$; add the result to (f_0, f_1, f_2, f_3) to obtain the desired $(f_0 + c_2f_2, f_1 + c_1f_1, f_0 - c_2f_2, f_1 - c_3f_3)$.

A remaining bottleneck is memory access. To minimize loads and stores, we merge levels 0,1,2, levels 3,4,5 and levels 6,7,8. The idea is that on one level two pairs of coefficients are interacting; through two levels it is 4-tuples of coefficients that interact and through 3 levels it is 8-tuples of coefficients that interact. On levels 0,1 and 2 we load these 8 coefficients; perform all transformations through the 3 levels and store them again, then proceed to the next 8 coefficients. On higher levels we load 32 coefficients, perform all transformations through 3 levels on them, store them and then proceed to the next 32 coefficients.

In total, one NTT transformation takes 4484 cycles on the Ivy Bridge processor. This includes about 500 cycles for the initial coefficient permutation. We are continuing to investigate the difference between the lower bound on cycles dictated by vector additions and the cycles actually taken by our software.

Addition and Subtraction. Addition and subtraction of polynomials simply means loading coefficients, performing double-precision floating-point addition or subtraction, and storing the result coefficient. This is completely parallel, so we do this in 256 vector loads, 128 vector additions or subtractions, and 128 vector stores.

Higher-Order Transformation. The higher-order transformation described in Algorithm 4 is a nice example of the power of representing polynomial coefficients as double-precision floats: The only operation required is the multiplication by the precomputed value $\overline{(2(k-32)+1)^{-1}}$ (a double-precision approximation of $(2(k-32)+1)^{-1}$) and a subsequent rounding towards the nearest integer. As for the coefficient reduction we perform these computations using the vmulpd and vroundpd instructions.

4   Performance Analysis and Benchmarks

In this section we analyze the performance of our software and report benchmarks for key generation (crypto_keypair), as well as the signing (crypto_sign) and verification (crypt_sign_open) algorithm. Our software implements the eBATS API [4] for signature software, but we did *not* use SUPERCOP for benchmarking. The reason is that SUPERCOP reports the *median* of multiple runs to filter out benchmarks that are polluted by, for example, an interrupt that occurred during some of the computations. Considering the median of timings when signing would be overly optimistic and cut off legitimate benchmarks of signature generations that took very long because they required many attempts. Therefore, for signing we report the *average* of 100000 signature generations; for key-pair generation, verification and lower-level functions we report the median of 1000 benchmarks. However, we will submit our software to eBACS for public benchmarking and discuss the issue with the editors of eBACS. Note that our software for signing is obviously not running in constant time but the timing variation is independent of secret data; our software is fully protected against timing attacks.

We performed benchmarks on two different machines:

- a machine called h9ivy at the University of Illinois at Chicago with an Intel Core i5-3210M CPU (Ivy Bridge) at 2500 MHz and 4 GB of RAM; and
- a machine called h6sandy at the University of Illinois at Chicago with an Intel Core i3-2310M CPU (Sandy Bridge) at 2100 MHz and 4 GB of RAM.

All software was compiled with gcc-4.7.2 and compiler flags -O3 -msse2avx -march=corei7-avx -fomit-frame-pointer. During the benchmarks Turbo-Boost and hyperthreading were switched off. The performance results for the most important operations are given in Table 1. The message length was 59 bytes for the benchmarking of crypto_sign and crypto_sign_open.

Table 1. Cycle counts of our software; $n = 512$ and $p = 8383489$

Operation	Sandy Bridge cycles	Ivy Bridge cycles
crypto_sign_keypair	33894	31140
crypto_sign	681500	634988
crypto_sign_open	47636	45036
ntt	4480	4484
poly_mul	16052	16096
poly_mul_a	11100	11044
poly_setrandom_maxk	12788	10824
poly_setrandom_max1	6072	5464

Polynomial-Multiplication Performance. The multiplication of two polynomials (poly_mul) takes 16096 cycles on the Ivy Bridge. Out of those, $3 \cdot 4484 = 13452$ cycles are for 3 NTT transformations (ntt).

Key-Generation Performance. Generating a key pair takes 31140 cycles on the Ivy Bridge. Out of those, $2 \cdot 5464 = 10928$ cycles are required to generate two random polynomials (poly_setrandom_max1); 11044 cycles are required for a multiplication by the constant system parameter a (poly_mul_a); the remaining

9168 cycles are required for one polynomial addition, compression of the two private-key polynomials and packing of the public-key polynomial into a byte array.

Signing Performance. Signing takes 634988 cycles on average on the Ivy Bridge. Each signing *attempt* takes 85384 cycles. We need 7 attempts on average, so those attempts account for about $7 \cdot 85384 = 597688$ cycles; the remaining cycles are required for constant overhead for extracting the private key from the byte array, copying the message to the signed message etc. *Some* of the remaining cycles may also be due to some measurements being polluted as explained above.

Out of the 85384 cycles for each signing attempt, $2 \cdot 10824 = 21648$ cycles are required to generate two random polynomials (`poly_setrandom_maxk`); $2 \cdot 16096 = 32192$ cycles are required for two polynomial multiplications; 11084 cycles are required for a multiplication with the system parameter a; the remaining 20460 cycles are required for hashing, the higher order transformation, four polynomial additions, one polynomial subtraction and testing whether the polynomial can be compressed.

Verification Performance. Verifying a signature takes 45036 cycles on the Ivy Bridge. Out of those, 16096 cycles are required for a polynomial multiplication; 11084 cycles are required for a multiplication with a; the remaining 17856 cycles are required for hashing, the high-order transformation, a polynomial addition and a polynomial subtraction, decompression of the signature, and unpacking of the public key from a byte array.

Comparison. As we provide the first software implementation of the signature scheme we cannot compare our result to other software implementations. In [14] only a hardware implementation is given which is naturally hard to compare to. For different types of FPGAs and parallelism, an implementation of sign/verify of 931/998 (Spartan-6 LX16) up to 12627/14580 (Virtex-6 LX130) messages/signatures per second is reported. However, the architecture is quite different; in particular it uses a configurable number of high-clock-frequency schoolbook multipliers instead of an NTT multiplier. The explanation for the low verification performance on the FPGA, compared with the software implementation, is that only one such multiplier is used in the verification engine.

Another target for comparison is a recently reported implementation of an ideal lattice-based encryption system in soft- and hardware [12]. In software, the necessary polynomial arithmetic relies on Shoup's NTL library [22]. Measurements confirmed that our basic arithmetic is faster than their prototype implementation (although their parameters are smaller) as we can rely on AVX, a hand-crafted NTT implementation and optimized modular reduction.

Various other implementations of post-quantum signature schemes have been described in the literature and many of them have been submitted to eBACS [4]. In Table 2 we compare our software in terms of security, speed, key sizes and signature size to the Rainbow, TTS, and C^* (pFLASH) software presented in [8], and the MQQ-Sig software presented in [11]. The cycle counts of these implementations are obtained from the eBACS website and have been measured on the same Intel Ivy Bridge machine that we used for benchmarking (h9ivy). We reference these implementations by their names in eBACS (in typewriter font) and their corresponding paper. For most of these multivariate schemes, the signing performance is much better, verification performance is somewhat better, but they suffer from excessive public-key sizes.

We furthermore compare to software described in the literature that has not been submitted to eBACS, specifically the implementation of the parallel-CFS code-based signature scheme presented in [16], the implementation of the treeless signature scheme TSS12 presented in [23], and the implementation of the hash-based signature scheme XMSS [6]. For those implementations we give the performance numbers from the respective paper and indicate the CPU used for benchmarking. Parallel-CFS not only has much larger keys, signing is also several orders of magnitude slower than with the lattice-based signature software presented in this paper. However, we expect that verification with parallel-CFS is very fast, but [16] does not give performance numbers for verification. The TSS software is using the scheme originally proposed in [17]. It makes an interesting target for comparison as it is similar to our scheme but relies on weaker assumptions. However, the software is much slower for both signing and verification. Hash-based signature schemes are also an interesting post-quantum signature alternative due to their well understood security properties and relatively small keys. However, the XMSS software presented in [6] is still an order of magnitude slower than our implementation and produces considerably larger signatures.

Finally we include two non-post-quantum signature schemes in the comparison in Table 2. First, the Ed25519 elliptic-curve signature scheme [3] and second, RSA-2048 signatures based on the OpenSSL implementation (ronald2048). Comparing to those schemes shows that our implementation and also most of the multivariate-signature software can even be faster or at least quite comparable to established schemes in terms of performance. However, the key and signature sizes of those two non-post-quantum signature are not beaten by any post-quantum proposal, yet.

Other lattice-based signature schemes that have a security reduction in the standard model are given in [7] and [5]. However, those papers do not give concrete parameters, security estimates or describe an implementation.

5   Future Work

As the initial implementation work has been carried out it is now necessary in future work to evaluate the security claims of the scheme by careful cryptanalysis and development of potential attacks. Especially, as the implemented scheme relaxes some assumptions that are required for connection to worst-case lattice problems more confidence is needed for real world usage. Other future work is the investigation of efficiency on more constrained devices like ARM (which, in some versions, also feature a SIMD unit) or even low-cost 8-bit processors.

Table 2. Comparison of different post-quantum signature software; **pk** stands for public key; **sk** stands for private key. The sizes are given in bytes. All software was benchmarked on **h9ivy** if not indicated otherwise.

Software	Security	Cycles		Sizes	
This work	100 bits	sign:	634988	pk:	1536
		verify:	45036	sk:	256
				sig:	1184
mqqsig160 [12]	80 bits	sign:	1996	pk:	206112
		verify:	33220	sk:	401
				sig:	20
mqqsig192 [12]	96 bits	sign:	3596	pk:	333540
		verify:	63488	sk:	465
				sig:	24
mqqsig224 [12]	112 bits	sign:	3836	pk:	529242
		verify:	65988	sk:	529
				sig:	28
mqqsig256 [12]	128 bits	sign:	4560	pk:	789552
		verify:	87904	sk:	593
				sig:	32
rainbow5640 [9]	80 bits	sign:	53872	pk:	44160
		verify:	34808	sk:	86240
				sig:	37
rainbowbinary16242020 [9]	80 bits	sign:	29364	pk:	102912
		verify:	17900	sk:	94384
				sig:	40
rainbowbinary256181212 [9]	80 bits	sign:	33396	pk:	30240
		verify:	27456	sk:	23408
				sig:	42
pflash1 [9]	80 bits	sign:	1473364	pk:	72124
		verify:	286168	sk:	5550
				sig:	37
tts6440 [9]	80 bits	sign:	33728	pk:	57600
		verify:	49248	sk:	16608
				sig:	43
Parallel-CFS [17]	80 bits	sign:	4200000000^a	pk:	20968300
$(20, 8, 10, 3)$		verify:	-	sk:	4194300
				sig:	75
TSS12 [24]	80 bits	sign:	93633000^b	pk:	13087
$(n = 512)$		verify:	13064000^b	sk:	13240
				sig:	8294
XMSS [7]	82 bits	sign:	7261100^c	pk:	912
$(H = 20, w = 4, \text{AES-128})$		verify:	556600^c	sk:	19
				sig:	2451
ed25519 [4]	128 bits	sign:	67564	pk:	32
		verify:	209328	sk:	64
				sig:	64
ronald2048	112 bits	sign:	5768360	pk:	256
(RSA-2048 based on		verify:	77032	sk:	2048
OpenSSL)				sig:	256

[a] Benchmarked on an Intel Xeon W3670 (3.20 GHz)
[b] Benchmarked on an AMD Opteron 8356 (2.3 GHz)
[c] Benchmarked on an Intel i5-M540 (2.53 GHz)

Acknowledgments. We would like to thank Michael Schneider, Vadim Lyubashevsky, and the anonymous reviewers for their helpful comments.

References

1. Arora, S., Ge, R.: New Algorithms for Learning in Presence of Errors. In: Aceto, L., Henzinger, M., Sgall, J. (eds.) ICALP 2011, Part I. LNCS, vol. 6755, pp. 403–415. Springer, Heidelberg (2011)
2. Bernstein, D.J.: The Salsa20 Family of Stream Ciphers. In: Robshaw, M., Billet, O. (eds.) New Stream Cipher Designs. LNCS, vol. 4986, pp. 84–97. Springer, Heidelberg (2008)
3. Bernstein, D.J., Duif, N., Lange, T., Schwabe, P., Yang, B.-Y.: High-speed high-security signatures. J. Cryptographic Engineering 2(2), 77–89 (2012)
4. Bernstein, D.J., Lange, T.: eBACS: ECRYPT benchmarking of cryptographic systems, http://bench.cr.yp.to (accessed January 25, 2013)
5. Boyen, X.: Lattice Mixing and Vanishing Trapdoors: A Framework for Fully Secure Short Signatures and More. In: Nguyen, P.Q., Pointcheval, D. (eds.) PKC 2010. LNCS, vol. 6056, pp. 499–517. Springer, Heidelberg (2010)
6. Buchmann, J., Dahmen, E., Hülsing, A.: XMSS - A Practical Forward Secure Signature Scheme Based on Minimal Security Assumptions. In: Yang, B.-Y. (ed.) PQCrypto 2011. LNCS, vol. 7071, pp. 117–129. Springer, Heidelberg (2011)
7. Cash, D., Hofheinz, D., Kiltz, E., Peikert, C.: Bonsai Trees, or How to Delegate a Lattice Basis. In: Gilbert, H. (ed.) EUROCRYPT 2010. LNCS, vol. 6110, pp. 523–552. Springer, Heidelberg (2010)
8. Chen, A.I.-T., Chen, M.-S., Chen, T.-R., Cheng, C.-M., Ding, J., Kuo, E.L.-H., Lee, F.Y.-S., Yang, B.-Y.: SSE Implementation of Multivariate PKCs on Modern x86 CPUs. In: Clavier, C., Gaj, K. (eds.) CHES 2009. LNCS, vol. 5747, pp. 33–48. Springer, Heidelberg (2009)
9. Cormen, T.H., Leiserson, C.E., Rivest, R.L., Stein, C.: Introduction to Algorithms, 3rd edn. MIT Press (2009)
10. Fog, A.: The microarchitecture of Intel, AMD and VIA CPUs: An optimization guide for assembly programmers and compiler makers (2010), http://www.agner.org/optimize/microarchitecture.pdf (version February 29, 2012)
11. Gligoroski, D., Ødegård, R.S., Jensen, R.E., Perret, L., Faugère, J.-C., Knapskog, S.J., Markovski, S.: MQQ-SIG – an ultra-fast and provably CMA resistant digital signature scheme. In: Chen, L., Yung, M., Zhu, L. (eds.) INTRUST 2011. LNCS, vol. 7222, pp. 184–203. Springer, Heidelberg (2012)
12. Göttert, N., Feller, T., Schneider, M., Buchmann, J., Huss, S.: On the Design of Hardware Building Blocks for Modern Lattice-Based Encryption Schemes. In: Prouff, E., Schaumont, P. (eds.) CHES 2012. LNCS, vol. 7428, pp. 512–529. Springer, Heidelberg (2012)
13. Grover, L.K.: A fast quantum mechanical algorithm for database search. In: Miller, G.L. (ed.) Proceedings of the Twenty-Eighth Annual ACM Symposium on the Theory of Computing, Philadelphia, Pennsylvania, USA, May 22-24, pp. 212–219. ACM (1996)
14. Güneysu, T., Lyubashevsky, V., Pöppelmann, T.: Practical Lattice-Based Cryptography: A Signature Scheme for Embedded Systems. In: Prouff, E., Schaumont, P. (eds.) CHES 2012. LNCS, vol. 7428, pp. 530–547. Springer, Heidelberg (2012)

15. Hoffstein, J., Howgrave-Graham, N., Pipher, J., Silverman, J.H., Whyte, W.: NTRUSIGN: Digital Signatures Using the NTRU Lattice. In: Joye, M. (ed.) CT-RSA 2003. LNCS, vol. 2612, pp. 122–140. Springer, Heidelberg (2003)

16. Landais, G., Sendrier, N.: Implementing CFS. In: Galbraith, S., Nandi, M. (eds.) INDOCRYPT 2012. LNCS, vol. 7668, pp. 474–488. Springer, Heidelberg (2012)

17. Lyubashevsky, V.: Lattice Signatures without Trapdoors. In: Pointcheval, D., Johansson, T. (eds.) EUROCRYPT 2012. LNCS, vol. 7237, pp. 738–755. Springer, Heidelberg (2012)

18. Lyubashevsky, V., Peikert, C., Regev, O.: On Ideal Lattices and Learning with Errors over Rings. In: Gilbert, H. (ed.) EUROCRYPT 2010. LNCS, vol. 6110, pp. 1–23. Springer, Heidelberg (2010)

19. Nguyên, P.Q., Regev, O.: Learning a Parallelepiped: Cryptanalysis of GGH and NTRU Signatures. In: Vaudenay, S. (ed.) EUROCRYPT 2006. LNCS, vol. 4004, pp. 271–288. Springer, Heidelberg (2006)

20. Petzoldt, A., Thomae, E., Bulygin, S., Wolf, C.: Small Public Keys and Fast Verification for Multivariate Quadratic Public Key Systems. In: Preneel, B., Takagi, T. (eds.) CHES 2011. LNCS, vol. 6917, pp. 475–490. Springer, Heidelberg (2011)

21. John, M.: Pollard. The Fast Fourier Transform in a finite field. Mathematics of Computation 25(114), 365–374 (1971)

22. Shoup, V.: NTL: A library for doing number theory, http://www.shoup.net/ntl/ (accessed March 18, 2013)

23. Weiden, P., Hülsing, A., Cabarcas, D., Buchmann, J.: Instantiating treeless signature schemes. IACR Crptology ePrint archive report 2013/065 (2013), http://eprint.iacr.org/2013/065

24. Winkler, F.: Polynomial Algorithms in Computer Algebra (Texts and Monographs in Symbolic Computation). Springer, 1st edn. (1996)

Solving the Shortest Vector Problem in Lattices Faster Using Quantum Search

Thijs Laarhoven[1], Michele Mosca[2], and Joop van de Pol[3]

[1] Dept. of Mathematics and Computer Science, Eindhoven University of Technology
t.m.m.laarhoven@tue.nl
[2] Institute for Quantum Computing and Dept. of C&O, University of Waterloo
and
Perimeter Institute for Theoretical Physics
michele.mosca@uwaterloo.ca
[3] Dept. of Computer Science, University of Bristol
joop.vandepol@bristol.ac.uk

Abstract. By applying Grover's quantum search algorithm to the lattice algorithms of Micciancio and Voulgaris, Nguyen and Vidick, Wang et al., and Pujol and Stehlé, we obtain improved asymptotic quantum results for solving the shortest vector problem. With quantum computers we can provably find a shortest vector in time $2^{1.799n+o(n)}$, improving upon the classical time complexity of $2^{2.465n+o(n)}$ of Pujol and Stehlé and the $2^{2n+o(n)}$ of Micciancio and Voulgaris, while heuristically we expect to find a shortest vector in time $2^{0.312n+o(n)}$, improving upon the classical time complexity of $2^{0.384n+o(n)}$ of Wang et al. These quantum complexities will be an important guide for the selection of parameters for post-quantum cryptosystems based on the hardness of the shortest vector problem.

Keywords: lattices, shortest vector problem, sieving, quantum algorithms, quantum search.

1 Introduction

Large-scale quantum computers will redefine the landscape of computationally secure cryptography, including breaking public-key cryptography based on integer factorization or the discrete logarithm problem [57] or the Principle Ideal Problem in real quadratic number fields [25], providing sub-exponential attacks for some systems based on elliptic curve isogenies [16], speeding up exhaustive searching [9,23], counting [12] and (with appropriate assumptions about the computing architecture) finding collisions and claws [4, 11, 13], among many other quantum algorithmic speed-ups [15, 42, 58].

Currently, a small set of systems [8] are being studied intensely as possible systems to replace those broken by large-scale quantum computers. These systems can be implemented with conventional technologies and to date seem resistant to substantial quantum attacks. It is critical that these systems receive intense

P. Gaborit (Ed.): PQCrypto 2013, LNCS 7932, pp. 83–101, 2013.
© Springer-Verlag Berlin Heidelberg 2013

scrutiny for possible quantum or classical attacks. This will boost confidence in the resistance of these systems to (quantum) attacks, and allow us to fine-tune secure choices of parameters in practical implementations of these systems.

One such set of systems bases its security on the computational hardness of certain lattice problems. Since the late 1990s, there has been a lot of research into the area of lattice-based cryptography, resulting in encryption schemes [27, 50], digital signature schemes [20, 39] and even fully homomorphic encryption schemes [10, 21]. Each of the lattice problems that underpin the security of these systems can be reduced to the shortest vector problem. Conversely, the decisional variant of the shortest vector problem can be reduced to the average case of such lattice problems. For a more detailed summary on the security of lattice-based cryptography, see [35, 45].

In this paper, we closely study the best-known algorithms for solving the shortest vector problem on a lattice, and how quantum algorithms may speed up these algorithms. By challenging and improving the best asymptotic complexities of these algorithms, we increase the confidence in the security of lattice-based schemes. Understanding these algorithms is critical when selecting key-sizes and other security parameters.

1.1 Lattices

Lattices are discrete subgroups of $\mathbb{R}^n$. Given a set of n linearly independent vectors $B = \{\mathbf{b}_1, \ldots, \mathbf{b}_n\}$ in $\mathbb{R}^n$, we define the lattice generated by these vectors as $L = \{\sum_{i=1}^n \lambda_i \mathbf{b}_i : \lambda_i \in \mathbb{Z}\}$. We call the set B a basis of the lattice L. This basis is not unique; applying a unimodular matrix transformation to the vectors of B leads to a new basis B' of the same lattice L.

In lattices, we generally work with the Euclidean or ℓ_2-norm, which we will denote by $\|\cdot\|$. For bases B, we write $\|B\| = \max_i \|\mathbf{b}_i\|$. We refer to a vector $\mathbf{s} \in L \setminus \{\mathbf{0}\}$ such that $\|\mathbf{s}\| \leq \|\mathbf{v}\|$ for any $\mathbf{v} \in L \setminus \{\mathbf{0}\}$ as a shortest (non-zero) vector of the lattice. Its length is denoted by $\lambda_1(L)$. Given a basis B, we write $\mathcal{P}(B) = \{\sum_{i=1}^n \lambda_i \mathbf{b}_i : 0 \leq \lambda_i < 1\}$ for the fundamental domain of B.

One of the most important hard problems in the theory of lattices is the Shortest Vector Problem (SVP). Given a basis of a lattice, the Shortest Vector Problem consists of finding a shortest vector in this lattice. In many applications, finding a short vector instead of a shortest vector is also sufficient. The Approximate Shortest Vector Problem with approximation factor γ (SVP$_\gamma$) asks to find a non-zero lattice vector $\mathbf{v} \in L$ with length bounded from above by $\|\mathbf{v}\| \leq \gamma \lambda_1(L)$.

1.2 Related Work

The Approximate Shortest Vector problem is integral in the cryptanalysis of lattice-based cryptography [18]. For small values of γ, this problem is known to be NP-hard [2, 31], while for certain exponentially large γ, polynomial time algorithms exist, such as the LLL algorithm of Lenstra, Lenstra and Lovász [37]. Other algorithms trade extra running time for a better γ, such as LLL with deep insertions [55] and the BKZ algorithm of Schnorr and Euchner [55].

The current state-of-the-art for classically finding short vectors is BKZ 2.0 [14], which is essentially the original BKZ algorithm with the improved SVP sub-routine of Gama et al. [19]. Implementations of this algorithm, due to Chen and Nguyen [14], and Aono and Naganuma [5], currently dominate the Lattice Challenge Hall of Fame [36]. The BKZ algorithm and its variants require a low-dimensional exact SVP solver as a subroutine. In theory, any of the known methods for finding a shortest vector could be used. For SVP solvers there is a similar online challenge [59], where the record is currently held by Kuo et al. [32].

In 2003, Ludwig [38] used quantum algorithms to speed up one such basis reduction algorithm, Random Sampling Reduction (RSR), which is due to Schnorr [56]. By replacing a random sampling from a big list by a quantum search, Ludwig achieves a quantum algorithm that is asymptotically faster than previous results. Ludwig also details the effect that this faster quantum algorithm would have had on the practical security of the lattice-based encryption scheme NTRU [27], had there been a quantum computer in 2005.

Enumeration. The classical method for finding shortest vectors is enumeration, dating back to work by Pohst [44], Kannan [30] and Fincke and Pohst [17] in the first half of the 1980s. In order to find a shortest vector, one enumerates all lattice vectors inside a giant ball around the origin. If the input basis is only LLL-reduced, enumeration runs in $2^{O(n^2)}$ time, where n is the lattice dimension. The algorithm by Kannan uses a stronger preprocessing of the input basis, and runs in $2^{O(n \log n)}$ time. Both approaches use only polynomial space in n.

Sieving/Saturation. In 2001, Ajtai et al. [3] introduced a technique called sieving, leading to the first probabilistic algorithm to solve SVP in time $2^{O(n)}$. Starting with a huge list of short vectors, the algorithm repeatedly applies a sieve to this list to end up with a smaller list of shorter lattice vectors. Eventually, we hope to be left with a list of lattice vectors of length $O(\lambda_1(L))$. Due to the size of the list, the space requirement of sieving is $2^{O(n)}$. Later work [26,41,43,48] investigated the constants in both exponents and ways to reduce these.

Recently, in 2009, Micciancio and Voulgaris [41] started a new branch of sieving algorithms, which may be more appropriately called saturation algorithms. While sieving starts out with a long list and repeatedly applies a sieve to reduce its length, saturation algorithms iteratively add vectors to an initially empty list, hoping that at some point the space of short lattice vectors is "saturated", and two of the vectors in the list are at most $\lambda_1(L)$ apart. The time and space requirements of these algorithms are also $2^{O(n)}$. In 2009, Pujol and Stehlé [46] showed that with this method, SVP can provably be solved in time $2^{2.465n+o(n)}$.

Voronoi. In 2010, Micciancio and Voulgaris presented a deterministic algorithm for solving SVP based on constructing the Voronoi cell of the lattice [40]. In time $2^{2n+o(n)}$ and space $2^{n+o(n)}$, this algorithm is able to find a shortest vector in any lattice. Currently this is the best provable asymptotic result for classical SVP solvers.

Practice. While many methods have surpassed the enumeration algorithms in terms of classical provable asymptotic time complexities, in practice the enumeration methods still dominate the field. The version of enumeration that is currently used in practice is due to Schnorr and Euchner [55] with improvements by Gama et al. [19]. It does not incorporate the stronger version of preprocessing of Kannan [30] and hence has an asymptotic time complexity of $2^{O(n^2)}$. However, due to the small hidden constants in the exponents and the exponential space complexity of the other algorithms, enumeration is actually faster than other methods for common values of n. That said, the other methods are still quite new, so a further study of these other methods may tip the balance.

1.3   Quantum Search

In this paper we will study how quantum algorithms can be used to speed up the SVP algorithms outlined above. For this, we will make use of Grover's quantum search algorithm [23], which considers the following problem:

Given a list L of length N and a function $f : L \to \{0, 1\}$, such that the number of elements $e \in L$ with $f(e) = 1$ is small. Construct an algorithm "search" that, given L and f as input, returns an $e \in L$ with $f(e) = 1$, or determines that (with high probability) no such e exists. We assume for simplicity that f can be evaluated in unit time.

Classical Algorithm. With classical computers, the natural way to find such an element is to go through the whole list, until one of these elements is found. This takes on average $O(N)$ time. This is also optimal up to a constant factor; no classical algorithm can find such an element in less than $\Omega(N)$ time.

Quantum Algorithm. Using quantum search [9, 12, 23], we can find such an element in time $O(\sqrt{N})$. This is optimal up to a constant factor, as any quantum algorithm needs at least $\Omega(\sqrt{N})$ evaluations of f [6].

Throughout the paper, we will write $x \leftarrow \text{search}_{e \in L}(f(e) = 1)$ to highlight subroutines that perform a search in a long list. This assignment returns true if an element $e \in L$ with $f(e) = 1$ exists (and assigns such an element to x), and returns false if no such e exists. This allows us to give one description for both the classical and quantum versions of each algorithm, as the only difference between the two versions is which version of the subroutine is used.

For both of these classical and quantum algorithms, we assume a RAM model of computation where the jth entry of the list L can be looked up in constant time (or polylogarithmic time). In the case that L is a virtual list where the jth element can be computed in time polynomial in the length of j (thus polylogarithmic in the length of the list L), then look-up time is not an issue. When L is indeed an unstructured list of values, for classical computation, the assumption of a RAM-like model has usually been valid in practice. However, there are fundamental reasons for questioning it [7], and there are practical computing architectures where the assumption does not apply. In the case of quantum computation, a practical RAM-like quantum memory (e.g. [22]) looks particularly challenging, especially for first generation quantum computers. Some authors have studied the limitations of quantum algorithms in this context [7, 24, 28].

Some algorithms (e.g. [4]) must store a large database of information in regular quantum memory (that is, memory capable of storing quantum superpositions of states). In contrast, quantum searching an actual list of N (classical) strings requires the N values to be stored in quantumly addressable classical memory (e.g. as Kuperberg discusses in [34]) and $O(\log N)$ regular qubits. Quantumly addressable classical memory in principle could be much easier to realize in practice than regular qubits. Furthermore, quantum searching for a value $x \in \{0, 1\}^n$ satisfying $f(x) = 1$ for a function $f : \{0, 1\}^n \to \{0, 1\}$ which can be implemented by a circuit on $O(n)$ qubits only requires $O(n)$ regular qubits, and there is no actual list to be stored in memory. In this paper, the quantum search algorithms used require the lists of size N to be stored in quantumly addressable classical memory and use $O(\log N)$ regular qubits and $O(\sqrt{N})$ queries into the list of numbers.

In this work, we consider (conventional) classical RAM memories for the classical algorithms, and RAM-like quantumly addressable classical memories for the quantum search algorithms. This is both a first step for future studies in assessing the impact of more practical quantum architectures, and also represents a more conservative approach in determining parameter choices for lattice-based cryptography that should be resistant against the potential power of quantum algorithmic attacks. Future work may also find ways to take advantage of advanced quantum search techniques, such as those surveyed in [51].

1.4 Contributions and Outline

In this paper, we show that quantum algorithms can significantly speed up sieving and saturation algorithms. The constant in the exponent decreases by approximately 25% in all cases, leading to an improvement upon both provable and heuristic asymptotic results for solving the Shortest Vector Problem:

- Provably, we can find a shortest vector in any lattice in time $2^{1.799n+o(n)}$.
- Heuristically, we can find a shortest vector in any lattice in time $2^{0.312n+o(n)}$.
- Extrapolating from classical experiments, with quantum computers we expect to be able to find a shortest vector in any lattice in time about $2^{0.39n}$.

Table 1 contains a comparison between our contributions and previous results, in both the classical and quantum setting. While the Voronoi Cell algorithm is asymptotically the best algorithm in the provable classical setting, our quantum saturation algorithm has better asymptotics in the provable quantum setting.

Why do we only consider sieving and saturation algorithms, and not the more practical enumeration or the theoretically faster Voronoi cell algorithms? It turns out that it is not as simple to significantly speed up these algorithms using similar techniques. For some intuition why this is the case, see Appendix C.

Table 1. A comparison of the results as expressed in logarithmic leading order terms, with provable results above and heuristic results below

Algorithm	Classical		Quantum		
	Time	Space	Time	Space	
(Enumeration)	$O(n \log n)$	$O(\log n)$	-	-	(Appendix C)
Pujol and Stehlé [46]	$2.47n$	$1.24n$	$1.80n$	$1.29n$	(Section 3.1)
(Voronoi)	$2.00n$	$1.00n$	-	-	(Appendix C)
Micciancio and Voulgaris [41]	$0.52n$	$0.21n$	$0.39n$	$0.21n$	(Section 3.2)
Nguyen and Vidick [43]	$0.42n$	$0.21n$	$0.32n$	$0.21n$	(Section 2.1)
Wang et al. [60]	$0.39n$	$0.26n$	$0.32n$	$0.21n$	(Section 2.2)

The outline of this paper is as follows. In Section 2 we look at sieving algorithms, and how quantum algorithms lead to speed-ups. In Section 3, we look at saturation algorithms, and their estimated time and space complexities on a quantum computer. Technical details regarding some of these results can be found in Appendices A and B.

2   Sieving Algorithms

Sieving was first introduced by Ajtai et al. [3] and later improved theoretically [26, 41, 43, 48] and practically [43, 60] in various papers. In these algorithms, first an exponentially long list of lattice vectors is generated. Then, by iteratively applying a sieve to this list, the size of the list, as well as the lengths of the vectors in the list are reduced. After a polynomial number of applications of the sieve, we hope to be left with a short but non-empty list of very short vectors, from which we can then obtain a shortest vector of the lattice with high probability.

2.1   The Heuristic Algorithm of Nguyen and Vidick

Nguyen and Vidick [43] considered a heuristic, practical variant of the sieve algorithm of Ajtai et al. [3], which provably returns a shortest vector under a certain natural, heuristic assumption. A slightly modified but equivalent version of this algorithm is given in Algorithm 1.

Description of the Algorithm. The algorithm starts by generating a big list S of random lattice vectors with length at most $n\|B\|$. Then, by repeatedly applying a sieve to this list, shorter lists of shorter vectors are obtained, until the list is completely depleted. In that case, we go back one step, and look for the closest pair of lattice vectors in the last non-empty list.

The sieving step consists of splitting the previous list S_{prev} in a set of 'centers' C and a new list of vectors S that will be used for the next sieve. For each vector

---
Algorithm 1 The Heuristic Sieve Algorithm of Nguyen and Vidick
---
Input: An LLL-reduced basis B of L, and constants $\gamma \in (\frac{2}{3}, 1)$ and $N = 2^{O(n)}$
Output: A short non-zero lattice vector $\mathbf{s}$
1: $S \leftarrow \emptyset$
2: **for** $i \leftarrow 1$ to N **do**
3:    $\mathbf{v} \in_R B_n(\mathbf{0}, \|B\|) \cap L$
4:    $S \leftarrow S \cup \{\mathbf{v}\}$
5: **while** $S \setminus \{\mathbf{0}\} \neq \emptyset$ **do**
6:    $S_{\text{prev}} \leftarrow S \setminus \{\mathbf{0}\}$
7:    $R \leftarrow \max_{\mathbf{v} \in S_{\text{prev}}} \|\mathbf{v}\|$
8:    $C \leftarrow \{\mathbf{0}\}$
9:    $S \leftarrow \emptyset$
10:    **for all** $\mathbf{v} \in S_{\text{prev}}$ **do**
11:        **if** $\mathbf{c} \leftarrow \text{search}_{\mathbf{c} \in C}(\|\mathbf{v} - \mathbf{c}\| \leq \gamma R)$ **then**
12:            $S \leftarrow S \cup \{\mathbf{v} - \mathbf{c}\}$
13:        **else**
14:            $C \leftarrow C \cup \{\mathbf{v}\}$
15: $\mathbf{s} \leftarrow \text{argmin}_{\mathbf{v} \in S_{\text{prev}}} \|\mathbf{v}\|$
16: **return s**

$\mathbf{v}$ in S_{prev}, the algorithm first checks if a vector $\mathbf{c}$ in C exists that is close to $\mathbf{v}$. If this is the case, then we add the difference $\mathbf{v} - \mathbf{c}$ to S_{prev}. If this is not the case, then $\mathbf{v}$ is added to C. Since the set C consists of vectors with a bounded norm and a specified minimum distance between any two points, one can bound the size of C from above using a result of Kabatiansky and Levenshtein [29] regarding sphere packings. In other words, C will be sufficiently small, so that the list S will be sufficiently large. After applying the sieve, we discard all vectors in C and apply the sieve again to the vectors in $S_{\text{prev}} = S$.

At each iteration of the sieve, the maximum norm of the vectors in the list decreases from some constant R to at most γR, where γ is some geometric factor smaller than 1. Nguyen and Vidick conjecture that throughout the algorithm, the longest vectors in S are uniformly distributed over the space of all n-dimensional vectors with norms between γR and R.

Heuristic 1. *[43] At any stage of Algorithm 1, the vectors in $S \cap C_n(\gamma R, R)$ are uniformly distributed in $C_n(\gamma R, R)$, where $C_n(r_1, r_2) = \{\mathbf{x} \in \mathbb{R}^n : r_1 \leq \|\mathbf{x}\| \leq r_2\}$.*

Classical Complexities. In Line 11 of Algorithm 1, we have highlighted an application of a search subroutine that could be replaced by a quantum search. Using a standard classical search algorithm for this subroutine, under this heuristic assumption Nguyen and Vidick give the following estimate for the time and space complexity of their algorithm.

Lemma 1. *[43] On a classical computer, assuming that Heuristic 1 holds, Algorithm 1 will return a shortest vector of a lattice in time at most $2^{0.415n+o(n)}$ and space at most $2^{0.208n+o(n)}$.*

Quantum Complexities. If we use a quantum search subroutine in Line 11, the complexity of this subroutine decreases from $\tilde{O}(|C|)$ to $\tilde{O}(\sqrt{|C|})$. Since this search is part of the bottleneck for the time complexity, applying a quantum search here will decrease the running time significantly. Note that in Line 15, it also seems like a search of a list is performed. In reality, this final search of S_{prev} can be done in constant time by using appropriate data structures, e.g., by keeping the vectors in S and S_{prev} sorted from short to long, or by manually keeping track of the shortest vector in S.

Since replacing the classical search by a quantum search does not change the internal behaviour of the algorithm, the estimates and heuristics are as valid as they were in the classical setting. The time complexity does change, as the following theorem explains. For details, see Appendix A.

Theorem 1. *On a quantum computer, assuming that Heuristic 1 holds, Algorithm 1 will return a shortest vector of a lattice in time $2^{0.312n+o(n)}$ and space $2^{0.208n+o(n)}$.*

In other words, applying quantum search to Nguyen and Vidick's sieve algorithm leads to a 25% decrease in the exponent of the runtime.

2.2 The Heuristic Algorithm of Wang et al.

To improve upon the time complexity of the algorithm of Nguyen and Vidick, Wang et al. [60] introduced a further trade-off between the time complexity and the space complexity. Their algorithm uses two lists of centers C_1 and C_2 and two geometric factors γ_1 and γ_2, instead of the single list C and single geometric factor γ in the algorithm of Nguyen and Vidick. For details, see [60].

Classical Complexities. The classical time complexity of this algorithm is bounded from above by $\tilde{O}(|S| \cdot (|C_1| + |C_2|))$, while the space required is at most $O(|S| + |C_1| + |C_2|)$. Optimizing the constants γ_1 and γ_2 leads to $\gamma_1 = 1.0927$ and $\gamma_2 \to 1$, with an asymptotic time complexity of less than $2^{0.384n+o(n)}$ and a space complexity of about $2^{0.256n+o(n)}$.

Quantum Complexities. By using the quantum search algorithm for searching the lists C_1 and C_2, the time complexity is reduced to $\tilde{O}(|S| \cdot (\sqrt{|C_1|} + \sqrt{|C_2|}))$, while the space complexity remains $O(|S| + |C_1| + |C_2|)$. Re-optimizing the constants for a minimum time complexity leads to $\gamma_1 \to \sqrt{2}$ and $\gamma_2 \to 1$, leading to the same time and space complexities as the quantum-version of the algorithm of Nguyen and Vidick. Due to the simpler algorithm and smaller constants, a quantum version of the algorithm of Nguyen and Vidick will most likely be more efficient than a quantum version of the algorithm of Wang et al.

3 Saturation Algorithms

Saturation algorithms were only recently introduced by Micciancio and Voulgaris [41], and further studied by Pujol and Stehlé [46] and Schneider [52].

Algorithm 2 The Provable Saturation Algorithm of Pujol and Stehlé

Input: An LLL-reduced basis B of L, $\mu \simeq \lambda_1(L)$, $\xi > \frac{1}{2}$, $R > 2\xi$, $N_1^{\max}, N_2 = 2^{O(n)}$
Output: A non-zero lattice vector $\mathbf{s}$ of norm less than μ
1: $\gamma \leftarrow 1 - \frac{1}{n}$
2: $T \leftarrow \emptyset$
3: $N_1 \in_R [0, N_1^{\max} - 1]$
4: **for** $i \leftarrow 1$ to N_1 **do**
5:     $\mathbf{x} \in_R B_n(\mathbf{0}, \xi\mu)$
6:     $\mathbf{v'} \leftarrow \mathbf{x} \bmod \mathcal{P}(B)$
7:     **while** $\mathbf{t} \leftarrow \text{search}_{t \in T}(\|\mathbf{v'} - \mathbf{t}\| < \gamma \|\mathbf{v'}\|)$ **do**
8:         $\mathbf{v'} \leftarrow \mathbf{v'} - \mathbf{t}$
9:     $\mathbf{v} \leftarrow \mathbf{v'} - \mathbf{x}$
10:    **if** $\|\mathbf{v}\| \geq R\mu$ **then**
11:       $T \leftarrow T \cup \{\mathbf{v}\}$
12: $S \leftarrow \emptyset$
13: **for** $i \leftarrow 1$ to N_2 **do**
14:    $\mathbf{x} \in_R B_n(\mathbf{0}, \xi\mu)$
15:    $\mathbf{v'} \leftarrow \mathbf{x} \bmod \mathcal{P}(B)$
16:    **while** $\mathbf{t} \leftarrow \text{search}_{t \in T}(\|\mathbf{v'} - \mathbf{t}\| < \gamma \|\mathbf{v'}\|)$ **do**
17:       $\mathbf{v'} \leftarrow \mathbf{v'} - \mathbf{t}$
18:    $\mathbf{v} \leftarrow \mathbf{v'} - \mathbf{x}$
19:    $S \leftarrow S \cup \{\mathbf{v}\}$
20: $\{\mathbf{s_1}, \mathbf{s_2}\} \leftarrow \text{search}_{\{s_1, s_2\} \in S \times S}(0 < \|\mathbf{s_1} - \mathbf{s_2}\| < \mu)$
21: **return** $\mathbf{s_1} - \mathbf{s_2}$

Instead of starting with a huge list and making the list smaller and smaller, this method starts with a small or empty list, and keeps adding more and more vectors to the list. Building upon the same result of Kabatiansky and Levenshtein about sphere packings [29], we know that if the list reaches a certain size and all vectors have a norm bounded by a sufficiently small constant, two of the vectors in the list must be close to one another. Thus, if we can guarantee that new short lattice vectors keep getting added to the list, then at some point, with high probability, we can find a shortest vector as the difference between two of the list vectors.

3.1 The Provable Algorithm of Pujol and Stehlé

Using the Birthday paradox, Pujol and Stehlé [46] showed that the constant in the exponent of the time complexity of the original algorithm of Micciancio and Voulgaris [41, Section 3.1] can be reduced by almost 25%. The algorithm is presented in Algorithm 2.

Description of the Algorithm. The algorithm can roughly be divided in three stages, as follows.

First, the algorithm generates a long list T of lattice vectors with norms between $R\mu$ and $\|B\|$. This 'dummy' list is only used for technical reasons, and in practice one does not seem to need such a list. Note that besides the actual lattice vectors $\mathbf{v}$, to generate this list we also consider slightly perturbed vectors $\mathbf{v}'$ which are not in the lattice, but are at most $r\mu$ away from $\mathbf{v}$. This is purely a technical modification to make the proofs work, as experiments show that without such perturbed vectors, saturation algorithms also work fine [40, 46, 52].

After generating T, we generate a fresh list of short lattice vectors S. The procedure for generating these vectors is similar to that of generating T, with two exceptions: (i) now all sampled lattice vectors are added to S (regardless of their norms), and (ii) the vectors are reduced with the dummy list T rather than with vectors in S. The latter guarantees that the vectors in S are i.i.d.

Finally, when S has been generated, we hope that it contains two distinct lattice vectors $\mathbf{s}_1$, $\mathbf{s}_2$ that are at most μ apart. So we search $S \times S$ for a pair $\{\mathbf{s}_1, \mathbf{s}_2\}$ of close, distinct lattice vectors, and return their difference.

Classical Complexities. With a classical search applied to the subroutines in Lines 7, 16, and 20, Pujol and Stehlé obtained the following results.

Lemma 2. *[46] Let $\xi \approx 0.9476$ and $R \approx 3.0169$. Then, using polynomially many queries to Algorithm 2, we can find a shortest vector in a lattice with probability exponentially close to 1, using time at most $2^{2.465n+o(n)}$ and space at most $2^{1.233n+o(n)}$.*

Quantum Complexities. Applying a quantum search algorithm to the search-subroutines in Lines 7, 16, and 20 leads to the following result. Details are given in Appendix B.

Theorem 2. *Let $\xi \approx 0.9086$ and $R \approx 3.1376$. Then, using polynomially many queries to the quantum version of Algorithm 2, we can find a shortest vector in a lattice with probability exponentially close to 1, using time at most $2^{1.799n+o(n)}$ and space at most $2^{1.286n+o(n)}$.*

So the constant in the exponent of the time complexity decreases by about 27% when using quantum search.

Remark. If we generate S in parallel, we can potentially achieve a time complexity of $2^{1.470n+o(n)}$, by setting $\xi \approx 1.0610$ and $R \approx 4.5166$. However, it would require exponentially many parallel quantum computers of size $O(n)$ to achieve a substantial theoretical speed-up over the $2^{1.799n+o(n)}$ of Theorem 2. (Recall that quantum searching a list of c^n elements (with $c > 1$) requires the list to be stored in quantumly addressable classical memory (versus regular quantum memory) and otherwise can be searched using only $O(n)$ qubits and $O(c^{n/2})$ queries to the list.)

3.2   The Heuristic Algorithm of Micciancio and Voulgaris

In practice, just like sieving algorithms, saturation algorithms are much faster than their worst-case running times and provable time complexities suggest.

Algorithm 3 The Heuristic Saturation Algorithm of Micciancio and Voulgaris

Input: An LLL-reduced basis B of L, and a constant C_0
Output: A short non-zero lattice vector **s**
1: $S \leftarrow \{\mathbf{0}\}$
2: $Q \leftarrow \emptyset$
3: $c \leftarrow 0$
4: **while** $c < C_0$ **do**
5:      **if** $Q \neq \emptyset$ **then**
6:          $\mathbf{v} \in_R Q$
7:          $Q \leftarrow Q \setminus \{\mathbf{v}\}$
8:      **else**
9:          $\mathbf{v} \in_R B_n(\mathbf{0}, \|B\|) \cap L$
10:      **while** $\mathbf{s} \leftarrow \text{search}_{\mathbf{s} \in S}(\max\{\|\mathbf{s}\|, \|\mathbf{v} - \mathbf{s}\|\} \leq \|\mathbf{v}\|)$ **do**
11:          $\mathbf{v} \leftarrow \mathbf{v} - \mathbf{s}$
12:      **while** $\mathbf{s} \leftarrow \text{search}_{\mathbf{s} \in S}(\max\{\|\mathbf{v}\|, \|\mathbf{v} - \mathbf{s}\|\} \leq \|\mathbf{s}\|)$ **do**
13:          $S \leftarrow S \setminus \{\mathbf{s}\}$
14:          $Q \leftarrow Q \cup \{\mathbf{v} - \mathbf{s}\}$
15:      **if** $\mathbf{v} = 0$ **then**
16:          $c \leftarrow c + 1$
17:      **else**
18:          $S \leftarrow S \cup \{\mathbf{v}\}$
19: $\mathbf{s} \leftarrow \text{argmin}_{\mathbf{v} \in S \setminus \{\mathbf{0}\}} \|\mathbf{v}\|$
20: **return** $\mathbf{s}$

Micciancio and Voulgaris [41] gave a heuristic variant of their saturation algorithm, for which they could not give a (heuristic) bound on the time complexity, but with a better bound on the space complexity, and a better practical time complexity. The algorithm is given in Algorithm 3.

Description of the Algorithm. The algorithm is similar to Algorithm 2, with the following main differences: (i) we do not explicitly generate two lists S, T to apply the birthday paradox; (ii) we do not use the geometric factor $\gamma < 1$ but always reduce a vector if it can be reduced; (iii) we also reduce the existing list vectors with newly sampled vectors, so that each two vectors in the list are pairwise Gauss-reduced; and (iv) instead of specifying the number of iterations, we run the algorithm until we reach a predefined number of collisions C_0.

Classical Complexities. Micciancio and Voulgaris state that the algorithm above has an experimental time complexity of about $2^{0.52n}$ and a space complexity which is most likely bounded from above by $2^{0.208n}$ due to the kissing constant [41, Section 5]. This is much faster than the theoretical time complexity of $2^{1.799n}$ of the quantum-enhanced saturation algorithm discussed in Section 3.1.

Remark 1. In practice, the algorithm of Micciancio and Voulgaris is faster than the one of Nguyen and Vidick of Section 2.1, even though the leading term in the exponent is larger. So asymptotically, this algorithm is dominated by the algorithm of Nguyen and Vidick, but in practice and for small dimensions, the algorithm of Micciancio and Voulgaris seems to perform better.

Remark 2. Schneider states [52] that the time complexity roughly scales like $2^{0.57n-23.5}$, instead of the $2^{0.52n}$ claimed by Micciancio and Voulgaris. Although asymptotically this time complexity is worse than the one of Micciancio and Voulgaris, the cross-over point of these rough approximations is around $n \approx 470$. So for most values of n that SVP solvers handle in practice, the term -23.5 is more significant than the small increase caused by n, and the conjectured time complexity of Schneider is better than that of Micciancio and Voulgaris.

Quantum Complexities. To this heuristic algorithm, the quantum speed-ups can also be applied. Generally, these saturation algorithms generate a list S of reasonably short lattice vectors by (i) first sampling a long, random lattice vector $\mathbf{v} \in L$; (ii) reducing the vector $\mathbf{v}$ with lattice vectors already in S; (iii) possibly reducing the vectors in S with this new vector $\mathbf{v}$; and (iv) finally adding $\mathbf{v}$ to S. The total classical time complexity of these algorithms is of the order $|S|^2$ due to (ii) and (iii), but by applying quantum speed-ups to these steps, this becomes $|S|^{3/2}$. This means that the exponent in the time complexity is generally reduced by about 25%, which is comparable to the improvement in Section 3.1. In practice, we therefore expect a time complexity of about $2^{0.39n}$ for the heuristic algorithm of Micciancio and Voulgaris with quantum search speed-ups, with constants that may make this algorithm faster than the sieving algorithm of Section 2.1.

Acknowledgments. This report is partly a result of fruitful discussions at the Lorentz Center Workshop on Post-Quantum Cryptography and Quantum Algorithms, Nov. 5–9, Leiden, The Netherlands. In particular, we would like to thank Felix Fontein, Nadia Heninger, Stacey Jeffery, Stephen Jordan, Michael Schneider, Damien Stehlé and Benne de Weger for the valuable discussions there. Finally, we thank the anonymous reviewers for their helpful comments and suggestions.

The first author is supported by DIAMANT and ECRYPT II (ICT-2007-216676). The second author is supported by Canada's NSERC (Discovery, SPG FREQUENCY, and CREATE CryptoWorks21), MPrime, CIFAR, ORF and CFI; IQC and Perimeter Institute are supported in part by the Government of Canada and the Province of Ontario. The third author is supported in part by EPSRC via grant EP/I03126X.

References

1. Aharonov, D., Regev, O.: A Lattice Problem in Quantum NP. In: 44th Annual IEEE Symposium on Foundations of Computer Science (FOCS), pp. 210–219. IEEE Press, New York (2003)

2. Ajtai, M.: The Shortest Vector Problem in L_2 is NP-hard for Randomized Reductions. In: 30th Annual ACM Symposium on Theory of Computing (STOC), pp. 10–19. ACM, New York (1998)

3. Ajtai, M., Kumar, R., Sivakumar, D.: A Sieve Algorithm for the Shortest Lattice Vector Problem. In: 33rd Annual ACM Symposium on Theory of Computing (STOC), pp. 601–610. ACM, New York (2001)

4. Ambainis, A.: Quantum Walk Algorithm for Element Distinctness. In: 45th Annual IEEE Symposium on Foundations of Computer Science (FOCS), pp. 22–31. IEEE Press, New York (2003)

5. Aono, Y., Naganuma, K.: Heuristic Improvements of BKZ 2.0. IEICE Tech. Rep. 112(211), 15–22 (2012)

6. Bennett, C.H., Bernstein, E., Brassard, G., Vazirani, V.: Strengths and Weaknesses of Quantum Computing. SIAM J. Comput. 26(5), 1510–1523 (1997)

7. Bernstein, D.J.: Cost analysis of hash collisions: Will quantum computers make SHARCs obsolete? In: SHARCS 2009: Special-purpose Hardware for Attacking Cryptographic Systems (2009)

8. Buchmann, J., Ding, J. (eds.): PQCrypto 2008. LNCS, vol. 5299. Springer, Heidelberg (2008)

9. Boyer, M., Brassard, G., Høyer, P., Tapp, A.: Tight Bounds on Quantum Searching. Fortschritte der Physik 46, 493–505 (1998)

10. Brakerski, Z., Gentry, C., Vaikuntanathan, V.: Fully homomorphic encryption without bootstrapping. In: Goldwasser, S. (ed.) Innovations in Theoretical Computer Science, ITCS 2012, pp. 309–325. ACM (2012)

11. Brassard, G., Høyer, P., Tapp, A.: Quantum cryptanalysis of hash and claw-free functions. In: Lucchesi, C.L., Moura, A.V. (eds.) LATIN 1998. LNCS, vol. 1380, pp. 163–169. Springer, Heidelberg (1998)

12. Brassard, G., Høyer, P., Mosca, M., Tapp, A.: Quantum Amplitude Amplification and Estimation. AMS Contemporary Mathematics Series Millennium Vol. entitled Quantum Computation & Information, vol. 305 (2002)

13. Buhrman, B., Dürr, C., Heiligman, M., Høyer, P., Magniez, F., Santha, M., de Wolf, R.: Quantum Algorithms for Element Distinctness. SIAM J. Comput. 34(6), 1324–1330 (2005)

14. Chen, Y., Nguyen, P.Q.: BKZ 2.0: Better Lattice Security Estimates. In: Lee, D.H., Wang, X. (eds.) ASIACRYPT 2011. LNCS, vol. 7073, pp. 1–20. Springer, Heidelberg (2011)

15. Childs, A., Van Dam, W.: Quantum algorithms for algebraic problems. Rev. Mod. Phys. 82, 1–52 (2010)

16. Childs, A.M., Jao, D., Soukharev, V.: Constructing elliptic curve isogenies in quantum subexponential time. arXiv:1012.4019 (2010)

17. Fincke, U., Pohst, M.: Improved methods for calculating vectors of short length in a lattice, including a complexity analysis. Math. Comp. 44, 463–471 (1985)

18. Gama, N., Nguyen, P.Q.: Predicting lattice reduction. In: Smart, N.P. (ed.) EUROCRYPT 2008. LNCS, vol. 4965, pp. 31–51. Springer, Heidelberg (2008)

19. Gama, N., Nguyen, P.Q., Regev, O.: Lattice Enumeration Using Extreme Pruning. In: Gilbert, H. (ed.) EUROCRYPT 2010. LNCS, vol. 6110, pp. 257–278. Springer, Heidelberg (2010)

20. Gentry, C., Peikert, C., Vaikuntanathan, V.: Trapdoors for hard lattices and new cryptographic constructions. In: Dwork, C. (ed.) STOC 2008, pp. 197–206. ACM (2008)

21. Gentry, C.: A fully homomorphic encryption scheme (Doctoral dissertation, Stanford University) (2009)

22. Giovannetti, V., Lloyd, S., Maccone, L.: Quantum Random Access Memory. Phys. Rev. Lett. 100, 160501 (2008)
23. Grover, L.K.: A Fast Quantum Mechanical Algorithm for Database Search. In: 28th Annual ACM Symposium on Theory of Computing (STOC), pp. 212–219. ACM, New York (1996)
24. Grover, L., Rudolph, T.: How significant are the known collision and element distinctness quantum algorithms? Quantum Info. Comput. 4(3), 201–206 (2004)
25. Hallgren, S.: Polynomial-time quantum algorithms for Pell's equation and the principal ideal problem. J. ACM. 54(1), 653–658 (2007)
26. Hanrot, G., Pujol, X., Stehlé, D.: Algorithms for the Shortest and Closest Lattice Vector Problems. In: Chee, Y.M., Guo, Z., Ling, S., Shao, F., Tang, Y., Wang, H., Xing, C. (eds.) IWCC 2011. LNCS, vol. 6639, pp. 159–190. Springer, Heidelberg (2011)
27. Hoffstein, J., Pipher, J., Silverman, J.H.: NTRU: A ring-based public key cryptosystem. In: Buhler, J.P. (ed.) ANTS 1998. LNCS, vol. 1423, pp. 267–288. Springer, Heidelberg (1998)
28. Jeffery, S.: Collision Finding with Many Classical or Quantum Processors. Master's thesis, University of Waterloo (2011)
29. Kabatiansky, G., Levenshtein, V.I.: On Bounds for Packings on a Sphere and in Space. Problemy Peredachi Informacii 14(1), 3–25 (1978)
30. Kannan, R.: Improved Algorithms for Integer Programming and Related Lattice Problems. In: 15th Annual ACM Symposium on Theory of Computing (STOC), pp. 193–206. ACM, New York (1983)
31. Khot, S.: Hardness of approximating the shortest vector problem in lattices. Journal of the ACM 52(5), 789–808 (2005)
32. Kuo, P.C., Schneider, M., Dagdelen, Ö., Reichelt, J., Buchmann, J., Cheng, C.M., Yang, B.Y.: Extreme Enumeration on GPU and in Clouds. In: Preneel, B., Takagi, T. (eds.) CHES 2011. LNCS, vol. 6917, pp. 176–191. Springer, Heidelberg (2011)
33. Kuperberg, G.: A Subexponential-Time Quantum Algorithm for the Dihedral Hidden Subgroup Problem. SIAM J. Comput. 35(1), 170–188 (2005)
34. Kuperberg, G.: Another Subexponential-Time Quantum Algorithm for the Dihedral Hidden Subgroup Problem. arXiv, Report 1112/3333, pp. 1–10 (2011)
35. Laarhoven, T., van de Pol, J., de Weger, B.: Solving Hard Lattice Problems and the Security of Lattice-Based Cryptosystems. Cryptology ePrint Archive, Report 2012/533, pp. 1–43 (2012)
36. TU Darmstadt Lattice Challenge, http://www.latticechallenge.org/
37. Lenstra, A.K., Lenstra, H., Lovász, L.: Factoring Polynomials with Rational Coefficients. Math. Ann. 261(4), 515–534 (1982)
38. Ludwig, C.: A Faster Lattice Reduction Method Using Quantum Search. In: Ibaraki, T., Katoh, N., Ono, H. (eds.) ISAAC 2003. LNCS, vol. 2906, pp. 199–208. Springer, Heidelberg (2003)
39. Lyubashevsky, V.: Lattice signatures without trapdoors. In: Pointcheval, D., Johansson, T. (eds.) EUROCRYPT 2012. LNCS, vol. 7237, pp. 738–755. Springer, Heidelberg (2012)
40. Micciancio, D., Voulgaris, P.: A Deterministic Single Exponential Time Algorithm for Most Lattice Problems based on Voronoi Cell Computations. In: 42nd Annual ACM Symposium on Theory of Computing (STOC), pp. 351–358. ACM, New York (2010)
41. Micciancio, D., Voulgaris, P.: Faster Exponential Time Algorithms for the Shortest Vector Problem. In: 21st Annual ACM Symposium on Discrete Algorithms (SODA), pp. 1468–1480. ACM, New York (2010)

42. Mosca, M.: Quantum Algorithms. In: Meyers, R. (ed.) Encyclopedia of Complexity and Systems Science (2009)
43. Nguyen, P.Q., Vidick, T.: Sieve Algorithms for the Shortest Vector Problem are Practical. J. Math. Crypt. 2(2), 181–207 (2008)
44. Pohst, M.: On the computation of lattice vectors of minimal length, successive minima and reduced bases with applications. ACM SIGSAM Bulletin 15(1), 37–44 (1981)
45. van de Pol, J.: Lattice-based cryptography. Master's thesis. Eindhoven University of Technology (2011)
46. Pujol, X., Stehlé, D.: Solving the Shortest Lattice Vector Problem in Time $2^{2.465n}$. Cryptology ePrint Archive, Report 2009/605, pp. 1–7 (2009)
47. Regev, O.: A Subexponential Time Algorithm for the Dihedral Hidden Subgroup Problem with Polynomial Space. arXiv, Report 0405/151, pp. 1–7 (2004)
48. Regev, O.: Lattices in Computer Science. Lecture Notes for a Course at the Tel Aviv University (2004)
49. Regev, O.: Quantum Computation and Lattice Problems. SIAM J. Comput. 33(3), 738–760 (2004)
50. Regev, O.: On lattices, learning with errors, random linear codes, and cryptography. In: 37th Annual ACM Symposium on Theory of Computing (STOC), pp. 84–93 (2005)
51. Santha, M.: Quantum Walk Based Search Algorithms. In: Agrawal, M., Du, D.-Z., Duan, Z., Li, A. (eds.) TAMC 2008. LNCS, vol. 4978, pp. 31–46. Springer, Heidelberg (2008)
52. Schneider, M.: Analysis of Gauss-Sieve for Solving the Shortest Vector Problem in Lattices. In: Katoh, N., Kumar, A. (eds.) WALCOM 2011. LNCS, vol. 6552, pp. 89–97. Springer, Heidelberg (2011)
53. Schneider, M.: Sieving for Short Vectors in Ideal Lattices. Cryptology ePrint Archive, Report 2011/458, pp. 1–19 (2011)
54. Schnorr, C.P.: A Hierarchy of Polynomial Time Lattice Basis Reduction Algorithms. Theoretical Computer Science 53(2-3), 201–224 (1987)
55. Schnorr, C.P., Euchner, M.: Lattice Basis Reduction: Improved Practical Algorithms and Solving Subset Sum Problems. Mathematical Programming 66(2-3), 181–199 (1994)
56. Schnorr, C.P.: Lattice reduction by random sampling and birthday methods. In: Alt, H., Habib, M. (eds.) STACS 2003. LNCS, vol. 2607, pp. 145–156. Springer, Heidelberg (2003)
57. Shor, P.W.: Polynomial-Time Algorithms for Prime Factorization and Discrete Logarithms on a Quantum Computer. SIAM J. Comput. 26(5), 1484–1509 (1997)
58. Smith, J.: Mosca. M.: Algorithms for Quantum Computers. In: Handbook of Natural Computing, pp. 1451–1492. Springer (2012)
59. SVP Challenge, http://latticechallenge.org/svp-challenge/
60. Wang, X., Liu, M., Tian, C., Bi, J.: Improved Nguyen-Vidick Heuristic Sieve Algorithm for Shortest Vector Problem. In: 6th ACM Symposium on Information, Computer and Communications Security (ASIACCS), pp. 1–9. ACM, New York (2011)

A   Analysis of the Sieve Algorithm of Nguyen and Vidick

Nguyen and Vidick showed that if their heuristic assumption holds, the time and space complexities of their algorithm can be bounded from above as follows.

Lemma 3. *[43] On a classical computer, assuming Heuristic 1 holds, Algorithm 1 will return a shortest vector of a lattice in time $2^{2c_h n + o(n)}$ and space $2^{c_h n + o(n)}$, where $\frac{2}{3} < \gamma < 1$ and*

$$c_h = -\log_2(\gamma) - \frac{1}{2}\log_2\left(1 - \frac{\gamma^2}{4}\right). \tag{1}$$

To obtain a minimum time complexity, γ should be chosen as close to 1 as possible. Letting $\gamma \to 1$ leads to an asymptotic time complexity of less than $2^{0.415n + o(n)}$ and an asymptotic space complexity of less than $2^{0.208n + o(n)}$.

To obtain these estimates, it is first noted that the sizes of S and C are bounded from above by $2^{c_h n + o(n)}$. The space complexity is therefore bounded from above by $O(|S| + |C|) = 2^{c_h n + o(n)}$, and since for every element in S the algorithm has to search the list C, the time complexity is bounded from above by $\tilde{O}(|S| \cdot |C|) = 2^{2c_h n + o(n)}$.

Using quantum search on the list C, the time complexity decreases to $\tilde{O}(|S| \cdot \sqrt{|C|}) = 2^{\frac{3}{2}c_h n + o(n)}$, while the space complexity remains the same. This leads to the following result.

Lemma 4. *On a quantum computer, assuming Heuristic 1 holds, Algorithm 1 will return a shortest vector of a lattice in time $2^{\frac{3}{2}c_h n + o(n)}$ and space $2^{c_h n + o(n)}$.*

Optimizing γ to obtain a minimum time complexity again corresponds to letting γ tend to 1 from below, leading to an asymptotic time complexity of $2^{0.312n + o(n)}$ and space complexity of $2^{0.208n + o(n)}$, as stated in Theorem 1.

B   Analysis of the Saturation Algorithm of Pujol and Stehlé

In the classical setting, the time complexities of the different parts of the algorithm are as follows. The constants are explained in the lemma below.

- Cost of generating T: $\tilde{O}(N_1^{\max} \cdot |T|) = 2^{(c_g + 2c_t)n + o(n)}$.
- Cost of generating S: $\tilde{O}(N_2 \cdot |T|) = 2^{(c_g + c_b/2 + c_t)n + o(n)}$.
- Cost of searching S for a pair of close vectors: $\tilde{O}(|S|^2) = 2^{(2c_g + c_b)n + o(n)}$.

The space complexity is at most $O(|T| + |S|) = 2^{\max(c_t, c_g + c_b/2)n + o(n)}$. In [46], this lead to the following lemma.

Lemma 5. *[46] Let $\xi > \frac{1}{2}$ and $R > 2\xi$, and suppose $\mu > \lambda_1(L)$. Then, with c_b, c_t, c_g, N_B, N_V, N_G, $N_1^{\max}$, N_2 chosen according to:*

$$c_b = \log_2(R) + 0.401, \qquad\qquad N_B = 2^{c_b n + o(n)}, \qquad (2)$$

$$c_t = \frac{1}{2}\log_2\left(1 + \frac{2\xi}{R - 2\xi}\right) + 0.401, \qquad N_T = 2^{c_t n + o(n)}, \qquad (3)$$

$$c_g = \frac{1}{2}\log_2\left(\frac{4\xi^2}{4\xi^2 - 1}\right), \qquad\qquad N_G = 2^{c_g n + o(n)}, \qquad (4)$$

$$N_1^{\max} = 2^{(c_g + c_t)n + o(n)}, \qquad\qquad N_2 = 2^{(c_g + c_b/2)n + o(n)}, \qquad (5)$$

with probability at least $\frac{1}{16}$, Algorithm 2 returns a lattice vector $\mathbf{s} \in L \setminus \{\mathbf{0}\}$ with $\|\mathbf{s}\| < \mu$, in time at most $2^{tn + o(n)}$ and space at most $2^{sn + o(n)}$, where t and s are given by

$$t = \max\left(c_g + 2c_t, c_g + \frac{c_b}{2} + c_t, 2c_g + c_b\right), \quad s = \max\left(c_t, c_g + \frac{c_b}{2}\right). \quad (6)$$

In the quantum setting, the costs are as follows.

- Cost of generating T: $\tilde{O}(N_1^{\max} \cdot \sqrt{|T|}) = 2^{(c_g + 3c_t/2)n + o(n)}$.
- Cost of generating S: $\tilde{O}(N_2 \cdot \sqrt{|T|}) = 2^{(c_g + c_b/2 + c_t/2)n + o(n)}$.
- Cost of searching S for a pair of close vectors: $\tilde{O}(\sqrt{|S|^2}) = 2^{(c_g + c_b/2)n + o(n)}$.

The total space complexity is still the same as in the classical setting, i.e., at most $O(|T| + |S|) = 2^{\max(c_t, c_g + c_b/2)n + o(n)}$. This leads to the following lemma.

Lemma 6. *Let $\xi > \frac{1}{2}$ and $R > 2\xi$, and suppose $\mu > \lambda_1(L)$. Then, with c_b, c_t, c_g, N_B, N_V, N_G, $N_1^{\max}$, N_2 chosen according to Equations (2) to (5), with probability at least $\frac{1}{16}$, Algorithm 2 returns a lattice vector $\mathbf{s} \in L \setminus \{\mathbf{0}\}$ with $\|\mathbf{s}\| < \mu$ on a quantum computer in time at most $2^{\tilde{t}n + o(n)}$ and space at most $2^{\tilde{s}n + o(n)}$, where $\tilde{t}$ and $\tilde{s}$ are given by*

$$\tilde{t} = \max\left(c_g + \frac{3c_t}{2}, c_g + \frac{c_b}{2} + \frac{c_t}{2}, c_g + \frac{c_b}{2}\right), \quad \tilde{s} = \max\left(c_t, c_g + \frac{c_b}{2}\right). \quad (7)$$

Optimizing ξ and R for the minimum time complexity, we get $\xi \approx 0.9086$ and $R \approx 3.1376$ as in Theorem 2. Note that if S is generated in parallel with exponentially many quantum computers, the cost of the second part of the algorithm becomes negligible, and the exponent in the time complexity changes to

$$\tilde{t}' = \max\left(c_g + \frac{3c_t}{2}, c_g + \frac{c_b}{2}\right). \quad (8)$$

In that case, the optimal choice of ξ and R (with respect to minimizing the time complexity) would be $\xi \approx 1.0610$ and $R \approx 4.5166$, leading to a time complexity of less than $2^{1.470n + o(n)}$.

C   Other SVP Algorithms

C.1   Enumeration

Recall that enumeration considers all lattice vectors inside a giant ball around the origin that is known to contain at least one lattice vector. Let L be a lattice with basis $\{\mathbf{b}_1, \ldots, \mathbf{b}_n\}$. Consider each lattice vector $\mathbf{u} \in L$ as a linear combination of the basis vectors, i.e., $\mathbf{u} = \sum_i u_i \mathbf{b}_i$. Now, we can represent each lattice vector by its coefficient vector $(u_1, \ldots, u_n)$. We would like to have all combinations of values for $(u_1, \ldots, u_n)$ such that the corresponding vector $\mathbf{u}$ lies in the ball. We could try any combination and see if it lies within the ball by computing the norm of the corresponding vector, but there is a smarter way that ensures we only consider vectors that lie within the ball and none that lie outside.

To this end, enumeration algorithms search from right to left, by identifying all values for u_n such that there might exist $u'_1, \ldots, u'_{n-1}$ such that the vector corresponding to $(u'_1, \ldots, u'_{n-1}, u_n)$ lies in the ball. To identify these values $u'_1, \ldots, u'_{n-1}$, enumeration algorithms use the Gram-Schmidt orthogonalization of the lattice basis as well as the projection of lattice vectors. Then, for each of these possible values for u_n, the enumeration algorithm considers all possible values for u_{n-1} and repeats the process until it reaches possible values for u_1. This leads to a search which is serial in nature, as each value of u_n will lead to different possible values for u_{n-1} and so forth. Unfortunately, we can only really apply the quantum search algorithm to problems where the list of objects to be searched is known in advance.

One might suggest to forego the smart way to find short vectors and just search all combinations of $(u_1, \ldots, u_n)$ with appropriate upper and lower bounds on the different u_i's. Then it becomes possible to apply quantum search, since we now have a predetermined list of vectors and just need to compute the norm of each vector. However, it is doubtful that this will result in a faster algorithm, because the recent heuristic changes by Gama et al. [19] have reduced the running time of enumeration dramatically (roughly by a factor $2^{n/2}$) and these changes only complicate the search area further by changing the ball to an ellipsoid. There seems to be no simple way to apply quantum search to the enumeration algorithms that are currently used in practice, but perhaps the algorithms can be modified in some way.

C.2   Voronoi Cell

Consider a set of points in the Euclidean space. For any given point in this set, its Voronoi cell is the region that contains all vectors that lie closer to this point than to any of the other points in the set. Now, given a Voronoi cell, we define a relevant vector to be any vector in the set whose removal from the set will change this particular Voronoi cell. If we pick our lattice as the set and we consider the Voronoi cell around the zero vector, then any shortest vector is also a relevant vector. Furthermore, given the relevant vectors of the Voronoi cell we can solve the closest vector problem in $2^{2n+o(n)}$ time.

So how can we compute the relevant vectors of the Voronoi cell of a lattice L? Micciancio and Voulgaris [40] show that this can be done by solving $2^n - 1$ instances of CVP in the lattice $2L$. However, in order to solve CVP we would need the relevant vectors which means we are back to our original problem. However, Micciancio and Voulgaris show that these instances of CVP can also be solved by solving several related CVP instances in a lattice of lower rank. They give a basic and an optimized version of the algorithm. The basic version only uses LLL as preprocessing and solves all these related CVP instances in the lower rank lattice separately. As a consequence, the basic algorithm runs in time $2^{3.5n+o(n)}$ and in space $2^{n+o(n)}$. The optimized algorithm uses a stronger preprocessing for the lattice basis, which takes exponential time. But since the most expensive part is the computation of the Voronoi relevant vectors, this extra preprocessing time does not increase the asymptotic running time as it is executed only once. In fact, having the reduced basis decreases the asymptotic running time to $\tilde{O}(2^{3n})$. Furthermore, the optimized algorithm employs a trick that allows it to reduce 2^k CVP instances in a lattice of rank k to a single instance of an enumeration problem related to the same lattice. The optimized algorithm solves CVP in time $\tilde{O}(2^{2n})$ using $\tilde{O}(2^n)$ space.

Now, in the basic algorithm, it would be possible to speed up the routine that solves the CVP given the Voronoi relevant vectors using a quantum computer. It would also be possible to speed up the routine that removes non-relevant vectors from the list of relevant vectors using a quantum computer. Combining these two changes gives a quantum algorithm with an asymptotic running time $\tilde{O}(2^{2.5n})$, which is still slower than the optimized classical algorithm. It is not possible to apply these same speedups to the optimized algorithm due to the aforementioned trick with the enumeration problem. The algorithm to solve this enumeration problem makes use of a priority queue, which means the search is not trivially parallellized. Once again, there does not seem to be a simple way to apply quantum search to this special enumeration algorithm. However, it may be possible that the algorithm can be modified in such a way that quantum search can be applied.

An Efficient Attack of a McEliece Cryptosystem Variant Based on Convolutional Codes

Grégory Landais and Jean-Pierre Tillich

SECRET Project - INRIA Rocquencourt
Domaine de Voluceau, B.P. 105 78153 Le Chesnay Cedex, France
{gregory.landais,jean-pierre.tillich}@inria.fr

Abstract. Löndahl and Johansson proposed last year a variant of the McEliece cryptosystem which replaces Goppa codes by convolutional codes. This modification is supposed to make structural attacks more difficult since the public generator matrix of this scheme contains large parts that are generated completely at random. They proposed two schemes of this kind, one of them consists in taking a Goppa code and extending it by adding a generator matrix of a time varying convolutional code. We show here that this scheme can be successfully attacked by looking for low-weight codewords in the public code of this scheme and using it to unravel the convolutional part. It remains to break the Goppa part of this scheme which can be done in less than a day of computation in the case at hand.

Keywords: Code-based cryptography, McEliece cryptosystem, convolutional codes, cryptanalysis.

1 Introduction

In [Sho97], Peter Shor showed that all cryptosystems based on the hardness of factoring or taking a discrete logarithm can be attacked in polynomial time with a quantum computer (see [BBD09] for an extensive report). This threatens most if not all public-key cryptosystems deployed in practice, such as RSA [RSA78] or DSA [Kra91]. Cryptography based on the difficulty of decoding a linear code, on the other hand, is believed to resist quantum attacks and is therefore considered as a viable replacement for those schemes in future applications. Yet, independently of their so-called "post-quantum" nature, code-based cryptosystems offer other benefits even for present-day applications due to their excellent algorithmic efficiency, which is up to several orders of complexity better than traditional schemes.

The first code-based cryptosystem is the McEliece cryptosystem [McE78], originally proposed using Goppa codes. Afterwards several code families have been suggested to replace the Goppa codes in this scheme: generalized Reed–Solomon codes (GRS) [Nie86] or subcodes of them [BL05], Reed–Muller codes [Sid94], algebraic geometry codes [JM96], LDPC codes [BBC08], MDPC codes

P. Gaborit (Ed.): PQCrypto 2013, LNCS 7932, pp. 102–117, 2013.

[MTSB12] or more recently convolutional codes [LJ12]. Some of these schemes allow to reduce the public key size compared to the original McEliece cryptosystem while presumably keeping the same level of security against generic decoding algorithms.

However, for several of the aforementioned schemes it has been shown that a description of the underlying code suitable for decoding can be obtained- this breaks the corresponding scheme. This has been achieved for generalized Reed-Solomon codes in [SS92] and for subcodes of generalized Reed-Solomon codes in [Wie10]. In this case, the attack takes polynomial time and recovers the complete structure of the underlying generalized Reed–Solomon code from the public key G'. The Reed-Muller code scheme has also been attacked, but this time the algorithm recovering the secret description of the permuted Reed-Muller code has sub-exponential complexity [MS07] which is enough for attacking the scheme with the parameters proposed in [Sid94] but which is not sufficient to break the scheme completely. Algebraic geometry codes are broken in polynomial time but only for low genus hyperelliptic curves [FM08]. Finally, it should be mentioned that a first version of the scheme based on LDPC codes proposed in [BC07] has been successfully attacked in [OTD10] (but the new scheme proposed in [BBC08] seems to be immune to this kind of attack), that a variant [BBC+11] of the generalized Reed-Solomon scheme which was supposed to resist to the attack of [SS92] has recently been broken in [GOT12], that another variant of the Generalized Reed-Solomon scheme [Wie06] has been broken in [CGG+13], both by an approach that is related to the distinguisher of Goppa codes which is proposed in [FGO+11] (see [FGO+10] for the full version).

The original McEliece cryptosystem with Goppa codes is still unbroken. It was modified in [BCGO09, MB09] by considering quasi-cyclic or quasi-dyadic versions of Goppa codes (or more generally of alternant codes in [BCGO09]) in order to reduce significantly the key size. However, in this case it was shown in [FOPT10, UL09] that the added structure allows a drastic reduction of the number of unknowns in algebraic attacks and most of the schemes proposed in [BCGO09, MB09] were broken by this approach. This kind of attack has exponential complexity and it can be thwarted by choosing smaller cyclic or dyadic blocks in this approach in order to increase the number of unknowns of the algebraic system. When the rate of the Goppa code is close to 1 (as is the case in signature schemes for instance [CFS01]) then it has been shown in [FGO+11] that the public key can be distinguished from a random public key. This invalidates all existing security proofs of the McEliece cryptosystem when the code rate is close to 1 since they all rely on the hardness of two problems: the hardness of decoding a generic linear code on one hand and the indistinguishability of the Goppa code family on the other hand.

These algebraic attacks motivate the research of alternatives to Goppa codes in the McEliece cryptosystem and it raises the issue of what kind of codes can be chosen in the McEliece cryptosystem. The proposal with convolutional codes made in [LJ12] falls into this thread of research. What makes this new scheme interesting is the fact that its secret generator matrix contains large parts which are generated completely at random and has no algebraic structure as in other schemes such as generalized Reed-Solomon codes, algebraic geometry codes, Goppa codes or Reed-Muller codes.

In [LJ12] two schemes are given. The first one simply considers as the secret key the generator matrix of a time varying tail-biting convolutional code. A scheme for which it is supposed to resist to attacks of time complexity of about 2^{80} elementary operations is suggested and has reasonable decoding complexity. This construction presents however the drawback that the complexity of decoding scales exponentially with the security level measured in bits. The authors give a second scheme which is scalable and which is built upon a Goppa code and extends it by adding a generator matrix of a time varying convolutional code.

We study the security of this second scheme in this article. It was advocated that the convolutional structure of the code can not be recovered due to the fact that the dual code has large enough minimum distance. However, we show here that this extra defense can be successfully attacked by looking for low-weight codewords in the public code of this scheme. By a suitable filtering procedure of these low weight codewords we can unravel the convolutional part.

The main point that makes this attack feasible is the following phenomenon: the public code of this scheme contains subcodes of much smaller support but whose rate is not much smaller than the rate of the public code. The support of such codes can be easily found by low weight codewords algorithms. It is worthwhile to notice that the code-based KKS signature scheme [KKS97] was broken with exactly the same approach [OT11]. It turns out that the support of these subcodes reveals the convolutional structure. By suitably puncturing the public code, only the Goppa part remains. Deciphering an encrypted message can then be done because for the concrete parameters example provided in [LJ12], algorithms for decoding general linear codes can be used in this case to decode the Goppa code successfully. This attack works successfully on the parameters proposed in [LJ12] and needs only a few hours of computation. It should be possible to change the parameters of the scheme to avoid this kind of attack. In order to do so an improved attack is suggested in Subsection 5.1, its complexity is analyzed in Section 5. This suggests that it should be possible to repair the scheme by fixing the parameters in a more conservative way. Some indications about how to perform such a task are given in Subsection 5.3.

2   The McEliece Scheme Based on Convolutional Codes

The scheme can be summarized as follows.

Secret key.

- G_{sec} is a $k \times n$ generator matrix which has a block form specified in Figure 1;
- P is an $n \times n$ permutation matrix;
- S is a $k \times k$ random invertible matrix over $\mathbb{F}_2$.

Public key. $G_{\text{pub}} \stackrel{\text{def}}{=} SG_{\text{sec}}P$.

Encryption. The ciphertext $c \in \mathbb{F}_2^n$ of a plaintext $m \in \mathbb{F}_2^k$ is obtained by drawing at random e in $\mathbb{F}_2^n$ of weight equal to some quantity t and computing $c \stackrel{\text{def}}{=} mG_{\text{pub}} + e$.

Decryption. It consists in performing the following steps

1. Calculating $c' \stackrel{\text{def}}{=} cP^{-1} = mSG_{\text{sec}} + eP^{-1}$ and using the decoding algorithm of the code with generator matrix G_{sec} to recover mS from the knowledge of c';
2. Multiplying the result of the decoding by S^{-1} to recover m.

The point of the whole construction is that if t is well chosen, then with high probability the Goppa code part can be decoded, and this allows a sequential decoder of the time varying convolutional code to decode the remaining errors. From now on we will denote by $\mathscr{C}_{\text{pub}}$ the code with generator matrix G_{pub} and by $\mathscr{C}_{\text{sec}}$ the code with generator matrix G_{sec}.

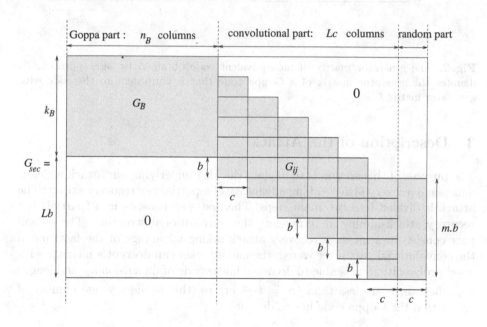

Fig. 1. The secret generator matrix. The areas in light pink indicate the only non zero parts of the matrix. G_B is a generator matrix of a binary Goppa code of length n_B and dimension k_B. This matrix is concatenated with a matrix of a time varying binary convolutional code where b bits of information are transformed into c bits of data (the corresponding G_{ij} blocks are therefore all of size $b \times c$) and terminated with c random columns at the end. The dimension of the corresponding code is $k \stackrel{\text{def}}{=} k_B + Lb$ and the length is $n \stackrel{\text{def}}{=} n_B + (L+1)c$ where L is the time duration of the convolutional code.

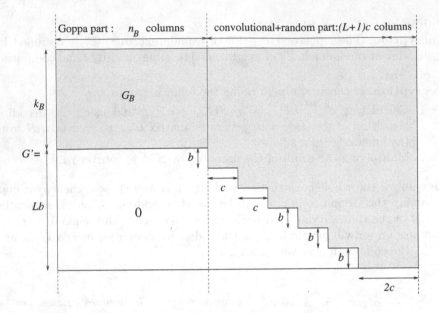

Fig. 2. The generator matrix of an equivalent code obtained by our approach. G'_B denotes the generator matrix of a Goppa code that is equivalent to the code with generator matrix G_B

3   Description of the Attack

The purpose of this section is to explain the idea underlying our attack which is a message recovery attack taking advantage of a partial key recovery attack. The attack is divided into two main steps. The first step consists in a (partial) key recovery attack aiming at unraveling the convolutional structure. The second part consists in a message recovery attack taking advantage of the fact that if the convolutional part is recovered, then an attacker can decrypt a message with good probability if he is able to decode a linear code of dimension k_B and length n_B when there are less than $t_B \stackrel{\text{def}}{=} t\frac{n_B}{n}$ errors (this is the average number of errors that the Goppa code has to decode).

3.1   Unraveling the Convolutional Structure

The authors of [LJ12] have chosen the parameters of their scheme so that it remains hard to find low-weight codewords in the dual of the public code $\mathscr{C}_{\text{pub}}$. It is advocated in [LJ12] that in their case the only deviation from a random code is the convolutional structure in terms of low weight parity-checks. For instance, the following parameters are suggested $(n, k, c, b, t) = (1800, 1200, 30, 20, 45)$ and in the construction phase the authors propose to throw away any code that would have parity-checks of weight less than 125. However, the fact that the structure of $\mathscr{C}_{\mathrm{pub}}$ leads in a natural way to low weight codewords is not taken into account. Indeed, we expect many (i.e. about 2^{b-1}) codewords of weight less than or equal to c. This comes from the fact that the subcode of $\mathscr{C}_{\mathrm{pub}}$ generated by the last b rows of $\boldsymbol{G}_{\mathrm{sec}}$ (and permuted by $\boldsymbol{P}$) has support of size $2c$ and dimension b. Therefore any algorithm aiming at finding codewords of weight less than c say should output such codewords. Looking at the support of such codewords reveals the $2c$ last columns of $\boldsymbol{G}_{\mathrm{sec}}$. By puncturing these columns and starting this process again but this time by looking for codewords of weight less than $c/2$ (since this time the punctured code contains a subcode of dimension b and support of size c arising from the penultimate block of rows of $\boldsymbol{G}_{\mathrm{sec}}$) will reveal the following block of c columns of the matrix. In other words we expect to capture by these means a first subcode of dimension b and support the $2c$ last positions of $\mathscr{C}_{\mathrm{sec}}$. Then we expect a second subcode of dimension b with support the $3c$ last positions of $\mathscr{C}_{\mathrm{pub}}$ and so on and so forth. Finally we expect to be able after suitable column swapping to obtain the generator matrix $\boldsymbol{G}'$ of an equivalent code to $\mathscr{C}_{\mathrm{pub}}$ that would have the form indicated in Figure 2.

More precisely the algorithm for finding a generator matrix of a code equivalent to $\mathscr{C}_{\mathrm{pub}}$ is given by Algorithm 1 given below.

Algorithm 1. An algorithm for finding $\boldsymbol{G}'$.

---

input: $\boldsymbol{G}_{\mathrm{pub}}$ the public generator matrix
output: a generator matrix $\boldsymbol{G}'$ of a code equivalent to $\mathscr{C}_{\mathrm{pub}}$ which has the form indicated in Fig. 2

$\mathcal{L} \leftarrow []$
for $i = L, \dots, 1$ do
    $\boldsymbol{G} \leftarrow$ GeneratorMatrixPuncturedCode$(\mathscr{C}_{\mathrm{pub}}, \mathcal{L})$
    $\boldsymbol{G} \leftarrow$ LowWeight$(\boldsymbol{G}, w)$
    $w \leftarrow$ Function(i)
    $\boldsymbol{G}_i \leftarrow$ ExtendedGeneratorMatrix$(\boldsymbol{G}, \mathcal{L}, \mathscr{C}_{\mathrm{pub}})$
    $\mathcal{L} \leftarrow$ Support$(\boldsymbol{G}) \| \mathcal{L}$
end for
$\boldsymbol{G} \leftarrow$ GeneratorMatrixPuncturedCode$(\mathscr{C}_{\mathrm{pub}}, \mathcal{L})$
$\boldsymbol{G}_0 \leftarrow$ ExtendedGeneratorMatrix$(\boldsymbol{G}, \mathcal{L}, \mathscr{C}_{\mathrm{pub}})$
$\boldsymbol{G}'$ is the concatenation of the rows of $\boldsymbol{G}_0, \boldsymbol{G}_1, \dots, \boldsymbol{G}_L$.
return $\boldsymbol{G}'$

We assume here that:

- the function GeneratorMatrixPuncturedCode takes as input a code $\mathscr{C}$ of length n and an ordered set of positions $\mathcal{L}$ which is a sublist of $[1, 2, \dots, n]$ and outputs a generator matrix of $\mathscr{C}$ punctured in the positions belonging to $\mathcal{L}$;
- Function will be a certain function which will be specified later on;
- Support$(\mathscr{C})$ yields the (ordered) support of $\mathscr{C}$ and $\|$ is the concatenation of lists;

- the function LowWeight takes as input a code $\mathscr{C}$ and a weight w. It outputs a generator matrix of a subcode of $\mathscr{C}$ obtained by looking for codewords of weight less than or equal to w. Basically a certain number of codewords of weight $\leq w$ are produced and the positions that are involved in at least t codewords are put in a list $\mathcal{L}$ (where t is some threshold depending on the weight w, the length n of the code, its dimension k and the number of codewords produced by the previous call of the function), which means that i is taken as soon as there are at least c elements in $\mathscr{C}$ for which $c_i = 1$. Then a generator matrix for the subcode of $\mathscr{C}$ formed by the codewords of $\mathscr{C}$ whose coordinates outside $\mathcal{L}$ are all equal to 0 is returned. See Algorithm 2 for further details.
- the function ExtendedGeneratorMatrix takes as input a generator matrix of some code $\mathscr{C}'$, an ordered set of positions $\mathcal{L}$ and a code $\mathscr{C}$ such that $\mathscr{C}'$ is the result of the puncturing of $\mathscr{C}$ in the positions belonging to $\mathcal{L}$. It outputs a generator matrix of the permuted subcode $\mathscr{C}''$ of $\mathscr{C}$ whose positions are reordered in such a way that the first positions correspond to the positions of $\mathscr{C}'$ and the remaining positions to the ordered list $\mathcal{L}$. This code $\mathscr{C}''$ corresponds to the codewords of $\mathscr{C}'$ that are extended as codewords of $\mathscr{C}$ over the positions belonging to $\mathcal{L}$ in an arbitrary linear way.

3.2   Finishing the Job: Decoding the Code with Generator Matrix G'_B

If we are able to decode the code with generator matrix G'_B, then standard sequential decoding algorithms for convolutional codes will allow to decode the last $(L + 1)c$ positions. Let G'_B be the generator matrix of a code equivalent to the secret Goppa code chosen for the scheme specified in Figure 2. Decoding such a code can be done by algorithms aiming at decoding generic linear codes such as Stern's algorithm [Ste88] and its subsequent improvements [Dum91, BLP11, MMT11, BJMM12]. This can be done for the parameters suggested in [LJ12].

4   Implementation of the Attack for the Parameters Suggested in [LJ12]

We have carried out the attack on the parameters suggested in [LJ12]. They are provided in Table 1.

Table 1. Parameters for the second scheme suggested in [LJ12]

n	n_B	k	k_B	b	c	L	m	t (number of errors)
1800	1020	1160	660	20	30	25	12	45

Setting the weight parameter w accurately when calling the function LowWeight is the key for finding the 60 last positions. If w is chosen to be too large, for

---

Algorithm 2. LowWeight(G, w)

---

input:

- G a certain $k \times n$ generator matrix of a code $\mathscr{C}$;
- w a certain weight

output: a generator matrix G' of a subcode of $\mathscr{C}$ obtained from the supports of a certain subset of codewords of weight w in $\mathscr{C}$.

$\mathscr{C} \leftarrow$ LowWeightCodewordSearch(G, w) {Produces a set of linear combinations of rows of G of weight $\leq w$}
Initialize an array tab of length n to zero
$t \leftarrow$ Threshold($w, n, k, |\mathscr{C}|$)
for all $c \in \mathscr{C}$ do
  for $i \in [1..n]$ do
    if $c_i = 1$ then
      tab[i] $\leftarrow$ tab[i] $+ 1$
    end if
  end for
end for
$\mathcal{L} \leftarrow []$
for $i \in [1..n]$ do
  if tab[i] $\geq$ t then
    $\mathcal{L} \leftarrow \mathcal{L} || \{i\}$
  end if
end for
$G' \leftarrow$ ShortenedCode($G, \mathcal{L}$) {Produces a generator matrix for the subcode of $\mathscr{C}$ formed by the codewords of $\mathscr{C}$ whose coordinates outside $\mathcal{L}$ are all equal to 0.}
return G'.

---

instance when $w = 22$, running Dumer's low weight codeword search algorithm [Dum91] gave the result given in Figure 3 concerning the frequencies of the code positions involved in the codewords of weight less than 22 output by the algorithm and stored in table *tab* during the execution of the algorithm.

We see in Figure 3 that this discriminates the 90 last code positions and not as we want the 60 last code positions. However choosing w to be equal to 18 enables to discriminate the 60 last positions as shown in Figure 4.

Data used in Figure 4 come from 3900 codewords generated in one hour and a half on an Intel Xeon W3550 (3 GHz) CPU by a monothread implementation in C of Dumer's algorithm. The message recovery part of the attack involving the Goppa code consists in decoding 25.5 errors on average in a linear code of dimension 660 and length 1020. The time complexity is about 2^{42}. This second part of the attack could be achieved using the previous program on the same computer in about 6.5 hours on average.

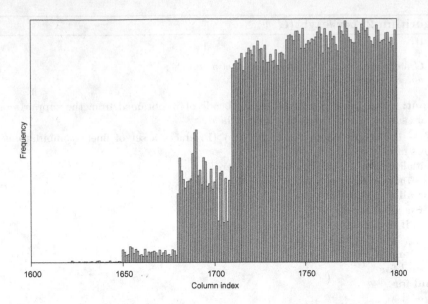

Fig. 3. The frequencies of the code positions involved in codewords of weight ≤ 22 output by Dumer's algorithm

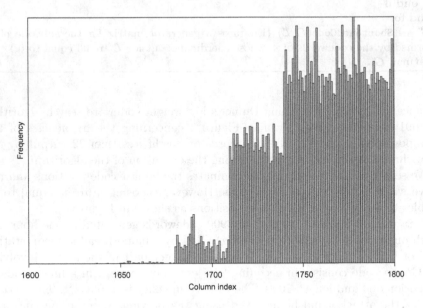

Fig. 4. The frequencies of the code positions involved in codewords of weight ≤ 18 output by Dumer's algorithm

5 Analysis of the Security of the Scheme

5.1 An Improved Attack

The purpose of this section is to provide a very crude analysis of the security of the scheme. We will not analyze our attack detailed in Section 3, since even though it was enough to break the second scheme suggested in [LJ12] it is not the most efficient one. We will give a sketch of a better attack and a rough analysis for it. Basically, the real threat on this scheme comes from the fact that there exists a subcode $\mathscr{C}$ of $\mathscr{C}_{\text{pub}}$ of very small support (of size $2c$ here), namely the code generated by the last b rows of G_{sec} permuted by the secret permutation matrix P. For instance, there are about 2^{b-1} codewords of weight less than or equal to c that should be found by a low weight codeword searching algorithm and that should reveal the support of $\mathscr{C}$. This is basically the idea underlying our attack. However there are other subcodes of rather small support that yield low weight codewords, namely the codes $\mathscr{C}_s$ generated by the $s \times b$ last rows of G for s ranging between 2 and L. The support of $\mathscr{C}_s$ is of size $(s+1)c$. Notice that its rate gets closer and closer to the rate $\frac{2}{3}$ (which is more or less the rate of the final code) as s increases. This is a phenomenon that helps low weight codeword algorithms as will be explained later on.

An improvement of our attack would consist in using a low weight codeword algorithm in order to find one of the codewords of $\mathscr{C}_s$ and to use this codeword c to bootstrap from here to find the whole support of $\mathscr{C}_s$. This is very much in the spirit of the attack against the KKS scheme which is explained in Algorithm 2 which can be found in Subsection 4.4 of [OT11]. With this approach, by using the codeword that has already been found, it is much easier to find new ones belonging to the same subcode with small support by imposing that the information set used for finding low weight codewords is chosen outside the positions belonging to the support of c. The complexity of the whole attack is dominated in this case by the complexity of finding just one codeword in $\mathscr{C}$ when there is a good way to identify the candidates in $\mathscr{C}$ (which can be done by checking the weight of c). Notice that it is very likely that $\mathscr{C}$ is actually the sub code of $\mathscr{C}$ of dimension b that has the smallest support. Recall here that this is precisely the notion captured by the generalized Hamming weights of a code [WY93], w_i being defined as the smallest support of a subcode of dimension i. In other words w_1 is nothing but the minimum distance of the code and in our case it is likely that $w_b = 2c$ (and more generally $w_{sb} = (s+1)c$ for $s = 1..L$). In other words, the problem which should be difficult to solve is the following one

Problem 1. Find one of the subcodes of dimension $s \times b$ whose support size is the $s \times b$-th generalized Hamming weight of $\mathscr{C}_{\text{pub}}$.

We will focus now on the following approach to solve this problem. Consider a low weight codeword algorithm that aims at finding low weight codewords in a code of dimension k by picking up a random set of positions $\mathcal{I}$ of size slightly larger than k, say $k+l$ and that looks for all (or at least a non-negligible fraction) of codewords that have weight equal to some small quantity p over these positions. These quantities are very good candidates for having low weight over the whole support. This is precisely the approach that is followed in the best low weight codeword search algorithms such as [Ste88, Dum91, MMT11, BJMM12]. We run such an algorithm for several different sets $\mathcal{I}$ and will be interested in the complexity of outputting at least one codeword that belongs to $\mathscr{C}$. This is basically the approach that has been very successful to break the KKS scheme [OT11] and that is the natural candidate to break the [LJ12] scheme.

To analyze such an algorithm we will make some simplifying assumptions

- The cost of checking one of those $\mathcal{I}$ is of order $O\left(\mathscr{L} + \frac{\mathscr{L}^2}{2^l}\right)$ where $\mathscr{L} = \sqrt{\binom{k+l}{p}}$. We neglect here the cost coming from writing the parity-check matrix in systematic form and this does not really cover the recent improvements in [MMT11, BJMM12]. We have made here such an approximation for sake of simplicity. We refer to [FS09] for an explanation of this cost.
- Denote by k' the dimension of the subcode $\mathscr{C}$, by n' the size of its support $\mathcal{J}$. We assume that the result of the puncturing of $\mathscr{C}$ by all positions that do no belong to $\mathcal{J}$ behaves like a random code of dimension k' and length n'.

Our main result to analyze such an algorithm consists in the following proposition.

Proposition 1. *Let*

- *$f(x)$ be the function defined over the positive reals by $f(x) \stackrel{def}{=} \max\left(x(1 - x/2), 1 - \frac{1}{x}\right)$;*
- *$\pi(s) \stackrel{def}{=} \frac{\binom{n'}{s}\binom{n-n'}{k+l-s}}{\binom{n}{k+l}}$;*
- *$\lambda(s) \stackrel{def}{=} \binom{s}{p}2^{k'-s}$.*
- *$C(k,l,p) \stackrel{def}{=} \mathscr{L} + \frac{\mathscr{L}^2}{2^l}$ where $\mathscr{L} \stackrel{def}{=} \sqrt{\binom{k+l}{p}}$;*
- *$\Pi \stackrel{def}{=} \sum_{s=1}^{n'} \pi(s) f(\lambda(s))$.*

Then the complexity that the low weight codeword search algorithm outputs an element in $\mathscr{C}$ is of order

$$O\left(\frac{C(k,l,p)}{\Pi}\right).$$

5.2 Proof of Proposition 1

Our first ingredient is a lower bound on the probability that a given set $X \subseteq \mathbb{F}_2^n$ intersects a random linear code $\mathscr{C}_{\mathrm{rand}}$ of dimension k and length n picked up uniformly at random. This lemma gives a sharp lower bound even when X is very large and when there is a big gap between the quantities $\mathbf{prob}(X \cap \mathscr{C}_{\mathrm{rand}} \neq \emptyset) = \mathbf{prob}(\cup_{x \in X}\{x \in \mathscr{C}_{\mathrm{rand}}\})$ and $\sum_{x \in X} \mathbf{prob}(x \in \mathscr{C}_{\mathrm{rand}})$.

Lemma 1. *Let X be some subset of $\mathbb{F}_2^n$ of size m and let f be the function defined over the positive reals by $f(x) \overset{def}{=} \max\left(x(1 - x/2), 1 - \frac{1}{x}\right)$. We denote by x the quantity $\frac{m}{2^{n-k}}$, then*

$$\mathbf{prob}(X \cap \mathscr{C}_{rand} \neq \emptyset) \geq f(x).$$

This lemma can be found in [OT11] and it is proved there.

Let us finish now the proof of Proposition 1. Denote by $\mathcal{J}$ the support of $\mathscr{C}$:

$$\mathcal{J} \overset{def}{=} \mathsf{supp}(\mathscr{C}).$$

Let us first calculate the expected number of sets $\mathcal{I}$ we have to consider before finding an element of $\mathscr{C}$. Such an event happens precisely when there is a nonzero word in $\mathscr{C}$ whose restriction to $\mathcal{I} \cap \mathcal{J}$ is of weight equal to p. Let $\mathscr{C}_{\mathcal{I} \cap \mathcal{J}}$ be the restriction of the codewords of $\mathscr{C}$ to the positions that belong to $\mathcal{I} \cap \mathcal{J}$, that is

$$\mathscr{C}_{\mathcal{I} \cap \mathcal{J}} \overset{def}{=} \{(c_i)_{i \in \mathcal{I} \cap \mathcal{J}} : (c_i)_{1 \leq i \leq n} \in \mathscr{C}\}.$$

Let X be the set of non-zero binary words of support $\mathcal{I} \cap \mathcal{J}$ that have weight equal to p. Denote by W the size of $\mathcal{I} \cap \mathcal{J}$. The probability that W is equal to s is precisely

$$\mathbf{prob}(W = s) = \frac{\binom{n'}{s}\binom{n-n'}{k+l-s}}{\binom{n}{k+l}} = \pi(s).$$

Then the probability Π that a certain choice of $\mathcal{I}$ gives among the codewords considered by the algorithm a codeword of $\mathscr{C}$ can be expressed as

$$\Pi = \sum_{s=1}^{n'} \mathbf{prob}(W = s)\mathbf{prob}(X \cap \mathscr{C}_{\mathcal{I} \cap \mathcal{J}} \neq \emptyset) \tag{1}$$

$$\geq \sum_{s=1}^{n'} \pi(s)f(\lambda(s)) \tag{2}$$

by using Lemma 1 with $\mathscr{C}_{\mathcal{I} \cap \mathcal{J}}$ and the aforementioned X. Therefore the average number of iterations that have to be performed before finding an element in $\mathscr{C}$ is equal to $\frac{1}{\Pi}$ and this yields immediately Proposition 1.

5.3 Repairing the Parameters and a Pitfall

A possible way to repair the scheme consists in increasing the size of the random part (which corresponds to the last c columns in G_{sec} here). Instead of choosing this part to be of size c as suggested in [LJ12], its size can be increased in order to thwart the algorithm of Subsection 5.1. Let r be the number of random columns we add at the end of the convolutional part, so that the final length of the code is now $n_B + Lc + r$ instead of $n_B + (L + 1)c$ as before. If we choose r to be equal to 140, then the aforementioned attack needs about 2^{80} operations before outputting an element of $\mathscr{C}$ which is the (permuted) subcode corresponding to the last b rows of $\boldsymbol{G}_{\text{sec}}$. As before, let us denote by $\mathscr{C}_s$ the permuted (by $\boldsymbol{P}$) subcode of $\mathscr{C}_{\text{pub}}$ generated by the last $s \times b$ rows of $\boldsymbol{G}_{\text{sec}}$ permuted by $\boldsymbol{P}$. We can use the previous analysis to estimate the complexity of obtaining an element of $\mathscr{C}_s$ by the previous algorithm. We have gathered the results in Table 2.

Table 2. Complexity of obtaining at least one element of $\mathscr{C}_s$ by the algorithm of Subsection 5.1

s	1	5	10	15	20	21	22	25
complexity (bits)	80.4	72.1	65.1	61.0	59.4	59.3	59.4	59.8

We see from this table that in this case the most important threat does not come from finding low weight codewords arising from codewords in $\mathscr{C}_1$, but codewords of moderate weight arising from codewords in $\mathscr{C}_{20}$ for instance. Codewords in this code have average weight $\frac{r+20c}{2} = 370$. This implies that a simple policy for detecting such candidates which consists in keeping all the candidates in the algorithm of Subsection 5.1 that have weight less than this quantity is very likely to filter out the vast majority of bad candidates and keep with a good chance the elements of $\mathscr{C}_{20}$. Such candidates can then be used as explained in Subsection 5.1 to check whether or not they belong to a subcode of large dimension and small support.

There is a simple way for explaining what is going on here. Notice that the rate of $\mathscr{C}$ is equal to $\frac{b}{c+r}$, which is much smaller than the rate of the overall scheme that is close to $\frac{b}{c}$ in this case by the choice of the parameters of the Goppa code. However as s increases, the rate of $\mathscr{C}_s$ gets closer and closer to $\frac{b}{c}$, since its rate is given by $\frac{sb}{sc+r} = \frac{b}{c+r/s}$. Assume for one moment that the rate of $\mathscr{C}_s$ is equal to $\frac{b}{c}$. Then putting $\boldsymbol{G}_{\text{pub}}$ in systematic form (which basically means that we run the aforementioned algorithm with $p = 1$ and $l = 0$) is already likely to reveal most of the support of $\mathscr{C}_s$ by looking at the support of the rows that have weight around $\frac{sc+r}{2}$ (notice that this phenomenon was already observed in [Ove07]). This can be explained like this. We choose $\mathcal{I}$ to be of size k, the dimension of $\mathscr{C}_{\text{pub}}$, and to be an information set for $\mathscr{C}_{\text{pub}}$. Then, because the rate of $\mathscr{C}_s$ is equal to the rate of $\mathscr{C}_{\text{pub}}$, we expect that the size of $\mathcal{I} \cap \mathcal{J}$ (where $\mathcal{J}$ is the support of $\mathscr{C}_s$) has a rather good chance to be of size smaller than or equal to the dimension of $\mathscr{C}_s$. This in turn implies that it is possible to get codewords from $\mathscr{C}_s$ by any choice over the information set $\mathcal{I}$ of weight 1 which is non zero over $\mathcal{I} \cap \mathcal{J}$ (and therefore of weight 1 there). More generally, even if $\mathcal{I} \cap \mathcal{J}$ is slightly bigger than the dimension of $\mathscr{C}_s$ we expect to be able to get codewords in $\mathscr{C}_s$ a soon as p is greater than the Gilbert-Varshamov distance of the restriction $\mathscr{C}_s'$ of $\mathscr{C}_s$ to $\mathcal{I} \cap \mathcal{J}$, because there is in this case a good chance that this punctured code has codewords of weight p. This Gilbert-Varshamov distance will be very small in this case, because the rate of $\mathscr{C}_s$ is very close to 1 (it is expected to be equal to $\frac{\dim(\mathscr{C}_s)}{|\mathcal{I} \cap \mathcal{J}|}$).

Nevertheless, it is clear that it should be possible to set up the parameters (in particular increasing r should do the job) so that existing low weight codeword algorithms should be unable to find these subcodes $\mathscr{C}_s$ with complexity less than some fixed threshold. However, all these codes $\mathscr{C}_s$ have to be taken into account and the attacks on the dual have also to be reconsidered carefully ([LJ12] considered only attacks on the dual aiming at finding the codewords of lowest weight, but obviously the same technique used for finding some of the $\mathscr{C}_s$ will also work for the dual). Moreover, even if by construction the restriction of $\mathscr{C} = \mathscr{C}_1$ to its support should behave as a random code, this is not true anymore for $\mathscr{C}_s$ with s greater than one, due to the convolutional structure. The analysis sketched in Subsection 5.1 should be adapted a little bit for this case and should take into account the improvements over low weight searching algorithms [MMT11, BJMM12]. Finally, setting up the parameters also requires a careful study of the error probability that sequential decoding fails. This whole thread of work is beyond the scope of the present paper.

Acknowledgements. We thank the reviewers for a careful reading of this manuscript which helped us to improve its editorial quality.

References

[BBC08]   Baldi, M., Bodrato, M., Chiaraluce, F.G.: A new analysis of the McEliece cryptosystem based on QC-LDPC codes. In: Ostrovsky, R., De Prisco, R., Visconti, I. (eds.) SCN 2008. LNCS, vol. 5229, pp. 246–262. Springer, Heidelberg (2008)

[BBC+11]   Baldi, M., Bianchi, M., Chiaraluce, F., Rosenthal, J., Schipani, D.: Enhanced public key security for the McEliece cryptosystem (2011) (submitted), arxiv:1108.2462v2[cs.IT]

[BBD09]   Bernstein, D.J., Buchmann, J., Dahmen, E. (eds.): Post-Quantum Cryptography. Springer (2009)

[BC07]   Baldi, M., Chiaraluce, G.F.: Cryptanalysis of a new instance of McEliece cryptosystem based on QC-LDPC codes. In: IEEE International Symposium on Information Theory, Nice, France, pp. 2591–2595 (March 2007)

[BCGO09]   Berger, T.P., Cayrel, P.-L., Gaborit, P., Otmani, A.: Reducing key length of the McEliece cryptosystem. In: Preneel, B. (ed.) AFRICACRYPT 2009. LNCS, vol. 5580, pp. 77–97. Springer, Heidelberg (2009)

[BJMM12]   Becker, A., Joux, A., May, A., Meurer, A.: Decoding random binary linear codes in $2^{n/20}$: How $1 + 1 = 0$ improves information set decoding. In: Pointcheval, D., Johansson, T. (eds.) EUROCRYPT 2012. LNCS, vol. 7237, pp. 520–536. Springer, Heidelberg (2012)

[BL05]   Berger, T.P., Loidreau, P.: How to mask the structure of codes for a cryptographic use. Designs Codes and Cryptography 35(1), 63–79 (2005)

[BLP11]   Bernstein, D.J., Lange, T., Peters, C.: Smaller decoding exponents: ball-collision decoding. In: Rogaway, P. (ed.) CRYPTO 2011. LNCS, vol. 6841, pp. 743–760. Springer, Heidelberg (2011)

[CFS01]   Courtois, N., Finiasz, M., Sendrier, N.: How to achieve a McEliece-based digital signature scheme. In: Boyd, C. (ed.) ASIACRYPT 2001. LNCS, vol. 2248, pp. 157–174. Springer, Heidelberg (2001)

[CGG+13] Couvreur, A., Gaborit, P., Gauthier, V., Otmani, A., Tillich, J.P.: Distinguisher-based attacks on public-key cryptosystems using Reed-Solomon codes. In: Proceedings of WCC 2013 (to appear, April 2013); see also arxiv

[Dum91]   Dumer, I.: On minimum distance decoding of linear codes. In: Proc. 5th Joint Soviet-Swedish Int. Workshop Inform. Theory, Moscow, pp. 50–52 (1991)

[FGO+10] Faugère, J.-C., Gauthier, V., Otmani, A., Perret, L., Tillich, J.-P.: A distinguisher for high rate McEliece cryptosystems. Cryptology ePrint Archive, Report 2010/331 (2010), http://eprint.iacr.org/

[FGO+11] Faugère, J.-C., Gauthier, V., Otmani, A., Perret, L., Tillich, J.-P.: A distinguisher for high rate McEliece cryptosystems. In: Proceedings of the Information Theory Workshop 2011, ITW 2011, Paraty, Brasil, pp. 282–286 (2011)

[FM08]   Faure, C., Minder, L.: Cryptanalysis of the McEliece cryptosystem over hyperelliptic curves. In: Proceedings of the eleventh International Workshop on Algebraic and Combinatorial Coding Theory, Pamporovo, Bulgaria, pp. 99–107 (June 2008)

[FOPT10] Faugère, J.-C., Otmani, A., Perret, L., Tillich, J.-P.: Algebraic cryptanalysis of McEliece variants with compact keys. In: Gilbert, H. (ed.) EURO-CRYPT 2010. LNCS, vol. 6110, pp. 279–298. Springer, Heidelberg (2010)

[FS09]   Finiasz, M., Sendrier, N.: Security bounds for the design of code-based cryptosystems. In: Matsui, M. (ed.) ASIACRYPT 2009. LNCS, vol. 5912, pp. 88–105. Springer, Heidelberg (2009)

[GOT12]   Gauthier, V., Otmani, A., Tillich, J.-P.: A distinguisher-based attack on a variant of McEliece's cryptosystem based on Reed-Solomon codes. CoRR, abs/1204.6459 (2012)

[JM96]   Janwa, H., Moreno, O.: McEliece public key cryptosystems using algebraic-geometric codes. Designs Codes and Cryptography 8(3), 293–307 (1996)

[KKS97]   Kabatianskii, G., Krouk, E., Smeets, B.J.M.: A digital signature scheme based on random error-correcting codes. In: Darnell, M.J. (ed.) Cryptography and Coding 1997. LNCS, vol. 1355, pp. 161–167. Springer, Heidelberg (1997)

[Kra91]   Kravitz, D.: Digital signature algorithm. US patent 5231668 (July 1991)

[LJ12]   Löndahl, C., Johansson, T.: A new version of McEliece PKC based on convolutional codes. In: Chim, T.W., Yuen, T.H. (eds.) ICICS 2012. LNCS, vol. 7618, pp. 461–470. Springer, Heidelberg (2012)

[MB09]   Misoczki, R., Barreto, P.S.L.M.: Compact mcEliece keys from goppa codes. In: Jacobson Jr., M.J., Rijmen, V., Safavi-Naini, R. (eds.) SAC 2009. LNCS, vol. 5867, pp. 376–392. Springer, Heidelberg (2009)

[McE78]   McEliece, R.J.: A Public-Key System Based on Algebraic Coding Theory, pp. 114–116. Jet Propulsion Lab. (1978); DSN Progress Report 44

[MMT11]   May, A., Meurer, A., Thomae, E.: Decoding random linear codes in $O(2^{0.054n})$. In: Lee, D.H., Wang, X. (eds.) ASIACRYPT 2011. LNCS, vol. 7073, pp. 107–124. Springer, Heidelberg (2011)

[MS07]   Minder, L., Shokrollahi, A.: Cryptanalysis of the Sidelnikov cryptosystem. In: Naor, M. (ed.) EUROCRYPT 2007. LNCS, vol. 4515, pp. 347–360. Springer, Heidelberg (2007)

[MTSB12] Misoczki, R., Tillich, J.-P., Sendrier, N., Barreto, P.S.L.M.: MDPC-McEliece: New McEliece variants from moderate density parity-check codes. IACR Cryptology ePrint Archive, 2012:409 (2012)

[Nie86]   Niederreiter, H.: Knapsack-type cryptosystems and algebraic coding theory. Problems of Control and Information Theory 15(2), 159–166 (1986)

[OT11]   Otmani, A., Tillich, J.-P.: An efficient attack on all concrete KKS proposals. In: Yang, B.-Y. (ed.) PQCrypto 2011. LNCS, vol. 7071, pp. 98–116. Springer, Heidelberg (2011)

[OTD10]   Otmani, A., Tillich, J.P., Dallot, L.: Cryptanalysis of two McEliece cryptosystems based on quasi-cyclic codes. Special Issues of Mathematics in Computer Science 3(2), 129–140 (2010)

[Ove07]   Overbeck, R.: Recognizing the structure of permuted reducible codes. In: Tillich, J.P., Augot, D., Sendrier, N. (eds.) Proceedings of WCC 2007, pp. 269–276 (2007)

[RSA78]   Rivest, R.L., Shamir, A., Adleman, L.M.: A method for obtaining digital signatures and public-key cryptosystems. Commun. ACM 21(2), 120–126 (1978)

[Sho97]   Shor, P.W.: Polynomial-time algorithms for prime factorization and discrete logarithms on a quantum computer. SIAM J. Comput. 26(5), 1484–1509 (1997)

[Sid94]   Sidelnikov, V.M.: A public-key cryptosytem based on Reed-Muller codes. Discrete Mathematics and Applications 4(3), 191–207 (1994)

[SS92]   Sidelnikov, V.M., Shestakov, S.O.: On the insecurity of cryptosystems based on generalized Reed-Solomon codes. Discrete Mathematics and Applications 1(4), 439–444 (1992)

[Ste88]   Stern, J.: A method for finding codewords of small weight. In: Wolfmann, J., Cohen, G. (eds.) Coding Theory 1988. LNCS, vol. 388, pp. 106–113. Springer, Heidelberg (1989)

[UL09]   Umana, V.G., Leander, G.: Practical key recovery attacks on two McEliece variants, IACR Cryptology ePrint Archive 509 (2009)

[Wie06]   Wieschebrink, C.: Two NP-complete problems in coding theory with an application in code based cryptography. In: 2006 IEEE International Symposium on Information Theory, pp. 1733–1737 (2006)

[Wie10]   Wieschebrink, C.: Cryptanalysis of the Niederreiter public key scheme based on GRS subcodes. In: Sendrier, N. (ed.) PQCrypto 2010. LNCS, vol. 6061, pp. 61–72. Springer, Heidelberg (2010)

[WY93]   Wei, V.K.-W., Yang, K.: On the generalized Hamming weights of product codes. Trans. Inf. Theory 39(5), 1709–1713 (1993)

Extended Algorithm for Solving Underdefined Multivariate Quadratic Equations

Hiroyuki Miura[1], Yasufumi Hashimoto[2], and Tsuyoshi Takagi[3]

[1] Graduate School of Mathematics, Kyushu University,
744, Motooka, Nishi-ku, Fukuoka, 819-0395, Japan
[2] Department of Mathematical Sciences, University of the Ryukyus,
1, Senbaru, Nishihara, Okinawa 903-0213, Japan
[3] Institute of Mathematics for Industry, Kyushu University,
744, Motooka, Nishi-ku, Fukuoka, 819-0395, Japan

Abstract. It is well known that solving randomly chosen Multivariate Quadratic equations over a finite field (MQ-Problem) is NP-hard, and the security of Multivariate Public Key Cryptosystems (MPKCs) is based on the MQ-Problem. However, this problem can be solved efficiently when the number of unknowns n is sufficiently greater than that of equations m (This is called "Underdefined"). Indeed, the algorithm by Kipnis et al. (Eurocrypt'99) can solve the MQ-Problem over a finite field of even characteristic in a polynomial-time of n when $n \geq m(m+1)$. Therefore, it is important to estimate the hardness of the MQ-Problem to evaluate the security of Multivariate Public Key Cryptosystems. We propose an algorithm in this paper that can solve the MQ-Problem in a polynomial-time of n when $n \geq m(m+3)/2$, which has a wider applicable range than that by Kipnis et al. We will also compare our proposed algorithm with other known algorithms. Moreover, we implemented this algorithm with Magma and solved the MQ-Problem of $m = 28$ and $n = 504$, and it takes 78.7 seconds on a common PC.

Keywords: Multivariate Public Key Cryptosystems (MPKCs), Multivariate Quadratic Equations, MQ-Problem.

1 Introduction

Multivariate Public Key Cryptosystems (MPKCs) are cryptosystems whose security depends on the hardness of solving Multivariate Quadratic equations over a finite field (MQ-Problem). It is known that the MQ-Problem over a finite field is NP-hard [13] when the coefficients are randomly chosen, and no quantum algorithm efficiently solving the MQ-Problem has been presented. Therefore, MPKCs are one of candidates for post quantum cryptographies. For example, the Matsumoto-Imai cryptosystem [16], Hidden Field Equation (HFE) [18], Unbalanced Oil and Vinegar (UOV) [15], and Rainbow [7] are MPKCs. However, the MQ-Problem is efficiently solved under special n and m conditions. In particular, the algorithm by Kipnis et al. [15] can solve the MQ-Problem over a finite field of even characteristic in a polynomial-time of n when $n \geq m(m+1)$. It is also

P. Gaborit (Ed.): PQCrypto 2013, LNCS 7932, pp. 118–135, 2013.
© Springer-Verlag Berlin Heidelberg 2013

known that the Gröbner basis algorithms [5,10,11] solve the MQ-Problem, and these algorithms are more effective in the Overdefined ($n \ll m$) MQ-Problem [1,2]. Thus, estimating the hardness of the MQ-Problem is important for the security of MPKCs.

The approach by Kipnis et al. diagonalizes the upper left $m \times m$ part of the coefficient matrices, solves linear equations, and reduces the MQ-Problem to find square roots over a finite field. Courtois et al. [6] and Hashimoto [14] modified this algorithm. Although the algorithm by Courtois et al. [6] has a much smaller applicable range, it can solve MQ-Problems over all the finite fields in polynomial-time. Hashimoto's algorithm presented a polynomial-time algorithm that solves those over all finite fields when $n \geq m^2 - 2m^{3/2} + 2m$, which extended the applicable range of that of Kipnis et al. [15]. However, we point out that Hashimoto's algorithm doesn't work efficiently due to some unsolved multivariate equations arisen from the linear transformation (See Appendix A). Recently, Thomae et al. [20] made n smaller than the algorithm by Kipnis et al. by using the Gröbner basis. This algorithm can be used when $n > m$, but it is an exponential-time algorithm.

We will present an algorithm in this paper solves the Underdefined ($n \gg m$) MQ-Problem in a polynomial-time when $n \geq m(m + 3)/2$, which is wider than $n \geq m(m + 1)$. Moreover, we implemented this algorithm on Magma [4] and solved an MQ-Problem with (n, m) which the algorithm by Kipnis et al. can't be used. We will compare these results with the algorithm by Kipnis et al. [15] and that by Courtois et al. [6].

2   MQ-Problem and Its Known Solutions

In this section we introduce the MQ-Problem and explain some algorithms to solve the Underdefined MQ-Problems.

2.1   MQ-Problem

Let q be a power of prime and k be a finite field of order q. For integers $n, m \geq 1$, denoted by $f_1(\boldsymbol{x}), f_2(\boldsymbol{x}), \ldots, f_m(\boldsymbol{x})$ quadratic polynomials of $\boldsymbol{x} = {}^t(x_1, x_2, \ldots, x_n)$ over k.

$$f_1(x_1, ..., x_n) = \sum_{1 \leq i \leq j \leq n} a_{1,i,j} x_i x_j + \sum_{1 \leq i \leq n} b_{1,i} x_i + c_1$$

$$f_2(x_1, ..., x_n) = \sum_{1 \leq i \leq j \leq n} a_{2,i,j} x_i x_j + \sum_{1 \leq i \leq n} b_{2,i} x_i + c_2$$

$$\vdots$$

$$f_m(x_1, ..., x_n) = \sum_{1 \leq i \leq j \leq n} a_{m,i,j} x_i x_j + \sum_{1 \leq i \leq n} b_{m,i} x_i + c_m,$$

where $a_{l,i,j}, b_{l,i}, c_l \in k; l = 1, ..., m$. We call it "the MQ-Problem of m equations and n unknowns over finite field k", that the problem tries to find one solution $(x_1, ..., x_n) \in k^n$ such that $f_i(x_1, ..., x_n) = 0$ for all $i = 1, \ldots, m$ among the many ones that exist.

2.2   Kipnis-Patarin-Goubin's Algorithm

We explain Kipnis-Patarin-Goubin's Algorithm [15].

Let $n, m \geq 1$ be integers with $n \geq m(m+1)$ and $f_1(\boldsymbol{x}), f_2(\boldsymbol{x}), \ldots, f_m(\boldsymbol{x})$ be the quadratic polynomials of $\boldsymbol{x} = {}^t(x_1, x_2, \ldots, x_n)$ over k. Our goal is to find a solution $x_1, x_2, \ldots, x_n$ such that $f_1(\boldsymbol{x}) = 0, f_2(\boldsymbol{x}) = 0, \ldots, f_m(\boldsymbol{x}) = 0$. For $i = 1, \ldots, n$ the polynomials $f_1(\boldsymbol{x}), f_2(\boldsymbol{x}), \ldots, f_m(\boldsymbol{x})$ are denoted by

$$f_i(x_1, x_2, \ldots, x_n) = {}^t\boldsymbol{x}F_i\boldsymbol{x} + (\text{linear.})$$

where $F_1, \ldots, F_m$ are $n \times n$ matrices over k.

We also use an $n \times n$ matrix T_t over k to transform all the unknowns, and T_t has the following form.

$$T_t = \begin{pmatrix} 1 & 0 & \cdots & 0 & a_{1,t} & 0 & \cdots & \cdots & 0 \\ 0 & 1 & \ddots & \vdots & a_{2,t} & \vdots & & & \vdots \\ \vdots & \ddots & \ddots & 0 & \vdots & \vdots & & & \vdots \\ \vdots & & \ddots & 1 & a_{t-1,t} & \vdots & & & \vdots \\ \vdots & & & 0 & 1 & 0 & & & \vdots \\ \vdots & & & \vdots & a_{t+1,t} & 1 & \ddots & & \vdots \\ \vdots & & & \vdots & \vdots & 0 & \ddots & \ddots & \vdots \\ \vdots & & & \vdots & \vdots & \vdots & \ddots & \ddots & 0 \\ 0 & \cdots & \cdots & 0 & a_{n,t} & 0 & \cdots & 0 & 1 \end{pmatrix} \tag{1}$$

where $a_{1,t}, \ldots, a_{t-1,t}, a_{t+1,t}, \ldots, a_{n,t} \in k$.

We want to obtain quadratic equations of the following form.

$$\begin{cases} \sum_{i=1}^{m} \beta_{1,i} x_i^2 - \lambda_1 = 0 \\ \qquad \vdots \\ \sum_{i=1}^{m} \beta_{m,i} x_i^2 - \lambda_m = 0, \end{cases} \tag{2}$$

where $\beta_{l,i}$ and $\lambda_l \in k$ ($l = 1, \ldots, m$).

Step 1. Transform $\boldsymbol{x} \mapsto T_2\boldsymbol{x}$ so that the coefficients of x_1x_2 in f_j ($j = 1, \ldots, m$) are zero.

$$F_j \mapsto \left(\begin{array}{c|c} * & 0 \\ \hline 0 & * \end{array} \right)_* \quad (j = 1, \ldots, m)$$

Step 2. Transform $x \mapsto T_3 x$ so that the coefficients of $x_1 x_3, x_2 x_3$ in f_j ($j = 1, \ldots, m$) are zero.

$$\begin{pmatrix} \begin{array}{c|c} * & 0 \\ \hline 0 & * \end{array} & \\ & * \end{pmatrix} \mapsto \begin{pmatrix} \begin{array}{c|c|c} * & 0 & 0 \\ \hline 0 & * & 0 \\ \hline 0 & 0 & * \end{array} & \\ & * \end{pmatrix}$$

$$\vdots$$

(We continue similar operations to "**Step $m - 1$.**".)

From "**Step 1.**" to "**Step $m - 1$.**", we require the condition $n - 1 \geq m(m-1)$, i.e., $n \geq m^2 - m + 1$.

Then we can obtain the coefficient matrices of the form

$$\begin{pmatrix} \begin{array}{ccc} * & & 0 \\ & \ddots & \\ 0 & & * \end{array} & \\ & * \end{pmatrix}$$

for each $i = 1, \ldots, m$, and the following quadratic equations.

$$\begin{cases} \sum_{i=1}^{m} \beta_{1,i} x_i^2 + \sum_{i=1}^{m} x_i L_{1,i}(x_{m+1}, \ldots, x_n) + Q_1(x_{m+1}, \ldots, x_n) = 0 \\ \qquad\qquad\qquad \vdots \\ \sum_{i=1}^{m} \beta_{m,i} x_i^2 + \sum_{i=1}^{m} x_i L_{m,i}(x_{m+1}, \ldots, x_n) + Q_m(x_{m+1}, \ldots, x_n) = 0 \end{cases} \tag{3}$$

where L's are linear polynomials and Q's are quadratic polynomials in these variables.

Step m. Solve linear equations $\{L_{i,j}(x_{m+1}, \ldots, x_n) = 0\}$ for $i = 1, \ldots, m$, and $j = 1, \ldots, m$, and substitute the solutions $x_{m+1}, \ldots, x_n$ into (3). This system of linear equations has $n - m$ unknowns and m^2 equations, so we can solve if n and m satisfy $n - m \geq m^2$ i.e. $n \geq m(m + 1)$. Finally, we obtain quadratic equations of the form (2). Then we can compute the $x_1^2, \ldots, x_m^2$ values easily. The complexity of this algorithm is

$$\begin{cases} O(n^w m (\log q)^2) & \text{(char } k \text{ is 2)} \\ O(2^m n^w m (\log q)^2) & \text{(char } k \text{ is odd)}, \end{cases}$$

where $2 \leq w \leq 3$ is the exponent of the Gaussian elimination. This is because this algorithm computes $n \times n$ matrices over finite field $k = \mathrm{GF}(q)$ and solves linear equations to obtain the $x_1^2, \ldots, x_m^2$ values. The complexity of these operations is $O(n^w (\log q)^2)$. When the characteristic of k is odd, the probability of existence of square roots is approximately $1/2$, and we can find a solution with probability of 2^{-m}. Therefore, when the characteristic of k is odd, the complexity of this algorithm is $O(2^m n^w (\log q)^2)$.

2.3   Courtois et al.'s Algorithm

Courtois et al. proposed an algorithm [6] which extend Kipnis-Patarin-Goubin's algorithm when char k is odd, and this algorithm can be applied when the number of equations m and the number of unknowns n satisfy $n \geq 2^{\frac{m}{7}}(m+1)$. This algorithm and Kipnis-Patarin-Goubin's algorithm are very similar until obtain quadratic equations of the form (2). Main idea of this algorithm is to reduce the number of equations and unknowns after they obtain the quadratic equations of the form (2). This algorithm can solve the MQ-Problem of m equations and n unknowns over k in time about $2^{40}(40 + 40/\log q)^{m/40}$.

2.4   Thomae et al.'s Algorithm

Thomae et al. proposed an algorithm [20] which extend Kipnis-Patarin-Goubin's algorithm, and this algorithm can be applied when the number of equations m and the number of unknowns n satisfy $n > m$. Main idea of this algorithm is to make more zero part by using more linear transformations than Kipnis-Patarin-Goubin's algorithm in order to reduce the number of equations and unknowns. This algorithm reduces the MQ-Problem of m equations and n unknowns over finite field k into the MQ-Problem of $(m - \lfloor n/m \rfloor)$ equations and $(m - \lfloor n/m \rfloor)$ unknowns over finite field k. Then this algorithm uses Gröbner basis algorithm, so the complexty of this algorithm exponential-time. Thomae et al. [20] claimed that the MQ-Problem of 28 equations and 84 unknowns over $GF(2^8)$ has 80-bit security.

3   Proposed Algorithm

We propose an algorithm in this section that solves the MQ-Problem with $n \geq m(m+3)/2$, and explain the analysis of this algorithm.

3.1   Proposed Algorithm

Let $n, m \geq 1$ be integers with $n \geq m(m+3)/2$ and $f_1(\boldsymbol{x}), f_2(\boldsymbol{x}), \ldots, f_m(\boldsymbol{x})$ be the quadratic polynomials of $\boldsymbol{x} = {}^t(x_1, x_2, \ldots, x_n)$ over k. Our goal is to find a solution $x_1, x_2, \ldots, x_n$ such that $f_1(\boldsymbol{x}) = 0, f_2(\boldsymbol{x}) = 0, \ldots, f_m(\boldsymbol{x}) = 0$. For $i = 1, \ldots, n$ the polynomials $f_1(\boldsymbol{x}), f_2(\boldsymbol{x}), \ldots, f_m(\boldsymbol{x})$ are denoted by

$$f_i(x_1, x_2, \ldots, x_n) = {}^t\boldsymbol{x} F_i \boldsymbol{x} + (\text{linear.})$$

where $F_1, \ldots, F_m$ are $n \times n$ matrices over kD We also use an $n \times n$ matrix T_t over k of the form (1) to transform all the unknowns in "**Step t.**" $(t = 2, \ldots, m)$.

Step 1. Choose $c_i^{(1)} \in k$ $(i = 1, \ldots, m-1)$ so that the $(1,1)$-elements of $F_i - c_i^{(1)} F_m$ are zero, and replace F_i with $F_i - c_i^{(1)} F_m$. If the $(1,1)$-element of F_m is zero, exchange F_m for one of $F_1, \ldots, F_{m-1}$ that satisfies the $(1,1)$-element is not zero.

$$F_1, F_2, \ldots, F_m \mapsto \underbrace{\begin{pmatrix} 0 \\ \ast \end{pmatrix}, \ldots, \begin{pmatrix} 0 \\ \ast \end{pmatrix}}_{m-1}, \begin{pmatrix} \ast \\ \ast \end{pmatrix}$$

Step 2. (i) Transform x to $T_2 x$ so that the coefficients of $x_1 x_2$ in $f_1, f_2, \ldots, f_m$ are zero.

$$\underbrace{\begin{pmatrix} 0 \\ \ast \end{pmatrix}, \ldots, \begin{pmatrix} 0 \\ \ast \end{pmatrix}}_{m-1}, \begin{pmatrix} \ast \\ \ast \end{pmatrix} \mapsto \underbrace{\left(\begin{array}{cc|c} 0 & 0 \\ 0 & \ast \\ \hline & & \ast \end{array}\right), \ldots, \left(\begin{array}{cc|c} 0 & 0 \\ 0 & \ast \\ \hline & & \ast \end{array}\right)}_{m-1}, \left(\begin{array}{cc|c} \ast & 0 \\ 0 & \ast \\ \hline & & \ast \end{array}\right)$$

After the linear transformation $x \mapsto T_2 x$, the coefficient matrices are denoted as

$$^t T_2 F_i T_2 \ (i = 1, 2, \ldots, m).$$

We determine the $a_{1,2}, a_{3,2}, \ldots, a_{n,2}$ values in T_2 by solving the linear equations of coefficients we want to make zero. Note that $(1,2)$-elements and $(2,1)$-elements of F_i are not always equal to zero. The picture means that sum of $(1,2)$-element and $(2,1)$-element of F_i is equal to zero for each $i = 1, \ldots, m$.

(ii) Choose $c_i^{(2)} \in k$ $(i = 1, \ldots, m-2)$ so that the $(2,2)$-elements of $F_i - c_i^{(2)} F_{m-1}$ are zero, and replace F_i with $F_i - c_i^{(2)} F_{m-1}$. If the $(2,2)$-element of F_{m-1} is zero, exchange F_{m-1} for one of $F_1, \ldots, F_{m-2}$ that satisfies the $(2,2)$-element is not zero.

$$\underbrace{\left(\begin{array}{cc|c} 0 & 0 \\ 0 & \ast \\ \hline & & \ast \end{array}\right), \ldots, \left(\begin{array}{cc|c} 0 & 0 \\ 0 & \ast \\ \hline & & \ast \end{array}\right)}_{m-1}, \left(\begin{array}{cc|c} \ast & 0 \\ 0 & \ast \\ \hline & & \ast \end{array}\right)$$

$$\mapsto \underbrace{\left(\begin{array}{cc|c} 0 & 0 \\ 0 & 0 \\ \hline & & \ast \end{array}\right), \ldots, \left(\begin{array}{cc|c} 0 & 0 \\ 0 & 0 \\ \hline & & \ast \end{array}\right)}_{m-2}, \left(\begin{array}{cc|c} 0 & 0 \\ 0 & \ast \\ \hline & & \ast \end{array}\right), \left(\begin{array}{cc|c} \ast & 0 \\ 0 & \ast \\ \hline & & \ast \end{array}\right)$$

Step 3. (i) Transform x to $T_3 x$ so that the coefficients of $x_1 x_3$ and $x_2 x_3$ in $f_1, f_2, \ldots, f_{m-1}$ and the coefficient of $x_1 x_3$ in f_m are zero.

$$\underbrace{\left(\begin{array}{cc|c} 0 & 0 \\ 0 & 0 \\ \hline & & \ast \end{array}\right), \ldots, \left(\begin{array}{cc|c} 0 & 0 \\ 0 & 0 \\ \hline & & \ast \end{array}\right)}_{m-2}, \left(\begin{array}{cc|c} 0 & 0 \\ 0 & \ast \\ \hline & & \ast \end{array}\right), \left(\begin{array}{cc|c} \ast & 0 \\ 0 & \ast \\ \hline & & \ast \end{array}\right)$$

$$\mapsto \underbrace{\left(\begin{array}{ccc|c} 0 & 0 & 0 \\ 0 & 0 & 0 \\ 0 & 0 & \ast \\ \hline & & & \ast \end{array}\right), \ldots, \left(\begin{array}{ccc|c} 0 & 0 & 0 \\ 0 & 0 & 0 \\ 0 & 0 & \ast \\ \hline & & & \ast \end{array}\right)}_{m-2}, \left(\begin{array}{ccc|c} 0 & 0 & 0 \\ 0 & \ast & 0 \\ 0 & 0 & \ast \\ \hline & & & \ast \end{array}\right), \left(\begin{array}{ccc|c} \ast & 0 & 0 \\ 0 & \ast & \ast \\ 0 & \ast & \ast \\ \hline & & & \ast \end{array}\right)$$

(ii) Choose $c_i^{(3)} \in k$ $(i = 1, \ldots, m-3)$ so that the $(3,3)$-elements of $F_i - c_i^{(3)} F_{m-2}$ are zero, and replace F_i with $F_i - c_i^{(3)} F_{m-2}$. If the $(3,3)$-element of F_{m-2} is zero, exchange F_{m-2} for one of $F_1, \ldots, F_{m-3}$ that satisfies the $(3,3)$-element is not zero.

$$\underbrace{\begin{pmatrix} \begin{array}{|c|c|c|} \hline 0 & 0 & 0 \\ \hline 0 & 0 & 0 \\ \hline 0 & 0 & * \\ \hline \end{array} & \\ & * \end{pmatrix}, \ldots, \begin{pmatrix} \begin{array}{|c|c|c|} \hline 0 & 0 & 0 \\ \hline 0 & 0 & 0 \\ \hline 0 & 0 & * \\ \hline \end{array} & \\ & * \end{pmatrix}}_{m-2}, \begin{pmatrix} \begin{array}{|c|c|c|} \hline 0 & 0 & 0 \\ \hline 0 & * & 0 \\ \hline 0 & 0 & * \\ \hline \end{array} & \\ & * \end{pmatrix}, \begin{pmatrix} \begin{array}{|c|c|c|} \hline * & 0 & 0 \\ \hline 0 & * & * \\ \hline 0 & * & * \\ \hline \end{array} & \\ & * \end{pmatrix} \mapsto$$

$$\underbrace{\begin{pmatrix} \begin{array}{|c|c|c|} \hline 0 & 0 & 0 \\ \hline 0 & 0 & 0 \\ \hline 0 & 0 & 0 \\ \hline \end{array} & \\ & * \end{pmatrix}, \ldots, \begin{pmatrix} \begin{array}{|c|c|c|} \hline 0 & 0 & 0 \\ \hline 0 & 0 & 0 \\ \hline 0 & 0 & 0 \\ \hline \end{array} & \\ & * \end{pmatrix}}_{m-3}, \begin{pmatrix} \begin{array}{|c|c|c|} \hline 0 & 0 & 0 \\ \hline 0 & 0 & 0 \\ \hline 0 & 0 & * \\ \hline \end{array} & \\ & * \end{pmatrix}, \begin{pmatrix} \begin{array}{|c|c|c|} \hline 0 & 0 & 0 \\ \hline 0 & * & 0 \\ \hline 0 & 0 & * \\ \hline \end{array} & \\ & * \end{pmatrix}, \begin{pmatrix} \begin{array}{|c|c|c|} \hline * & 0 & 0 \\ \hline 0 & * & * \\ \hline 0 & * & * \\ \hline \end{array} & \\ & * \end{pmatrix}$$

$$\vdots$$

(We continue similar operations to "**Step m.**".)

Then we can obtain the coefficient matrices of the form

$$\begin{pmatrix} \begin{smallmatrix} 0 \\ & 0 \\ & & \ddots \\ & & & 0 \\ & & & & * \end{smallmatrix} \\ * \end{pmatrix}, \begin{pmatrix} \begin{smallmatrix} 0 \\ & \ddots \\ & & 0 \\ & & & * \\ & & & & * \end{smallmatrix} \\ * \end{pmatrix}, \begin{pmatrix} \begin{smallmatrix} 0 \\ & \ddots \\ & & 0 \\ & & & * \\ & & & & * \end{smallmatrix} \\ * \end{pmatrix}, \ldots, \begin{pmatrix} \begin{smallmatrix} * \\ & * \end{smallmatrix} \\ * \end{pmatrix}$$

for each $i = 1, \ldots, m$, and the following quadratic equations.

$$\begin{cases} x_m^2 + \sum_{1 \leq i \leq m} x_i L_{1,i}(x_{m+1}, \ldots, x_n) + Q_{1,2}(x_{m+1}, \ldots, x_n) = 0 \\ x_{m-1}^2 + x_m^2 + \sum_{1 \leq i \leq m} x_i L_{2,i}(x_{m+1}, \ldots, x_n) + Q_{2,2}(x_{m+1}, \ldots, x_n) = 0 \\ x_{m-2}^2 + Q_{3,1}(x_{m-1}, x_m) + \sum_{1 \leq i \leq m} x_i L_{3,i}(x_{m+1}, \ldots, x_n) + Q_{3,2}(x_{m+1}, \ldots, x_n) = 0 \\ \quad\quad\quad\quad\quad \vdots \\ x_1^2 + Q_{m,1}(x_2, \ldots, x_m) + \sum_{1 \leq i \leq m} x_i L_{m,i}(x_{m+1}, \ldots, x_n) + Q_{m,2}(x_{m+1}, \ldots, x_n) = 0 \end{cases} \tag{4}$$

where L's are linear polynomials and Q's are quadratic polynomials in these variables.

Step $m + 1$. Solve linear equations $\{L_{i,j}(x_{m+1}, \ldots, x_n) = 0\}$ of $x_{m+1}, \ldots, x_n$ for $i = 1, \ldots, m$ and $j = 1, \ldots, m-i+1$, and substitute the solutions $x_{m+1}, \ldots, x_n$ into (4). If there exists $t = 1, \ldots, m$ such that the (t, t)-elements of $F_1, \ldots, F_{m-t+1}$ are zero, remove $L_{m-t+1,t} = 0$ from the linear systems and choose the $x_{m+1}, \ldots, x_n$ that satisfies $L_{m-t+1,t} \neq 0$.

Finally, we obtain the following quadratic equations.

$$\begin{cases} x_m^2 - \lambda_1 = 0 \\ x_{m-1}^2 + \tilde{Q}_2(x_m) - \lambda_2 = 0 \\ x_{m-2}^2 + \tilde{Q}_3(x_{m-1}, x_m) - \lambda_3 = 0 \\ \qquad \vdots \\ x_2^2 + \tilde{Q}_{m-1}(x_3, \ldots, x_m) - \lambda_{m-1} = 0 \\ x_1^2 + \tilde{Q}_m(x_2, \ldots, x_m) - \lambda_m = 0 \end{cases}$$

where $\lambda_1, \ldots, \lambda_m \in k$ and $\tilde{Q}$'s are quadratic polynomials in these variables.

We can find a solution for the quadratic equations in the following way. First, we solve the first equation and substitute the solution x_m into the others. Next, we solve the second equation and substitute the solution x_{m-1} into the remaining equations $\cdots$. If there exists $t = 1, \ldots, m$ such that (t, t)-elements of $F_1, \ldots, F_{m-t+1}$ are zero, the $(m - t + 1)$-th equation takes the form of $x_t + Q(x_{t+1}, \ldots, x_m) - \lambda_{m-t+1} = 0$.

3.2 Analysis of Proposed Algorithm

We will explain the required conditions and complexity of the proposed algorithm in this section.

Theorem 3.1. The proposed algorithm works when $n \geq m(m + 3)/2$.

Proof. Our algorithm works if we can solve the linear equations.

In "**Step t.**" $(t = 2, \ldots, m)$, the number of linear equations to be solved is

$$(m - t + 1)(t - 1) + \sum_{i=1}^{t-1} i = -\frac{1}{2}\left\{t - \left(m + \frac{3}{2}\right)\right\}^2 + \frac{1}{2}m^2 + \frac{1}{2}m + \frac{1}{8},$$

and the number of unknowns is $n - 1$. Thus, we require $n \geq m(m + 1)/2$ until "**Step m.**".

In "**Step $m + 1$.**", the number of linear equations to be solved is

$$\sum_{t=1}^{m}(m - t + 1) = \frac{1}{2}m(m + 1),$$

and the number of unknowns is $n - m$. Thus, we require $n \geq m(m + 3)/2$.

For these reasons, we found that the proposed algorithm can be applied when $n \geq m(m + 3)/2$. □

Lemma 3.2. For $n = m(m + 3)/2$, the proposed algorithm succeeds in finding a solution of the MQ-Problem of m equations, n unknowns with probability of approximately

$$\begin{cases} 1 - q^{-1} & (\text{char } k \text{ is } 2) \\ 2^{-m}(1 - q^{-1}) & (\text{char } k \text{ is odd}). \end{cases}$$

Proof. When $n = m(m + 3)/2$, we must solve the linear equations that are not underdefined in "**Step $m + 1$.**". Then, we fail to solve linear equations with probability of q^{-1}. When the characteristic of k is odd, the probability of existence of square roots over k is approximately $1/2$. Therefore, the success probability of this algorithm is approximately $2^{-m}(1 - q^{-1})$ when the characteristic of k is odd. □

Moreover, the proposed algorithm uses only $n \times n$ matrix operations and the calculation of square roots over finite field k, so we obtain the following result concerning the complexity of the proposed algorithm.

Theorem 3.3. The complexity of the proposed algorithm is

$$\begin{cases} O(n^w m(\log q)^2) & (\text{char } k \text{ is } 2) \\ O(2^m n^w m(\log q)^2) & (\text{char } k \text{ is odd}), \end{cases}$$

where $2 \le w \le 3$ is the exponent of the Gaussian elimination.

Proof. In this algorithm, we calculate $n \times n$ matrices over finite field $k = \mathrm{GF}(q)$ for about m times. The complexity of this operation is $O(n^w(\log q)^2)$. When the characteristic of k is odd, the probability of existence of square roots is approximately $1/2$, and we can find a solution with probability of 2^{-m}. Therefore, when the characteristic of k is odd, the complexity of this algorithm is $O(2^m n^w m(\log q)^2)$. □

4 Implementations

We implemented the proposed algorithm using Magma [4], and compare the proposed algorithm and other known algorithms in this section. The results depend on the characteristic of k, and we will explain two cases, when the characteristic of k is 2 and an odd prime.

4.1 Parameters and Computational Environments

We chose the n and m parameters in which other algorithms can't be applied, and used homogeneous quadratic polynomials to experiment. We also chose $m = 28$ the same as Thomæ et al. [20], and n so that the proposed algorithm can apply. The computer specification and software are listed in Table 1.

Table 1. Computer specifications

OS	CPU	RAM	Software
Windows 7 (64bit)	Intel Core i3 (1.33GHz)	4.00 GB	Magma V2.17-9

4.2   When char k Is 2

These algorithms have the same complexity $O(n^w m(\log q)^2)$, but the proposed algorithm has a wider applicable range than the others. The applicable ranges of the algorithms are drawn in Fig. 1.

Table 2. Applicable ranges of the proposed algorithm and other known algorithms (char k is 2)

	Applicable range	Complexity
Proposed	$n \geq m(m+3)/2$	(poly.)
Kipnis et al. [15]	$n \geq m(m+1)$	(poly.)
Courtois et al. [6]	$n \geq m(m+1)$	(poly.)

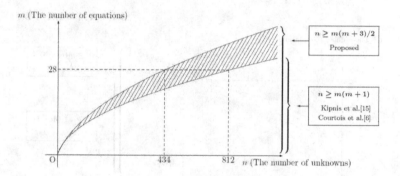

Fig. 1. Applicable range of proposed algorithm and other known algorithms

When $m = 28$, we can reduce the number of unknowns n from 812 to 434. The experimental results in our implementation are in Table. 3.

Table 3. Experimental results (char k is 2)

Field	n	m	Time / a try	Success probability
	16	4	8.76 (msec.)	99.99 %
GF(2^8)	84	11	506.83 (msec.)	100.0 %
	504	28	78.71 (sec.)	100.0 %

4.3   When char k Is Odd

We consider the algorithms by Courtois et al. [6]. Although the former one is polynomial-time of n, but it is not practical because the applicable range is too small. Thus, we compare the proposed algorithm and the latter one by Courtois et al. These algorithms are exponential-time. The applicable ranges of the algorithms are drawn in Fig. 2.

Table 4. Applicable ranges of the proposed algorithm and other known algorithms (char k is odd)

	Applicable range	Complexity
Proposed	$n \geq m(m+3)/2$	(exp.)
Courtois et al. [6]	$n \geq 2^{\frac{m}{7}} m(m+1)$	(poly.)
	$n \geq 2^{\frac{m}{7}}(m+1)$	(exp.)

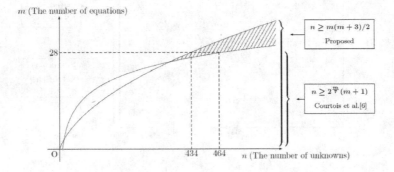

Fig. 2. Applicable range of proposed algorithm and algorithm by Courtois et al.

If $m \geq 27$, we can reduce the number of unknowns n to smaller than that of the algorithm by Courtois et al. The experimental results in our implementation are in Table. 5.

Table 5. Experimental results (char k is odd)

Field	n	m	Time / a try	Success probability
GF(7)	16	4	3.99 (msec.)	11.83 %
	84	11	259.28 (msec.)	0.22 %
	434	28	39.99 (sec.)	0.00 %

The proposed algorithm succeeds in solving the MQ-Problem with probability of 11.83% when $n = 16$ and $m = 4$, and 0.22% when $n = 84$ and $m = 11$. These results follow our success probability estimation, and we can get a similar result when $n = 434$ and $m = 28$ which can't use the algorithm by Courtois et al., and then, the success probability is $(4/7)^{28} \times (6/7) \approx 10^{-6.87} \approx 2^{-22.83}$. We estimate that it takes 1-core PC 9.44 years to solve the MQ-Problem of $n = 434$ and $m = 28$.

5 Conclusion

We presented an algorithm in this paper that can solve the MQ-Problem when $n \geq m(m + 3)/2$, where n is the number of unknowns and m is the number of equations. This algorithm makes the range of solvable MQ-Problems wider than that by Kipnis et al. Moreover, we compared this algorithm and other known algorithms, and found that the proposed algorithm is easier to use than the others. In order to demonstrate the effectiveness of the proposed algorithm we implemented it using Magma on a PC. We were able to solve the MQ-Problem of $m = 28$ and n = 504 in 78.7 seconds.

Two open problems remain. The first is to make the applicable range wider and the second is to apply the proposed algorithm to the algorithm developed by Thomae et al. [20].

References

1. Bardet, M., Faugère, J.-C., Salvy, B., Yang, B.-Y.: Asymptotic Behaviour of the Degree of Regularity of Semi-Regular Polynomial Systems, MEGA 2005 (2005), http://www-polsys.lip6.fr/~jcf/Papers/BFS05b.pdf
2. Bettale, L., Faugère, J.-C., Perret, L.: Hybrid Approach for Solving Multivariate Systems over Finite Fields. Journal of Mathematical Cryptology 3, 177–197 (2009)
3. Braeken, A., Wolf, C., Preneel, B.: A Study of the Security of Unbalanced Oil and Vinegar Signature Schemes. In: Menezes, A. (ed.) CT-RSA 2005. LNCS, vol. 3376, pp. 29–43. Springer, Heidelberg (2005)
4. Computational Algebra Group, University of Sydney. The MAGMA Computational Algebra System for Algebra, Number Theory, and Geometry
5. Courtois, N.T., Klimov, A.B., Patarin, J., Shamir, A.: Efficient Algorithms for Solving Overdefined Systems of Multivariate Polynomial Equations. In: Preneel, B. (ed.) EUROCRYPT 2000. LNCS, vol. 1807, pp. 392–407. Springer, Heidelberg (2000), http://www.minrank.org/xlfull.pdf
6. Courtois, N.T., Goubin, L., Meier, W., Tacier, J.-D.: Solving Underdefined Systems of Multivariate Quadratic Equations. In: Naccache, D., Paillier, P. (eds.) PKC 2002. LNCS, vol. 2274, pp. 211–227. Springer, Heidelberg (2002)
7. Ding, J., Schmidt, D.: Rainbow, a New Multivariate Polynomial Signature Scheme. In: Ioannidis, J., Keromytis, A.D., Yung, M. (eds.) ACNS 2005. LNCS, vol. 3531, pp. 164–175. Springer, Heidelberg (2005)
8. Ding, J., Gower, J.E., Schmidt, D.S.: Multivariate Public Key Cryptosystems. Springer (2006)

9. Ding, J., Hodges, T.J.: Inverting HFE Systems Is Quasi-Polynomial for All Fields. In: Rogaway, P. (ed.) CRYPTO 2011. LNCS, vol. 6841, pp. 724–742. Springer, Heidelberg (2011)
10. Faugère, J.-C.: "A New Efficient Algorithm for Computing Gröbner bases (F_4)". Journal of Pure and Applied Algebra 139, 61–88 (1999)
11. Faugère, J.-C.: A New Efficient Algorithm for Computing Gröbner bases without reduction to zero (F_5). In: Proceedings of ISSAC 2002, pp. 75–83. ACM Press (2002)
12. Faugère, J.-C., Perret, L.: On the Security of UOV. In: Proceedings of SCC 2008, pp. 103–109 (2008)
13. Garey, M.R., Johnson, D.S.: Computers and Intractability: A Guide to the Theory of NP-Completeness. W.H.Freeman (1979)
14. Hashimoto, Y.: Algorithms to Solve Massively Under-Defined Systems of Multivariate Quadratic Equations. IEICE Trans. Fundamentals E94-A(6), 1257–1262 (2011)
15. Kipnis, A., Patarin, J., Goubin, L.: Unbalanced Oil and Vinegar Signature Schemes. In: Stern, J. (ed.) EUROCRYPT 1999. LNCS, vol. 1592, pp. 206–222. Springer, Heidelberg (1999)
16. Matsumoto, T., Imai, H.: Public Quadratic Polynomial-Tuples for Efficient Signature-Verification and Message-Encryption. In: Günther, C.G. (ed.) EUROCRYPT 1988. LNCS, vol. 330, pp. 419–453. Springer, Heidelberg (1988)
17. Patarin, J.: Cryptanalysis of the Matsumoto and Imai Public Key Scheme of Eurocrypt '88. In: Coppersmith, D. (ed.) CRYPTO 1995. LNCS, vol. 963, pp. 248–261. Springer, Heidelberg (1995)
18. Patarin, J.: Hidden Field Equations (HFE) and Isomorphisms of Polynomials (IP): Two New Families of Asymmetric Algorithms. In: Maurer, U.M. (ed.) EUROCRYPT 1996. LNCS, vol. 1070, pp. 33–48. Springer, Heidelberg (1996)
19. Shor, P.W.: Algorithms for Quantum Computation: Discrete Logarithms and Factoring. In: Proceedings of 35th Annual Symposium on Foundations of Computer Science, pp. 124–134. IEEE Computer Society Press (1994)
20. Thomae, E., Wolf, C.: Solving Underdetermined Systems of Multivariate Quadratic Equations Revisited. In: Fischlin, M., Buchmann, J., Manulis, M. (eds.) PKC 2012. LNCS, vol. 7293, pp. 156–171. Springer, Heidelberg (2012)

Appendix A: Hashimoto's Algorithm

In this appendix we explain Hashimoto's algorithm [14], which claimed that the MQ-Problem of $n \geq m^2 - 2m^{3/2} + 2m$ over all finite fields can be solved in a polynomial-time. The applicable range of Hashimoto's algorithm is wider than that of the algorithm by Kipnis et al. [15]. However, we point out that Hashimoto's algorithm doesn't work efficiently due to some unsolved multivariate equations arisen from the linear transformation.

A.1   Outline

In the following we describe Hashimoto's algorithm which consists of Algorithm A and Algorithm B.

Algorithm A

Let $g(\boldsymbol{x})$ be a quadratic form of unknowns $\boldsymbol{x} = {}^t(x_1, \ldots, x_n)$ over finite field k. We transform $\boldsymbol{x}$ by a linear matrix $U \in k^{n \times n}$. For $a_{2,1}, a_{3,1}, a_{3,2}, \ldots, a_{n,n-1} \in k$ we define U as follows :

$$
U = \begin{pmatrix}
1 & 0 & 0 & \cdots & & \cdots & 0 \\
a_{2,1} & 1 & 0 & & & & \vdots \\
a_{3,1} & a_{3,2} & 1 & \ddots & & & \vdots \\
0 & 0 & a_{4,3} & \ddots & & \ddots & \vdots \\
\vdots & \vdots & & \ddots & \ddots & 1 & 0 \\
0 & 0 & \cdots & & 0 & a_{n,n-1} & 1
\end{pmatrix}.
$$

We determine the linear transformation U such that the coefficients of $x_1^2, x_1 x_2$, $x_1 x_3, \ldots, x_1 x_{n-1}$ in $g(U\boldsymbol{x})$ are all zero in the following way.
Step 1. Calculate $a_{2,1}, a_{3,1}$ such that the coefficient of x_1^2 in $g(U\boldsymbol{x})$ is zero.
Step 2. Calculate $a_{3,2}$ such that the coefficient of $x_1 x_2$ in $g(U\boldsymbol{x})$ is zero.
Step 3. Calculate $a_{4,3}$ such that the coefficient of $x_1 x_3$ in $g(U\boldsymbol{x})$ is zero.

$$\vdots$$

Step $n-1$. Calculate $a_{n,n-1}$ such that the coefficient of $x_1 x_{n-1}$ in $g(U\boldsymbol{x})$ is zero.

Algorithm B

Let $n, L, M \geq 1$ be integers that satisfy the following condition :

$$
n \geq \begin{cases}
2L & (M = 1) \\
ML - M + L & (1 < M < L) \\
L^2 + 1 & (M = L)
\end{cases} .
\tag{5}
$$

Let $g_1(\boldsymbol{x}), \ldots, g_M(\boldsymbol{x})$ be quadratic forms of $\boldsymbol{x}$ over k such that the coefficients of $x_i x_j$ $(1 \leq i, j \leq L)$ in $g_1(\boldsymbol{x}), \ldots, g_{M-1}(\boldsymbol{x})$ are all zero. Then we can find an invertible linear transformation U such that the coefficients of $x_i x_j$ $(1 \le i, j \le L)$ in $g_1(U\boldsymbol{x}), \ldots, g_M(U\boldsymbol{x})$ are all zero.

$$\underbrace{{}^t\boldsymbol{x} \begin{pmatrix} O_L & * \\ * & * \end{pmatrix} \boldsymbol{x}, \ldots, {}^t\boldsymbol{x} \begin{pmatrix} O_L & * \\ * & * \end{pmatrix} \boldsymbol{x}, {}^t\boldsymbol{x} \begin{pmatrix} * & * \\ * & * \end{pmatrix} \boldsymbol{x}}_{M-1} \mapsto \underbrace{{}^t\boldsymbol{x} \begin{pmatrix} O_L & * \\ * & * \end{pmatrix} \boldsymbol{x}, \ldots, {}^t\boldsymbol{x} \begin{pmatrix} O_L & * \\ * & * \end{pmatrix} \boldsymbol{x}}_{M}$$

where O_L is $L \times L$ zero matrix. **Step 1.** (i) Using Algorithm A, find a transformation $T_{1,1}$ such that the coefficients of $x_1 x_j$ $(j = 1, \ldots, L-1)$ in $g_M(\boldsymbol{x})$ are zero, and transform $\boldsymbol{x} \mapsto T_{1,1}\boldsymbol{x}$.
(ii) Transform $\boldsymbol{x} \mapsto T_{2,1}\boldsymbol{x}$ such that the coefficients of $x_1 x_L$ in $g_M(\boldsymbol{x})$ and $x_i x_L$ $(i = 1, \ldots, L)$ in $g_1(\boldsymbol{x}), \ldots, g_{M-1}(\boldsymbol{x})$ are all zero.
Step 2. (i) Using Algorithm A, find a transformation $T_{1,2}$ such that the coefficients of $x_2 x_j$ $(j = 2, \ldots, L-1)$ in $g_M(\boldsymbol{x})$ are all zero, and transform $\boldsymbol{x} \mapsto T_{1,2}\boldsymbol{x}$.
(ii) Transform $\boldsymbol{x} \mapsto T_{2,2}\boldsymbol{x}$ such that the coefficients of $x_2 x_L$ in $g_M(\boldsymbol{x})$ and $x_i x_L$ $(i = 2, \ldots, L)$ in $g_1(\boldsymbol{x}), \ldots, g_{M-1}(\boldsymbol{x})$ are all zero.

$$\vdots$$

(We continue similar operations to "**Step $L-1$.**".)
In "**Step t.**-(i), (ii)" $(t = 1, \ldots, L-1)$, we use $n \times n$ matrices $T_{1,t}, T_{2,t}$ which have the following form :

$$T_{1,t} = \begin{pmatrix} 1 & 0 & \cdots & \cdots & & \cdots & 0 \\ a_{2,1}^{(t)} & 1 & & & & & \vdots \\ a_{3,1}^{(t)} & a_{3,2}^{(t)} & 1 & & & & \vdots \\ 0 & 0 & a_{4,3}^{(t)} & \ddots & \ddots & & \vdots \\ \vdots & \vdots & & \ddots & 1 & 0 \\ 0 & 0 & \cdots & 0 & a_{n,n-1}^{(t)} & 1 \end{pmatrix}, \quad T_{2,t} = \begin{pmatrix} 1 & 0 & \cdots & 0 & b_{1,L}^{(t)} & 0 & \cdots & \cdots & 0 \\ 0 & \ddots & \ddots & \vdots & b_{2,L}^{(t)} & & \vdots & & \vdots \\ \vdots & \ddots & \ddots & 0 & \vdots & & \vdots & & \vdots \\ \vdots & & \ddots & 1 & b_{L-1,L}^{(t)} & \vdots & & \vdots \\ \vdots & & & 0 & 1 & 0 & & \vdots \\ \vdots & & & \vdots & b_{L+1,L}^{(t)} & 1 & \ddots & \vdots \\ \vdots & & & \vdots & \vdots & 0 & \ddots & \ddots & 0 \\ \vdots & & & \vdots & \vdots & \vdots & & \ddots & 0 \\ 0 & \cdots & \cdots & 0 & b_{n,L}^{(t)} & 0 & \cdots & 0 & 1 \end{pmatrix}.$$

Step L. Transform $\boldsymbol{x} \mapsto T_L \boldsymbol{x}$ such that the coefficients of $x_i x_L$ $(i = 1, \ldots, L)$ in $g_1(\boldsymbol{x}), \ldots, g_M(\boldsymbol{x})$ are all zero, where

$$T_L = \begin{pmatrix} 1 & 0 & \cdots & 0 & a_{1,L}^{(L)} & 0 & \cdots & \cdots & 0 \\ 0 & \ddots & \ddots & \vdots & a_{2,L}^{(L)} & \vdots & & & \vdots \\ \vdots & \ddots & \ddots & 0 & \vdots & \vdots & & & \vdots \\ \vdots & & \ddots & 1 & a_{L-1,L}^{(L)} & \vdots & & & \vdots \\ \vdots & & & 0 & a_{L,L} & 0 & & & \vdots \\ \vdots & & & & a_{L+1,L}^{(L)} & 1 & \ddots & & \vdots \\ \vdots & & & & \vdots & 0 & \ddots & \ddots & \vdots \\ \vdots & & & \vdots & \vdots & \vdots & & \ddots & 0 \\ 0 & \cdots & \cdots & 0 & a_{n,L}^{(L)} & 0 & \cdots & 0 & 1 \end{pmatrix}.$$

If there is no such transformation, then go back to "**Step $L-1$.**".

Step $L+1$. Return $U = T_L T_{2,L-1} T_{1,L-1} \cdots T_{2,1} T_{1,1}$.

A.2   Analysis of Algorithm B

We find the following facts about Algorithm B.

Lemma A.1. Suppose $L \geq 3$. In "**Step t.**" $t = 1, \ldots, L-2$) of Algorithm B,

$${}^t T_{1,t} \begin{pmatrix} O_{L,L} & * \\ * & * \end{pmatrix} T_{1,t} = \begin{pmatrix} O_{L,L} & * \\ * & * \end{pmatrix}.$$

This lemma shows that the $L \times L$ upper left part of $g_1(\boldsymbol{x}), \ldots, g_{M-1}(\boldsymbol{x})$ remains zero by linear transformation $T_{1,t}$.

Lemma A.2. In "**Step t.-(ii)**" $(t = 1, \ldots, L-1)$, the coefficient of x_L^2 in $g_i(\boldsymbol{x})$ $(i = 1, \ldots, M-1)$ is

$$\sum_{1 \leq j \leq L-1} a_{j,L} L_{i,j}(a_{L+1,L}^{(t)}, \ldots, a_{n,L}^{(t)}) + Q_i(a_{L+1,L}^{(t)}, \ldots, a_{n,L}^{(t)}).$$

Theorem A.3. In "**Step t.-(ii)**" $(t = 1, \ldots, L-1)$, the coefficient of $x_j x_L$ in $g_i(\boldsymbol{x})$ is equal to $L_{i,j}(a_{L+1,L}^{(t)}, \ldots, a_{n,L}^{(t)})$ $(i = 1, \ldots, M-1; j = 1, \ldots, L-1)$.

Observation A.4. In "**Step t.-(ii)**" $(t = 1, \ldots, L-1)$, we must solve equations

$$\begin{cases} \text{(The coefficient of } x_1 x_L \text{ in } g_1(\boldsymbol{x})) = 0 \\ \quad \vdots \\ \text{(The coefficient of } x_{L-1} x_L \text{ in } g_1(\boldsymbol{x})) = 0 \\ \text{(The coefficient of } x_L^2 \text{ in } g_1(\boldsymbol{x})) = 0 \\ \text{(The coefficient of } x_1 x_L \text{ in } g_2(\boldsymbol{x})) = 0 \\ \quad \vdots \\ \text{(The coefficient of } x_L^2 \text{ in } g_{M-1}(\boldsymbol{x})) = 0 \\ \text{(The coefficient of } x_t x_L \text{ in } g_M(\boldsymbol{x})) = 0, \end{cases}$$

i.e.,

$$
\begin{cases}
L_{1,1}(a^{(t)}_{L+1,L}, \ldots, a^{(t)}_{n,L}) = 0 \\
\quad\vdots \\
L_{1,L-1}(a^{(t)}_{L+1,L}, \ldots, a^{(t)}_{n,L}) = 0 \\
\displaystyle\sum_{1 \le j \le L-1} a_{j,L} L_{i,j}(a^{(t)}_{L+1,L}, \ldots, a^{(t)}_{n,L}) + Q_i(a^{(t)}_{L+1,L}, \ldots, a^{(t)}_{n,L}) = 0 \\
L_{2,1}(a^{(t)}_{L+1,L}, \ldots, a^{(t)}_{n,L}) = 0 \\
\quad\vdots \\
L_{M-1,L-1}(a^{(t)}_{L+1,L}, \ldots, a^{(t)}_{n,L}) = 0 \\
(\text{The coefficient of } x_t x_L \text{ in } g_M(\boldsymbol{x})) = 0
\end{cases}
$$

Note that we can solve linear equations without the L-th equation

$$
\begin{cases}
L_{1,1}(a^{(t)}_{L+1,L}, \ldots, a^{(t)}_{n,L}) = 0 \\
\quad\vdots \\
L_{1,L-1}(a^{(t)}_{L+1,L}, \ldots, a^{(t)}_{n,L}) = 0 \\
L_{2,1}(a^{(t)}_{L+1,L}, \ldots, a^{(t)}_{n,L}) = 0 \\
\quad\vdots \\
L_{M-1,L-1}(a^{(t)}_{L+1,L}, \ldots, a^{(t)}_{n,L}) = 0 \\
(\text{The coefficient of } x_t x_L \text{ in } g_M(\boldsymbol{x})) = 0
\end{cases} \tag{6}
$$

under the condition (5). However, $Q_i(a^{(t)}_{L+1,L}, \ldots, a^{(t)}_{n,L})$ is not equal to zero in general for the solution of equations (6). It means that **Step t.**-(ii) fails in the case of $Q_i(a^{(t)}_{L+1,L}, \ldots, a^{(t)}_{n,L}) \neq 0$.

A.3   Example of Algorithm B

Let $k = \mathrm{GF}(7), n = 7, M = 2, L = 3$. We consider quadratic forms represented by the following matrices.

$$
G_1 = \begin{pmatrix}
0 & 0 & 0 & 5 & 2 & 5 & 1 \\
0 & 0 & 0 & 2 & 3 & 1 & 2 \\
0 & 0 & 0 & 4 & 1 & 6 & 2 \\
6 & 5 & 5 & 4 & 1 & 6 & 1 \\
1 & 5 & 6 & 5 & 2 & 1 & 3 \\
2 & 3 & 5 & 1 & 5 & 3 & 1 \\
1 & 4 & 4 & 1 & 4 & 1 & 5
\end{pmatrix}, G_2 = \begin{pmatrix}
1 & 3 & 4 & 1 & 5 & 2 & 3 \\
6 & 5 & 2 & 1 & 4 & 3 & 0 \\
4 & 6 & 4 & 1 & 5 & 0 & 2 \\
6 & 1 & 0 & 0 & 3 & 2 & 4 \\
4 & 2 & 6 & 6 & 0 & 1 & 3 \\
1 & 5 & 4 & 6 & 6 & 3 & 4 \\
2 & 5 & 2 & 4 & 6 & 3 & 2
\end{pmatrix}.
$$

Step 1.-(i) Using Algorithm A, we solve the equations

$$
\begin{cases}
(\text{The coefficient of } x_1^2 \text{ in } g_2(\boldsymbol{x})) = 0 \\
(\text{The coefficient of } x_1 x_2 \text{ in } g_2(\boldsymbol{x})) = 0,
\end{cases}
$$

i.e.,

$$\begin{cases} 1 + 2a_{2,1}^{(1)} + a_{3,1}^{(1)} + 5a_{2,1}^{(1)^2} + a_{2,1}^{(1)}a_{3,1}^{(1)} + 4a_{3,1}^{(1)^2} = 0 \\ 6 + a_{3,2}^{(1)} = 0 \end{cases}$$

From these equations, we obtain $(a_{2,1}^{(1)}, a_{3,1}^{(1)}, a_{3,2}^{(1)}) = (2, 5, 1)$. Then,

$$G_1 \mapsto \begin{pmatrix} 0 & 0 & 0 & 1 & 6 & 2 & 1 \\ 0 & 0 & 0 & 6 & 4 & 0 & 4 \\ 0 & 0 & 0 & 4 & 1 & 6 & 2 \\ 6 & 3 & 5 & 4 & 1 & 6 & 1 \\ 6 & 4 & 6 & 5 & 2 & 1 & 3 \\ 5 & 1 & 5 & 1 & 5 & 3 & 1 \\ 1 & 1 & 4 & 1 & 4 & 1 & 5 \end{pmatrix}, G_2 \mapsto \begin{pmatrix} 0 & 1 & 0 & 1 & 3 & 1 & 6 \\ 6 & 3 & 6 & 2 & 2 & 3 & 2 \\ 1 & 3 & 4 & 1 & 5 & 0 & 2 \\ 1 & 1 & 0 & 0 & 3 & 2 & 4 \\ 3 & 1 & 6 & 6 & 0 & 1 & 3 \\ 3 & 2 & 4 & 6 & 6 & 3 & 4 \\ 1 & 0 & 2 & 4 & 6 & 3 & 2 \end{pmatrix}.$$

Step 1.-(ii)

$$\begin{cases} (\text{The coefficient of } x_1 x_3 \text{ in } g_1(\boldsymbol{x})) = 0 \\ (\text{The coefficient of } x_2 x_3 \text{ in } g_1(\boldsymbol{x})) = 0 \\ (\text{The coefficient of } x_1 x_3 \text{ in } g_2(\boldsymbol{x})) = 0 \\ (\text{The coefficient of } x_3^2 \text{ in } g_1(\boldsymbol{x})) = 0, \end{cases}$$

i.e.,

$$\begin{cases} 5b_{5,3}^{(1)} + 2b_{7,3}^{(1)} = 0 \\ 2b_{4,3}^{(1)} + b_{5,3}^{(1)} + b_{6,3}^{(1)} + 5b_{7,3}^{(1)} = 0 \\ 1 + 2b_{4,3}^{(1)} + 6b_{5,3}^{(1)} + 4b_{6,3}^{(1)} = 0 \\ b_{1,3}^{(1)}(5b_{5,3}^{(1)} + 2b_{7,3}^{(1)}) + b_{2,3}^{(1)}(2b_{4,3}^{(1)} + b_{5,3}^{(1)} + b_{6,3}^{(1)} + 5b_{7,3}^{(1)}) + 4b_{4,3}^{(1)^2} \\ \quad + 6b_{4,3}^{(1)}b_{5,3}^{(1)} + 2b_{4,3}^{(1)}b_{7,3}^{(1)} + 2b_{5,3}^{(1)^2} + 6b_{5,3}^{(1)}b_{6,3}^{(1)} + 3b_{6,3}^{(1)^2} + 2b_{6,3}^{(1)}b_{7,3}^{(1)} + 5b_{7,3}^{(1)^2} = 0 \end{cases}$$

These multivariate equations are hard to solve.

Quantum Key Distribution in the Classical Authenticated Key Exchange Framework

Michele Mosca[1,2], Douglas Stebila[3], and Berkant Ustaoğlu[4]

[1] Institute for Quantum Computing and Dept. of Combinatorics & Optimization
University of Waterloo, Waterloo, Ontario, Canada
[2] Perimeter Institute for Theoretical Physics, Waterloo, Ontario, Canada
mmosca@uwaterloo.ca
[3] Information Security Discipline, Queensland University of Technology, Brisbane,
Queensland, Australia
stebila@qut.edu.au
[4] Department of Mathematics, Izmir Institute of Technology, Urla, Izmir, Turkey
bustaoglu@uwaterloo.ca

Abstract. Key establishment is a crucial primitive for building secure channels in a multi-party setting. Without quantum mechanics, key establishment can only be done under the assumption that some computational problem is hard. Since digital communication can be easily eavesdropped and recorded, it is important to consider the secrecy of information anticipating future algorithmic and computational discoveries which could break the secrecy of past keys, violating the secrecy of the confidential channel.

Quantum key distribution (QKD) can be used generate secret keys that are secure against any future algorithmic or computational improvements. QKD protocols still require authentication of classical communication, although existing security proofs of QKD typically assume idealized authentication. It is generally considered folklore that QKD when used with computationally secure authentication is still secure against an unbounded adversary, provided the adversary did not break the authentication during the run of the protocol.

We describe a security model for quantum key distribution extending classical authenticated key exchange (AKE) security models. Using our model, we characterize the long-term security of the BB84 QKD protocol with computationally secure authentication against an eventually unbounded adversary. By basing our model on traditional AKE models, we can more readily compare the relative merits of various forms of QKD and existing classical AKE protocols. This comparison illustrates in which types of adversarial environments different quantum and classical key agreement protocols can be secure.

Keywords: quantum key distribution, authenticated key exchange, cryptographic protocols, security models.

P. Gaborit (Ed.): PQCrypto 2013, LNCS 7932, pp. 136–154, 2013.
© Springer-Verlag Berlin Heidelberg 2013

1 Introduction

Quantum key distribution (QKD) promises new security properties compared to cryptography based on computational assumptions: two parties can establish a key using a pair of quantum and classical channels, secure against any adversary who is limited solely by the laws of quantum mechanics. Most information-theoretically secure classical[1] cryptographic tasks have limited practicality, so many schemes' security rely on computational assumptions, the most widely used of which—factoring, discrete logarithms—could be efficiently solved by a large-scale quantum computer. As a result, QKD could be an important primitive for cryptography secure against advances in computing technology, provided quantum mechanics remains an accurate description of the laws of nature.

The classical cryptographic literature has extensively studied *authenticated key exchange* (AKE) since the founding of public key cryptography in 1976. After a period of ad hoc security analysis, protocols are now generally analyzed in a security model where an active attacker controls communication and can possibly compromise certain private information; proofs usually consist of probabilistic reductions to computationally hard problems. The seminal work in this area by Bellare and Rogaway [1] was followed by the more modern CK01 [2] and eCK [3] models; an alternative approach to this family of security models is given by Canetti's *universal composability (UC) framework* [4]. Typically in AKE protocols, calculating a secret key is relatively easy, but authentication—ensuring that the key is shared only with only the intended party—requires greater care.

There are many types of QKD protocols, but for our purposes we will divide them into 3 classes: prepare-send-measure protocols, measure-only protocols, and prepare-send-only protocols. The first QKD protocol, now called BB84 [5], is an example of a prepare-send-measure protocol in which Alice randomly prepares one of several quantum states, sends it to Bob, and Bob randomly measures in one of several settings. Ekert [6] proposed an entanglement-based protocol, which is an example of a measure-only protocol: Alice and Bob only randomly measure in one of several settings; the state itself can be prepared by Eve entirely untrusted. Biham et al. [7] proposed a prepare-send-only protocol, in which Alice and Bob each randomly prepare one of several quantum states and send them to Eve, who measures and sends back a classical result. Different versions can be appealing due to ease of implementation, resistance to side-channel attacks on preparing or measuring, or device independence.

Research on QKD security has largely proceeded independent of the afore-mentioned classical AKE security models. Various proofs of QKD have been given in a stand-alone 2-party setting [8,9,10,11,12,13,14]. This contrasts with the aforementioned security models used in classical AKE protocols, which consider the multi-party, multi-session setting, and consider various types of information leakage or compromise. Existing QKD proofs typically take place under the assumption that classical communication happens over on authentic public

---

[1] We use the adjective "classical" to mean "non-quantum", so "classical cryptography" means "non-quantum cryptography", not "historical cryptography".

channel. It is generally considered folklore [15,16,17,18] that if QKD was performed using a computationally secure message authentication scheme (such as public key digital signatures), then messages encrypted under the keys output by QKD would be secure provided that the adversary could not break the authentication scheme *before or during* the QKD protocol. This result has only been justified formally in this paper and in our concurrent work by Unruh in the universal composability setting [19].

Contributions. Our goal is to describe the security of quantum key distribution in a security model similar to existing classical authenticated key exchange protocols and compare the relative security properties of various QKD and classical AKE protocols. Our model is explicitly a multi-party model, includes authentication, and allows for either computationally secure or information theoretically secure authentication. We aim to capture two properties: (1) QKD is *immediately secure* against an active adversary who is restricted such that he is unable to break the authentication scheme, and (2) QKD is *long-term secure*, meaning that, if it is secure against an active adversary who is restricted during the run of the protocol to be unable to break the authentication scheme, then it remains secure even when the (classical and quantum) data obtained by the active bounded adversary are later given to an unbounded quantum adversary.

Security model for classical-quantum AKE protocols. We first introduce in Section 2 a multi-party model for analyzing the security of QKD protocols. In our model, which adopts the formalism of Goldberg et al.'s framework for AKE [20], parties consist of a pair of classical and quantum Turing machines, each of which is capable of sending and receiving messages. The adversary controls all communications between parties, but is restricted in its ability to affect communication between a party's classical and quantum devices. The adversary also has the ability to compromise various values used by parties before, during, or after the run of the protocol. As is typical, the adversary's goal is to distinguish the session key of a completed session from a random string of the same length. A novelty of our approach is a new technique for defining matching sessions.

Having defined the adversarial model, we then introduce our two security definitions, *immediate security* against an active, potentially bounded adversary, and *long-term security*, against an adversary who during the run of the protocol may be bounded, but after the protocol completes is unbounded (except by the laws of quantum mechanics). Our model is generic enough to allow the bound on the adversary to be computational—assuming that a particular computational problem is hard—or run-time or memory-bounded [21]. We adapt the long-term security notion of Müller-Quade and Unruh [22] from the classical universal composability framework to our classical-quantum model.

Security of BB84. We then proceed in Section 3 to show that the BB84 protocol, when used with a computationally secure classical authentication scheme such as a digital signature, is secure in this model. For the quantum aspects of the proof, we rely on existing proof techniques. This is next extended to provide a proof of the folklore theorem that QKD, when used with computationally secure authentication in a multi-party setting, is information theoretically secure, provided the adversary did not break the authentication during the run of the protocol. Our argument explicitly identifies which secret information leakage does not affect security either before or after the run of the protocol.

Comparison of quantum and classical AKE protocols. Finally, we use our generic security model to compare in Section 4 the security properties of classical key exchange protocols and examples from each of the three classes of QKD protocols (prepare-send-measure, measure-only, prepare-send-only). This comparison is facilitated by our phrasing of QKD in a security model more closely related to traditional AKE security models, which we can then use to compare the relative powers afforded to the adversary under those models. In particular, our model allows us to compare how different protocols react when the randomness used in the protocol is revealed—or if it is later discovered that bad randomness was used. For example, some classical AKE protocols such as UP [23] are secure even if the randomness used for either a party's long-term secret key or ephemeral secret key is revealed *before* the run of the protocol, but the same is not true for the randomness used to pick basis choices in BB84. And the EPR protocol of Ekert is secure even if all of the randomness used by the parties is leaked after the protocol completes, unlike BB84 where data bit choices must remain secret. Since obtaining high quality randomness can be very challenging in practice—requiring either a separate, tested quantum source, or relying on a pseudorandom number generator seeded from a high quality source of entropy—it may be desirable to select a protocol based on the quality of randomness available, and our framework provides a method for comparing protocols along these lines.

Comparison with other frameworks. Our approach to defining security differs from existing work in several essential ways. Stand-alone QKD security definitions do not consider the security in a *multi-party setting*, and also tend to ignore entirely the question of *explicit authentication*, instead assuming an authentic classical channel. It is widely recognized that the authentication can be secure against an unbounded adversary if all classical communication is protected by information-theoretically secure message authentication codes, such as the Wegman-Carter 2-universal hash function [24,25]. However, as mentioned above, the classical AKE experience suggests that it is the authentication part of the overall security definition that is often violated; more so when there is *information leakage* to adversary. With a few exceptions (e.g., [13]), stand-alone definitions also exclude the possibility of the adversary learning private information. The universal composability definition of QKD security of Ben-Or et al. [26] (which is an adaptation of Canetti's UC framework [4] to the quantum setting), notably referenced by Renner in his thesis [14], also brushes aside the possibility of any information being leaked to the adversary and focuses solely on information-theoretic authentication. Other frameworks for composability of quantum protocols have been given [27,28,29,30,31] and applied to other types of cryptographic protocols, but not QKD. Our model, then, is the first to define QKD security in the multi-party setting, with explicit consideration of authentication, allowing leakage of information the adversary. Moreover, it defines both short-term and long-term security; last but not least our definitions paves way for formally analyzing and comparing both classical and quantum AKE protocols within the same framework. In work concurrent with our, Unruh [19] analyzes the long-term security of QKD in the UC framework.

2   QKD Model

Our model begins as an enhancement to the eCK model [3], following the notation of Goldberg et al. [20]. In our model, each party has access to a quantum device. The quantum device may be viewed as limited based on for example current hardware limitations. As usual we consider interactive protocols within a multi-party multi-session setting, where communication is controlled by the adversary. The adversary controls the quantum communication channel between parties, subject to the laws of quantum physics. We also describe how, if at all, the adversary may gain access to secrets used by the parties. We then define secrecy against bounded adversaries and long-term security against unbounded adversaries: the long-term security definition is achieved by having the active bounded short-term adversary output a classical and quantum transcript upon which the unbounded quantum adversary may operate.

We next formally describe the model. We use k to denote a security parameter. Our description uses qubits but can be generalized to arbitrary-dimension quantum systems.

2.1   Parties and Protocols

A *party* (see also [32, Def. 1.1, bullet 2]) is an interactive classical Turing machine with access to a quantum Turing machine. We refer to this pair jointly as the party.

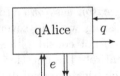

(a) Quantum Turing machine

The classical machine can activate the quantum device via a special activation request or receive (via designated activation routines) measurement outcomes from the quantum device. The communication is delivered over a two way classical communication tape (the e-channel in Figure 1(b)). The classical Turing machine has also access to a sequence of random bits – the r-tape in Figure 1(b) – and a separate c-tape over which the party can receive and send other

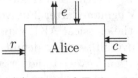

(b) Classical Turing machine

Fig. 1. A party's classical and quantum Turing machines

activation requests and messages as specified by designated routines. Similarly, the quantum Turing device can be activated by the classical Turing machine and can receive and send qubits over a designated quantum channel q as in Figure 1(a).

Each party can have associated authenticated public strings (e.g., public keys or identifiers), which are assumed to be distributed over an authenticated channel to other parties. Furthermore, pairs of parties may possess shared secrets that were distributed confidentially a priori.

A *protocol* is a collection of interactive classical and quantum subroutines that produce a shared secret key between two (or more parties) or output an indicator of an error. The interactions may use messages received on either the classical or quantum channels. The final output of the protocol is made via the classical Turing machine.

A *session* is an execution of the protocol. Sessions are initiated via a special incoming request and upon initiation each one is identified with a unique[2] *session identifier* Ψ chosen by the party at which the session is executed (in which case we say the party *owns* the session). A session that has been initiated but is not yet completed is called *active*. Since sessions are interactive procedures a party may own more than one active session at a given point of time. Each active session has a separate *session state* that stores session-specific classical data.[3]

Upon receiving and sending all protocol messages and performing the required measurements and computations specified by the protocol, the session *completes* by having the classical Turing machine output either an error symbol $\perp$ or a tuple $(sk, pid, \mathbf{v}, \mathbf{u})$. The tuple consists of:

- sk: a session key;
- pid: a party identifier;
- $\mathbf{v}$: a vector $(\mathbf{v}_0, \mathbf{v}_1, \dots)$ where each $\mathbf{v}_i$ is a vector of public values or labels; (For example, $\mathbf{v}_1$ may consist of the public values contributed by party P_1. Including $\mathbf{v}$ in the session output binds the session with the various values used by the parties to compute the session key.)
- $\mathbf{u}$: a vector $(\mathbf{u}_0, \mathbf{u}_1, \dots)$ where each $\mathbf{u}_i$ is vector of a public values or labels; $\mathbf{u}$ is called the *authentication* vector and indicates what the session owner uses to identify its peer pid.

The vectors $\mathbf{v}$ and $\mathbf{u}$ will play an important role in defining freshness.

Definition 1 (Correctness). *A key exchange protocol π is correct if, when all protocol messages are relayed faithfully, without changes to content or ordering, the peer parties output the same session key k and the same vector $\mathbf{v}$.*

Memory. A party may hold in its memory several *value pairs* of the form (x, X), generated by some algorithm specified by the protocol, where x is a private value and X is a public value or label. The pair may be a *public key pair*, such as private key x and public key X, or a *labelled private value*, such as a private value x and a unique public label $X = \ell(x)$.

---

[2] With this definition uniqueness is guaranteed only within a party; globally uniqueness can be guaranteed by requiring the session identifier is the concatenation of the unique party identifier and the party's own session identifier.

[3] While quantum protocols in general may make use of quantum memory for storing quantum states during a session, the current QKD protocols we consider in this paper, such as BB84 or EPR, do not, so we omit this from our model.

There are two classifications of value pairs: *ephemeral* value pairs, which are associated with a particular session Ψ, and *static* value pairs, which can be used across multiple sessions. The party may also have value pairs that have been generated but not yet used. If necessary, different types of key pairs may be permitted, for example, if a protocol uses one type of key pair for digital signatures and another type of key pair for public-key encryption. The protocol specifies an algorithm for generating new pairs.

Classical Turing machine communication. As described above each classical Turing machine has two incoming-outgoing classical communication channels, denoted by e and c in Figure 1(b), over which the classical Turing machine receives activations and submits responses. The responses themselves can be activation requests. Furthermore the classical Turing machine has an input of classical random bits which can be read at will by the Turing machine, denoted by r in Figure 1(b). The following activations of the classical Turing machine are allowed:

- SendC($params, pid$): This activation is received via channel c and directs the party to begin a new key exchange session. A new session is initiated and assigned a unique session identifier Ψ based on protocol-specific public parameters $params$ and an identifier pid of the party with whom to establish the session. The response to this query includes the session identifier Ψ and any protocol-specific outgoing classical message msg' that are sent via the outgoing channel c. If required by the protocol, the Turing machine can send an activation request C2Q(m) over the e outgoing channel, which may in turn cause that quantum Turing machine to write an output to its q channel as well, or to prepare its measurement device to receive quantum messages.
- SendC(Ψ, msg): This query models the delivery of classical messages over c-channel. The party's classical Turing machine is activated with session Ψ and classical message msg. It returns any outgoing classical message msg' over the c-channel. If required by the protocol, the Turing machine can send an activation request C2Q(m) over the e outgoing channel, which may in turn cause that quantum Turing machine to write an output to its q channel as well, or to prepare its measurement device to receive quantum messages.
- Q2C(m): Upon activation with this query the classical Turing machine activates its most recent session with input m. This query may cause the classical Turing machine to output to its c channel, or send another activation over the e channel.

A protocol may request that the classical Turing machine acts probabilistically, in which case it reads random bits from the r-channel.

Quantum Turing machine communication. Each party's quantum Turing machine has a two-way quantum communication channel, denoted by q in Figure 1(a), over which the machine receives and submits quantum information. The responses themselves can be activation requests. Furthermore the quantum Turing machine has a two-way classical control channel (denoted by e in Figure 1(a)) with which it communicates with the classical Turing machine.

The following activations of the quantum Turing machine are allowed:

- SendQ(ρ): This query activates the quantum Turing machine with quantum message ρ; it returns any outgoing quantum message ρ' over the q-channel. If required by the protocol, the quantum Turing machine can send an activation request C2Q(m) over the e outgoing channel (for example, to report any measurement results obtained from measuring ρ), which may in turn cause that classical Turing machine to write an output to its c channel as well.
- C2Q(m): This query activates the quantum Turing machine with classical control message m, for example to prepare the quantum circuit for execution due to an anticipated SendQ activation. The activation may cause a quantum state to be output over the outgoing quantum channel q as well as a classical message to be returned over classical control channel e.

2.2 Adversarial Model

The *adversary* is, similar to a party, a pair of interactive classical and quantum Turing machines. The adversary's classical Turing machine runs in time at most $t_c(k)$ and has access to a quantum Turing machine with runtime bounded by $t_q(k)$ and memory bounded by $m_q(k)$ qubits; bounds may be unlimited. The adversary takes as its input all public information and may interact with the (honest) parties. Furthermore the adversary can establish corrupted (dishonest) parties which it fully controls. Honest parties cannot distinguish between honest and dishonest parties.

Communication over the parties' classical c-channels is controlled by the adversary. On the classical channels, the adversary can read, copy, reorder, insert, delay, modify, drop or forward messages at will. The sending and receiving parties have no intrinsic mechanism to detect which actions, if any, the adversary performed on the classical messages.

Communication over the parties' quantum q channels is also controlled by the adversary. The adversary's operations on the quantum channels are bound by the laws of quantum mechanics: the delivery of quantum messages can be delayed, modified in order, forwarded, or dropped; the adversary can create new quantum states and perform joint quantum operations on quantum messages received from the parties as well as on the adversary's state. However, due to the laws of quantum mechanics, the adversary cannot necessarily obtain full information about quantum messages from the parties; for example, measurements may irrevocably disturb the state of messages transmitted by the parties, and the adversary may be unable to precisely copy a message due to the no-cloning theorem. We assume communication between the adversary's quantum machine and party's quantum machines is perfect: the adversary can simulate any environmental effect or noise on qubits sent by a party.

Queries. The adversary can direct a party to perform certain actions by sending any of the aforementioned activation queries over party's the c and q channels.

The adversary has neither immediate control and cannot observe the content exchanged between the classical and quantum subcomponents of a party over the e channel, nor has information about the bits obtained from the r-channel. Furthermore, to allow for information leakage the adversary may issue the following queries to parties:

- RevealNext $\to X$: This query allows the adversary to activate the classical Turing machine to read input from the r-channel and learn future public values. The activated party generates a new value pair (x, X), records it as unused, and returns the public value X. (This query may be specialized if there are multiple value pair types specified by the protocol.)
- Reveal$(X) \to x$: This query allows the adversary to compromise secret values used in the protocol computation.[4] If the party has a value pair (x, X) in its memory, it returns the private value x. Reveal(Ψ) returns the secret key sk for session Ψ, if it exists; this is often referred to as a RevealSessionKey query.

Where necessary to avoid ambiguity, we use a superscript to indicate the party to whom the query is directed, for example SendC$^{P_i}(\Psi, msg)$.

Revealing. If (x, X) is a value pair, with public key value or public label X, then the adversary is said to have *revealed the secret for* X if the adversary issued the query Reveal(X) to a party holding that value pair in its memory. In general, the adversary can reveal the secret for any value X, though this may affect which sessions are fresh.

2.3   Security Definition

For the purpose of defining session key security, the adversary has access to the following additional oracle:

- Test$(i, \Psi) \to \kappa$: If party P_i has not output a session key, return $\perp$. Otherwise, choose $b \xleftarrow{\$} \{0, 1\}$. If $b = 1$, then return the session key sk from the output for session Ψ at party P_i. If $b = 0$, return a random bit string of length equal to the length of the session key sk in session Ψ at party P_i. Only one call to the Test query is allowed.

Definition 2 (Fresh session). *A session Ψ owned by an honest party P_i is fresh if all of the following occur:*

1. *For every vector $\mathbf{v}_j$ in P_i's output for session Ψ, there is at least one element X in $\mathbf{v}_j$ for which the adversary has not revealed the secret.*
2. *The adversary did not issue Reveal(Ψ') to any honest party P_j for which Ψ' has the same public output vector as Ψ (including the case where $\Psi' = \Psi$ and $P_j = P_i$).*

---

[4] Our notation here is altered from that of Goldberg et al. [20], in that we call this query Reveal instead of their original term Partner.

3. At the time of session completion, *for every vector* $\mathbf{u}_j$, $j \geq 1$, *in P_i's output for session Ψ, there was at least one element X in $\mathbf{u}_j$ for which the adversary has not revealed the secret.*

The difference between the first condition (involving $\mathbf{v}$) and the third condition (involving $\mathbf{u}$) is that there are some values ($\mathbf{u}$) that are okay for the adversary to learn after the session completes but not before, whereas there may be other values ($\mathbf{v}$) that he can never learn.

Definition 3 (Security). *Let k be a security parameter. An authenticated key exchange protocol is secure if, for all adversaries $\mathcal{A}$ with classical runtime bounded by $t_c(k)$, quantum runtime bounded by $t_q(k)$, and quantum memory bounded by $m_q(k)$, the advantage of $\mathcal{A}$ in guessing the bit b used in the Test query of a fresh session is negligible in k; in other words, the probability that $\mathcal{A}$ can distinguish the session key of a fresh session from a random string of the same length is negligible.*

Output vectors. One of the key differences between our model and traditional AKE security models is how we phrase restrictions on what secret values the adversary can learn and when. In the eCK model, for example, a fresh session is defined as one in which the adversary has not learned (a) both the session owner's ephemeral secret key x and long-term secret key a, and (b) both the peer's ephemeral secret key y and long-term secret key b (or just the peer's long-term key if no matching peer session exists). In our model, this could be specified as $\mathbf{v} = (\mathbf{v}_0 = (a, x), \mathbf{v}_1 = (b, y))$.

Since in traditional AKE security models the restriction on values learned is specified in the security model, a new security model is required for each differing combination of learnable values. Though models may often appear similar, they sometimes contain subtle but important formal differences and thus become formally incomparable [33]. The traditional approach of specifying the values that can or cannot be learned in the security definition itself contrasts with our approach—building on that of Goldberg et al. [20]—where the vectors $\mathbf{v}$ and $\mathbf{u}$ in the session output specify what can or cannot be learned. As a result, two protocols with differing restrictions on values that can be learned could both be proven secure in our model and then compared based on which values can or cannot be revealed.

2.4 Long-Term Security

One of the main benefits of quantum key distribution is that it can be secure against unbounded adversaries, but this comes at the cost of being unable to use computationally secure cryptographic primitives such as public key digital signatures for authentication. Definition 3 can be used to analyze QKD when computationally secure cryptographic primitives are used by choosing a $t_c(k)$, $t_q(k)$, and $m_q(k)$ such that the cryptographic primitive is believed secure against such an adversary. The particular values may be chosen based on known classical algorithms for factoring or discrete logarithms and on present-day limits of quantum devices.

Regardless of the bound on the active adversary, we can still recover a very strong form of long-term security by considering an unbounded quantum Turing machine acting after the protocol has completed. In other words, during the run of the protocol, we assume a bounded adversary as in Definition 3; this bounded active adversary produces some classical and quantum transcript which it provides to the unbounded adversary. This models the real-world scenario of an adversary being somewhat limited by its classical and quantum computing equipment now but later having much more powerful equipment or making an algorithmic breakthrough.

Definition 4 (Long-term security). *An AKE protocol is* long-term secure *if, for all unbounded quantum Turing machines $\mathcal{M}$ acting on a classical and quantum transcript produced by a (bounded) adversary $\mathcal{A}$ in Definition 3, the advantage of $\mathcal{M}$ in guessing the bit b used in the* Test *query of a fresh session is negligible in the security parameter.*

Bounds on devices. If $t_q(k) = m_q(k) = 0$, and Definition 4 is omitted, the model reduces to a classical definition for secure session key establishment. It refines the idea of authentication as the session output can explicitly identify how peers were identified and authenticated. Thus any classical protocol analyzed in [20] can also be analyzed in this model.

This model can be used in conjunction with present limitations of quantum devices. While there are ongoing improvements in controlling quantum systems, at present the number of qubits a device can work with is essentially a small constant compared to classical computers. Thus, using our model with appropriate values of $t_q(k)$ and $m_q(k)$, one can devise efficient protocols that are easy to implement but guarantee unconditional future secrecy. An appropriate assumption on $t_c(k)$—for example that all adversaries with polynomial running time $t_c(k)$ cannot solve a particular hard problem—allow the model to be used as existing classical reductionist security models are used.

Of course, the devices available to the adversary can be made unbounded essentially allowing a complete quantum world. Thus the definitions presented here are suitable for analyzing novel QKD protocols. These alternatives show the wide range of scenarios our definitions incorporate. Due to the unified underlying framework it is easier to compare various protocols and decide which one is the best for the task at hand.

3   BB84

We now turn to the BB84 protocol [5]. We first specify the protocol in the language of the model of Section 2, discuss some aspects of our formulation, and complete the section with a security analysis. Our presentation of BB84 explicitly includes the authentication operations. We choose to focus on authentication using digital signatures, rather than authentication using symmetric key primitives, for several reasons: first, establishment of shared secret keys for authentication is in practice harder than authentic distribution of public keys; and second, the short-term and long-term security properties resulting from the use of public key authentication with QKD are not yet understood.

Definition 5. *Let k be a security parameter. The* BB84 *protocol is defined by having parties responding to activations as follows:*

1. Upon activation SendC(start, initiator, B) the classical Turing machine A does the following:
   (a) create a new session Ψ^A with peer identifier B;
   (b) read n_1 (random) data bits Ψ^A_{dAB} and n_1 (random) basis bits Ψ^A_{bA} from its r-tape;
   (c) send activation C2Q($\Psi^A_{bA}, \Psi^A_{dAB}$) on its e-tape, which indicates that the quantum device should encode each data bit from Ψ^A_{dAB} as $|0\rangle$ or $|1\rangle$ if the corresponding basis bit Ψ^A_{bA} is 0, or as $|+\rangle$ or $|-\rangle$ if the corresponding basis bit Ψ^A_{bA} is 1;
   (d) send activation SendC(Ψ^A, start, responder, A) on its c-tape to B.
2. Upon activation SendC(Ψ^A, start, responder, A) the classical Turing machine B does the following:
   (a) create a new session Ψ^B with peer identifier A;
   (b) read n_1 (random) basis bits Ψ^B_{bB} from its r-tape;
   (c) send activation C2Q(Ψ^B_{bB}) on its e-tape, which indicates the quantum device should measure the ith qubit in the $|0\rangle/|1\rangle$ if the ith bit of Ψ^B_{bB} is 0, or in the $|+\rangle/|-\rangle$ basis if ith bit of Ψ^B_{bB} is 1.
3. Upon activation Q2C(m), the classical Turing machine B does the following:
   (a) set Ψ^B_{dAB} equal to m;
   (b) compute $\sigma \leftarrow \text{Sign}_{pk_B}(\Psi^A, \Psi^B, \Psi^B_{bB}, B)$;
   (c) send activation SendC($\Psi^A, \Psi^B, \Psi^B_{bB}, \sigma$) on its c-tape to A.
4. Upon activation SendC($\Psi^A, \Psi^B, \Psi^B_{bB}, \sigma$), the classical Turing machine A does the following:
   (a) verify σ with pk_B;
   (b) discard all bit positions from Ψ^A_{dAB} for which Ψ^A_{bA} is not equal to Ψ^B_{bB}; assume n_2 such positions remain;
   (c) read n_2 (random) bits Ψ^A_{indAB} from its r-tape; set Ψ^A_{chkAB} to be the substring of Ψ^A_{dAB} for which the bits of Ψ^A_{indAB} are 1, and set Ψ^A_{kAB} to be the substring of Ψ^A_{dAB} for which the bits of Ψ^A_{indAB} are 0; let n_3 denote the length of Ψ^A_{kAB}
   (d) compute $\sigma \leftarrow \text{Sign}_{pk_A}(\Psi^A, \Psi^B, \Psi^A_{bA}, \Psi^A_{indAB}, \Psi^A_{chkAB}, A)$;
   (e) send activation SendC($\Psi^A, \Psi^B, \Psi^A_{bA}, \Psi^A_{indAB}, \Psi^A_{chkAB}, \sigma$) on its c-tape to B.
5. Upon activation SendC($\Psi^A, \Psi^B, \Psi^A_{indAB}, \Psi^A_{chkAB}, \sigma$), the classical Turing machine B does the following:
   (a) verify σ with pk_A;
   (b) discard all bit positions from Ψ^B_{dAB} for which Ψ^A_{bA} is not equal to Ψ^B_{bB}
   (c) set Ψ^B_{chkAB} to be the substring of Ψ^B_{dAB} for which the bits of Ψ^A_{indAB} are 1, and set Ψ^B_{kAB} to be the substring of Ψ^B_{dAB} for which the bits of Ψ^A_{indAB} are 0
   (d) let ϵ be the proportion of bits of Ψ^A_{chkAB} that do not match Ψ^B_{chkAB}; if $\epsilon > 0.061$ then abort;
   (e) compute $\sigma \leftarrow \text{Sign}_{pk_B}(\Psi^A, \Psi^B, \epsilon, B)$;

(f) send activation $\mathsf{SendC}(\Psi^A, \Psi^B, \epsilon, \sigma)$ on its c-tape to A.

6. Upon activation $\mathsf{SendC}(\Psi^A, \Psi^B, \epsilon, \sigma)$, the classical Turing machine A does the following:

   (a) verify σ with pk_B;

   (b) read (random) bits Ψ^A_F from its r-tape to construct a random a 2-universal hash function $F : \{0,1\}^{n_3} \to \{0,1\}^{r'}$ (where $r' = n_3 h(\epsilon) + o(n_3)$) for information reconciliation[5] and compute $F' = F(\Psi^A_{kAB})$;

   (c) read (random) bits $\Psi^A_{P,G}$ from its r-tape to generate a random permutation P on n_3 elements and a 2-universal hash function $G : \{0,1\}^{n_3} \to \{0,1\}^{s'}$ (where $s' = n_3(1 - 3h(\epsilon)) + o(n_3)$) for privacy amplification, respectively; compute $\Psi^A_{skAB} \leftarrow G(P(\Psi^A_{kAB}))$;

   (d) compute $\sigma \leftarrow \mathsf{Sign}_{pk_A}(\Psi^A, \Psi^B, F, F', P, G, A)$;

   (e) send activation $\mathsf{SendC}(\Psi^A, \Psi^B, F, F', P, G, \sigma)$ on its c-tape to B;

   (f) output $(sk = \Psi^A_{skAB}, pid = B, \mathbf{v} = (\mathbf{v}_0 = (\ell(\Psi^A_{dAB})), \mathbf{v}_1 = (\ell(\Psi^A_{bAB})), \mathbf{v}_2 = (\ell(\Psi^B_{dAB})), \mathbf{v}_3 = (\ell(\Psi^B_{bAB})), \mathbf{v}_4 = (\ell(\Psi^A_F)), \mathbf{v}_5 = (\ell(\Psi^A_{P,G}))), \mathbf{u} = (\mathbf{u}_1 = (pk_B)))$ (recall $\ell(\cdot)$ denotes the label describing the corresponding secret value).

7. Upon activation $\mathsf{SendC}(\Psi^A, \Psi^B, F, F', P, G, \sigma)$, the classical Turing machine B does the following:

   (a) verify σ with pk_A;

   (b) use F and F' to correct Ψ^B_{kAB} to $\Psi^B_{kAB'}$;

   (c) compute $\Psi^B_{skAB} \leftarrow G(P(\Psi^B_{kAB'}))$;

   (d) output $(sk = \Psi^B_{skAB}, pid = A, \mathbf{v} = (\mathbf{v}_0 = (\ell(\Psi^A_{dAB})), \mathbf{v}_1 = (\ell(\Psi^A_{bAB})), \mathbf{v}_2 = (\ell(\Psi^B_{dAB})), \mathbf{v}_3 = (\ell(\Psi^B_{bAB})), \mathbf{v}_4 = (\ell(\Psi^A_F)), \mathbf{v}_5 = (\ell(\Psi^A_{P,G})),), \mathbf{u} = (\mathbf{u}_1 = (pk_A)))$.

Remark 1. In the output vector $\mathbf{v}$, the values $\ell(\Psi^A_{bAB})$, $\ell(\Psi^B_{bAB})$, $\ell(\Psi^A_F)$, and $\ell(\Psi^A_{P,G})$ appear as single component vectors. But in step 6(e) the values are broadcast in the clear. This may seem a bit contradictory since, if the adversary has revealed the secret for either of those values (and therefore learns their content), the session is not fresh, but because of the broadcast the adversary *does* in fact learn the values corresponding to the aforementioned labels. The important distinction is *when* the adversary obtains these values, either before or after the protocol commences and measurements are performed. For the adversary to learn these values before parties' measurements, it must reveal the secret for these values, violating session freshness. Learning the values after the session completes is not an issue and the values are given to the adversary "for free", without the need for revealing the secrets.

Remark 2. The output vector $\mathbf{u}$ represents the values which the session owner uses to authenticate its peer. Similar to $\ell(\Psi^A_{bAB})$ the authentication information has to be exclusively available to the alleged peer, but only at the time of protocol execution: they may subsequently be revealed.

Observe that for the BB84 protocol above, Alice's own authentication secret pk_A is not included in her $\mathbf{u}$ or $\mathbf{v}$ vectors. This implies that the protocol is resilient to *key compromise impersonation (KCI) attacks* [35, §2.4.2]: even with Alice's authentication keys no party is able to pretend to be someone other than Alice to Alice.

---

[5] For details on information reconciliation and privacy amplification, see the full version [34, Appendix A].

3.1   Security of BB84

We now show that the BB84 protocol stated above is a secure (Theorem 1) and long-term-secure (Theorem 2) AKE protocol assuming that the bounded active adversary cannot break the signature scheme.

Theorem 1 (Security of BB84). *Let k be a security parameter. Suppose that the probability ϵ_{sig} that any probabilistic polynomial time classical Turing machine with oracle access to a $(t_q(k), m_q(k))$-bounded quantum Turing machine can break the signature scheme is negligible in k. Then the BB84 protocol is a secure AKE protocol (Definition 3).*

Proof sketch. Our proof combines an existing proof of security by Christandl et al. [36] for the BB84 protocol with the sequence-of-games technique of Shoup [37]. First we show—using techniques from classical reductionist security—that no bounded adversary can (except with negligible probability) successfully tamper with the classical authenticated communication. Then we show—using techniques from QKD security proofs—that the adversary cannot distinguish the key from random. Details appear in the full version [34].

Theorem 2 (Long-term security of BB84). *Let k be a security parameter. Suppose the signature scheme is secure against all bounded adversaries as specified in Theorem 1. Then the BB84 protocol is a long-term secure authenticated key exchange protocol (Definition 4).*

Proof. The argument in fact appears in the proof of Theorem 1. In its proof, the bounds on $t_c(k)$, $t_q(k)$, and $m_q(k)$ and on the adversary are required only for guaranteeing the authenticity and origin of messages in a game hop that assures that the classical authentic communication has not been tampered with. The remainder of the argument is a typical argument for a quantum key distribution scheme, which does not require any bounds on the adversarial power. Since the unbounded adversary runs after the protocol completes, meaning it cannot inject reorder or modify messages in the transcript, therefore the past classical communication remains authentic and the result follows.

4   Comparing Classical and Quantum Key Exchange Protocols

Given the similarity of our model for both classical and quantum AKE protocols to existing classical AKE security models and our model's flexibility in analyzing the security of a variety of protocols, we can use our model to identify qualitative differences between classes of protocols.

One of the key differences between existing AKE security models such as CK01 and eCK is what randomness the adversary is allowed reveal—and when—yet still have the protocol be secure. Our framework is more generic: it is not the *model* that specifies which randomness can be revealed but the *protocol itself* in its output vectors **v** and **u**. As a result, we can "compare" protocols by viewing

Table 1. Comparison of security properties of various classical and quantum AKE protocols

Protocol	Signed Diffie–Hellman [2]	UP [23]	BB84 [5]	EPR [6]	BHM96 [7,12]
Protocol type	classical	classical	quantum prepare-send-measure	quantum measure-only	quantum prepare-send-only
Security model	CK01 [2]	eCK [3], this paper	this paper	this paper	this paper
Randomness revealable **before** protocol run?	× static key × ephemeral key	at most 1 of static key, ephemeral key	× static key × basic choice × data bits × info. recon. × priv. amp.	× static key × basis choice × info. recon. × priv. amp.	× static key × basis choice × data bits × info. recon. × priv. amp.
Randomness revealable **after** protocol run?	✓ static key × ephemeral key	at most 1 of static key, ephemeral key	✓ static key ✓ basis choice × data bits ✓ info. recon. ✓ priv. amp.	✓ static key ✓ basis choice ✓ info. recon. ✓ priv. amp.	✓ static key ✓ basis choice × data bits ✓ info. recon. ✓ priv. amp.
Short-term security	computational assumption	computational assumption	computational or inf.-th.	computational or inf.-th.	computational or inf.-th.
Long-term security w/short-term-secure authentication	×	×	✓	✓	✓

them all within our model and then comparing which values are included in the output vector.[6]

Table 1 summarizes the observations of this section. We compare two qualitatively different classical AKE protocols and three qualitatively different QKD protocols: (1) the signed Diffie–Hellman protocol [2] (which can be proven secure in the CK01 model), (2) the UP protocol [23], a variant of the MQV protocol [38] which can be proven secure in the eCK model, (3) the BB84 [5] prepare-send-measure QKD protocol, (4) the EPR [6] (entanglement-based) measure-only QKD protocol, and (5) the BHM96 [7,12] prepare-send-only QKD protocol. Our model is flexible enough to allow all these protocols to be proven secure in it, of course with different cryptographic assumptions, bounds on the adversary, and different output vectors, which we compare in Table 1.

Revealing randomness before the run of the protocol. Some classical AKE protocols, especially eCK-secure protocols such as UP and similar MQV-style protocols, remain secure even if the adversary learns either the ephemeral secret key or the long-term secret key, but not both, before the run of the protocol. This contrasts with all known QKD protocols, where none of the random values—the long-term secret key, the basis choices (for measure protocols), data bits (for prepare protocols), information reconciliation function, or privacy amplification function—can be revealed to the adversary in advance. (This is why all of these values are included individually in the output vector **v** in the BB84 specification in Section 3.)

Revealing randomness after the run of the protocol. For classical AKE protocols to remain secure, at least some secret values must not be revealed after the

---

[6] We note that it has been shown [33] that the CK01 and eCK models are *formally incomparable*, meaning neither can be shown to imply the other.

run of the protocol. For protocols with so-called perfect forward secrecy, such as signed Diffie–Hellman, the parties' long-term secret keys can be corrupted after the run of the protocol, but not the ephemeral secret keys. For eCK-secure protocols such as MQV-style protocols like UP, either the long-term or the ephemeral secret key, but not both, can be revealed before, during, or after the protocol run. For measure-only entanglement-based QKD protocols such as EPR, all random choices made by the parties can be revealed after the run of the protocol: this is because the key bits are not chosen by the parties, nor in fact by the adversary, but are the result of measurements and (after successful privacy amplification) are uncorrelated with any of the input bits of any of the parties, including the adversary. This is not the case for prepare-and-send protocols such as BB84 or BHM96, as the sender randomly chooses data bits which must remain secret.

Short-term and long-term security. Classical AKE protocols can be proven secure only under computational assumptions, and as such only offer short-term security in the sense of Definition 3. Even against an unbounded passive adversary they do not retain any of their secrecy properties. Thus classical AKE protocols are only secure against bounded short-term adversaries; however, they can be compared on the relative strength of the bound on the adversary. This contrasts with QKD protocols. QKD can be shown to be secure against either *unbounded* short-term adversaries, by using information-theoretic authentication, or secure against bounded short-term adversaries when using a computationally secure authentication scheme as we have shown for BB84 in Section 3.1. A key contribution of the model in Section 2 is a formalism which captures the notion that QKD can remain secure against an unbounded adversary after the protocol completes, provided the adversary at the time of the run of the protocol could not break the authentication scheme.

Applications wishing to achieve both long-term security (like QKD) and resistance to randomness revelation (like eCK-secure classical AKE protocols) could do so by running both protocols in parallel for each session, and then combining the keys output by the two protocols together; if combined correctly, the resulting key would provide strong short-term security and strong long-term security. This approach is being used by QKD implementers, such as commercial QKD vendor ID Quantique.[7]

5   Conclusions

We have presented a model for key establishment which incorporates both classical key agreement and quantum key distribution. Our model can accommodate a wide range of practical and theoretical scenarios and can serve as a common framework in which to compare relative security properties of different protocols. A key aspect of our model is that restrictions on values the adversary can compromise are not specified by the model but by the output of the protocol. Using our model, we were able to provide a formal argument for the short-term and

---

[7] http://www.idquantique.com/images/stories/PDF/cerberis-encryptor/
cerberis-specs.pdf

long-term security of BB84 in the multi-user setting while using computationally secure authentication.

The ability to compare various classical and quantum protocols in our model has allowed us to identify an important distinction between existing classical and quantum key exchange protocols. At a high level, classical protocols can provide more assurances against online adversaries who can leak or infiltrate in certain ways, but in the long run may be insecure against potential future advances. Current quantum protocols provide assurances against somewhat weaker online adversaries but retain secrecy indefinitely, even against future advances in computing technology.

Since in our model the relative strength of a fresh session is specified by the conditions given in the output vector, an interesting open problem would be to use our model develop a quantum key distribution protocol which does retain its security attributes in the short- and long-terms even if some random values were known before the run of the protocol. Also of interest is how to best combined keys from both quantum and classical key exchange protocols run in parallel.

Acknowledgements. The authors acknowledge helpful discussions with Norbert Lütkenhaus, Alfred Menezes, and Kenny Paterson.

MM is supported by NSERC (Discovery, SPG FREQUENCY, CREATE), QuantumWorks, MITACS, CIFAR, ORF. IQC and Perimeter Institute are supported in part by the Government of Canada and the Province of Ontario.

References

1. Bellare, M., Rogaway, P.: Entity authentication and key distribution. In: Stinson, D.R. (ed.) CRYPTO 1993. LNCS, vol. 773, pp. 232–249. Springer, Heidelberg (1994)
2. Canetti, R., Krawczyk, H.: Analysis of key-exchange protocols and their use for building secure channels. In: Pfitzmann, B. (ed.) EUROCRYPT 2001. LNCS, vol. 2045, pp. 453–474. Springer, Heidelberg (2001)
3. LaMacchia, B., Lauter, K., Mityagin, A.: Stronger security of authenticated key exchange. In: Susilo, W., Liu, J.K., Mu, Y. (eds.) ProvSec 2007. LNCS, vol. 4784, pp. 1–16. Springer, Heidelberg (2007)
4. Canetti, R.: Universally composable security: a new paradigm for cryptographic protocols (extended abstract). In: Proc. 42nd Annual IEEE Symposium on Foundations of Computer Science (FOCS), pp. 136–145. IEEE Press (2001)
5. Bennett, C.H., Brassard, G.: Quantum cryptography: public key distribution and coin tossing. In: Proc. IEEE International Conf. on Computers, Systems and Signal Processing, pp. 175–179. IEEE (December 1984)
6. Ekert, A.K.: Quantum cryptography based on Bell's theorem. Physical Review Letters 67, 661–663 (1991)
7. Biham, E., Huttner, B., Mor, T.: Quantum cryptographic network based on quantum memories. Physical Review A 54(4), 2651–2658 (1996)
8. Mayers, D.: Quantum key distribution and string oblivious transfer in noisy channels. In: Koblitz, N. (ed.) CRYPTO 1996. LNCS, vol. 1109, pp. 343–357. Springer, Heidelberg (1996)

9. Lo, H.K., Chau, H.F.: Unconditional security of quantum key distribution over arbitrarily long distances. Science 283(5410), 2050–2056 (1999)

10. Biham, E., Boyer, M., Boykin, P.O., Mor, T., Roychowdhury, V.: A proof of the security of quantum key distribution (extended abstract). In: Proc. 32nd Annual ACM Symposium on the Theory of Computing (STOC), pp. 715–724. ACM Press (2000)

11. Shor, P., Preskill, J.: Simple proof of security of the BB84 quantum key distribution protocol. Physical Review Letters 85(2), 441–444 (2000)

12. Inamori, H.: Security of practical time-reversed EPR quantum key distribution. Algorithmica 34(4), 340–365 (2002)

13. Gottesman, D., Lo, H.K., Lütkenhaus, N., Preskill, J.: Security of quantum key distribution with imperfect devices. Quantum Information and Computation 4(5), 325–360 (2004)

14. Renner, R.: Security of Quantum Key Distribution. PhD thesis, Swiss Federal Institute of Technology Zürich (2005)

15. Paterson, K.G., Piper, F., Schack, R.: Quantum cryptography: A practical information security perspective. In: Zukowski, M., Kilin, S., Kowalik, J. (eds.) Proc. NATO Advanced Research Workshop on Quantum Communication and Security. NATO Science for Peace and Security Series, Sub-Series D: Information and Communication Security, vol. 11. IOS Press (2007), http://arxiv.org/abs/quant-ph/0406147

16. Alléaume, R., Bouda, J., Branciard, C., Debuisschert, T., Dianati, M., Gisin, N., Godfrey, M., Grangier, P., Länger, T., Leverrier, A., Lütkenhaus, N., Painchault, P., Peev, M., Poppe, A., Pornin, T., Rarity, J., Renner, R., Ribordy, G., Riguidel, M., Salvail, L., Shields, A., Weinfurter, H., Zeilinger, A.: SECOQC white paper on quantum key distribution and cryptography (January 2007), http://www.arxiv.org/abs/quant-ph/0701168

17. Stebila, D., Mosca, M., Lütkenhaus, N.: The case for quantum key distribution. In: Sergienko, A., Pascazio, S., Villoresi, P. (eds.) QuantumComm 2009. LNICST, vol. 36, pp. 283–296. Springer, Heidelberg (2010)

18. Ioannou, L.M., Mosca, M.: A new spin on quantum cryptography: Avoiding trapdoors and embracing public keys. In: Yang, B.-Y. (ed.) PQCrypto 2011. LNCS, vol. 7071, pp. 255–274. Springer, Heidelberg (2011)

19. Unruh, D.: Everlasting quantum security. Cryptology ePrint Archive, Report 2012/177 (2012), http://eprint.iacr.org/

20. Goldberg, I., Stebila, D., Ustaoglu, B.: Anonymity and one-way authentication in key exchange protocols. Designs, Codes and Cryptography 67(2), 245–269 (2013)

21. Cachin, C., Maurer, U.: Unconditional security against memory-bounded adversaries. In: Kaliski Jr., B.S. (ed.) CRYPTO 1997. LNCS, vol. 1297, pp. 292–306. Springer, Heidelberg (1997)

22. Müller-Quade, J., Unruh, D.: Long-term security and universal composability. Journal of Cryptology 23(4), 594–671 (2010)

23. Ustaoglu, B.: Comparing SessionStateReveal and EphemeralKeyReveal for Diffie-Hellman protocols. In: Pieprzyk, J., Zhang, F. (eds.) ProvSec 2009. LNCS, vol. 5848, pp. 183–197. Springer, Heidelberg (2009)

24. Carter, J.L., Wegman, M.N.: Universal classes of hash functions. Journal of Computer and System Sciences 18(2), 143–154 (1979)

25. Wegman, M.N., Carter, J.L.: New hash functions and their use in authentication and set equality. Journal of Computer and System Sciences 22(3), 265–279 (1981)

26. Ben-Or, M., Horodecki, M., Leung, D.W., Mayers, D., Oppenheim, J.: The universal composable security of quantum key distribution. In: Kilian, J. (ed.) TCC 2005. LNCS, vol. 3378, pp. 386–406. Springer, Heidelberg (2005)

27. Ben-Or, M., Mayers, D.: General security definition and composability for quantum & classical protocols (2004); arXiv:quant-ph/0409062.

28. Fehr, S., Schaffner, C.: Composing quantum protocols in a classical environment. In: Reingold, O. (ed.) TCC 2009. LNCS, vol. 5444, pp. 350–367. Springer, Heidelberg (2009)

29. Unruh, D.: Simulatable security for quantum protocols arXiv:quant-ph/0409125. Extended abstract published as [31]

30. Unruh, D.: Universally composable quantum multi-party computation (full version) (October 2009); arXiv:0910.2912. Short version published as [31]

31. Unruh, D.: Universally composable quantum multi-party computation. In: Gilbert, H. (ed.) EUROCRYPT 2010. LNCS, vol. 6110, pp. 486–505. Springer, Heidelberg (2010)

32. Aharonov, D., Ben-Or, M., Eban, E.: Interactive proofs for quantum computations. In: Yao, A.C.C. (ed.) Proc. Innovations in Computer Science (ICS 2010), pp. 453–469 (October 2010)

33. Cremers, C.: Examining indistinguishability-based security models for key exchange protocols: the case of CK, CK-HMQV, and eCK. In: Proc. 6th ACM Symposium on Information, Computer and Communications Security (ASIACCS 2011), pp. 80–91. ACM (2011)

34. Mosca, M., Stebila, D., Ustaoğlu, B.: Quantum key distribution in the classical authenticated key exchange framework. Cryptology ePrint Archive, Report 2012/361 (2012), http://eprint.iacr.org/2012/361, http://arxiv.iacr.org/2012/361

35. Boyd, C., Mathuria, A.: Protocols for Authentication and Key Establishment. Springer (2003)

36. Christandl, M., Renner, R., Ekert, A.: A generic security proof for quantum key distribution (February 2004), http://arxiv.org/abs/quant-ph/0402131v2

37. Shoup, V.: Sequences of games: A tool for taming complexity in security proofs, http://www.shoup.net/papers/games.pdf (2006) (first version appeared in 2004)

38. Law, L., Menezes, A., Qu, M., Solinas, J., Vanstone, S.A.: An efficient protocol for authenticated key agreement. Designs, Codes and Cryptography 28(2), 119–134 (2003)

Cryptanalysis of Hash-Based Tamed Transformation and Minus Signature Scheme

Xuyun Nie[1,2,3,4], Zhaohu Xu[1,3], and Johannes Buchmann[2]

[1] School of Computer Science and Engineering,
University of Electronic Science and Technology of China, Chengdu 611731, China
[2] Technische Universität Darmstadt, Department of Computer Science,
Hochschulstraße 10, 64289 Darmstadt, Germany
[3] Network and Data Security Key Laboratory of Sichuan Province
[4] State Key Laboratory of Information Security, Institute of Information Engineering,
Chinese Academy of Sciences, Beijing 100093, China
xynie@uestc.edu.cn, xzh_tiger@yahoo.cn,
buchmann@cdc.informatik.tu-darmstadt.de

Abstract. In 2011, wang et al. proposed a security enhancement method of Multivariate Public Key Cryptosystems (MPKCs), named Extended Multivariate public key Cryptosystems (EMC). They introduced more variables in an original MPKC by a so-called Hash-based Tamed (HT) transformation in order to resist existing attack on the original MPKC. They proposed Hash-based Tamed Transformation and Minus (HTTM) signature scheme which combined EMC method with minus method. Through our analysis, the HTTM is not secure as they declared. If we can forge a valid signature of the original MPKC-minus signature scheme, we could forge a valid signature of HTTM scheme successfully.

Keywords: Multivariate public key cryptosystem, Minus method, Algebraic attack, Hash-based tamed transformation.

1 Introduction

For last three decades, due to the quantum computer attack [Sho99] on the traditional public key cryptosystems which based on the assumption about the difficulty of certain number theory problems, such as the Integer Prime Factorization Problem or the Discrete Logarithm Problem, people are constantly looking for cryptographic algorithms that can resist quantum computer algorithms attack. Multivariate public key cryptosystem (MPKC) is one of the promising alternatives to resist the quantum computer attack. The security of MPKC relies on the difficulty of solving systems of nonlinear multivariate quadratic (MQ) polynomial equations in a finite field, which is a NP-hard problem in general. However, this does not guarantee that these new cryptosystems are secure. Compared with RSA public key cryptosystems, the computation in MPKC can be very fast because it is operated on a small finite field. By now, there is no quantum computer algorithm to solve MQ problem in polynomial time.

P. Gaborit (Ed.): PQCrypto 2013, LNCS 7932, pp. 155–164, 2013.

The first promising construction of MPKC is the Matsumoto-Imai (MI) scheme [MI88] proposed in 1988. Unfortunately, it was defeated by Patarin in 1995 with the linearization equation method [Pat95]. Since then, many types of MPKCs were proposed such as HFE [Pat96], MFE [WYH06], TTM [Moh99], Rainbow [DS05], TTS [YC05] etc. Also, there are many attack methods proposed, for instance, linearization equation attack [Pat95] [DHN07], XL [CKPS00], Groebner basis [FJ03], differential attack [FGS05] [DFSS07] and so on. In addition to the design of the new systems, many security enhancement methods were proposed to resist existing attack. There are plus/minus, internal perturbation [Ding04], piece-in-hand [TTF04] etc. But some of them are not very successful.

In 2011, Wang et al. proposed a method named Extended Multivariate public key Cryptosystems (EMC)[WZW11]. Given an MPKC cryptosystem, they used a Hash-based Tame (HT) transformation working on the plaintext variables to introduce some new variables in public key to enhance the security of the original MPKC. This made the public key seems more complicated. Combined with HT transformation and minus method, they proposed Hash-based Tamed Transformation and Minus signature scheme. They claimed the HTTM is secure against the existing attacks without losing the efficiency of the original MPKC.

Through analysis, we found that the HT transformation can not really enhance the security of the original MPKC. Given a public key of HTTM signature scheme, if there were an algorithm $\mathcal{A}$ which can be used to forge a valid signature of the original MPKC combined with minus method, there would also exist an algorithm which could be used to forge a valid signature of HTTM. Firstly, we get a new public key which is equivalent to the original MPKC by setting all the new variables which were introduced by HT transformation equal to zero. This step can remove all the new variables and make the HT transformation change to be an affine map. And using the special structure of the HT transformation, we get the value of the matrix D which is a key parameter of the HT transformation. And then, we can derive the relationship between the inverses of two public keys on the same message. At last, given a message to be signed, we use algorithm $\mathcal{A}$ forge a valid signature under the new public key and then forge a valid signature under the HTTM scheme according the relationship derived above.

The paper is organized as follows. We introduce EMC and HTTM scheme in section 2 and present our cryptanalysis in section 3. In section 4, we present a practical attack on an instance of HTTM. Finally, in section 5, we conclude the paper.

2   Hash-Based Tamed Transformation and Minus Signature Scheme

Wang et al. introduced a function named Hash-based Tamed Transformation (HT for short) to enhance the security of MPKC. They used HT transformation on the plaintext variables and put the output into the original MPKC. They called the new scheme Extended Multivariate public key Cryptosystems (EMC).

We use the same notation as in [WZW11]. Let $\mathbb{F}_q$ be a degree k extension of the field $\mathbb{F}_2$, where $q = 2^k$, $\mathbb{F}_q^n$ be n-dimensional vector space over $\mathbb{F}_q$, $\mathbb{F}_{q^n}$ be a degree n extension of the field $\mathbb{F}_q$. Let $H(\cdot)$ be a standard hash function such as SHA-1, $H_k(\cdot)$ be an operation extracting the first k bits of $H(\cdot)$ and mapping the bit string into an element in $\mathbb{F}_q$. Let $a \parallel b$ be concatenation of variables a and b. let δ be the number of extended input variables of public key and μ $(0 \le \mu < \delta)$ be the number of deleted equations of the central map.

2.1 General Form of MPKC

The general form of MPKC : $P : \mathbb{F}_q^n \to \mathbb{F}_q^n$, $F : \mathbb{F}_q^n \to \mathbb{F}_q^n$

$$y = (y_1, \cdots, y_n) = P(x_1, \cdots, x_n) = T \circ F \circ U(x_1, \cdots, x_n)$$

The public key of MPKC is a set of quadratic polynomials $P(x_1, \ldots, x_n) = (P_1(x_1, \ldots, x_n), \ldots, P_n(x_1, \ldots, x_n))$. The private key are two invertible affine maps T and U. The function F is called the central map of MPKC.

2.2 HT Transformation and EMC

The form of HT transformation is described as follow. $L : F_q^{n+\delta} \to F_q^n$

$$\begin{cases} \begin{pmatrix} h_1 \\ \vdots \\ h_{n-\delta} \end{pmatrix} = A \cdot \begin{pmatrix} x_1 \\ \vdots \\ x_{n-\delta} \end{pmatrix} + \alpha_1 \\ \begin{pmatrix} h_{n-\delta+1} \\ \vdots \\ h_n \end{pmatrix} = \begin{pmatrix} x_{n-\delta+1} \\ \vdots \\ x_n \end{pmatrix} + D \cdot \begin{pmatrix} x_{n+1} \\ \vdots \\ x_{n+\delta} \end{pmatrix} + B \cdot \begin{pmatrix} x_1 \\ \vdots \\ x_{n-\delta} \end{pmatrix} + \alpha_2 \end{cases}$$

where α_1, α_2 are $n - \delta$-dimension vector and δ-dimension vector respectively; $(n - \delta) \times (n - \delta)$ invertible matrix A and full-rank $\delta \times \delta$ diagonal matrix D ; B is a $\delta \times (n - \delta)$ random matrix. The extended variables x_{n+i} $1 \le i \le \delta$ are defined by

$$x_{n+i} = H_k(x_1 \parallel x_2 \parallel \cdots \parallel x_{n-\delta+i-1})$$

Hence, $(h_1, \ldots, h_n) = L(x_1, \ldots, x_n, x_{n+1}, \ldots, x_{n+\delta})$. Due to its structure, the function L can be easy inverted.

The public key of EMC is the expression of function $\bar{P}$.

$$\bar{P} = (\bar{P}_1, \ldots, \bar{P}_n) = P \circ L = T \circ F \circ U \circ L.$$

To encrypt a plaintext $x = (x'_1, \ldots, x'_n)$, they can firstly computer $x'_{n+i} = H_k(x'_1 \parallel x'_2 \parallel \cdots \parallel x'_{n-\delta+i-1})$ and substitute $x'_1, \ldots, x'_{n+\delta}$ into public key. Then, the ciphertext $y' = (y'_1, \cdots, y'_n)$ can be derived.

To decrypt a valid ciphertext is to computer T^{-1}, F^{-1}, U^{-1}, L^{-1} in turn, that is

$$x = (x'_1, \ldots, x'_n) = L^{-1} \circ U^{-1} \circ F^{-1} \circ T^{-1}(y'_1, \cdots, y'_n).$$

2.3   HTTM Signature Scheme

Wang et al. combined EMC and minus method to construct the HTTM signature scheme. we used same notations above.

Private key. The private keys of the original MPKC scheme (T, U, F and their inverses) plus L and L^{-1}.

Public key

$$\bar{P}^-(x_1,\ldots,x_{n+\delta}) = (\bar{P}_1,\ldots,\bar{P}_{n-\mu})$$

which is derived by removing the last μ polynomials in the public key of EMC.

Signing. Let the message be $y' = (y'_1,\cdots,y'_{n-\mu}) \in \mathbb{F}_q^{n-\mu}$. Then a signer chooses μ random elements $y'_{n-\mu+1},\cdots,y'_n$, which are appended y' to $y' = (y'_1,\cdots,y'_n)$ $\in \mathbb{F}_q^n$. To obtain the valid signature x', he (or she) calculates

$$x = (x'_1,\ldots,x'_{n+\delta}) = L^{-1} \circ U^{-1} \circ F^{-1} \circ T^{-1}(y'_1,\cdots,y'_n).$$

Verification. After receiving the message $y' = (y'_1,\cdots,y'_{n-\mu})$ and its signature $(x'_1,\ldots,x'_{n+\delta})$, the verifier performs the following steps. Firstly, the verifier checks whether or not

$$x'_{n+i} = H_k(x'_1 \parallel x'_2 \parallel \cdots \parallel x'_{n-\delta+i-1}), 1 \le i \le \delta.$$

If they are true, then the verifier checks whether or not

$$\bar{P}^-(x'_1,\ldots,x'_{n+\delta}) = (y'_1,\cdots,y'_{n-\mu}).$$

Practical Parameters. They gave an practical scheme of HTTM, named HTTM^{v_1}, in which they chose MI as an original MPKC scheme and $n = 31$, $k = 6$, $\delta = 10$, $\mu = 5$.

See reference [WZW11] for more details.

3   Cryptanalysis of HTTM

Through theoretical analysis, we found that if there exists an algorithm $\mathcal{A}$ can forge a valid signature of the original MPKC-minus signature scheme, we could also forge a valid signature of HTTM. That is, HT transformation cannot enhance the security of MPKC signature scheme.

To show this, we need three propositions.

Proposition 1. Let all terms which contained $x_{n+i}(1 \le i \le \delta)$ equal to zero in public key $\bar{P}^-$ of a HTTM scheme, we can get a new public key $\bar{P}^-_{L'}$, which is equivalent to the public key of the original MPKC-minus signature scheme, where L' is the special case of the function L with matrix $D = 0$.

Proof. Let D be a zero matrix in L, we get the function $L' : F_q^n \to F_q^n$.

$$\begin{cases} \begin{pmatrix} h_1 \\ \vdots \\ h_{n-\delta} \end{pmatrix} = A \cdot \begin{pmatrix} x_1 \\ \vdots \\ x_{n-\delta} \end{pmatrix} + \alpha_1 \\ \begin{pmatrix} h_{n-\delta+1} \\ \vdots \\ h_n \end{pmatrix} = \begin{pmatrix} x_{n-\delta+1} \\ \vdots \\ x_n \end{pmatrix} + B \cdot \begin{pmatrix} x_1 \\ \vdots \\ x_{n-\delta} \end{pmatrix} + \alpha_2 \end{cases}$$

Namely,

$$L'(x_1, \cdots, x_n)^t = \begin{pmatrix} A & O \\ B & I \end{pmatrix} \begin{pmatrix} x_1 \\ \vdots \\ x_n \end{pmatrix} + \begin{pmatrix} \alpha_1 \\ \alpha_2 \end{pmatrix}$$

Note that the function L' is exactly an invertible affine map on $\mathbb{F}_q$. Denote $U_{L'} = U \circ L'$. The $U_{L'}$ is also an invertible affine map on $\mathbb{F}_q$.

So, if we set $D = 0$ in public key of $\bar{P}^-$, we could get a new public key, denoted by $\bar{P}_{L'}^-$:

$$\bar{P}_{L'}^- = M_\mu \circ T \circ F \circ U \circ L' = M_\mu \circ T \circ F \circ U_{L'},$$

where M_μ is the minus function which moves the last μ polynomials in the public key. Clearly, $\bar{P}_{L'}^-$ is equivalent to the public key of the original MPKC-minus signature scheme.

The expression of $\bar{P}_{L'}^-$ can be also derived by setting all terms which contained $x_{n+i}(1 \leq i \leq \delta)$ equal to zero in public key $\bar{P}^-$.

$$\begin{aligned} \bar{P}_{L'}^-(x_1, \cdots, x_n) &= M_\mu \circ T \circ F \circ U_{L'}(x_1, \cdots, x_n) \\ &= M_\mu \circ T \circ F \circ U \circ L'(x_1, \cdots, x_n) \\ &= M_\mu \circ T \circ F \circ U(\begin{pmatrix} A & O \\ B & I \end{pmatrix} \begin{pmatrix} x_1 \\ \vdots \\ x_n \end{pmatrix} + \begin{pmatrix} \alpha_1 \\ \alpha_2 \end{pmatrix})) \\ &= M_\mu \circ T \circ F \circ U(\begin{pmatrix} A & O & O \\ B & I & O \end{pmatrix} \begin{pmatrix} x_1 \\ \vdots \\ x_{n+\delta} \end{pmatrix} + \begin{pmatrix} \alpha_1 \\ \alpha_2 \end{pmatrix})) \\ &= M_\mu \circ T \circ F \circ U(\begin{pmatrix} A & O & O \\ B & I & D \end{pmatrix} \begin{pmatrix} x_1 \\ \vdots \\ x_n \\ 0 \\ \vdots \\ 0 \end{pmatrix} + \begin{pmatrix} \alpha_1 \\ \alpha_2 \end{pmatrix})) \\ &= M_\mu \circ T \circ F \circ U \circ L(x_1, \cdots, x_n, 0, \cdots, 0) \end{aligned}$$

$\square$

The signatures of a message under $\bar{P}^-$ and $\bar{P}_{L'}^-$, respectively, have following relationship.

Proposition 2. Given a message $y' = (y'_1, \cdots, y'_{n-\mu}) \in \mathbb{F}_q^{n-\mu}$, consider the signatures under $\bar{P}^-$ and $\bar{P}_{L'}^-$, denote them $x' = (x'_1, \cdots, x'_{n+\delta})$ and $x'' = (x''_1, \cdots, x''_n)$, respectively. If we choose the same values of $y'_{n-\mu+1}, \cdots, y'_n$, then $x' = (x'_1, \cdots, x'_{n+\delta})$ and $x'' = (x''_1, \cdots, x''_n)$ satisfy:

(1) $x'_i = x''_i, \quad i = 1, \cdots, n - \delta;$
(2) $x'_{n-\delta+i} = x''_{n-\delta+i} - D[i][i]x'_{n+i}, \quad i = 1, \cdots, \delta$

where $D[i][i]$ be the ith element in the diagonal of matrix D.

Proof. Given a message $y' = (y'_1, \cdots, y'_{n-\mu})$ and randomly chosen the value of $y'_{n-\mu+1}, \cdots, y'_n$, consider its corresponding signatures under $\bar{P}^-$ and $\bar{P}_{L'}^-$. Observing the signature generation process, we found that the only difference in two functions is the difference between L' and L.

Given $(h_1, \ldots, h_n) = (h'_1, \ldots, h'_n)$. Let $(x'_1, \cdots, x'_{n+\delta})$ and $(x''_1, \cdots, x''_n)$ be $(h'_1, \ldots, h'_n)$'s inverse under functions L and L', respectively. Then, we can easily check from the structure of functions L and L':

(1) $x'_i = x''_i, \quad i = 1, \cdots, n - \delta;$
(2) $x'_{n-\delta+i} = x''_{n-\delta+i} - D[i][i]x'_{n+i}, \quad i = 1, \cdots, \delta$

where $D[i][i]$ be the ith element in the diagonal of matrix D. $\square$

Hence, if we can get the value of matrix D, we could forge a valid signature of HTTM after forging a valid signature of $\bar{P}_{L'}^-$ by the algorithm $\mathcal{A}$.

The value of matrix D can be derived from the public key of HTTM efficiently.

Proposition 3. Given a public key of HTTM, $\bar{P}^- = M_\mu \circ T \circ F \circ U \circ L$, we can recover the value of matrix D from it.

Proof. Firstly, we derived the function $\bar{P}_{L'}^- = M_\mu \circ T \circ F \circ U \circ L'$ by setting all terms which contained $x_{n+i}(1 \le i \le \delta)$ equal to zero in public key $\bar{P}^-$.

Comparing $\bar{P}^-$ and $\bar{P}_{L'}^-$, if the inputs of map S in two functions are equal, the outputs are also equal. Hence, we focus on the outputs of L and L' in order to recover the value of matrix D.

Note that, let $x_1 = x_2 = \cdots = x_n = 0$ in L, the output of L will be

$$
\begin{cases}
\begin{pmatrix} h_1 \\ \vdots \\ h_{n-\delta} \end{pmatrix} = \alpha_1 \\
\begin{pmatrix} h_{n-\delta+1} \\ \vdots \\ h_n \end{pmatrix} = D \cdot \begin{pmatrix} x_{n+1} \\ \vdots \\ x_{n+\delta} \end{pmatrix} + \alpha_2 = \begin{pmatrix} D[1][1]x_{n+1} \\ \vdots \\ D[\delta][\delta]x_{n+\delta} \end{pmatrix} + \alpha_2
\end{cases}
$$

while let $x_1 = x_2 = \cdots = x_{n-\delta} = 0$ in L', the output of L' will be

$$\left\{ \begin{array}{l} \begin{pmatrix} h_1 \\ \vdots \\ h_{n-\delta} \end{pmatrix} = \alpha_1 \\ \begin{pmatrix} h_{n-\delta+1} \\ \vdots \\ h_n \end{pmatrix} = \begin{pmatrix} x_{n-\delta+1} \\ \vdots \\ x_n \end{pmatrix} + \alpha_2 \end{array} \right.$$

Thus, if

$$\begin{pmatrix} D[1][1]x_{n+1} \\ \vdots \\ D[\delta][\delta]x_{n+\delta} \end{pmatrix} = \begin{pmatrix} x_{n-\delta+1} \\ \vdots \\ x_n \end{pmatrix},$$

the outputs of L and L' will be equal. Thereby, the outputs of $\bar{P}^-$ and $\bar{P}_{L'}^-$ will be equal.

Due to the observation above, we can recover D by performing the following steps:

(1) For the function $\bar{P}^-$, let $x_1 = x_2 = \cdots = x_n = 0$ and $x_{n+2} = x_{n+3} = \cdots = x_{n+\delta} = 0$, thus the function changes into

$$\bar{P}^-(x_{n+1}) = T^- \circ F \circ S \circ L(\overbrace{0, \cdots, 0}^{n}, x_{n+1}, \overbrace{0, \cdots, 0}^{\delta-1})^T.$$

Taking x_{n+1} over the finite field $\mathbb{F}_q$ and storing all results of the function $\bar{P}^-(x_{n+1})$.

(2) For the function $\bar{P}_{L'}^-$, let $x_1 = x_2 = \cdots = x_{n-\delta} = 0$ and $x_{n-\delta+2} = x_{n-\delta+3} = \cdots = x_{n+\delta} = 0$, thus the function changes into

$$\bar{P}_{L'}^-(x_{n-\delta+1}) = M_\mu \circ T \circ F \circ S \circ L'(\overbrace{0, \cdots 0}^{n-\delta}, x_{n-\delta+1}, \overbrace{0, \cdots, 0}^{\delta-1})^T.$$

(3) Taking $x_{n-\delta+1}$ over the finite field $\mathbb{F}_q$ and comparing the results of $\bar{P}_{L'}^-(x_{n-\delta+1})$ to the results of the function $\bar{P}^-(x_{n+1})$, if there were x'_{n+1} and $x'_{n-\delta+1}$ satisfied $\bar{P}^-(x_{n+1}) = \bar{P}_{L'}^-(x_{n-\delta+1})$, then we have $D[1][1]x'_{n+1} = x'_{n-\delta+1}$, namely, $D[1][1] = x'_{n+1}{}^{-1}x'_{n-\delta+1}$. Similarly, we can get the values of $D[2][2], \cdots, D[\delta][\delta]$.

$\square$

The time-complexity of recovering D is $\delta|\mathbb{F}_q|$ and the space-complexity is $|\mathbb{F}_q|$.

Hence, if there exists an algorithm $\mathcal{A}$ can forge a valid signature of the original MPKC-minus signature scheme, we could also forge a valid signature of HTTM through following steps.

(1) Firstly, we derived the function $\bar{P}_{L'}^- = M_\mu \circ T \circ F \circ U \circ L'$ by setting all terms which contained $x_{n+i}(1 \leq i \leq \delta)$ equal to zero in public key $\bar{P}^-$.

(2) Recovering the value of matrix D following by proposition 3.
(3) Given a message $y' = (y'_1, \cdots, y'_{n-\mu})$, forging a valid signature of $\bar{P}_{L'}^-$ using algorithm $\mathcal{A}$.
(4) Deriving a valid signature corresponding to the message y' of HTTM according to proposition 2.

4   Practical Cryptanalysis of HTTM^{v_1}

In [WZW11], the authors gave a practical example of HTTM, namely HTTM^{v_1}. They chose MI scheme as the original MPKC, namely, the central map of HTTM^{v_1} is

$$Y = \hat{F}(X) = X^{1+q^\theta},$$

and they set $q = 2$, $n = 31$, $k = 6$, $\delta = 10$, $\mu = 5$. They did not give the value of θ HTTM^{v_1}. We set $\theta = 11$ in our cryptanalysis such that $\gcd(q^\theta + 1, q^n - 1) = 1$. The hash function in our experiments is SHA-1.

It is well-known that we can forge a valid signature of MI- signature scheme by differential attack [DFSS07].

After generating a public key of HTTM^{v_1}, we perform the following steps.

Firstly, we set all terms which contained $x_{n+i}(1 \leq i \leq \delta)$ equal to zero in public key $\bar{P}^-$ and get the function $\bar{P}_{L'}^- = M_\mu \circ T \circ F \circ U \circ L'$. The function $\bar{P}_{L'}^-$ is equivalent to MI- scheme.

And then, we recover the value of matrix D according to the proposition 3. We did many computer experiments to verify it. The complexity of this step is that the time-complexity is $\delta|\mathbb{F}_q| = 10 \times 2^6 < 2^{10}$ and the space-complexity is $|\mathbb{F}_q| = 2^6$.

Next, given a message $y' = (y'_1, \cdots, y'_{n-\mu})$, we forge a valid signature of it under the function $\bar{P}_{L'}^-$ by using the same technique as in [DFSS07].

At last, we derive a valid signature of the message y' under the public key of HTTM^{v_1} according to proposition 2.

5   Conclusion

In this paper, we gave a practical cryptanalysis of HTTM scheme. The EMC method did not enhance the security of original MPKC. For HTTM scheme, we could forge a valid signature of it if there were an algorithm that can forge a valid signature of the original MPKC. Although the EMC method did not work, it is an interesting method which is worth further studying.

Acknowledgements. The work of this paper was supported by the National Key Basic Research Program of China (2013CB834203), the Fundamental Research Funds for the Central Universities under Grant ZYGX2010J069, the National Natural Science Foundation of China (No. 61103205).

References

[CKPS00]   Courtois, N., Klimov, A., Patarin, J., Shamir, A.: Efficient algorithms for solving overdefined systems of multivariate polynomial equations. In: Preneel, B. (ed.) EUROCRYPT 2000. LNCS, vol. 1807, pp. 392–407. Springer, Heidelberg (2000)

[Ding04]   Ding, J.: A new variant of the Matsumoto-Imai cryptosystem through perturbation. In: Bao, F., Deng, R., Zhou, J. (eds.) PKC 2004. LNCS, vol. 2947, pp. 305–318. Springer, Heidelberg (2004)

[DS05]   Ding, J., Schmidt, D.: Rainbow, a new multivariable polynomial signature scheme. In: Ioannidis, J., Keromytis, A.D., Yung, M. (eds.) ACNS 2005. LNCS, vol. 3531, pp. 164–175. Springer, Heidelberg (2005)

[DHN07]   Ding, J., Hu, L., Nie, X., Li, J., Wagner, J.: High Order Linearization Equation (HOLE) Attack on Multivariate Public Key Cryptosystems. In: Okamoto, T., Wang, X. (eds.) PKC 2007. LNCS, vol. 4450, pp. 233–248. Springer, Heidelberg (2007)

[DFSS07]   Dubois, V., Fouque, P., Shamir, A., Stern, J.: Practical Cryptanalysis of SFLASH. In: Menezes, A. (ed.) CRYPTO 2007. LNCS, vol. 4622, pp. 1–12. Springer, Heidelberg (2007)

[FGS05]   Fouque, P.-A., Granboulan, L., Stern, J.: Differential Cryptanalysis for Multivariate Schemes. In: Cramer, R. (ed.) EUROCRYPT 2005. LNCS, vol. 3494, pp. 341–353. Springer, Heidelberg (2005)

[FJ03]   Faugère, J.-C., Joux, A.: Algebraic cryptanalysis of hidden field equation (HFE) cryptosystems using gröbner bases. In: Boneh, D. (ed.) CRYPTO 2003. LNCS, vol. 2729, pp. 44–60. Springer, Heidelberg (2003)

[GJ79]   Garey, M., Johnson, D.: Computers and intractability, A Guide to the theory of NP-compuleteness. W.H.Freeman (1979)

[MI88]   Matsumoto, T., Imai, H.: Public quadratic polynomial-tuples for efficient signature-verification and message-encryption. In: Günther, C.G. (ed.) EUROCRYPT 1988. LNCS, vol. 330, pp. 419–453. Springer, Heidelberg (1988)

[Moh99]   Moh, T.: A fast public key system with signature and master key functions. Lecture Notes at EE department of Stanford University (May 1999), http://www.usdsi.com/ttm.html

[Pat95]   Patarin, J.: Cryptanalysis of the Matsumoto and Imai Public Key Scheme of Eurocrypt '88. In: Coppersmith, D. (ed.) CRYPTO 1995. LNCS, vol. 963, pp. 248–261. Springer, Heidelberg (1995)

[Pat96]   Patarin, J.: Hidden fields equations (HFE) and isomorphisms of polynomials (IP): Two new families of asymmetric algorithms. In: Maurer, U.M. (ed.) EUROCRYPT 1996. LNCS, vol. 1070, pp. 33–48. Springer, Heidelberg (1996)

[Sho99]   Shor, P.: Polynomial-time algorithms for prime factorization and discrete logarithms on a quantum computer. SIAM Rev. 41(2), 303–332 (1999)

[TTF04]   Tsujii, S., Tadaki, K., Fujioka, R.: Piece in Hand concept for enhanceing the security of multivariate type pulic key cryptosystem: public key without containing all the information of secret key. IACR eprint 2004/366, http://eprint.iacr.org

[W07]   Wang, Z.: An Improved Medium-Field Equation (MFE) Multivariate Public Key Encryption Scheme. IIH-MISP (2007), http://bit.kuas.edu.tw/iihmsp07/acceptedlistgeneralsession.html

[WYH06]   Wang, L.-C., Yang, B.-Y., Hu, Y.-H., Lai, F.: A "Medium-Field" Multi-
           variate Public-Key Encryption Scheme. In: Pointcheval, D. (ed.) CT-RSA
           2006. LNCS, vol. 3860, pp. 132–149. Springer, Heidelberg (2006)
[WZW11]   Wang, H., Zhang, H., Wang, Z., Tang, M.: Extended multivariate public
           key cryptosystems with secure encryption function. SCIENCE CHINA
           Information Sciences 54(6), 1161–1171 (2011)
[YC05]   Yang, B.-Y., Chen, J.-M.: Building Secure Tame-like Multivariate Public-
           Key Cryptosystems: The New TTS. In: Boyd, C., González Nieto, J.M.
           (eds.) ACISP 2005. LNCS, vol. 3574, pp. 518–531. Springer, Heidelberg
           (2005)

A Classification of Differential Invariants for Multivariate Post-quantum Cryptosystems

Ray Perlner[1] and Daniel Smith-Tone[1,2]

[1] National Institute of Standards and Technology,
Gaithersburg, Maryland, USA
[2] Department of Mathematics, University of Louisville,
Louisville, Kentucky, USA
{daniel.smith,ray.perlner}@nist.gov

Abstract. Multivariate Public Key Cryptography(MPKC) has become one of a few options for security in the quantum model of computing. Though a few multivariate systems have resisted years of effort from the cryptanalytic community, many such systems have fallen to a surprisingly small pool of techniques. There have been several recent attempts at formalizing more robust security arguments in this venue with varying degrees of applicability. We present an extension of one such recent measure of security against a differential adversary which has the benefit of being immediately applicable in a general setting on unmodified multivariate schemes.

Keywords: Matsumoto-Imai, multivariate public key cryptography, differential, symmetry.

1 Introduction

Since Peter Shor's discovery of quantum algorithms for factoring and computing discrete logarithms quickly with quantum computers, there has been a growing community with the goal of establishing a replacement for RSA or Diffie-Hellman in the quantum realm. The last two decades have witnessed a great deal of progress towards realizing that quantum computing world, indicating that Shor's discovery is a great deal more than a mathematical curiosity; instead, his discovery marks the need for an eventual paradigm shift in our public key infrastructure.

Multivariate Public Key Cryptography(MPKC) has emerged as one of a few serious candidates for security in the post-quantum world. This emergence is due to several facts. First, the problem of solving a system of quadratic equations is known to be NP-hard, and seems to be hard even in the average case. No great reduction of the complexity of this problem has been found in the quantum model of computing, and, indeed, if this problem is discovered to be solvable in the quantum model, we can solve all NP problems, which seems particularly wishful. Second, multivariate systems are very efficient, often having speeds dozens of times faster than RSA, [1–3]. Finally, several theoretical

P. Gaborit (Ed.): PQCrypto 2013, LNCS 7932, pp. 165–173, 2013.
© Springer-Verlag Berlin Heidelberg 2013

advances have resulted in the development of modification techniques which allow multiple parameters to be hidden within a system which can be altered to achieve different performance or security properties.

One of the great challenges facing MPKC is the task of establishing reasonable security assurance. Though there have been some recent attempts at forming a new model in which to offer provable security for encryption and signatures, see for example [4, 5], it seems apparent that these models are not as general as we would like or require modifications of realistic protocols to carry their full meaning. The task of quantifying indistinguishability between general classes of systems of multivariate equations seems exceptionally difficult in light of the fact that even with a great deal of structure in the construction of a multivariate cryptosystem, the coefficients can appear to have a uniform distribution. Although history has shown that once a way to distinguish a class of systems of structured multivariate equations from a collection of randomly generated equations is discovered, a method of solving this system is often quickly developed, it is not clear that the techniques for distinguishing such systems are indicative of an underlying theme powerful enough to establish a general method of security proof.

The many cryptanalyses of various big field multivariate cryptosystems have, however, pointed out weaknesses in the predominant philosophy for the construction of such multivariate public key cryptosystems. Several systems, SFLASH, Square, for example, which are based on simple modifications of the prototypical Matsumoto-Imai public key cryptosystem, have been broken by very similar differential attacks exploiting some symmetry which is inherent to the field structure these systems utilize. See [6–9]. Even in the small field milieu, various attacks, for example the oil-vinegar attack, see [10], can be viewed as an attack on differential structure; specifically, discovering a differential invariant.

In [11], a measure of security against attacks exploiting differential symmetry was advanced. This methodology allows one to construct proofs that a cryptosystem is secure against a differential symmetry adversary by classifying the differential symmetric structure of the cryptosystem. By identifying all possible initial general linear differential symmetries possessed by a field map, one can determine which linear relations involving the differential of a public key are accessible to any adversary, and thus guarantee security against such an attack model. Although this result is not as robust as a reduction theoretic proof of security, it has the benefit, first, of being far stronger than the traditional model of checking the vulnerability of new schemes against old attacks, second, of being immediately applicable in the design of cryptosystems, and third, of perhaps being a more realistic goal than that of reduction theoretic proof.

In this article, we introduce a technique which is dual to that of [11] in the sense that it assures security against any first-order differential invariant adversary. Specifically, we establish a model for classifying first-order differential invariants of a field map and apply the model, providing classifications of such invariants for specific cryptosystems. This characterization, in conjunction with an analogous classification in the symmetric setting, provides a model for security against any first-order differential adversary, and is the first step towards establishing general differential security via an existence criterion. We suggest such an analysis of differential invariant security as a reasonable criterion and pragmatic tool for cryptographers in the development of future multivariate schemes.

The paper is organized as follows. The next section illustrates the ubiquitous nature of the differential attack by recasting the attack on the balanced oil and vinegar scheme in the differential setting. In the following section, we focus on differential invariants, presenting the first-order differential invariant and discussing the technique for realizing the theoretical differential invariant structure of any class of MPKC. The subsequent section restricts the analysis of this space to the case in which the hidden field map of the cryptosystem is a C^* monomial. The differential invariant structure is then determined for projected systems such as the projected SFLASH analogue, pSFLASH. Finally, we review these results and suggest a general model for differential security.

2   Differential Symmetries and Invariants

Differential attacks play a crucial role in multivariate public key cryptography. Such attacks have not only broken many of the so called "big field" schemes, they have directed the further development of the field by inspiring modifiers — Plus (+), Minus (-), Projection (p), Perturbation (P), Vinegar (v) — and the creation of newer more robust techniques.

The differential of a field map, f, is defined by $Df(a,x) = f(a+x) - f(a) - f(x) + f(0)$. The use of this discrete differential appears to occur in very many cryptanalyses of post-quantum multivariate schemes. In fact, we can even consider Patarin's initial attack, in [12], on Imai and Matsumoto's C^* scheme, see [13], as the exploitation of a trivial differential symmetry. Suppose $f(x) = x^{q^\theta+1}$ and let $y = f(x)$. Since the differential of f, Df, is a symmetric bilinear function, $0 = Df(y,y) = Df(y,x^{q^\theta+1}) = yx^{q^{2\theta}+q^\theta} + y^{q^\theta}x^{q^\theta+1} = x^{q^\theta}(yx^{q^{2\theta}} + y^{q^\theta}x)$. Dividing by x^{q^θ} we have Patarin's linear relation, $yx^{q^{2\theta}} = y^{q^\theta}x$; see [12] for details.

Differential methods provide powerful tools for decomposing a multivariate scheme. To illustrate the nearly universal nature of differential attacks, we review the attack of Kipnis and Shamir, see [10], on a non-big-field system, the oil and vinegar scheme. Though they use differing terminology, the attack exploits a symmetry hidden in the differential structure of the scheme.

Recall that the oil and vinegar scheme is based on a hidden quadratic system of equations, $f : k^n \to k^o$, in two types of variables, $x_1, ..., x_o$, the oil variables, and $x_{o+1}, ..., x_{o+v=n}$, the vinegar variables. We focus on the balanced oil and vinegar scheme, in which $o = v$. Let $c_1, ..., c_v$ be random constants. The map f has the property that $f(x_1, ..., x_v, c_1, ..., c_v)$ is affine in $x_1, ..., x_v$. The encryption map, $\overline{f}$ is the composition of f with an n-dimensional invertible affine map, L.

Let O represent the subspace generated by the first v basis vectors, and let V denote the cosummand of O. Notice that the discrete differential given by $Df(a,x) = f(x+a) - f(x) - f(a) + f(0)$ has the property that for all a and x

in O, $Df(a, x) = 0$. Thus for each coordinate, i, the differential coordinate form Df_i can be represented:

$$Df_i = \begin{bmatrix} 0 & Df_{i1} \\ Df_{i1}^T & Df_{i2} \end{bmatrix}.$$

Let M_1 and M_2 be two invertible matrices in the span of the Df_i. Then $M_1^{-1}M_2$ is an O-invariant transformation of the form:

$$\begin{bmatrix} A & B \\ 0 & C \end{bmatrix}.$$

Now the Df_i are not known, but $D(f \circ L)_i = L^T Df_i L$, so the $L^T Df_i L$ are known. Notice that if M is in the span of the Df_i, then $L^T ML$ is in the span of the $L^T Df_i L$. Also, since $(L^T M_1 L)^{-1}(L^T M_2 L) = L^{-1}M_1^{-1}M_2 L$, there is a large space of matrices leaving $L^{-1}O$ invariant, which Kipnis and Shamir are able to exploit to effect an attack against the balanced oil and vinegar scheme; see [10] for details. Making the oil and vinegar scheme unbalanced, see [14], corrects this problem by making any subspace which is invariant under a general product $M_1^{-1}M_2$ very small, see [15].

3   First-Order Differential Invariants

Let $f : k \to k$ be an arbitrary fixed function on k, a degree n extension of the Galois field $\mathbb{F}_q$. Consider the differential $Df(a, x) = f(a+x) - f(a) - f(x) + f(0)$. We can express the differential as an n-tuple of differential coordinate forms in the following way:

$$[Df(a, x)]_i = a^T Df_i x,$$

where Df_i is a symmetric matrix representation of the action on the ith coordinate of the bilinear differential. A first-order differential invariant of f is a subspace $V \subseteq k$ with the property that there exists a $W \subseteq k$ of dimension at most $dim(V)$ for which simultaneously $AV \subseteq W$ for all $A \in Span_i(Df_i)$.

We note that any simultaneous invariant of all $Span_i(Df_i)$ satisfies the above definition, as well the situation for balanced oil and vinegar, in which the invariant was found in the product of an element and an inverse of an element in $Span_i(Df_i)$. A first-order differential invariant is thus a more general construct than a simultaneous invariant among all differential coordinate forms. We present a proof theoretic technique for classifying the first-order differential invariants of such a multivariate map $f : k \to k$ which can specify parameters admitting such invariant structure.

Suppose f has a first-order differential invariant V. Let $V^\perp$ represent the set of all elements x in k such that the dot product $< x, Ay > = 0$ for all $y \in V$ and for all $A \in Span_i(Df_i)$. We should note that in positive characteristic there is a great deal of freedom in membership in $V^\perp$; there is no reason that $V \cap V^\perp$ should be empty in general or even that $V \oplus V^\perp$ be contained in k. Let $M : k \to V$ be an arbitrary linear map. Choosing an arbitrary linear map $M^\perp : k \to V^\perp$

we have the following (non-linear) symmetric relation, a dual expression of the differential invariance:

$$[Df(M^\perp a, Mx)]_i = a^T (M^\perp)^T Df_i Mx = 0,$$

for all i. Thus $Df(M^\perp a, Mx)$ is identically zero for all $a, x \in k$.

Consequently, the existence of a first-order differential invariant for a map f implies the existence of a nonlinear symmetry on f, that is, a symmetry induced by linear maps such that the system of equations expressing the symmetric relation are nonlinear in the coefficients of the maps. Note that the converse implication is false, so that having a first-order differential invariant is a stronger property than having this manner of nonlinear differential symmetry. By explicitly constructing the polynomial map $\overline{f}(a, x) = Df(M^\perp a, Mx) \equiv 0$ over k^2, we can derive relations permitting the existence of this nonlinear symmetry, and hence the first-order differential invariant.

4 Invariants in the Prototypical Case

As an illustration of this technique we examine the case when $f : k \to k$ is a C^* monomial map. Specifically, we let $f(x) = x^{q^\theta + 1}$ where $(\theta, [k : \mathbb{F}_q]) = 1$. This case in particular applies to the famously broken, see [9], SFLASH signature scheme, which was constructed by composing f with two affine transformations: $P = T \circ f \circ U$, where T is singular and U is of full rank.

Theorem 1. *Let $f : k \to k$ be a C^* monomial map. Then f has no nontrivial first-order differential invariant.*

Proof. Suppose by way of contradiction that f has a first-order differential invariant $\{0\} \subsetneq V \subsetneq k$. Define $V^\perp = \{x| < x, Ay > = 0, \forall y \in V$ and $\forall A \in Span_i(Df_i)\}$. Then f satisfies the relation $Df(M^\perp a, Mx) = 0$ for all $a, x \in k$.

$$
\begin{aligned}
Df(M^\perp a, Mx) &= f(M^\perp a + Mx) - f(M^\perp a) - f(Mx) + f(0) \\
&= f(\sum_{i=0}^{n-1} m_i^\perp a^{q^i} + \sum_{i=0}^{n-1} m_i x^{q^i}) - f(\sum_{i=0}^{n-1} m_i^\perp a^{q^i}) - f(\sum_{i=0}^{n-1} m_i x^{q^i}) + f(0) \\
&= (\sum_{i=0}^{n-1} m_i^\perp a^{q^i} + \sum_{i=0}^{n-1} m_i x^{q^i})^{q^\theta + 1} - (\sum_{i=0}^{n-1} m_i^\perp a^{q^i})^{q^\theta + 1} - (\sum_{i=0}^{n-1} m_i x^{q^i})^{q^\theta + 1} \\
&= \sum_{i=0}^{n-1} \sum_{j=0}^{n-1} (m_j (m_{i-\theta}^\perp)^{q^\theta} + m_i^\perp m_{j-\theta}^{q^\theta}) a^{q^i} x^{q^j}.
\end{aligned}
$$

(1)

Since the collection of monomials $\{a^{q^i} x^{q^j}\}$ are algebraically independent, the fact that the above function is identically zero implies that,

$$m_j (m_{i-\theta}^\perp)^{q^\theta} + m_i^\perp m_{j-\theta}^{q^\theta} = 0,$$

for all $0 \leq i, j \leq n - 1$. This fact implies that all 2×2 minors of the following matrix are zero:

$$\begin{bmatrix} m_0 & m_0^\perp & m_1 & \cdots & m_{n-1} & m_{n-1}^\perp \\ m_{-\theta}^{q^\theta} & (m_{-\theta}^\perp)^{q^\theta} & m_{1-\theta}^{q^\theta} & \cdots & m_{n-1-\theta}^{q^\theta} & (m_{n-1-\theta}^\perp)^{q^\theta} \end{bmatrix}.$$

Thus, the rank of this matrix is one, and we have that the second row is a multiple of the first, say $m_i^* = r(m_{i-\theta}^*)^{q^\theta}$, as well as the fact that each column is a multiple of the first, implying, for example, $m_0^\perp = sm_0$.

Consequently, for all $0 \leq i \leq n - 1$, $m_{i\theta}^* = r^{\frac{q^{i\theta}-1}{q^\theta-1}}(m_0^*)^{q^{i\theta}}$. Moreover, we can specify that $m_{i\theta} = r^{\frac{q^{i\theta}-1}{q^\theta-1}} m_0^{q^{i\theta}}$ and $m_{i\theta}^\perp = r^{\frac{q^{i\theta}-1}{q^\theta-1}} s^{q^{i\theta}} m_0^{q^{i\theta}}$, which implies that $m_i^\perp = s^{q^i} m_i$ for all $0 \leq i \leq n - 1$. Thus

$$M^\perp x = \sum_{i=0}^{n-1} m_i^\perp x^{q^i}$$

$$= \sum_{i=0}^{n-1} m_i s^{q^i} x^{q^i} \qquad (2)$$

$$= \sum_{i=0}^{n-1} m_i (sx)^{q^i}$$

$$= M(sx).$$

Hence, the fact that $Df(M(sa), Mx) = 0$ for all $a, x \in k$ implies that $Df(Ma, Mx) = 0$ for all $a, x \in k$. This result implies that $dim(Mk) \leq 1$, that is, the dimension of the image of M in k is one, by the following argument.

If $Df(\overline{a}, \overline{x}) = 0$, then $\overline{a}\overline{x}\left(\overline{x}^{q^\theta-1} + \overline{a}^{q^\theta-1}\right) = 0$, and $\overline{a}^{q^\theta-1} = -\overline{x}^{q^\theta-1}$ implies that $\overline{a}^{q-1} = -\overline{x}^{q-1}$ since $(q^\theta - 1, q^n - 1) = q - 1$. This equation is satisfied exactly when there exists $\alpha \in \mathbb{F}_q$ such that $\overline{a} = \alpha\overline{x}$.

Since this nonlinear differential symmetry exists for any map $g : k \to k$, there exists no nontrivial differential invariant of f.

We can therefore conclude that C^* has no first-order differential invariant weaknesses, even though it is fraught with linear differential symmetric weaknesses. The significance of this result is that we can prove that the cryptosystem in question is secure against all first-order differential invariant adversaries, even those employing attacks yet undiscovered.

5   Invariant Properties under Projection

After SFLASH was broken, it was suggested in [16] that the affine map U be made singular. We continue, establishing security bounds for this suggestion, one of the last unbroken C^* variants, pC^{*-}, or pSFLASH. We recall that in [11] it was established that pSFLASH with appropriately chosen parameters has no general linear differential symmetries and is thus immune to any type of differential attack relying on the accumulation of linear equations involving the differential of the public key. While it has been established in [17] that the projection in pSFLASH can be removed, the structure when the projection modifier is removed is no longer that of a C^* function; rather, it is an HFE^- scheme. Thus pSFLASH is no more secure than HFE^-, which remains unbroken. For the security details of HFE^-, please see [18].

Theorem 2. *Let $f : k \to k$ be a C^* monomial, and let $\pi : k \to k$ be a linear projection onto a codimension r subspace. Then every nontrivial first-order differential invariant V satisfies $dim(V) \leq dim(V \cap ker(\pi)) + 1$. Consequently, if $r = 1$, there is no nontrivial first-order differential invariant structure beyond the obvious $ker(\pi)$.*

Proof. Let V be a first-order differential invariant of $f \circ \pi$, and let $M : k \to V$ be an arbitrary linear map. Then $\pi \circ M$ is a first-order differential invariant of f, and there exist maps $\overline{M} = \pi \circ M$ and $\overline{M}^{\perp}$ such that:

$$D(f \circ \pi)(M^{\perp}a, Mx) = Df(\pi M^{\perp}a, \pi Mx) = Df(\overline{M}^{\perp}a, \overline{M}x) = 0,$$

for all $a, x \in k$. We note that there are exactly as many possible maps $\overline{M}^{\perp}$ as maps $\pi \circ M^{\perp}$; indeed, the proof of Theorem 1 shows us that $\overline{M}^{\perp}x = \pi \circ M^{\perp}(sx)$ for some s. As in the proof of Theorem 1, $dim(\overline{M}k) \leq 1$, and since π is of codimension r, $dim(Mk) \leq dim(Mk \cap ker(\pi)) + 1$. We note that since any map $g : k \to k$ has this property, $f \circ \pi$ has no nontrivial first-order differential invariant structure beyond $ker(\pi)$.

We can conclude from the above theorem that pSFLASH is secure against any first-order differential invariant adversary.

6 Conclusion

Multivariate public key cryptography has several desirable traits as a potential candidate for post-quantum security. Unfortunately, a standard metric by which we can judge the security of a multivariate scheme has yet to be determined. One consequence of this current status of the field is the similar cryptanalyses of several promising ideas.

We suggest the classification of first-order differential invariants as a second benchmark for the determination of differential security for multivariate public key cryptosystems. We note that while the lack of the symmetric and invariant differential security argument does not imply that a cryptosystem is insecure against a differential adversary, the presence of such an assurance guarantees the resistance against any future first-order differential attack.

The case of pSFLASH is particularly interesting because while retaining the prototypical C^* underlying structure which plagued other variants, the modifications implemented in the scheme seem to perform their intended tasks perfectly.

Most significantly, the projection modifier has provably removed the linear symmetric differential structure, as shown in [11], while retaining the flawless differential invariant structure. On the other hand, the reduction provided by the algorithm in [17] to remove the projection modifier succeeds in transforming pSFLASH into an HFE^- scheme. Although the transformation removes the C^* properties of the core map, it may well prove to be the case that the extra structure the resultant particular HFE^- scheme retains may reveal a weakness. Any new attack on this system will be very exciting, as it will indicate a fundamentally new cryptanalytic technique.

References

1. Chen, A.I.T., Chen, M.S., Chen, T.R., Cheng, C.M., Ding, J., Kuo, E.L.H., Lee, F.Y.S., Yang, B.Y.: Sse implementation of multivariate pkcs on modern x86 cpus. In: Clavier, C., Gaj, K. (eds.) CHES 2009. LNCS, vol. 5747, pp. 33–48. Springer, Heidelberg (2009)
2. Chen, A.I.-T., Chen, C.-H.O., Chen, M.-S., Cheng, C.-M., Yang, B.-Y.: Practical-sized instances of multivariate pKCs: Rainbow, TTS, and ℓIC-derivatives. In: Buchmann, J., Ding, J. (eds.) PQCrypto 2008. LNCS, vol. 5299, pp. 95–108. Springer, Heidelberg (2008)
3. Yang, B.-Y., Cheng, C.-M., Chen, B.-R., Chen, J.-M.: Implementing minimized multivariate PKC on low-resource embedded systems. In: Clark, J.A., Paige, R.F., Polack, F.A.C., Brooke, P.J. (eds.) SPC 2006. LNCS, vol. 3934, pp. 73–88. Springer, Heidelberg (2006)
4. Sakumoto, K., Shirai, T., Hiwatari, H.: On provable security of uov and hfe signature schemes against chosen-message attack. In: [19], pp. 68–82.
5. Huang, Y.J., Liu, F.H., Yang, B.Y.: Public-key cryptography from new multivariate quadratic assumptions. In: Fischlin, M., Buchmann, J., Manulis, M. (eds.) PKC 2012. LNCS, vol. 7293, pp. 190–205. Springer, Heidelberg (2012)
6. Clough, C., Baena, J., Ding, J., Yang, B.-Y., Chen, M.-S.: Square, a New Multivariate Encryption Scheme. In: Fischlin, M. (ed.) CT-RSA 2009. LNCS, vol. 5473, pp. 252–264. Springer, Heidelberg (2009)
7. Baena, J., Clough, C., Ding, J.: Square-vinegar signature scheme. In: Buchmann, J., Ding, J. (eds.) PQCrypto 2008. LNCS, vol. 5299, pp. 17–30. Springer, Heidelberg (2008)
8. Billet, O., Macario-Rat, G.: Cryptanalysis of the square cryptosystems. In: Matsui, M. (ed.) ASIACRYPT 2009. LNCS, vol. 5912, pp. 451–468. Springer, Heidelberg (2009)
9. Dubois, V., Fouque, P.-A., Shamir, A., Stern, J.: Practical Cryptanalysis of SFLASH. In: Menezes, A. (ed.) CRYPTO 2007. LNCS, vol. 4622, pp. 1–12. Springer, Heidelberg (2007)
10. Kipnis, A., Shamir, A.: Cryptanalysis of the oil & vinegar signature scheme. In: Krawczyk, H. (ed.) CRYPTO 1998. LNCS, vol. 1462, pp. 257–266. Springer, Heidelberg (1998)
11. Smith-Tone, D.: On the differential security of multivariate public key cryptosystems. In: [19], pp. 130–142.
12. Patarin, J.: Cryptanalysis of the Matsumoto and Imai public key scheme of Eurocrypt '88. In: Coppersmith, D. (ed.) CRYPTO 1995. LNCS, vol. 963, pp. 248–261. Springer, Heidelberg (1995)

13. Matsumoto, T., Imai, H.: Public quadratic polynomial-tuples for efficient signature-verification and message-encryption. In: Günther, C.G. (ed.) EUROCRYPT 1988. LNCS, vol. 330, pp. 419–453. Springer, Heidelberg (1988)
14. Kipnis, A., Patarin, J., Goubin, L.: Unbalanced oil and vinegar signature schemes. In: Stern, J. (ed.) EUROCRYPT 1999. LNCS, vol. 1592, pp. 206–222. Springer, Heidelberg (1999)
15. Patarin, J.: The oil and vinegar algorithm for signatures. In: Presented at the Dagsthul Workshop on Cryptography (1997)
16. Ding, J., Dubois, V., Yang, B.-Y., Chen, O.C.-H., Cheng, C.-M.: Could SFLASH be repaired? In: Aceto, L., Damgård, I., Goldberg, L.A., Halldórsson, M.M., Ingólfsdóttir, A., Walukiewicz, I. (eds.) ICALP 2008, Part II. LNCS, vol. 5126, pp. 691–701. Springer, Heidelberg (2008)
17. Bettale, L., Faugère, J.C., Perret, L.: Cryptanalysis of multivariate and odd-characteristic hfe variants. In: Catalano, D., Fazio, N., Gennaro, R., Nicolosi, A. (eds.) PKC 2011. LNCS, vol. 6571, pp. 441–458. Springer, Heidelberg (2011)
18. Ding, J., Kleinjung, T.: Degree of regularity for hfe-. IACR Cryptology ePrint Archive 2011, 570 (2011)
19. Yang, B.-Y. (ed.): PQCrypto 2011. LNCS, vol. 7071. Springer, Heidelberg (2011)

Secure and Anonymous Hybrid Encryption
from Coding Theory*

Edoardo Persichetti

University of Warsaw

Abstract. Cryptographic schemes based on coding theory are one of the most accredited choices for cryptography in a post-quantum scenario. In this work, we present a hybrid construction based on the Niederreiter framework that provides IND-CCA security in the random oracle model. In addition, the construction satisfies the IK-CCA notion of anonymity whose importance is ever growing in the cryptographic community.

1 Introduction

A *Hybrid Encryption* scheme is a cryptographic protocol that features both a public-key encryption scheme and a symmetric encryption scheme, the former with the task of encrypting a key for the latter, in charge of encrypting the actual body of the message. The first component is therefore known as *Key Encapsulation Mechanism (KEM)* while the second is called *Data Encapsulation Mechanism (DEM)*. Key feature is that the two parts are independent of one another. The framework was first introduced in a seminal work by Cramer and Shoup [6], along with the corresponding notions of security and an example of a scheme based on the DDH assumptions. In a subsequent work [12], Shoup presents a proposal for an ISO standard on public-key encryption including many different schemes based on the RSA assumptions (RSA-OAEP, RSA-KEM), elliptic curves (ECIES) and Diffie-Hellman (PSEC, ACE). Other schemes based on integer factorization such as EPOC or HIME are also mentioned.

In this paper we present a new KEM construction, based on the Niederreiter framework [9]. The work follows up a suggestion from Bernstein [4] and stems from the RSA-KEM scheme (also known as "Simple RSA" in earlier versions of the paper), and as far as we know is the first proposal for a KEM based on coding theory assumptions. The construction is proved to be CCA secure; moreover, it is shown that, for the resulting Hybrid Niederreiter encryption scheme, it is possible to achieve key-privacy in the IK-CCA sense, as formalized by Bellare et al. in [1]. Key-privacy for coding theory schemes has been studied by Yamakawa et al. in [15], where it is proved that the IND-CPA variant of McEliece by Nojima et al. [10] satisfies the weaker anonymity notion of IK-CPA. To the best of our knowledge, our work is the first code-based construction achieving IK-CCA security.

---

* European Research Council has provided financial support under the European Community's Seventh Framework Programme (FP7/2007-2013) / ERC grant agreement no CNTM-207908.

P. Gaborit (Ed.): PQCrypto 2013, LNCS 7932, pp. 174–187, 2013.

The paper is organized as follows: first, we briefly recall the basic notions of coding theory and the Niederreiter cryptosystem. We then introduce all the definitions and notions of security for KEMs and DEMs, plus other cryptographic tools that we will need for our scheme, such as KDFs and MACs. In Section 3 we introduce the construction and prove its security, then show how to realize an efficient DEM and how to compose the two parts. Anonymity notions and the corresponding result about the Hybrid Niederreiter scheme are presented in Section 4. We conclude in Section 5.

2   Preliminaries

2.1   The Niederreiter Cryptosystem

This cryptosystem was introduced by H. Niederreiter in 1985 [9]. Since it makes use of the parity-check matrix rather than the generator matrix, it is often considered as a "dual" version of the McEliece cryptosystem [7]. Due to space limitations, we leave a detailed description to Appendix A.

The security of the scheme follows from the two following computational assumptions.

Assumption 1 (Indistinguishability). *The $(n - k) \times k$ matrix M output by KeyGen is computationally indistinguishable from a uniformly chosen matrix of the same size.*

Assumption 2 (Syndrome Decoding Problem (SDP)). *Let H be a parity-check matrix for a random $[n, k]$ linear code over $\mathbb{F}_q$ and s be chosen uniformly at random in $\mathbb{F}_q^{(n-k)}$. Then it is hard to find a vector $e \in \mathbb{F}_q^n$ with $\mathsf{wt}(e) \leq w$ such that $He^\top = s$.*

SDP was proved to be NP-complete in [3].

2.2   Encapsulation Mechanisms and the Hybrid Framework

A key encapsulation mechanism is essentially a public-key encryption scheme (PKE), with the exception that the encryption algorithm takes no input apart from the public key, and returns a pair (K, ψ_0). The string K has fixed length ℓ_K, specified by the KEM, and ψ_0 is an "encryption" of K in the sense that $\mathsf{Dec}_{\mathsf{sk}}(\psi_0) = K$. Formally, a KEM consists of the following three algorithms.

A KEM is required to be *sound* for at least all but a negligible portion of public key/private key pairs, that is, if $\mathsf{Enc}_{\mathsf{pk}}(\ ) = (K, \psi_0)$ then $\mathsf{Dec}_{\mathsf{sk}}(\psi_0) = K$ with overwhelming probability.

The data encapsulation mechanism is a (possibly labeled) symmetric encryption scheme (SE) that uses as a key the string K output by the KEM. In what follows we only discuss, for simplicity, un-labeled DEMs.

Formally, a DEM consists of the following two algorithms.

Table 1. Key Encapsulation Mechanism

KeyGen	A probabilistic key generation algorithm that takes as input a security parameter 1^λ and outputs a public key pk and a private key sk.
Enc	A probabilistic encryption algorithm that receives as input a public key pk and returns a key/ciphertext pair (K, ψ_0).
Dec	A deterministic decryption algorithm that receives as input a private key sk and a ciphertext ψ_0 and outputs either a key K or the failure symbol $\perp$.

Table 2. Data Encapsulation Mechanism

Enc	A deterministic encryption algorithm that receives as input a key K and a plaintext ϕ and returns a ciphertext ψ_1.
Dec	A deterministic decryption algorithm that receives as input a key K and a ciphertext ψ_1 and outputs either a plaintext ϕ or the failure symbol $\perp$.

The security notions are similar to their corresponding ones for PKE and SE schemes (see Appendix B). We present them below.

Definition 1. *The adaptive chosen-ciphertext attack game for a KEM proceeds as follows:*

1. *Query a key generation oracle to obtain a public key pk.*
2. *Make a sequence of calls to a decryption oracle, submitting any string ψ_0 of the proper length. The oracle will respond with $Dec_{sk}^{KEM}(\psi_0)$.*
3. *Query an encryption oracle. The oracle runs Enc_{pk}^{KEM} to generate a pair $(\tilde{K}, \tilde{\psi}_0)$, then chooses a random $b \in \{0, 1\}$ and replies with the "challenge" ciphertext $(K^*, \tilde{\psi}_0)$ where $K^* = \tilde{K}$ if $b = 1$ or K^* is a random string of length ℓ_K otherwise.*
4. *Keep performing decryption queries. If the submitted ciphertext is ψ_0^*, the oracle will return $\perp$.*
5. *Output $b^* \in \{0, 1\}$.*

The adversary succeeds if $b^* = b$. More precisely, we define the advantage *of $\mathcal{A}$ against KEM as*

$$Adv_{KEM}(\mathcal{A}, \lambda) = \left| Pr[b^* = b] - \frac{1}{2} \right|. \tag{1}$$

We say that a KEM is secure if the advantage Adv_{KEM} of any polynomial-time adversary $\mathcal{A}$ in the above CCA attack model is negligible.

Definition 2. *The attack game for a DEM proceeds as follows:*

1. *Choose two plaintexts ϕ_0, ϕ_1 and submit them to an encryption oracle. The oracle will choose a random key K and a random bit $b \in \{0, 1\}$ and reply with the "challenge" ciphertext $\psi_1^* = Enc_K^{DEM}(\phi_b)$.*

2. *Make a sequence of calls to a decryption oracle, submitting any string ψ_1 of the proper length. The oracle will respond with $Dec_K^{DEM}(\psi_1)$. If the submitted ciphertext is ψ_1^*, the oracle will return $\bot$.*
3. *Output $b^* \in \{0, 1\}$.*

The adversary succeeds if $b^* = b$. As above, we define the advantage *of $\mathcal{A}$ against DEM as*

$$Adv_{DEM}(\mathcal{A}, \lambda) = \left| Pr[b^* = b] - \frac{1}{2} \right|. \tag{2}$$

We say that a DEM is secure if the advantage Adv_{DEM} of any polynomial-time adversary $\mathcal{A}$ in the above attack model is negligible.

We require that the key K used in Enc^{DEM} and Dec^{DEM} has the same length ℓ_K as in the KEM. In this case, the mechanisms are said to be *compatible*, and can be composed in the canonical way as shown in Table 3.

Remark 1. An alternative definition of advantage against KEM, or more in general any PKE scheme, is the following:

$$Adv'_{KEM}(\mathcal{A}, \lambda) = \left| Pr[b^* = 1 | b = 1] - Pr[b^* = 1 | b = 0] \right|. \tag{3}$$

The two notions are related in the sense that, for any adversary $\mathcal{A}$, we have $Adv'_{KEM}(\mathcal{A}, \lambda) = 2 \cdot Adv_{KEM}(\mathcal{A}, \lambda)$. However, as we will see, the above expression is often more convenient for interpreting the behavior of an adversary in two different attack games, where b is always equal to 0 in one game, and to 1 in the other. This is usually accomplished by replacing a honest encryption with the encryption of a "rubbish" message (commonly a randomly generated string of the proper length), and then analyzing the behavior of the adversary.

Table 3. Hybrid Encryption scheme

K	K_{publ} the *public key space.*
	K_{priv} the *private key space.*
P	The set of messages to be encrypted, or *plaintext space.*
C	The set of the messages transmitted over the channel, or *ciphertext space.*
KeyGen	A probabilistic key generation algorithm that takes as input a security parameter 1^λ and outputs a public key $pk \in K_{publ}$ and a private key $sk \in K_{priv}$.
Enc	A probabilistic encryption algorithm that receives as input a public key $pk \in K_{publ}$ and a plaintext $\phi \in P$. The algorithm invokes $Enc_{pk}^{KEM}(\ )$ and obtains a key/ciphertext pair (K, ψ_0), then runs $Enc_K^{DEM}(\phi)$ and gets a ciphertext ψ_1. Finally, it outputs the ciphertext $\psi = (\psi_0 || \psi_1)$.
Dec	A deterministic decryption algorithm that receives as input a private key $sk \in K_{priv}$ and a ciphertext $\psi \in C$. The algorithm parses ψ as $(\psi_0 || \psi_1)$, then decrypts the left part by running $Dec_{sk}^{KEM}(\psi_0)$; it either gets $\bot$ or a key K. In the first case, the algorithm returns $\bot$, otherwise it runs $Dec_K^{DEM}(\psi_1)$ and returns either the resulting plaintext ϕ or the failure symbol $\bot$.

It has then been proved that, given a CCA adversary $\mathcal{A}$ for the hybrid scheme (HY), there exist an adversary $\mathcal{A}_1$ for KEM and an adversary $\mathcal{A}_2$ for DEM running in roughly the same time as $\mathcal{A}$, such that for any choice of the security parameter λ we have $\mathsf{Adv}_{\mathsf{HY}}(\mathcal{A}, \lambda) \leq \mathsf{Adv}'_{\mathsf{KEM}}(\mathcal{A}_1, \lambda) + \mathsf{Adv}_{\mathsf{DEM}}(\mathcal{A}_2, \lambda)$. See Cramer and Shoup [6, Th. 5] for a complete proof.

2.3   Other Cryptographic Tools

In this section we introduce other cryptographic tools that we need for our construction. We start with key derivation functions.

Definition 3. *A* Key Derivation Function (KDF) *is a function that takes as input a string* x *of arbitrary length and an integer* $\ell \geq 0$ *and outputs a bit string of length* ℓ.

A KDF is modelled as a random oracle, and it satisfies the *entropy smoothing* property, that is, if x is chosen at random from a high entropy distribution, the output of KDF should be computationally indistinguishable from a random length-ℓ bit string.

Intuitively, a good choice for a KDF could be a hash function with a variable (arbitrary) length output, such as the new SHA-3, Keccak [5].

Definition 4. *A* Message Authentication Code (MAC) *is an algorithm that produces a short piece of information* (tag) *used to authenticate a message. A MAC is defined by a function* Ev *that takes as input a key* K *of length* ℓ_{MAC} *and an arbitrary string* T *and returns a tag to be appended to the message, that is, a string* τ *of fixed length* ℓ_{TAG}.

Informally, a MAC is similar to a signature scheme, with the difference that the scheme makes use of private keys both for evaluation and verification; in this sense, it could be seen as a "symmetric encryption equivalent" of a signature scheme. The usual desired security requirement is existential unforgeability under chosen message attacks (see Appendix B).

3   The Hybrid Encryption Scheme

3.1   The KEM Construction

The KEM we present here follows closely the Niederreiter framework, and is thus based on the hardness of SDP. Note that, compared to the original Niederreiter scheme, a slight modification is introduced in the decryption process. As we will see later, this is necessary for the proof of security.

If the ciphertext is correctly formed, decoding will always succeed, hence the KEM is perfectly sound. Furthermore, we will see in Section 3.2 that, even if with this formulation $\mathsf{Dec}^{\mathsf{KEM}}$ never fails, there is no integrity loss in the hybrid encryption scheme thanks to the check given by the MAC.

We prove the security of the KEM in the following theorem.

---

[1] A natural suggestion is for example to set $K = \mathsf{KDF}(\psi_0, \ell_K)$.

Table 4. The Niederreiter KEM

Setup	Fix public system parameters $q, n, k, w \in \mathbb{N}$, then choose a family $\mathcal{F}$ of w-error-correcting $[n, k]$ linear codes over $\mathbb{F}_q$.

KeyGen	Generate at random a code $\mathcal{C} \in \mathcal{F}$ given by its code description Δ and compute its parity-check matrix in systematic form $H = (M|I_{n-k})$. Publish the public key M and store the private key Δ.
Enc	On input a public key M choose a random $e \in \mathbb{W}_{q,n,w}$, set $H = (M|I_{n-k})$, then compute $K = \mathsf{KDF}(e, \ell_K)$, $\psi_0 = He^\mathsf{T}$ and return the key/ciphertext pair (K, ψ_0).
Dec	On input a private key Δ and a ciphertext ψ_0, compute $\mathsf{Decode}_\Delta(\psi_0)$. If the decoding succeeds, use its output e to compute $K = \mathsf{KDF}(e, \ell_K)$. Otherwise, set K to be a string of length ℓ_K determined as a pseudorandom function[1] of ψ_0. Return K.

Theorem 1. *Let $\mathcal{A}$ be an adversary in the random oracle model for the Niederreiter KEM as in Definition 1. Let θ be the running time of $\mathcal{A}$, n_{KDF} and n_{Dec} be two bounds on, respectively, the total number of random oracle queries and the total number of decryption queries performed by $\mathcal{A}$, and set $N = |\mathbb{W}_{q,n,w}|$. Then there exists an adversary $\mathcal{A}'$ for SDP such that $\mathsf{Adv}_{\mathsf{KEM}}(\mathcal{A}, \lambda) \leq \mathsf{Adv}_{\mathsf{SDP}}(\mathcal{A}', \lambda) + n_{\mathsf{Dec}}/N$. The running time of $\mathcal{A}'$ will be approximately equal to θ plus the cost of n_{KDF} matrix-vector multiplications and some table lookups.*

Proof. We replace KDF with a random oracle $\mathcal{H}$ mapping words in $\mathbb{W}_{q,n,w}$ to bit strings of length ℓ_K. To prove our claim, we proceed as follows. Let's call G_0 the original attack game played by $\mathcal{A}$, and S_0 the event that $\mathcal{A}$ succeeds in game G_0. We define a new game G_1 which is identical to G_0 except that the game is halted if the challenge ciphertext $\psi_0^* = He^{*\mathsf{T}}$ obtained when querying the encryption oracle had been previously submitted to the decryption oracle: we call this event F_1. Since the number of valid ciphertexts is N, we have $\Pr[\mathsf{F}_1] \leq n_{\mathsf{Dec}}/N$. It follows that $\left|\Pr[\mathsf{S}_0] - \Pr[\mathsf{S}_1]\right| \leq n_{\mathsf{Dec}}/N$, where S_1 is the event that $\mathcal{A}$ succeeds in game G_1. Next, we define game G_2 which is identical to G_1 except that we generate the challenge ciphertext ψ_0^* at the beginning of the game, and we halt if $\mathcal{A}$ ever queries $\mathcal{H}$ at e^*: we call this event F_2. By construction, since $\mathcal{H}(e^*)$ is undefined, it is not possible to tell whether $K^* = K$, thus we have $\Pr[\mathsf{S}_2] = 1/2$, where S_2 is the event that $\mathcal{A}$ succeeds in game G_2. We obtain that $\left|\Pr[\mathsf{S}_1] - \Pr[\mathsf{S}_2]\right| \leq \Pr[\mathsf{F}_2]$ and we just need to bound $\Pr[\mathsf{F}_2]$.

We now construct an adversary $\mathcal{A}'$ against SDP. $\mathcal{A}'$ interacts with $\mathcal{A}$ and is able to simulate the random oracle and the decryption oracle with the help of two tables T_1 and T_2, initially empty, as described below.

Key Generation: On input the instance (H, s^*, w) of SDP, return the public key $\mathsf{pk} = H$.

Challenge Queries: When $\mathcal{A}$ asks for the challenge ciphertext:

1. Generate a random string K^* of length ℓ_K.

2. Set $\psi_0^* = s^*$.

3. Return the pair (K^*, ψ_0^*).

Random Oracle Queries: Upon $\mathcal{A}$'s random oracle query $e \in \mathbb{W}_{q,n,w}$:

1. Look up e in T_1. If (e, s, K) is in T_1 for some s and K, return K.

2. Compute $s = He^\mathsf{T}$.

3. If $s = s^*$ then $\mathcal{A}'$ outputs e and the game ends.

4. Look up s in T_2. If (s, K) is in T_2 for some K (i.e. the decryption oracle has been evaluated at s), return K.

5. Set K to be a random string of length ℓ_K and place the triple (e, s, K) in table T_1.

6. Return K.

Decryption Queries: Upon $\mathcal{A}$'s decryption query $\psi_0 = s \in \mathbb{F}_q^{(n-k)}$:

1. Look up s in T_2. If (s, K) is in T_2 for some K, return K.

2. Look up s in T_1. If (e, s, K) is in T_1 for some e and K (i.e. the random oracle has been evaluated at e such that $s = He^\mathsf{T}$), return K.

3. Generate a random string K of length ℓ_K and place the pair (s, K) in T_2.

4. Return K.

Note that, in both random oracle and decryption queries, we added Step 1 to guarantee the integrity of the simulation, that is, if the same value is queried more than once, the same output is returned.

A fundamental issue is that it is impossible for the simulator to determine if a word is decodable or not. If the decryption algorithm returned $\perp$ if and only if a word was not decodable, then it would be impossible to simulate decryption properly. We have resolved this problem by insisting that the KEM decryption algorithm always outputs a hash value. With this formulation, the simulation is flawless and $\mathcal{A}'$ outputs a solution to the SDP instance with probability equal to $\Pr[\mathsf{F}_2]$. □

3.2   A Standard DEM

A standard way to construct a DEM by means of a SE scheme and a one-time MAC is shown in Table 5.

It is easy to prove that if the underlying components are secure, so is the resulting DEM. In particular it is possible to prove [6, Th. 4] that, for any DEM adversary $\mathcal{A}$, we have $\mathsf{Adv}_{\mathsf{DEM}}(\mathcal{A}, \lambda) \leq \mathsf{Adv}_{\mathsf{FG}}(\mathcal{A}_1, \lambda) + \mathsf{Adv}_{\mathsf{MAC}}(\mathcal{A}_2, \lambda)$, where $\mathcal{A}_1$ and $\mathcal{A}_2$ are, respectively, a find-guess adversary for SE and a one-time existential forgery adversary for MAC, both running in about the same time of $\mathcal{A}$.

Table 5. Standard DEM

Enc On input a key K and a plaintext ϕ, parse K as $(K_1\|K_2)$ then compute $\psi' = \mathsf{Enc}^{\mathsf{SE}}_{K_1}(\phi)$, set $T = \psi'$ and evaluate $\tau = \mathsf{Ev}(K_2, T)$. Return the ciphertext $\psi_1 = (\psi'\|\tau)$.

Dec On input a key K and a ciphertext ψ_1, parse ψ_1 as $(\psi'\|\tau)$ then parse K as $(K_1\|K_2)$, set $T = \psi'$ and apply the MAC algorithm to obtain $\tau' = \mathsf{Ev}(K_2, T)$. If $\tau' \neq \tau$ the verification fails, hence return $\bot$. Otherwise, compute $\phi = \mathsf{Dec}^{\mathsf{SE}}_{K_1}(\psi')$ and return the plaintext ϕ.

3.3 Hybrid Niederreiter

For our purposes, and throughout the rest of this paper, we will think at the DEM as a one-time pad with fixed input/output length m (e.g. 128 or 256 bits), together with a MAC (any of the ISO standards is acceptable). The Hybrid Niederreiter (HN) scheme is simply the composition of the two components, as described in Table 3. Details are presented in Table 6.

Table 6. Hybrid Niederreiter scheme

Setup Fix public system parameters $q, n, k, w \in \mathbb{N}$, then choose a family $\mathcal{F}$ of w-error-correcting $[n, k]$ linear codes over $\mathbb{F}_q$.

K $\quad$ $\mathsf{K}_{\mathsf{publ}}$ the set of $(n - k) \times k$ matrices over $\mathbb{F}_q$.

$\quad$ $\mathsf{K}_{\mathsf{priv}}$ the set of code descriptions for $\mathcal{F}$.

P $\quad$ The set of binary strings $\{0, 1\}^m$.

C $\quad$ The set of triples formed by a vector of $\mathbb{F}_q^{(n-k)}$, a bit string of length ℓ, and a tag.

KeyGen Generate at random a code $\mathcal{C} \in \mathcal{F}$ given by its code description Δ and compute its parity-check matrix in systematic form $H = (M|I_{n-k})$. Publish the public key $M \in \mathsf{K}_{\mathsf{priv}}$ and store the private key $\Delta \in \mathsf{K}_{\mathsf{publ}}$.

Enc $\quad$ On input a public key M and a plaintext $\phi \in \mathsf{P}$, choose a random $e \in \mathbb{W}_{q,n,w}$, set $H = (M|I_{n-k})$, then compute $K = \mathsf{KDF}(e, m + \ell_{\mathsf{MAC}})$ and $\psi_0 = He^{\mathsf{T}}$. Parse K as $(K_1\|K_2)$ then compute $\psi' = K_1 \oplus \phi$, set $T = \psi'$ and evaluate $\tau = \mathsf{Ev}(K_2, T)$. Return the ciphertext $\psi = (\psi_0\|\psi'\|\tau)$.

Dec $\quad$ On input a private key Δ and a ciphertext ψ, first parse ψ as $(\psi_0\|\psi_1)$, then compute $\mathsf{Decode}_\Delta(\psi_0)$. If the decoding succeeds, use its output e to compute $K = \mathsf{KDF}(e, m + \ell_{\mathsf{MAC}})$. Otherwise, determine K as a pseudorandom function of ψ_0. Parse ψ_1 as $(\psi'\|\tau)$ then parse K as $(K_1\|K_2)$, set $T = \psi'$ and apply the MAC algorithm to obtain $\tau' = \mathsf{Ev}(K_2, T)$. If $\tau' \neq \tau$ the verification fails, hence return $\bot$. Otherwise, compute $\phi = K_1 \oplus \psi'$ and return the plaintext ϕ.

4   Anonymity

Anonymity for public-key encryption schemes was first introduced by Bellare et al. in [1] and, as opposed to the classical notions of data-privacy such as indistinguishability, it captures the idea of *key-privacy*. This means that an anonymous PKE scheme does not disclose information about which key, among a set of valid keys, has been used to encrypt. We therefore speak about *indistinguishability of keys*.

Definition 5. *The Indistinguishability of Keys game in the adaptive chosen-ciphertext attack model (IK-CCA) is defined as follows:*

1. *Query a key generation oracle to obtain two public keys pk_0 and pk_1.*

2. *Make a sequence of calls to either of the two decryption oracles (corresponding to the two keys), submitting any string ψ of the proper length. The oracle will respond respectively with $Dec_{sk_0}^{PKE}(\psi)$ or $Dec_{sk_1}^{PKE}(\psi)$.*

3. *Choose a plaintext ϕ^* and submit it to an encryption oracle. The oracle chooses a random $b \in \{0, 1\}$ and replies with the "challenge" ciphertext $\psi^* = Enc_{pk_b}^{PKE}(\phi^*)$.*

4. *Keep performing decryption queries. If the submitted ciphertext (to any of the two decryption oracles) is ψ^*, return $\perp$.*

5. *Output $b^* \in \{0, 1\}$.*

The adversary succeeds if $b^* = b$. More precisely, we define the advantage of $\mathcal{A}$ against PKE as

$$Adv_{IK\text{-}CCA}(\mathcal{A}, \lambda) = \left| Pr[b^* = b] - \frac{1}{2} \right|. \tag{4}$$

We say that PKE is secure in this sense if the advantage $Adv_{IK\text{-}CCA}$ of any adversary $\mathcal{A}$ in the above CCA attack model is negligible.

If the attack model does not allow for decryption queries, the security notion is known as IK-CPA.

The above notions for PKE schemes apply also to hybrid encryption schemes. Unlike the case of data-privacy, though, it is not enough to have two anonymous components for the resulting hybrid encryption scheme to be anonymous: a counterexample is given by Mohassel in [8]. The author, however, shows how this can be fixed by using a KEM component that satisfies an additional property called *robustness*. In practice, this requires that a ciphertext does not decrypt to a valid plaintext under distinct private keys.

Now, it is easy to see that plain coding theory schemes are not anonymous: this is immediate for Niederreiter (being deterministic), and it was shown in [15] for McEliece. In the same paper, the authors prove that the randomized version of McEliece by Nojima et al. [10] is IK-CPA secure. We use a similar technique to prove that the Hybrid Niederreiter scheme described in Table 6 is IK-CCA secure.

Theorem 2. *Let $\mathcal{A}$ be an IK-CCA adversary in the random oracle model for the Hybrid Niederreiter scheme. Let θ be the running time of $\mathcal{A}$, n_{KDF} and n_{Dec} be two bounds on, respectively, the total number of random oracle queries and the total number of decryption queries performed by $\mathcal{A}$, and set $N = |\mathbb{W}_{q,n,w}|$. Then there exists an IND-CCA adversary $\mathcal{A}'$ against HN such that $\mathsf{Adv}_{IK\text{-}CCA}(\mathcal{A}, \lambda) \leq \mathsf{Adv}_{HN}(\mathcal{A}', \lambda) + n_{Dec}/2N$. The running time of $\mathcal{A}'$ will be approximately equal to θ plus n_{KDF} matrix-vector multiplications and at most n_{Dec} decryption operations.*

Proof. As in the proof of Theorem 1, we replace KDF with a random oracle $\mathcal{H}$ mapping words in $\mathbb{W}_{q,n,w}$ to bit strings of length ℓ_K. Let's call G_0 the original attack game played by $\mathcal{A}$, and S_0 the event that $\mathcal{A}$ succeeds in game G_0. We define a new game G_1 which is identical to G_0 except that the game is halted if the challenge ciphertext $\psi^* = \mathsf{Enc}_{\mathsf{pk}_b}(\phi^*)$ obtained when querying the encryption oracle had been previously submitted to the decryption oracle: we call this event F_1. Since, for any fixed plaintext, the number of valid ciphertexts is $2N$, we have $\Pr[\mathsf{F}_1] \leq n_{Dec}/2N$. It follows that $\left|\Pr[\mathsf{S}_1] - \Pr[\mathsf{S}_0]\right| \leq n_{Dec}/2N$, where S_1 is the event that $\mathcal{A}$ succeeds in game G_1. Next, we define game G_2 which is the same as G_1 apart from the following modification: when $\mathcal{A}$ queries the encryption oracle on an input ϕ^*, a random string $\phi' \in \mathsf{P}$ is generated and returned together with the challenge ψ^*. Since this carries no additional information, we have $\Pr[\mathsf{S}_2] = \Pr[\mathsf{S}_1]$, where S_2 is the event that $\mathcal{A}$ succeeds in game G_2. Finally, we define game G_3 by modifying the encryption oracle such that it replies instead with $\psi^* = \mathsf{Enc}_{\mathsf{pk}_b}^{HN}(\phi')$. Now, note that the success probability of $\mathcal{A}$ does not change unless it is able to distinguish which of the two plaintexts had been used by the encryption oracle. More precisely, there exists an IND-CCA adversary $\mathcal{A}'$ against HN that uses $\mathcal{A}$ as a subroutine. Since it plays the adaptive chosen-ciphertext game, $\mathcal{A}'$ has access to a decryption oracle $\mathcal{D}$ for its public key pk. The interaction with $\mathcal{A}$ is described below:

Key Generation: On input the public key $\mathsf{pk} = H$, generate a pair $(\mathsf{pk}', \mathsf{sk}') \in \mathsf{K}$ then set $\mathsf{pk}_0 = \mathsf{pk}$ and $\mathsf{pk}_1 = \mathsf{pk}'$ and send $(\mathsf{pk}_0, \mathsf{pk}_1)$ to $\mathcal{A}$.

Challenge Queries: When $\mathcal{A}$ asks for the challenge ciphertext:

1. Receive as input the string ϕ^* from $\mathcal{A}$.

2. Generate a random string $\phi' \in \mathsf{P}$.

3. Submit ϕ^* and ϕ' to the IND-CCA encryption oracle. The oracle will reply with $\psi^* = (\psi_0^* \| \psi_1^*) = \mathsf{Enc}_{\mathsf{pk}}^{HN}(\phi_\beta)$, where ϕ_β is equal to ϕ^* if $\beta = 1$ or to ϕ' otherwise. This is the challenge ciphertext for $\mathcal{A}'$.

4. Return ϕ' and ψ^* to $\mathcal{A}$.

Random Oracle Queries: Upon $\mathcal{A}$'s random oracle query $e \in \mathbb{W}_{q,n,w}$:

1. Submit e to $\mathcal{H}$ and get $K = \mathcal{H}(e)$.

2. If $H e^\mathsf{T} = \psi_0^*$ then $\mathcal{A}'$ decrypts ψ_1^* with K and the game ends.

3. Return $\mathcal{H}(e)$.

Decryption Queries: Upon $\mathcal{A}$'s decryption query ψ for sk_b:

1. If $b = 1$, return $\mathsf{Dec}_{\mathsf{sk}'}^{\mathsf{HN}}(\psi)$.
2. If $b = 0$, submit ψ to $\mathcal{D}$ and return $\mathcal{D}(\psi)$.

Observe that, in the attack game that $\mathcal{A}'$ is playing against HN, the value of β is equal to 1 in game G_2, and to 0 in game G_3. Following up from Remark 1, we conclude that $\left| \mathsf{Pr}[S_3] - \mathsf{Pr}[S_2] \right| \leq \mathsf{Adv}_{\mathsf{HN}}'(\mathcal{A}', \lambda)$.

Now, since ϕ' is chosen uniformly at random and Assumption 1 holds, the distributions $\{(\mathsf{pk}_0, \mathsf{pk}_1, \mathsf{Enc}_{\mathsf{pk}_0}^{\mathsf{HN}}(\phi')) | (\mathsf{pk}_0, \mathsf{sk}_0) \overset{\$}{\leftarrow} \mathsf{K}, (\mathsf{pk}_1, \mathsf{sk}_1) \overset{\$}{\leftarrow} \mathsf{K}\}$ and $\{(\mathsf{pk}_0, \mathsf{pk}_1, \mathsf{Enc}_{\mathsf{pk}_1}^{\mathsf{HN}}(\phi')) | (\mathsf{pk}_0, \mathsf{sk}_0) \overset{\$}{\leftarrow} \mathsf{K}, (\mathsf{pk}_1, \mathsf{sk}_1) \overset{\$}{\leftarrow} \mathsf{K}\}$ are computationally indistinguishable. It follows that $\mathsf{Pr}[S_3] = 1/2$; hence, $\mathsf{Adv}_{\mathsf{IK\text{-}CCA}}(\mathcal{A}, \lambda) = = \left| \mathsf{Pr}[S_0] - 1/2 \right| = \left| \mathsf{Pr}[S_0] - \mathsf{Pr}[S_1] \right| + \left| \mathsf{Pr}[S_1] - \mathsf{Pr}[S_2] \right| + \left| \mathsf{Pr}[S_2] - \mathsf{Pr}[S_3] \right| \leq \mathsf{Adv}_{\mathsf{HN}}'(\mathcal{A}', \lambda) + n_{\mathsf{Dec}}/2N$ as claimed.   □

5   Conclusions

In this paper, we have introduced a key encapsulation method based on the Niederreiter cryptosystem. This is the first KEM based directly on a coding theory problem and it enjoys a simple construction and a tight security proof. We have also shown that the Hybrid Niederreiter encryption scheme that makes use of our KEM satisfies the most important notion of anonymity, IK-CCA. Our work builds on the results of [15], and is the first code-based encryption scheme to enjoy IK-CCA security. Future work includes investigating practical applications of our construction, with the aim of an implementation.

References

1. Bellare, M., Boldyreva, A., Desai, A., Pointcheval, D.: Key-privacy in public-key encryption. In: Boyd, C. (ed.) ASIACRYPT 2001. LNCS, vol. 2248, pp. 566–582. Springer, Heidelberg (2001)
2. Bellare, M., Desai, A., Jokipii, E., Rogaway, P.: A Concrete Security Treatment of Symmetric Encryption. In: FOCS, pp. 394–403. IEEE Computer Society (1997)
3. Berlekamp, E., McEliece, R., van Tilborg, H.: On the inherent intractability of certain coding problems. IEEE Transactions on Information Theory 24(3), 384–386 (1978)
4. Bernstein, D.J.: Personal communication (May 2012)
5. Bertoni, G., Daemen, J., Peeters, M., Van Assche, G.: http://keccak.noekeon.org/
6. Cramer, R., Shoup, V.: Design and Analysis of Practical Public-Key Encryption Schemes Secure against Adaptive Chosen Ciphertext Attack. SIAM J. Comput. 33(1), 167–226 (2004)
7. McEliece, R.: A Public-Key Cryptosystem Based on Algebraic Coding Theory. Technical report, NASA (1978)

8. Mohassel, P.: A Closer Look at Anonymity and Robustness in Encryption Schemes. In: Abe, M. (ed.) ASIACRYPT 2010. LNCS, vol. 6477, pp. 501–518. Springer, Heidelberg (2010)
9. Niederreiter, H.: Knapsack-type cryptosystems and algebraic coding theory. Problems of Control and Information Theory 15(2), 159–166 (1986)
10. Nojima, R., Imai, H., Kobara, K., Morozov, K.: Semantic security for the McEliece cryptosystem without random oracles. Des. Codes Cryptography 49(1-3), 289–305 (2008)
11. Patterson, N.: The algebraic decoding of Goppa codes. IEEE Transactions on Information Theory 21(2), 203–207 (1975)
12. Shoup, V.: A proposal for an ISO standard for public key encryption (version 2.1). IACR Cryptology ePrint Archive, 112 (2001)
13. Strenzke, F.: A Timing Attack against the Secret Permutation in the McEliece PKC. In: Sendrier, N. (ed.) PQCrypto 2010. LNCS, vol. 6061, pp. 95–107. Springer, Heidelberg (2010)
14. Strenzke, F., Tews, E., Molter, H.G., Overbeck, R., Shoufan, A.: Side Channels in the McEliece PKC. In: Buchmann, J., Ding, J. (eds.) PQCrypto 2008. LNCS, vol. 5299, pp. 216–229. Springer, Heidelberg (2008)
15. Yamakawa, S., Cui, Y., Kobara, K., Hagiwara, M., Imai, H.: On the key-privacy issue of McEliece public-key encryption. In: Boztaş, S., Lu, H.-F(F.) (eds.) AAECC 2007. LNCS, vol. 4851, pp. 168–177. Springer, Heidelberg (2007)

A   The Niederreiter Cryptosystem

The scheme we present below is a generalization of the Niederreiter scheme in two senses: first, we extend it to any family of codes with an efficient decoding algorithm; second, we avoid using the outdated method of scrambling matrices S and P (see [9] for details). While the former allows for a wider and safer choice of codes, the latter provides a simpler formulation and resistance to side-channel attacks (Strenzke et al. [13,14]).

In what follows, we consider only families of codes to which is possible to associate an efficient decoding algorithm; we denote this with Decode_Δ, where Δ is a description of the selected code that depends on $\mathcal{F}$. For example, in case $\mathcal{F}$ is the family of binary Goppa codes, the associated algorithm is Patterson's algorithm [11] and Δ is given by a Goppa polynomial $g(x)$ and its support $(\alpha_1, \ldots, \alpha_n)$.

B   Standard Security Definitions

Definition 6 (IND). *An adversary $\mathcal{A}$ for the indistinguishability (IND) property is a two-stage polynomial-time algorithm. In the first stage, $\mathcal{A}$ takes as input a public key $\mathsf{pk} \in K_{publ}$, then outputs two arbitrary plaintexts ϕ_0, ϕ_1. In the second stage, it receives a ciphertext $\psi^* = \mathsf{Enc}_{pk}(\phi_b)$, for $b \in \{0, 1\}$, and returns a bit b^*. The adversary succeeds if $b^* = b$. More precisely, we define the* advantage *of $\mathcal{A}$ against PKE as*

$$Adv(\mathcal{A}, \lambda) = \left| Pr[b^* = b] - \frac{1}{2} \right|. \tag{5}$$

Table 7. The Niederreiter cryptosystem

Setup	Fix public system parameters $q, n, k, w \in \mathbb{N}$, then choose a family $\mathcal{F}$ of w-error-correcting $[n, k]$ linear codes over $\mathbb{F}_q$.

K	$\mathsf{K}_{\mathsf{publ}}$ the set of $(n - k) \times k$ matrices over $\mathbb{F}_q$.
	$\mathsf{K}_{\mathsf{priv}}$ the set of code descriptions for $\mathcal{F}$.
P	The set $\mathbb{W}_{q,n,w}$ of words of $\mathbb{F}_q^n$ with Hamming weight w.
C	The vector space $\mathbb{F}_q^{(n-k)}$.
KeyGen	Generate at random a code $\mathcal{C} \in \mathcal{F}$ given by its code description Δ and compute its parity-check matrix in systematic form $H = (M|I_{n-k})$. Publish the public key $M \in \mathsf{K}_{\mathsf{publ}}$ and store the private key $\Delta \in \mathsf{K}_{\mathsf{priv}}$.
Enc	On input a public key $M \in \mathsf{K}_{\mathsf{publ}}$ and a plaintext $\phi = e \in \mathsf{P}$, set $H = (M|I_{n-k})$, then compute the syndrome $s = He^{\mathsf{T}}$ and return the ciphertext $\psi = s \in \mathsf{C}$.
Dec	On input the private key $\Delta \in \mathsf{K}_{\mathsf{priv}}$ and a ciphertext $\psi \in \mathsf{C}$, compute $\mathsf{Decode}_\Delta(\psi)$. If the decoding succeeds, return its output $\phi = e$. Otherwise, output $\perp$.

We say that a PKE scheme enjoys Indistinguishability *if the advantage of any adversary $\mathcal{A}$ over all choices of pk, ψ^* and the randomness used by $\mathcal{A}$ is negligible in the security parameter.*

Definition 7 (IND-CCA). *The attack game for IND-CCA proceeds as follows:*

1. *Query a key generation oracle to obtain a public key pk.*

2. *Make a sequence of calls to a decryption oracle, submitting any string ψ of the proper length (not necessarily an element of C). The oracle will respond with $\mathsf{Dec}_{\mathsf{sk}}(\psi)$.*

3. *Choose $\phi_0, \phi_1 \in \mathsf{P}$ and submit them to an encryption oracle. The oracle will choose a random $b \in \{0, 1\}$ and reply with the "challenge" ciphertext $\psi^* = \mathsf{Enc}_{\mathsf{pk}}(\phi_b)$.*

4. *Keep performing decryption queries. If the submitted ciphertext is $\psi = \psi^*$, the oracle will return $\perp$.*

5. *Output $b^* \in \{0, 1\}$.*

We say that a PKE scheme has Indistinguishability against Adaptive Chosen Ciphertext Attacks (IND-CCA) *if the advantage $\mathsf{Adv}_{\mathsf{CCA}}$ of any IND adversary $\mathcal{A}$ in the CCA attack model is negligible.*

A similar but weaker notion is *Indistinguishability against Chosen Plaintext Attacks (IND-CPA)*. The game proceeds at above, but no decryption queries are allowed.

The equivalent scenario for symmetric schemes is a model called *find-guess* (Bellare et al., [2]). The definition is similar to IND, except that in this case some extra information is needed before producing the response bit. This replaces the role of the randomness in the adversary since we are now operating with symmetric encryption. The names "find" and "guess" refer to the two stages of the algorithm.

Definition 8 (FG). *An adversary $\mathcal{A}$ for the find-guess (FG) property is a two-stage polynomial-time algorithm. In the first stage (find), $\mathcal{A}$ takes as input a key $\kappa \in K$, then outputs two arbitrary plaintexts ϕ_0, ϕ_1 along with some extra information ι to be used later. In the second stage (guess), it receives a ciphertext $\psi^* = \mathsf{Enc}_\kappa(\phi_b)$ for $b \in \{0, 1\}$, and returns a bit $b^* = \mathcal{A}(\kappa, \psi^*, \iota)$. The adversary succeeds if $b^* = b$. More precisely, we define the* advantage *of $\mathcal{A}$ against SE as*

$$\mathsf{Adv}(\mathcal{A}, \lambda) = \left| \Pr[b^* = b] - \frac{1}{2} \right|. \tag{6}$$

We say that a SE enjoys Find-Guess *security if the probability of success of any adversary $\mathcal{A}$ over all choices of pk, ψ^* and ι is negligible in the security parameter.*

Finally, the following is the most desirable security properties for signature schemes. It challenges an adversary, equipped with a signing oracle, to reproduce at least one valid message/signature pair.

Definition 9 (EUF-CMA). *We define an adversary $\mathcal{A}$ as a polynomial-time algorithm that acts as follows:*

1. *Query a key generation oracle to obtain a verification key vk.*
2. *Make a sequence of calls to a signing oracle, submitting any message $\mu \in M$. The oracle will reply with $\sigma = \mathsf{Sign}_{\mathsf{sgk}}(\mu)$.*
3. *Output a pair (μ^*, σ^*).*

The adversary succeeds if $\mathsf{Ver}_{\mathsf{vk}}(\mu^*, \sigma^*) = 1$ and $(\mu^*, \sigma^*) \neq (\mu, \sigma)$ for any pair (μ, σ) previously obtained by querying the signing oracle. We say that a signature scheme is Existentially Unforgeable against Chosen Message Attacks (EUF-CMA) *if the probability of success of any adversary $\mathcal{A}$ is negligible in the security parameter.*

Fast Verification for Improved Versions
of the UOV and Rainbow Signature Schemes

Albrecht Petzoldt[1], Stanislav Bulygin, and Johannes Buchmann[1,2]

[1] Technische Universität Darmstadt, Department of Computer Science
Hochschulstraße 10, 64289 Darmstadt, Germany
{apetzoldt,buchmann}@cdc.informatik.tu-darmstadt.de

[2] Center for Advanced Security Research Darmstadt - CASED
Mornewegstraße 32, 64293 Darmstadt, Germany
johannes.buchmann@cased.de

Abstract. Multivariate cryptography is one of the main candidates to guarantee the security of communication in the post-quantum era. While multivariate signature schemes are fast and require only modest computational resources, the key sizes of such schemes are quite large. In [14] Petzoldt et al. proposed a way to reduce the public key size of certain multivariate signature schemes like UOV and Rainbow by a large factor. In this paper we show that by using this idea it is possible to speed up the verification process of these schemes, too. For example, we are able to speed up the verification process of UOV by a factor of 5.

Keywords: Multivariate Cryptography, UOV Signature Scheme, Rainbow Signature Scheme, Key Size Reduction, Fast Verification.

1 Introduction

When quantum computers arrive, classical public-key cryptosystems like RSA and ECC will be broken [1]. The reason for this is Shor's algorithm [18] which solves number theoretic problems like integer factorization and discrete logarithms in polynomial time on a quantum computer. So, to guarantee the security of communication in the post-quantum era, we need alternatives to those classical schemes. Besides lattice-, code-, and hash-based cryptosystems multivariate cryptography seems to be a candidate for this.

Additionally to its (believed) resistance against quantum computer attacks, multivariate cryptosystems are very fast, especially for signatures [2,3]. Furthermore they require only modest computational resources, which makes them appropriate for the use on low-cost devices like smartcards and RFID chips. However, multivariate schemes are not widely used yet, mainly because of the large size of their public and private keys.

In [14], [16] and [17] Petzoldt et al. showed different possibilities to decrease the public key size of the Unbalanced Oil and Vinegar (UOV) and Rainbow signature schemes. The key idea is it to insert a highly structured matrix into the coefficient matrix of the public key. Therefore, the coefficient matrix of the

P. Gaborit (Ed.): PQCrypto 2013, LNCS 7932, pp. 188–202, 2013.

public key has the form $M_P = (B|C)$, where B is a matrix of a very special form (e.g. partially circulant or generated by an LFSR) and C is a matrix without visible structure. By doing so, they were able to decrease the public key size of UOV by 86 %, namely from 99.9 kB to 13.4 kB.

In this paper we show that this idea can not only be used to decrease the size of the public key, but also to speed up the verification process. We use the rich structure of the matrix B to reduce the number of field multiplications needed during the verification process by a large factor (for cyclicUOV this factor is about 80 %). We derive our results both theoretically and show them using a C implementation of the schemes.

The structure of this paper is as follows: In Section 2 we give a short overview on multivariate signature schemes and describe the UOV and Rainbow signature schemes. Section 3 reviews the approach of [14] and [16] to create UOV and Rainbow schemes with structured public keys. In Section 4 we demonstrate how we can use this special structure to speed up the verification process of the schemes. In Subsection 4.2 we look hereby on partially cyclic UOV schemes, whereas Subsection 4.3 deals with cyclic versions of Rainbow. Section 5 presents the results of our experiments and Section 6 concludes the paper.

2   Multivariate Public Key Cryptography

The basic idea behind multivariate cryptography is to choose a system $\mathcal{F}$ of m quadratic polynomials in n variables which can be easily inverted (central map). After that one chooses two affine invertible maps $\mathcal{S}$ and $\mathcal{T}$ to hide the structure of the central map. The public key of the cryptosystem is the composed quadratic map $\mathcal{P} = \mathcal{S} \circ \mathcal{F} \circ \mathcal{T}$ which is difficult to invert. The private key consists of $\mathcal{S}$, $\mathcal{F}$ and $\mathcal{T}$ and therefore allows to invert $\mathcal{P}$.

Due to this construction, the security of multivariate cryptography is based on two mathematical problems:

Problem MQ. Solve the system $p^{(1)} = \cdots = p^{(m)} = 0$, where each $p^{(i)}$ is a quadratic polynomial in the n variables $x_1, \ldots, x_n$ with coefficients and variables in $GF(q)$.

The MQ-problem is proven to be NP-hard even for quadratic polynomials over $GF(2)$ [8].

Problem EIP (Extended Isomorphism of Polynomials). Given a class of central maps $\mathcal{C}$ and a map $\mathcal{P}$ expressible as $\mathcal{P} = \mathcal{S} \circ \mathcal{F} \circ \mathcal{T}$, where $\mathcal{S}$ and $\mathcal{T}$ are affine maps and $\mathcal{F} \in \mathcal{C}$, find a decomposition of $\mathcal{P}$ of the form $\mathcal{P} = \mathcal{S}' \circ \mathcal{F}' \circ \mathcal{T}'$, with affine maps $\mathcal{S}'$ and $\mathcal{T}'$ and $\mathcal{F}' \in \mathcal{C}$.

In this paper we concentrate on the case of multivariate signature schemes. The standard process for signature generation and verification works as shown in Figure 1.

Signature Generation. To sign a document d, we use a hash function $\mathcal{H} : \{0,1\}^* \rightarrow \mathbb{F}^m$ to compute the value $\mathbf{h} = \mathcal{H}(d) \in \mathbb{F}^m$. Then we compute $\mathbf{x} = \mathcal{S}^{-1}(\mathbf{h})$,

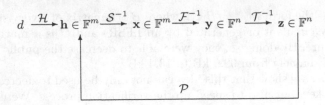

$$d \xrightarrow{\mathcal{H}} \mathbf{h} \in \mathbb{F}^m \xrightarrow{\mathcal{S}^{-1}} \mathbf{x} \in \mathbb{F}^m \xrightarrow{\mathcal{F}^{-1}} \mathbf{y} \in \mathbb{F}^n \xrightarrow{\mathcal{T}^{-1}} \mathbf{z} \in \mathbb{F}^n$$

$$\mathcal{P}$$

Fig. 1. Signature generation and verification

$\mathbf{y} = \mathcal{F}^{-1}(\mathbf{x})$ and $\mathbf{z} = \mathcal{T}^{-1}(\mathbf{y})$. The signature of the document is $\mathbf{z} \in \mathbb{F}^n$. Here, $\mathcal{F}^{-1}(\mathbf{x})$ means finding one (of the possibly many) pre-image of $\mathbf{x}$ under the central map $\mathcal{F}$.

Verification. To verify the authenticity of a document, one simply computes $\mathbf{h}' = \mathcal{P}(\mathbf{z})$ and the hash value $\mathbf{h} = \mathcal{H}(d)$ of the document. If $\mathbf{h}' = \mathbf{h}$ holds, the signature is accepted, otherwise rejected.

There are several ways to build the central map $\mathcal{F}$ of multivariate schemes. In this paper we concentrate on the so called SingleField constructions. In contrast to BigField schemes like Matsumoto-Imai [11] and MiddleField schemes like ℓiC [6], here all the computations are done in one (relatively small) field. In the following two subsections we describe two well known examples of these schemes in detail.

2.1 The Unbalanced Oil and Vinegar (UOV) Signature Scheme

One way to create an easily invertible multivariate quadratic system is the principle of Oil and Vinegar, which was first proposed by J. Patarin in [13].

Let $\mathbb{F}$ be a finite field. Let o and v be two integers and set $n = o + v$. We set $V = \{1, \ldots, v\}$ and $O = \{v+1, \ldots, n\}$. We call $x_1, \ldots, x_v$ the Vinegar variables and $x_{v+1}, \ldots, x_n$ Oil variables. We define o quadratic polynomials $f^{(k)}(\mathbf{x}) = f^{(k)}(x_1, \ldots, x_n)$ by

$$f^{(k)}(\mathbf{x}) = \sum_{i \in V,\, j \in O} \alpha_{ij}^{(k)} x_i x_j + \sum_{i,j \in V,\, i \le j} \beta_{ij}^{(k)} x_i x_j + \sum_{i \in V \cup O} \gamma_i^{(k)} x_i + \eta^{(k)} \ (1 \le k \le o).$$

$$(1)$$

Note that Oil and Vinegar variables are not fully mixed, just like oil and vinegar in a salad dressing.

The map $\mathcal{F} = (f^{(1)}(\mathbf{x}), \ldots, f^{(o)}(\mathbf{x}))$ can be easily inverted. First, we choose the values of the v Vinegar variables $x_1, \ldots, x_v$ at random. Therefore we get a system of o linear equations in the o variables $x_{v+1}, \ldots, x_n$ which can be solved e.g. by Gaussian Elimination. If the system does not have a solution, one has to choose other values of $x_1, \ldots, x_v$ and try again.

The public key of the scheme is given as $\mathcal{P} = \mathcal{F} \circ \mathcal{T}$, where $\mathcal{T}$ is an affine map from $\mathbb{F}^n$ to itself. The private key consists of the two maps $\mathcal{F}$ and $\mathcal{T}$ and therefore allows to invert the public key.

Remark. In opposite to other multivariate schemes the second affine map $\mathcal{S}$ is not needed for the security of UOV. So it can be omitted.

In his original paper [13] Patarin suggested to choose $o = v$ (Balanced Oil and Vinegar (OV)). After this scheme was broken by Kipnis and Shamir in [10], it was recommended in [9] to choose $v > o$ (Unbalanced Oil and Vinegar (UOV)). The UOV signature scheme over GF(256) is commonly believed to be secure for $o \geq 28$ equations [19] and $v = 2 \cdot o$ Vinegar variables. For UOV schemes over GF(31) we set $(o, v) = (33, 66)$.

2.2   The Rainbow Signature Scheme

In [4] J. Ding and D. Schmidt proposed a signature scheme called Rainbow, which is based on the idea of (Unbalanced) Oil and Vinegar [9].

Let $\mathbb{F}$ be a finite field and V be the set $\{1, \ldots, n\}$. Let $v_1, \ldots, v_{u+1}, u \geq 1$ be integers such that $0 < v_1 < v_2 < \cdots < v_u < v_{u+1} = n$ and define the sets of integers $V_i = \{1, \ldots, v_i\}$ for $i = 1, \ldots, u$. We set $o_i = v_{i+1} - v_i$ and $O_i = \{v_i + 1, \ldots, v_{i+1}\}$ $(i = 1, \ldots, u)$. The number of elements in V_i is v_i and we have $|O_i| = o_i$. For $k = v_1 + 1, \ldots, n$ we define multivariate quadratic polynomials in the n variables $x_1, \ldots, x_n$ by

$$f^{(k)}(\mathbf{x}) = \sum_{i \in O_l, \, j \in V_l} \alpha_{ij}^{(k)} x_i x_j + \sum_{i,j \in V_l, \, i \leq j} \beta_{ij}^{(k)} x_i x_j + \sum_{i \in V_l \cup O_l} \gamma_i^{(k)} x_i + \eta^{(k)}, \quad (2)$$

where l is the only integer such that $k \in O_l$. Note that these are Oil and Vinegar polynomials with x_i, $i \in V_l$ being the Vinegar variables and x_j, $j \in O_l$ being the Oil variables.

The map $\mathcal{F}(\mathbf{x}) = (f^{(v_1+1)}(\mathbf{x}), \ldots, f^{(n)}(\mathbf{x}))$ can be inverted as follows. First, we choose $x_1, \ldots, x_{v_1}$ at random. Hence we get a system of o_1 linear equations (given by the polynomials $f^{(k)}$ $(k \in O_1)$) in the o_1 unknowns $x_{v_1+1}, \ldots, x_{v_2}$, which can be solved by Gaussian Elimination. The so computed values of x_i $(i \in O_1)$ are plugged into the polynomials $f^{(k)}(\mathbf{x})$ $(k > v_2)$ and a system of o_2 linear equations (given by the polynomials $f^{(k)}$ $(k \in O_2)$) in the o_2 unknowns x_i $(i \in O_2)$ is obtained. By repeating this process we can get values for all the variables x_i $(i = 1, \ldots, n)$ [1].

The public key of the scheme is given as $\mathcal{P} = \mathcal{S} \circ \mathcal{F} \circ \mathcal{T}$ with two invertible affine maps $\mathcal{S} : \mathbb{F}^m \to \mathbb{F}^m$ and $\mathcal{T} : \mathbb{F}^n \to \mathbb{F}^n$. The private key consists of $\mathcal{S}$, $\mathcal{F}$ and $\mathcal{T}$ and therefore allows to invert te public key.

In the following, we restrict ourselves to Rainbow schemes with two layers (i.e. $u = 2$). For this, $\mathbb{F} = GF(256)$, $(v_1, o_1, o_2) = (17, 13, 13)$ provides 80-bit security under known attacks [15]. For Rainbow schemes over GF(31), we choose $(v_1, o_1, o_2) = (14, 19, 14)$.

---

[1] It may happen, that one of the linear systems does not have a solution. If so, one has to choose other values of $x_1, \ldots x_{v_1}$ and try again.

3   Improved Versions of UOV and Rainbow

In [14] and [16] Petzoldt et al. presented an approach to create UOV- and Rainbow-based schemes with structured public keys, by which they could reduce the public key size of these schemes by up to 83 %. Due to lack of space we give here only a very brief description and refer to [14] and [16] for the details.

The main idea of the approach is to insert a structured matrix B into the Macauley matrix M_P of the public key. In our case the matrix B is chosen partially circulant, i.e. its rows are given by

$$B[i] = \mathcal{R}^{i-1}(\mathbf{b}) \ (i = 1, \ldots, m), \tag{3}$$

where $\mathbf{b}$ is a randomly chosen vector and $\mathcal{R}^i$ denotes the cyclic right shift by i positions.

To insert this matrix B into M_P, the authors used the relation $\mathcal{P} = \mathcal{F} \circ \mathcal{T}$ between a UOV public and private key, which translates into the matrix equation

$$M_P = M_F \cdot A \tag{4}$$

with a transformation matrix A whose elements are given as quadratic functions in the coefficients of the affine map $\mathcal{T}$. If this matrix is invertible, one can compute the matrix M_F in such a way that M_P has the form $M_P = (B|C)$ with a partially circulant matrix B and a matrix C without visible structure. Figure 2 shows this key generation process graphical form.

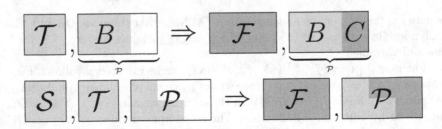

Fig. 2. Alternative key generation for UOV (above) and Rainbow. The light gray parts are chosen by the user, the dark gray parts are computed during the key generation process.

4   The Verification Process

The central part of the verification process for multivariate signature schemes is the evaluation of the public polynomials. Normally this is done as follows: For a given (valid or invalid) signature $\mathbf{z} = (z_1, \ldots, z_n) \in \mathbb{F}^n$ one first computes an $\frac{(n+1) \cdot (n+2)}{2}$ vector mon, which contains the values of all monomials of degree ≤ 2, i.e.

$$\text{mon} = (z_1^2, z_1 z_2, \ldots, z_n^2, z_1, \ldots, z_n, 1). \tag{5}$$

Then we have

$$\mathcal{P}(\mathbf{z}) = \begin{pmatrix} M_P[1] \cdot \mathrm{mon}^T \\ \vdots \\ M_P[m] \cdot \mathrm{mon}^T \end{pmatrix}, \tag{6}$$

with $M_P[i]$ being the i-th row of the Macauley matrix M_P and $\cdot$ being the standard scalar product.

For schemes with partially cyclic public key, the following strategy seems to be more promising:

4.1   Notations

Let $\mathbf{h} = (h_1, \ldots, h_m)$ be the hash value of the signed message.

The public polynomials can be written as

$$p^{(k)}(x_1, \ldots, x_n) = \sum_{i=1}^{n} \sum_{j=i}^{n} p_{ij}^{(k)} \cdot x_i x_j + \sum_{i=1}^{n} p_i^{(k)} \cdot x_i + p_0^{(k)} \quad (k = 1, \ldots, m). \tag{7}$$

For $k = 1, \ldots, m$ we define upper triangular matrices $MP^{(k)}$ by

$$MP^{(k)} = \begin{pmatrix} p_{11}^{(k)} & p_{12}^{(k)} & p_{13}^{(k)} & \cdots & p_{1n}^{(k)} & p_1^{(k)} \\ 0 & p_{22}^{(k)} & p_{23}^{(k)} & \cdots & p_{2n}^{(k)} & p_2^{(k)} \\ 0 & 0 & p_{33}^{(k)} & & p_{3n}^{(k)} & p_3^{(k)} \\ \vdots & & \ddots & & \vdots & \vdots \\ 0 & 0 & \cdots & 0 & p_{nn}^{(k)} & p_n^{(k)} \\ 0 & 0 & \cdots & 0 & 0 & p_0^{(k)} \end{pmatrix}. \tag{8}$$

For a (valid or invalid) signature $\mathbf{z} = (z_1, \ldots, z_n)$ of the message we define the extended signature vector

$$\mathrm{sign} = (z_1, \ldots, z_n, 1). \tag{9}$$

With this notation we can write the verification process in the following form

$$\text{accept the signature } \mathbf{z} \iff \mathrm{sign} \cdot MP^{(k)} \cdot \mathrm{sign}^T = h_k \; \forall k \in \{1, \ldots, m\}. \tag{10}$$

In the following two subsections we consider the question how we can evaluate this equation more efficiently for improved versions of UOV and Rainbow.

4.2   cyclicUOV

In the case of cyclicUOV [14], the matrices $MP^{(k)}$ are of the form shown in Figure 3. We have

$$MP_{ij}^{(k)} = MP_{i,j-1}^{(k-1)} \; \forall i = 1, \ldots, v, \; j = i+1, \ldots, n, \; k = 2, \ldots, o. \tag{11}$$

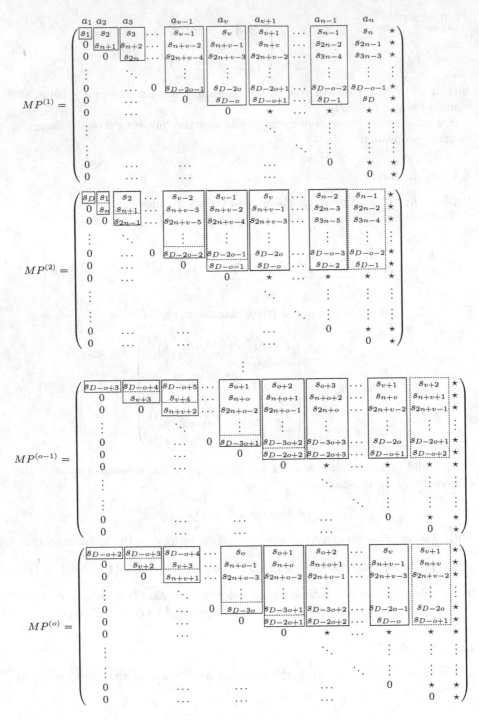

Fig. 3. Matrices $MP^{(i)}$ for cyclicUOV

Therefore we get

$$(\text{sign}_1, \ldots, \text{sign}_i) \cdot \begin{pmatrix} MP_{1,j}^{(k)} \\ MP_{2,j}^{(k)} \\ \vdots \\ MP_{i,j}^{(k)} \end{pmatrix} = (\text{sign}_1, \ldots, \text{sign}_i) \cdot \begin{pmatrix} MP_{1,j-1}^{(k-1)} \\ MP_{2,j-1}^{(k-1)} \\ \vdots \\ MP_{i,j-1}^{(k-1)} \end{pmatrix} \begin{array}{l} \forall i = 1, \ldots, v \\ j = i + 1, \ldots, n, \\ k = 2, \ldots, o. \end{array}$$

$$(12)$$

The boxes in Figure 3 illustrate this equation. Boxes with continuous lines show the vector $(MP_{1,j-1}^{(k-1)}, \ldots, MP_{i,j-1}^{(k-1)})^T$ on the right hand side of the equation, whereas the boxes with dashed lines represent the vector $(MP_{1,j}^{(k)}, \ldots, MP_{i,j}^{(k)})^T$ on the left hand side. As one can see, the dashed boxes in the matrix $MP^{(k)}$ are exactly the same as the boxes with continuous lines in the matrix $MP^{(k-1)}$ ($k = 2, \ldots, o$). We can use this fact to speed up the verification process of cyclicUOV by a large factor (see Algorithm 1).

Algorithm 1. Verification process for cyclicUOV

1: **for** $i = 1$ to $n - 1$ **do**   ▷ first polynomial
2:     $a_i \leftarrow \sum_{j=1}^{\min(i,v)} MP_{ji}^{(1)} \cdot \text{sign}_j$
3:     $\text{temp}_i \leftarrow a_i$
4: **end for**
5: **for** $i = v + 1$ to $n - 1$ **do**
6:     $\text{temp}_i \leftarrow a_i + \sum_{j=v+1}^{i} MP_{ji}^{(1)} \cdot \text{sign}_j$
7: **end for**
8: $\text{temp}_n \leftarrow \sum_{j=1}^{n} MP_{ji}^{(1)} \cdot \text{sign}_j$
9: $\text{temp}_{n+1} \leftarrow \sum_{j=1}^{n+1} MP_{j,n+1}^{(1)} \cdot \text{sign}_j$
10: $h_1' \leftarrow \sum_{j=1}^{n+1} \text{temp}_j \cdot \text{sign}_j$
11: **for** $l = 2$ to o **do**   ▷ polynomials $2, \ldots, o$
12:     $\text{temp}_{n+1} \leftarrow \sum_{j=1}^{n+1} MP_{j,n+1}^{(l)} \cdot \text{sign}_j$
13:     **for** $i = n$ to $v + 1$ by -1 **do**
14:         $a_i \leftarrow a_{i-1}$
15:         $\text{temp}_i \leftarrow a_i + \sum_{j=v+1}^{i} MP_{ji}^{(l)} \cdot \text{sign}_j$
16:     **end for**
17:     **for** $i = v$ to 2 by -1 **do**
18:         $a_i \leftarrow a_{i-1} + MP_{ii}^{(l)} \cdot \text{sign}_i$
19:         $\text{temp}_i \leftarrow a_i$
20:     **end for**
21:     $a_1 \leftarrow MP_{11}^{(l)} \cdot \text{sign}_1$
22:     $\text{temp}_1 \leftarrow a_1$
23:     $h_l' \leftarrow \sum_{j=1}^{n+1} \text{temp}_j \cdot \text{sign}_j$
24: **end for**
25: **if** $h_l = h_l' \ \forall l \in \{1, \ldots, o\}$ **then return** "ACCEPT"   ▷ TEST
26: **else return** "REJECT"
27: **end if**

Algorithm 1 works as follows. The first matrix-vector product $\text{sign} \cdot MP^{(1)} \cdot \text{sign}^T$ is computed as for a random polynomial: From step 1 to step 9 we compute the product $\text{sign} \cdot MP^{(1)}$ (the result is written into the vector temp) and step 10 computes the scalar product of temp and sign. Furthermore we compute the vector $a = (a_1, \ldots, a_{n-1})$ which can be used for the computation of $\text{sign} \cdot MP^{(2)}$. In the loop (step 11 to step 24 of the algorithm) we compute the matrix vector products $\text{sign} \cdot MP^{(k)} \cdot \text{sign}^T$ ($k = 2, \ldots, o$). Step 12 to step 22 computes the vector $\text{temp} = \text{sign} \cdot MP^{(k)}$. We begin with temp_{n+1} and go back to temp_1. In the computation of temp_i ($i = 2, \ldots, n$) we use the values a_i computed before, since, due to the cyclic structure of the public key, they appear in several of the products $\text{sign} \cdot MP^{(k)}$ (see equation (12)). Furthermore (step 18 and 21) we update the values of the a_i ($i = 1, \ldots, n-1$) for the use in the next iteration of the loop. Step 23 computes the scalar product of temp and sign. The last three steps (step 25 to 27) use the values h'_l ($l = 1, \ldots o$) computed in step 10 and 23 to verify the authenticity of the signature.

Computational Effort. Evaluating the system $\mathcal{P}$ in the standard way, one needs

- $\frac{n \cdot (n+1)}{2}$ field multiplications to compute the vector mon (c.f. equation (5))
- and $o \cdot \frac{(n+1) \cdot (n+2) - 2}{2}$ field multiplications to compute the scalar products of equation (6).

Altogether, we need

$$\frac{n+1}{2} \cdot (n \cdot (o+1) + 2 \cdot o) - o \tag{13}$$

field multiplications. Algorithm 1 needs

- in step 2 $\frac{v \cdot (v+1)}{2} + (o-1) \cdot v$ field multiplications,
- in step 6 $\frac{(o-1) \cdot o}{2}$ field multiplications,
- in step 8 n field multiplications,
- in step 9 $n+1$ field multiplications,
- and in step 10 again $n+1$ field multiplications.

Therefore, to compute the value of h'_1, the algorithm needs $\frac{(n+1) \cdot (n+4)}{2}$
In the loop (step 11 to 24) Algorithm 1 needs

- in step 12 $n+1$ field multiplications,
- in step 15 $\frac{o \cdot (o+1)}{2}$ field multiplications,
- in step 18 $v-1$ field multiplications,
- in step 21 1 field multiplication,
- and in step 23 $n+1$ field multiplications.

So, for every iteration of the loop the algorithm needs $2 \cdot (n+1) + v + \frac{o \cdot (o+1)}{2}$ field multiplications.

Altogether, we need therefore

$$(o-1) \cdot \left(2 \cdot (n+1) + v + \frac{o \cdot (o+1)}{2} \right) + \frac{(n+1) \cdot (n+4)}{2} \quad (14)$$

field multiplications to evaluate equation (10).

For $\mathbb{F} = $GF(256), $(o, v) = (28, 56)$ this means a reduction of the number of field multiplications needed during the verification process by 80 % or a factor of 5.0. For a UOV scheme over GF(31), $(o, v) = (33, 66)$, we get a reduction factor of 5.4.

4.3   cyclicRainbow

The verification process for cyclicRainbow is mainly done as for cyclicUOV. However we have to consider the different structure of the polynomials. For cyclicRainbow, the matrices $MP^{(k)}$ look as shown in Figure 4.

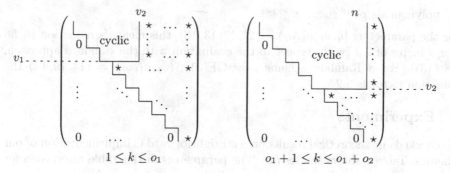

Fig. 4. Matrices $MP^{(k)}$ for cyclicRainbow

So we get for the polynomials $2, \ldots, o_1 + 1$

$$MP_{ij}^{(k)} = MP_{i,j-1}^{(k-1)} \ \forall i = 1, \ldots, v_1, \ j = i+1, \ldots, v_2, \ k = 2, \ldots, o_1 + 1 \quad (15)$$

or

$$(\mathrm{sign}_1, \ldots, \mathrm{sign}_i) \cdot \begin{pmatrix} MP_{1,j}^{(k)} \\ MP_{2,j}^{(k)} \\ \vdots \\ MP_{i,j}^{(k)} \end{pmatrix} = (\mathrm{sign}_1, \ldots, \mathrm{sign}_i) \cdot \begin{pmatrix} MP_{1,j-1}^{(k-1)} \\ MP_{2,j-1}^{(k-1)} \\ \vdots \\ MP_{i,j-1}^{(k-1)} \end{pmatrix} \begin{matrix} \forall i = 1, \ldots, v_1, \\ j = i+1, \ldots, v_2, \\ k = 2, \ldots, o_1 + 1. \end{matrix}$$

$$(16)$$

For the polynomials $o_1 + 2, \ldots, o_1 + o_2$ we get

$$MP_{ij}^{(k)} = MP_{i,j-1}^{(k-1)} \ \forall i = 1, \ldots, v_2, \ j = i+1, \ldots, n, \ k = o_1+2, \ldots, o_1+o_2 \quad (17)$$

or

$$(\text{sign}_1, \ldots, \text{sign}_i) \cdot \begin{pmatrix} MP_{1,j}^{(k)} \\ MP_{2,j}^{(k)} \\ \vdots \\ MP_{i,j}^{(k)} \end{pmatrix} = (\text{sign}_1, \ldots, \text{sign}_i) \cdot \begin{pmatrix} MP_{1,j-1}^{(k-1)} \\ MP_{2,j-1}^{(k-1)} \\ \vdots \\ MP_{i,j-1}^{(k-1)} \end{pmatrix} \quad \begin{array}{l} \forall i = 1, \ldots, v_2, \\ j = i+1, \ldots, n, \\ k = o_1 + 2, \ldots, o_1 + o_2. \end{array}$$

$$(18)$$

To cover this fact, we use Algorithm 1 for both groups of polynomials separately (see Algorithm 2 in Appendix A).

Computational Cost. Our algorithm needs

- $\frac{(n+1)\cdot(n+4)}{2}$ field multiplications to evaluate $p^{(1)}$,
- $o_1 \cdot \left(\frac{(n+1)\cdot(n+4)}{2} - \frac{v_2\cdot(v_2+1)}{2} + \frac{o_1\cdot(o_1+1)}{2} + v_1 \right)$ field multiplications to evaluate the polynomials $p^{(2)} \ldots, p^{(o_1+1)}$ and
- $(o_2 - 1) \cdot \left(2 \cdot (n+1) + v_2 + \frac{o_2\cdot(o_2+1)}{2} \right)$ field multiplications to evaluate the polynomials $p^{(o_1+2)}, \ldots, p^{(o_1+o_2)}$.

For the parameters $(q, v_1, o_1, o_2) = (2^8, 17, 13, 13)$, this means a reduction by 56 % or a factor of 2.3 (with respect to the evaluation with the standard approach, see (13)). For a Rainbow scheme over GF(31), $(v_1, o_1, o_1) = (14, 19, 14)$ the reduction factor is 2.2.

5   Experiments

We checked our theoretical results on a straightforward C implementation of our schemes. Table 1 shows the results. The parameters in this table are chosen for 80 bit security.

Table 1. Improved versions of UOV and Rainbow

Scheme	private key size (kB)	hash length (bit)	signature length (bit)	public key size (kB)	public key red. factor	verification time ms	verification time s. u. f.[1]
UOV$(31, 33, 66)$	102.9	160	528	108.5	-	1.75	-
cyclicUOV$(31, 33, 66)$	102.9	160	528	17.1	6.3	0.34	5.2
UOV$(256, 28, 56)$	95.8	224	672	99.9	-	0.98	-
cyclicUOV$(256, 28, 56)$	95.8	224	672	16.5	6.1	0.20	4.9
Rainbow$(31, 14, 19, 14)$	17.1	160	256	25.3	-	0.44	-
cyclicRainbow$(31, 14, 19, 14)$	17.1	160	256	12.0	2.1	0.21	2.1
Rainbow$(256, 17, 13, 13)$	19.1	208	344	25.1	-	0.26	-
cyclicRainbow$(256, 17, 13, 13)$	19.1	208	344	9.5	2.6	0.13	2.0

[1]speed up factor of the verification time

The differences between the results of our theoretical analysis (see Section 4) and the actual runtime of the verification process is mainly caused by the heavy use of control structures in Algorithms 1 and 2.

6   Conclusion

In this paper we show a way how the structure in the public keys of cyclic versions of UOV and Rainbow can be used to achieve a significant speed up of the verification process. We propose improved algorithms for the verification process of UOV and Rainbow which run up to 5 times faster than the standard verification algorithm. Future research includes:

- *Use of special processor instructions*
  Like in the paper of Chen et al. [3] we plan to use special processor instructions to speed up our implementations.
- *Implementation in hardware*
  We plan to implement our schemes in hardware (e.g. on FPGA and HSM), which should also decrease the verification time.

Acknowledgements. We want to thank the anonymous reviewers for their comments which helped to improve the paper. The first author is supported by the Horst Görtz Foundation.

References

1. Bernstein, D.J., Buchmann, J., Dahmen, E. (eds.): Post Quantum Cryptography. Springer, Heidelberg (2009)
2. Bogdanov, A., Eisenbarth, T., Rupp, A., Wolf, C.: Time-area optimized public-key engines: $\mathcal{MQ}$-cryptosystems as replacement for elliptic curves? In: Oswald, E., Rohatgi, P. (eds.) CHES 2008. LNCS, vol. 5154, pp. 45–61. Springer, Heidelberg (2008)
3. Chen, A.I.-T., Chen, M.-S., Chen, T.-R., Cheng, C.-M., Ding, J., Kuo, E.L.-H., Lee, F.Y.-S., Yang, B.-Y.: SSE implementation of multivariate PKCs on modern x86 CPUs. In: Clavier, C., Gaj, K. (eds.) CHES 2009. LNCS, vol. 5747, pp. 33–48. Springer, Heidelberg (2009)
4. Ding, J., Schmidt, D.: Rainbow, a new multivariable polynomial signature scheme. In: Ioannidis, J., Keromytis, A.D., Yung, M. (eds.) ACNS 2005. LNCS, vol. 3531, pp. 164–175. Springer, Heidelberg (2005)
5. Ding, J., Yang, B.-Y., Chen, C.-H.O., Chen, M.-S., Cheng, C.M.: New Differential-Algebraic Attacks and Reparametrization of Rainbow. In: Bellovin, S.M., Gennaro, R., Keromytis, A.D., Yung, M. (eds.) ACNS 2008. LNCS, vol. 5037, pp. 242–257. Springer, Heidelberg (2008)
6. Ding, J., Wolf, C., Yang, B.-Y.: ℓ-Invertible Cycles for $\mathcal{M}$ultivariate $\mathcal{Q}$uadratic ($\mathcal{MQ}$) Public Key Cryptography. In: Okamoto, T., Wang, X. (eds.) PKC 2007. LNCS, vol. 4450, pp. 266–281. Springer, Heidelberg (2007)
7. Faugère, J.C.: A new efficient algorithm for computing Groebner bases (F4). Journal of Pure and Applied Algebra 139, 61–88 (1999)
8. Garey, M.R., Johnson, D.S.: Computers and Intractability: A Guide to the Theory of NP-Completeness. W.H. Freeman and Company (1979)
9. Kipnis, A., Patarin, J., Goubin, L.: Unbalanced Oil and Vinegar Signature Schemes. In: Stern, J. (ed.) EUROCRYPT 1999. LNCS, vol. 1592, pp. 206–222. Springer, Heidelberg (1999)

10. Kipnis, A., Shamir, A.: Cryptanalysis of the Oil & Vinegar Signature Scheme. In: Krawczyk, H. (ed.) CRYPTO 1998. LNCS, vol. 1462, pp. 257–266. Springer, Heidelberg (1998)
11. Matsumoto, T., Imai, H.: Public Quadratic Polynomial-Tuples for Efficient Signature-Verification and Message-Encryption. In: Günther, C.G. (ed.) EURO-CRYPT 1988. LNCS, vol. 330, pp. 419–453. Springer, Heidelberg (1988)
12. Patarin, J.: Hidden Fields Equations (HFE) and Isomorphisms of Polynomials (IP): Two New Families of Asymmetric Algorithms. In: Maurer, U.M. (ed.) EU-ROCRYPT 1996. LNCS, vol. 1070, pp. 33–48. Springer, Heidelberg (1996)
13. Patarin, J.: The oil and vinegar signature scheme. Presented at the Dagstuhl Workshop on Cryptography (September 1997)
14. Petzoldt, A., Bulygin, S., Buchmann, J.: A Multivariate Signature Scheme with a partially cyclic public key. In: Proceedings of SCC 2010, pp. 229–235 (2010)
15. Petzoldt, A., Bulygin, S., Buchmann, J.: Selecting Parameters for the Rainbow Signature Scheme. In: Sendrier, N. (ed.) PQCrypto 2010. LNCS, vol. 6061, pp. 218–240. Springer, Heidelberg (2010)
16. Petzoldt, A., Bulygin, S., Buchmann, J.: CyclicRainbow – A Multivariate Signature Scheme with a Partially Cyclic Public Key. In: Gong, G., Gupta, K.C. (eds.) INDOCRYPT 2010. LNCS, vol. 6498, pp. 33–48. Springer, Heidelberg (2010)
17. Petzoldt, A., Bulygin, S., Buchmann, J.: Linear Recurring Sequences for the UOV Key Generation. In: Catalano, D., Fazio, N., Gennaro, R., Nicolosi, A. (eds.) PKC 2011. LNCS, vol. 6571, pp. 335–350. Springer, Heidelberg (2011)
18. Shor, P.: Polynomial-Time Algorithms for Prime Factorization and Discrete Logarithms on a Quantum Computer. SIAM J. Comput. 26(5), 1484–1509
19. Thomae, E., Wolf, C.: Solving Underdetermined Systems of Multivariate Quadratic Equations Revisited. In: Fischlin, M., Buchmann, J., Manulis, M. (eds.) PKC 2012. LNCS, vol. 7293, pp. 156–171. Springer, Heidelberg (2012)

A   Algorithm for the Verification Process of Cyclicrainbow

Algorithm 2 shows the improved verification process for Rainbow schemes with two layers and partially circulant public key. The algorithm can be extended to Rainbow schemes with more than two layers in a natural way.

Algorithm 2. Verification process for cyclicRainbow

1: **for** $i = 1$ to $v_2 - 1$ **do**   ▷ First polynomial
2:     $a_i \leftarrow \sum_{j=1}^{\min(i,v_1)} MP_{ji}^{(1)} \cdot \text{sign}_j$
3:     $\text{temp}_i \leftarrow a_i$
4: **end for**
5: **for** $i = v_1 + 1$ to $v_2 - 1$ **do**
6:     $\text{temp}_i \leftarrow a_i + \sum_{j=v_1+1}^{i} MP_{ji}^{(1)} \cdot \text{sign}_j$
7: **end for**
8: **for** $i = v_2$ to $n + 1$ **do**
9:     $\text{temp}_i \leftarrow \sum_{j=1}^{i} MP_{ji}^{(1)} \cdot \text{sign}_j$
10: **end for**
11: $h_1' \leftarrow \sum_{j=1}^{n+1} \text{temp}_j \cdot \text{sign}_j$
12: **for** $l = 2$ to o_1 **do**   ▷ Polynomials 2 to o_1
13:     **for** $i = v_2 + 1$ to $n + 1$ **do**
14:         $\text{temp}_i \leftarrow \sum_{j=1}^{i} MP_{ji}^{(l)} \cdot \text{sign}_j$
15:     **end for**
16:     **for** $i = v_2$ to $v_1 + 1$ by -1 **do**
17:         $a_i \leftarrow a_{i-1}$
18:         $\text{temp}_i \leftarrow a_i + \sum_{j=v+1}^{i} MP_{ji}^{(l)} \cdot \text{sign}_j$
19:     **end for**
20:     **for** $i = v_1$ to 2 by -1 **do**
21:         $a_i \leftarrow a_{i-1} + MP_{ii}^{(l)} \cdot \text{sign}_i$
22:         $\text{temp}_i \leftarrow a_i$
23:     **end for**
24:     $a_1 \leftarrow MP_{11}^{(l)} \cdot \text{sign}_1$
25:     $\text{temp}_1 \leftarrow a_1$
26:     $h_l' \leftarrow \sum_{j=1}^{n+1} \text{temp}_j \cdot \text{sign}_j$
27: **end for**

Algorithm 2. Verification process for cyclicRainbow (cont.)

28:   $\text{temp}_{n+1} \leftarrow \sum_{j=1}^{n+1} MP_{j,n+1}^{(o_1+1)} \cdot \text{sign}_j$ $\triangleright$ $(o_1 + 1)$-th polynomial

29: **for** $i = n$ to $v_2 + 1$ by -1 **do**

30:     $a_i \leftarrow \sum_{j=1}^{v_2} MP_{ji}^{(o_1+1)} \cdot \text{sign}_j$

31:     $\text{temp}_i \leftarrow a_i + \sum_{j=v_2+1}^{i} MP_{ji}^{(o_1+1)} \cdot \text{sign}_j$

32: **end for**

33: **for** $i = v_2$ to $v_1 + 1$ by -1 **do**

34:     $a_i \leftarrow a_{i-1} + \sum_{j=v_1+1}^{i} MP_{ji}^{(o_1+1)} \cdot \text{sign}_j$

35:     $\text{temp}_i \leftarrow a_i$

36: **end for**

37: **for** $i = v_1$ to 2 by -1 **do**

38:     $a_i \leftarrow a_{i-1} + MP_{ii}^{(o_1+1)} \cdot \text{sign}_i$

39:     $\text{temp}_i \leftarrow a_i$

40: **end for**

41: $a_1 \leftarrow MP_{11}^{(o_1+1)} \cdot \text{sign}_1$

42: $\text{temp}_1 \leftarrow a_1$

43: $h'_{o_1+1} \leftarrow \sum_{j=1}^{n+1} \text{temp}_j \cdot \text{sign}_j$

44: **for** $l = o_1 + 2$ to $o_1 + o_2$ **do** $\triangleright$ Polynomials $o_1 + 2$ to $o_1 + o_2$

45:     $\text{temp}_{n+1} \leftarrow \sum_{j=1}^{n+1} MP_{j,n+1}^{(l)} \cdot \text{sign}_j$

46:     **for** $i = n$ to $v_2 + 1$ by -1 **do**

47:        $a_i \leftarrow a_{i-1}$

48:        $\text{temp}_i \leftarrow a_i + \sum_{j=v_2+1}^{i} MP_{ji}^{(l)} \cdot \text{sign}_j$

49:     **end for**

50:     **for** $i = v_2$ to 2 by -1 **do**

51:        $a_i \leftarrow a_{i-1} + MP_{ii}^{(l)} \cdot \text{sign}_i$

52:        $\text{temp}_i \leftarrow a_i$

53:     **end for**

54:     $a_1 \leftarrow MP_{11}^{(l)} \cdot \text{sign}_1$

55:     $\text{temp}_1 \leftarrow a_1$

56:     $h'_l \leftarrow \sum_{j=1}^{n+1} \text{temp}_j \cdot \text{sign}_j$

57: **end for**

58: **if** $h_l = h'_l \ \forall l \in \{1, \ldots, m\}$ **then return** "ACCEPT" $\triangleright$ TEST

59: **else return** "REJECT"

60: **end if**

The Hardness of Code Equivalence over $\mathbb{F}_q$ and Its Application to Code-Based Cryptography

Nicolas Sendrier[1] and Dimitris E. Simos[1,2]

[1] INRIA Paris-Rocquencourt
Project-Team SECRET
78153 Le Chesnay Cedex, France
[2] SBA Research
1040 Vienna, Austria
{nicolas.sendrier,dimitrios.simos}@inria.fr,
dsimos@sba-research.org

Abstract. The code equivalence problem is to decide whether two linear codes over $\mathbb{F}_q$ are identical up to a linear isometry of the Hamming space. In this paper, we review the hardness of code equivalence over $\mathbb{F}_q$ due to some recent negative results and argue on the possible implications in code-based cryptography. In particular, we present an improved version of the three-pass identification scheme of Girault and discuss on a connection between code equivalence and the hidden subgroup problem.

Keywords: Code Equivalence, Isometry, Hardness, Zero-Knowledge Protocols, Quantum Fourier Sampling, Linear Codes.

1   Introduction

The purpose of this work is to examine the applications of the worst-case and average-case hardness of the CODE EQUIVALENCE problem to the field of code-based cryptography. The latter problem is, given the generator matrices of two q-ary linear codes, how hard is it to decide whether or not these codes are identical up to an isometry of Hamming space? The support splitting algorithm ($\mathcal{SSA}$) [28] runs in polynomial time for all but a negligible proportion of all linear codes, and solves the latter problem by recovering the isometry when it is just a permutation of the code support.

The McEliece public-key cryptosystem [23] and Girault's zero-knowledge protocol [17], both candidates for post-quantum cryptography, are related to the hardness of permutationally equivalent linear codes. For the McEliece cryptosystem, the $\mathcal{SSA}$ is able to detect some weak keys but a polynomial attack is infeasible due to the large number of possible private keys. However, the security of Girault's zero-knowledge protocol is severely weakened and cannot longer be used with random codes but only with weakly self-dual codes (the hard instances of $\mathcal{SSA}$).

P. Gaborit (Ed.): PQCrypto 2013, LNCS 7932, pp. 203–216, 2013.

Recently in [29], the worst-case and average-case hardness of code equivalence over $\mathbb{F}_q$ was studied and it was shown that in practice, $\mathcal{SSA}$ could be extended for $q \in \{3, 4\}$, and similarly solve all but an exponentially small proportion of the instances in polynomial time, when isometries are under consideration. However, for any fixed $q \geq 5$, the problem seems to be intractable for almost all instances.

In light of these new results, we repair Girault's zero-knowledge protocol over $\mathbb{F}_q$, when $q \geq 5$, by showing that random codes are again a viable option. Moreover, the context of the framework built in [11] suggests that codes with large automorphism groups resist quantum Fourier sampling as long as permutation equivalence is considered. We examine whether it is possible to extend these results, when a more general notion of code equivalence over $\mathbb{F}_q$ is taken into account, in particular when the equivalence mapping is an isometry and not just a permutation of the code support.

The paper is structured as follows. In section 2 we define the different notions of equivalence of linear codes over $\mathbb{F}_q$ when isometries are considered, while in section 3 we formally define the CODE EQUIVALENCE problem and present a thorough analysis of its hardness. In section 4 we review the protocol of Girault together with its weakness and repair its security using results based on the hardness of code equivalence, while in the last section we elaborate on the connection between code equivalence over $\mathbb{F}_q$ and the quantum Fourier sampling.

2   Equivalence of Linear Codes over $\mathbb{F}_q$

Code equivalence is a basic concept in coding theory with several applications in code-based cryptography; the McEliece public-key cryptosystem [23], Girault's identification scheme [17] and the CFS signature scheme [10], to name a few. The notion of equivalence of linear codes used in code-based cryptography usually involves only permutations as the code alphabet is the binary field. However, this is by far the case in coding theory where for a more general notion of equivalence all isometries of the Hamming space have to be included. In this section, we review the concept of what it means for codes to be "essentially different" by considering the metric Hamming space together with its isometries, which are the maps preserving the metric structure. This in turn will lead to a rigorous definition of equivalence of linear codes and as we shall see later on may provide additional applications in cryptography. In fact, we will call codes isometric if they are equivalent as subspaces of the Hamming space.

Let $\mathbb{F}_q$ be a finite field of cardinality $q = p^r$, where the prime number p is its characteristic, and r is a positive integer. As usual, a linear $[n, k]$ code C is a k-dimensional subspace of the finite vector space $\mathbb{F}_q^n$ and its elements are called codewords. We consider all vectors, as row vectors. Therefore, an element v of $\mathbb{F}_q^n$ is of the form $v := (v_1, \ldots, v_n)$. It can also be regarded as the mapping v from the set $\mathcal{I}_n = \{1, \ldots, n\}$ to $\mathbb{F}_q$ defined by $v(i) := v_i$. The Hamming distance (metric) on $\mathbb{F}_q^n$ is the following mapping,

$$d : \mathbb{F}_q^n \times \mathbb{F}_q^n \to \mathbb{N} : (x, y) \mapsto d(x, y) := \mid \{i \in \{1, 2, \ldots, n\} \mid x_i \neq y_i\} \mid.$$

The pair $(\mathbb{F}_q^n, d)$ is a metric space, called the Hamming space of dimension n over $\mathbb{F}_q$, denoted by $H(n, q)$. The Hamming weight $w(x)$ of a codeword $x \in C$ is simply the number of its non-zero coordinates, i.e. $w(x) := d(x, 0)$.

It is well-known due to a theorem of MacWilliams that any isometry between linear codes preserving the weight of the codewords induces an equivalence for codes [22]. Therefore, two codes C, C' are of the same quality if there exists a mapping $\iota : \mathbb{F}_q^n \mapsto \mathbb{F}_q^n$ with $\iota(C) = C'$ which preserves the Hamming distance, i.e. $d(v, v') = d(\iota(v), \iota(v'))$, for all $v, v' \in \mathbb{F}_q^n$. Mappings with the latter property are called the isometries of $H(n, q)$, and the two codes C and C' will be called isometric. Clearly, isometric codes have the same error-correction capabilities, and obvious permutations of the coordinates are isometries. We write $\mathcal{S}_n$ for the symmetric group acting on the set $\mathcal{I}_n$, equipped with the composition of permutations.

Definition 1. *Two linear codes $C, C' \subseteq \mathbb{F}_q^n$ will be called permutationally equivalent[1], and will be denoted as $C \overset{\text{PE}}{\sim} C'$, if there exists a permutation $\sigma \in \mathcal{S}_n$ that maps C onto C', i.e. $C' = \sigma(C) = \{\sigma(x) \mid x = (x_1, \ldots, x_n) \in C\}$ where $\sigma(x) = \sigma(x_1, \ldots, x_n) := (x_{\sigma^{-1}(1)}, \ldots, x_{\sigma^{-1}(n)})$.*

Note also that the use of σ^{-1} in the index is consisted as we have $\sigma(\pi(C)) = \sigma \circ \pi(C)$. This can easily be seen by considering $x \in C$, and $\sigma, \pi \in \mathcal{S}_n$ such that $\sigma(\pi(x)) = \sigma((x_{\pi^{-1}(i)})_{i \in \mathcal{I}_n})$. Let $y_i = x_{\pi^{-1}(i)}$, $i \in \mathcal{I}_n$. Then $\sigma(\pi(x)) = \sigma((y_i)_{i \in \mathcal{I}_n}) = (y_{\sigma^{-1}(i)})_{i \in \mathcal{I}_n} = (x_{\pi^{-1}\sigma^{-1}(i)})_{i \in \mathcal{I}_n} = (x_{(\sigma\pi)^{-1}(i)})_{i \in \mathcal{I}_n} = \sigma \circ \pi(x)$.

Moreover, there is a particular subgroup of $\mathcal{S}_n$ that maps C onto itself, the permutation group of C defined as $\text{PAut}(C) := \{C = \sigma(C) \mid \sigma \in \mathcal{S}_n\}$. $\text{PAut}(C)$ always contains the identity permutation. If it does not contain any other element, we will say that it is trivial.

Recall, that we defined two codes to be isometric if there exists an isometry that maps one into another. Isometries that are linear[2], are called linear isometries. Therefore, we can obtain a more general notion of equivalence for codes induced by linear isometries of $\mathbb{F}_q$. Moreover, it can be shown that any linear isometry between two linear codes $C, C' \subseteq \mathbb{F}_q^n$ can always be extended to an isometry of $\mathbb{F}_q^n$ [6].

The group of all linear isometries of $H(n, q)$ corresponds to the semidirect product of $\mathbb{F}_q^{*n}$ and $\mathcal{S}_n$, $\mathbb{F}_q^{*n} \rtimes \mathcal{S}_n = \{(v; \pi) \mid v : \mathcal{I}_n \mapsto \mathbb{F}_q^*, \pi \in \mathcal{S}_n\}$, called the monomial group of degree n over $\mathbb{F}_q^*$, where the multiplication within this group is defined by

$$(v; \pi)(v'; \pi') = (vv'_\pi, \pi\pi') \quad \text{and} \quad (vv'_\pi)_i := v_i v'_{\pi^{-1}(i)} \tag{1}$$

where $\mathbb{F}_q^*$ denotes the multiplicative group of $\mathbb{F}_q$. Hence, any linear isometry ι can be expressed as a pair of mappings $(v; \pi) \in \mathbb{F}_q^{*n} \rtimes \mathcal{S}_n$. Note that, some authors [6,14,16], describe this group as the wreath product $\mathbb{F}_q^* \wr_n \mathcal{S}_n$. The action

---

[1] This definition can also met as permutationally isometric codes in the literature, see [6].

[2] For all $u, v \in \mathbb{F}_q^n$ we have $\iota(u + v) = \iota(u) + \iota(v)$ and $\iota(0) = 0$.

of the latter group in an element of $\mathbb{F}_q^n$ is translated into an equivalence for linear codes.

Definition 2. *Two linear codes $C, C' \subseteq \mathbb{F}_q^n$ will be called linearly or monomially equivalent, and will be denoted as $C \overset{\mathbf{LE}}{\sim} C'$, if there exists a linear isometry $\iota = (v; \sigma) \in \mathbb{F}_q^{*n} \rtimes \mathcal{S}_n$ that maps C onto C', i.e. $C' = (v; \sigma)(C) = \{(v; \sigma)(x) \mid (x_1, \ldots, x_n) \in C\}$ where $(v; \sigma)(x_1, \ldots, x_n) := (v_1 x_{\sigma^{-1}(1)}, \ldots, v_n x_{\sigma^{-1}(n)}).$*

If $q = p^r$ is not a prime, then the Frobenius automorphism $\tau : \mathbb{F}_q \to \mathbb{F}_q, x \mapsto x^p$ applied on each coordinate of $\mathbb{F}_q^n$ preserves the Hamming distance, too. Moreover, for $n \geq 3$, the isometries of $\mathbb{F}_q^n$ which map subspaces onto subspaces are exactly the semilinear mappings[3] of the form $(v; (\alpha, \pi))$, where $(v; \pi)$ is a linear isometry and α is a field automorphism, i.e. $\alpha \in \text{Aut}(\mathbb{F}_q)$ (c.f. [6,21]). All these mappings form the group of semilinear isometries of $H(n, q)$ which is isomorphic to the semidirect product $\mathbb{F}_q^{*n} \rtimes (\text{Aut}(\mathbb{F}_q) \times \mathcal{S}_n)$, where the multiplication of elements is given by

$$(v; (\alpha, \pi))(\varphi; (\beta, \sigma)) := (v \cdot \alpha(\varphi_\pi); (\alpha\beta, \pi\sigma)) \tag{2}$$

Moreover, there is a description of $\mathbb{F}_q^{*n} \rtimes (\text{Aut}(\mathbb{F}_q) \times \mathcal{S}_n)$ as a generalized wreath product $\mathbb{F}_q^* \wr_n (\text{Aut}(\mathbb{F}_q) \times \mathcal{S}_n)$, see [6,15,21]. Clearly, the notion of semilinear isometry which can be expressed as a group action on the set of linear subspaces gives rise to the most general notion of equivalence for linear codes.

Definition 3. *Two linear codes $C, C' \subseteq \mathbb{F}_q^n$ will be called semilinearly equivalent, and will be denoted as $C \overset{\mathbf{SLE}}{\sim} C'$, if there exists a semilinear isometry $(v; (\alpha, \sigma)) \in \mathbb{F}_q^{*n} \rtimes (\text{Aut}(\mathbb{F}_q) \times \mathcal{S}_n)$ that maps C onto C', i.e. $C' = (v; (\alpha, \sigma))(C) = \{(v; (\alpha, \sigma))(x) \mid (x_i)_{i \in \mathcal{I}_n} \in C\}$ where $(v; (\alpha, \sigma))(x_1, \ldots, x_n) = (v_1 \alpha(x_{\sigma^{-1}(1)}), \ldots, v_n \alpha(x_{\sigma^{-1}(n)})).$*

Finally, we can define the monomial group of C as $\text{MAut}(C) := \{C = (v; \sigma)(C) \mid (v; \sigma) \in \mathbb{F}_q^{*n} \rtimes \mathcal{S}_n\}$ and the automorphism group of C as $\text{Aut}(C) := \{C = (v; (\alpha, \sigma))(C) \mid (v; (\alpha, \sigma)) \in \mathbb{F}_q^{*n} \rtimes (\text{Aut}(\mathbb{F}_q) \times \mathcal{S}_n)\}$ where their elements map each codeword of C to another codeword of C, under the respective actions of the involved groups. For more details, on automorphism groups of linear codes we refer to [20]. In addition, we remark the following:

1. When $\mathbb{F}_q = \mathbb{F}_2$ the group of linear isometries of $H(n, 2)$ is isomorphic to $\mathcal{S}_n$, therefore all notions of equivalence are the same.
2. The group of semilinear isometries of $H(n, q)$ is the same as the group of linear isometries if and only if q is a prime (since $\text{Aut}(\mathbb{F}_q)$ is trivial if and only if q is a prime). Therefore, semilinear equivalence reduces to linear equivalence for prime fields, and is different for all other cases.

---

[3] $\sigma : \mathbb{F}_q^n \to \mathbb{F}_q^n$ is semilinear if there exists $\alpha \in \text{Aut}(\mathbb{F}_q)$ such that for all $u, v \in \mathbb{F}_q^n$ and $k \in \mathbb{F}_q$ we have $\sigma(u + v) = \sigma(u) + \sigma(v)$ and $\sigma(ku) = \alpha(k)\sigma(u)$.

3   The Code Equivalence Problem

For efficient computation of codes we represent them with generator matrices. A $k \times n$ matrix G over $\mathbb{F}_q$, is called a generator matrix for the $[n, k]$ linear code C if the rows of G form a basis for C, so that $C = \{xG \mid x \in \mathbb{F}_q^k\}$. In general, a linear code possess many different bases, and it is clear from linear algebra that the set of all generator matrices for C can be reached by $\{SG \mid S \in \mathrm{GL}_k(q)\}$, where $\mathrm{GL}_k(q)$ is the group of all $k \times k$ invertible matrices over $\mathbb{F}_q$.

For any $\sigma \in \mathcal{S}_n$ associate by $P_\sigma = [p_{i,j}]$ the $n \times n$ matrix such that $p_{i,j} = 1$ if $\sigma(i) = j$ and $p_{i,j} = 0$ otherwise, therefore P_σ is a permutation matrix. Note that, the action of $\sigma \in \mathcal{S}_n$ on $x \in \mathbb{F}_q^n$ agrees with the ordinary matrix multiplication. The permutation matrices form a subgroup of $M_n(q)$, the set of all $n \times n$ monomial matrices over $\mathbb{F}_q$, that is, matrices with exactly one nonzero entry per row and column from $\mathbb{F}_q$. If $M = [m_{i,j}] \in M_n(q)$, then $M = DP$, where P is a permutation matrix and $D = [d_{i,j}] = \mathrm{diag}(d_1, \ldots, d_n)$ is a diagonal matrix with $d_i = d_{i,i} = m_{i,j}$ if $m_{i,j} \neq 0$ and $d_{i,j} = 0$ if $i \neq j$. There is an isomorphism between diagonal matrices and $\mathbb{F}_q^{*n}$, therefore we associate $D_v = \mathrm{diag}(v_1, \ldots, v_n)$ for $v = (v_i)_{i \in \mathcal{I}_n} \in \mathbb{F}_q^{*n}$. Hence, we can map any linear isometry $(v; \sigma) \in \mathbb{F}_q^{*n} \rtimes \mathcal{S}_n$ to a monomial matrix $M_{(v;\sigma)} = D_v P_\sigma \in M_n(q)$, and this mapping is an isomorphism between $\mathbb{F}_q^{*n} \rtimes \mathcal{S}_n$ and $M_n(q)$. Therefore, we can express the equivalence between linear codes in terms of their generator matrices.

Problem 1. Given two $k \times n$ matrices G and G' over $\mathbb{F}_q$, whose rows span two $[n, k]$ linear codes C and C' over $\mathbb{F}_q$, does there exist $S \in \mathrm{GL}_k(q)$ and a monomial matrix $M_{(v;\sigma)} = D_v P_\sigma \in M_n(q)$ such that $G' = SGD_v P_\sigma$?

We will refer to the decidability of the previous problem, as the LINEAR CODE EQUIVALENCE problem. The SEMI-LINEAR CODE EQUIVALENCE problem can be defined analogously by permitting the application of a field automorphism in the columns of the scrambled generator matrix. In particular, we define the following problem.

Problem 2. Given two $k \times n$ matrices G and G' over $\mathbb{F}_q$, whose rows span two $[n, k]$ linear codes C and C' over $\mathbb{F}_q$, does there exist $S \in \mathrm{GL}_k(q)$, a monomial matrix $M_{(v;\sigma)} = D_v P_\sigma \in M_n(q)$ and a field automorphism $\alpha \in \mathrm{Aut}(\mathbb{F}_q)$ such that $G' = S\alpha(GD_v P_\sigma)$?

Finally, we review the hardness of the code equivalence problem, therefore we deem necessary to briefly mention the most significant results in terms of complexity, for deciding it, and algorithms, for solving it.

When the linear isometry $(v; \sigma)$ is just a permutation, i.e. D_v is equal to I_n, we will call problem 1, as the PERMUTATION CODE EQUIVALENCE problem. The latter problem, was introduced in [26], who showed that if $\mathbb{F}_q = \mathbb{F}_2$ then it is harder than the GRAPH ISOMORPHISM, there exists a polynomial time reduction, but not NP-complete unless P = NP. A different proof of this reduction is also given in [21]. Recently, the reduction of [26] was generalized in [18] over any field $\mathbb{F}_q$, hence PERMUTATION CODE EQUIVALENCE is harder than the GRAPH

ISOMORPHISM, for any field $\mathbb{F}_q$. The latter problem, has been extensively studied for decades, but until now there is no polynomial-time algorithm for solving all of its instances. Clearly, (SEMI)-LINEAR CODE EQUIVALENCE for any $\mathbb{F}_q$ cannot be easier than the GRAPH ISOMORPHISM, since it contains the PERMUTATION CODE EQUIVALENCE as a subproblem.

Last but not least, we would like to mention that the McEliece public-key cryptosystem [23] is related to the hardness of permutationally equivalent binary linear codes. Towards this direction, another important complexity result was shown in [11], that the HIDDEN SUBGROUP problem also reduces to PERMUTATION CODE EQUIVALENCE for any field $\mathbb{F}_q$.

The Support Splitting Algorithm can be used as an oracle to decide whether two binary codes are permutationally equivalent [28], as well as to retrieve the equivalence mapping. Other notable algorithms for code equivalence can be found in [3,7,13]. The main idea of $\mathcal{SSA}$ is to partition the support $\mathcal{I}_n$ of a code $C \subseteq \mathbb{F}_2^n$, into small sets that are fixed under operations of PAut(C). The algorithm employs the concept of invariants and signatures, defined in [28]. Invariants are mappings such that any two permutationally equivalent codes take the same value, while signatures depends on the code and one of its positions.

Definition 4. *A signature S over a set F maps a code $C \subseteq \mathbb{F}_q^n$ and an element $i \in \mathcal{I}_n$ into an element of F and is such that for all $\sigma \in \mathcal{S}_n$, $S(C, i) = S(\sigma(C), \sigma(i))$. Moreover, S is called discriminant for C if there exist $i, j \in \mathcal{I}_n$ such that $S(C, i) \neq S(C, j)$ and fully discriminant if this holds $\forall\, i, j \in \mathcal{I}_n$.*

The fundamental idea of $\mathcal{SSA}$ is to be able to find a distinct property for the code and one of its positions, and thus by labeling them accordingly it is possible to recover the permutation between equivalent codes.

The main difficulty of the algorithm, is to obtain a fully discriminant signature, for as many codes as possible. In [28] it was shown that such a signature, can be built from the weight enumerator of the hull of a code C, denoted by $\mathcal{H}(C)$, and defined as the intersection of the code with its dual, $\mathcal{H}(C) = C \cap C^\perp$ [2], because the hull commutes with permutations[4], $\mathcal{H}(\sigma(C)) = \sigma(\mathcal{H}(C))$, and therefore is an invariant for permutation equivalence. The (heuristic) complexity of $\mathcal{SSA}$ for an $[n, k]$ code C is $\mathcal{O}(n^3 + 2^h n^2 \log n)$ where h is the dimension of the hull [24,28]. The first term is the cost of the Gaussian elimination needed to compute the hull. The second term is the (conjectured) number of refinements, $\log n$, multiplied by the cost one refinement (n weight enumerators of codes of dimension h and length n). Moreover, the cost of computing the weight enumerator of an $[n, h]$ code over $\mathbb{F}_q$ is proportional to nq^h operations in $\mathbb{F}_q$ [28].

In practice, for random codes, the hull has a small dimension with overwhelming probability [27] and the dominant cost for the average case is $\mathcal{O}(n^3)$. Note that, the worst case occurs when the hull dimension is maximal; weakly self-dual

---

[4] No such property exists in general for linear codes when (semi)-linear equivalence is considered, because the dual of equivalent codes do not remain equivalent with the same isometry as the original codes.

codes ($C \subset C^\perp$) are equal to their hulls. Then the algorithm becomes intractable with a complexity equal to $\mathcal{O}(2^k n^2 \log n)$.

Reduction of Linear Code Equivalence to Permutation Code Equivalence was made possible via the introduction of the closure of a linear code in [29]. A similar approach was given in [30].

Definition 5. *Let* $\mathbb{F}_q = \{a_0, a_1, \ldots, a_{q-1}\}$, *with* $a_0 = 0$, *and a linear code* $C \subseteq \mathbb{F}_q^n$. *Define* $\mathcal{I}_{q-1}^{(n)}$ *as the cartesian product of* $\mathcal{I}_{q-1} \times \mathcal{I}_n$. *The closure* $\widetilde{C}$ *of the code* C *is a code of length* $(q-1)n$ *over* $\mathbb{F}_q$ *where,*

$$\widetilde{C} = \{(a_k x_i)_{(k,i) \in \mathcal{I}_{q-1}^{(n)}} \mid (x_i)_{i \in \mathcal{I}_n} \in C\}.$$

Clearly, we see that every coordinate of the closure $\widetilde{C}$, corresponds to a coordinate position of a codeword of C multiplied by a nonzero element of $\mathbb{F}_q$. Since, the index $(k,i) \in \mathcal{I}_{q-1}^{(n)}$ of a position of a codeword of the closure means that $k \in \mathcal{I}_{q-1}$ and $i \in \mathcal{I}_n$, we have taken into account every possible multiplication of x_i with nonzero elements of $\mathbb{F}_q$. The fundamental property of the closure is realised in the following theorem, first given in [29].

Theorem 1. *Let* $C, C' \subseteq \mathbb{F}_q^n$. *Then* C *and* C' *are linearly equivalent, i.e.* $C \overset{\mathbf{LE}}{\sim} C'$, *if and only if* $\widetilde{C}$ *and* $\widetilde{C'}$ *are permutationally equivalent, i.e.* $\widetilde{C} \overset{\mathbf{PE}}{\sim} \widetilde{C'}$.

Theorem 1 is of great importance, because it realizes a reduction from the LINEAR CODE EQUIVALENCE problem to the PERMUTATION CODE EQUIVALENCE problem. Thus, we are able to decide if the codes C and C' are linearly equivalent by checking their closures for permutation equivalence. Moreover, if the closures are permutation equivalent then there exists an algorithmic procedure that allows the retrieval of the initial isometry between C and C' by considering that a signature for an extension of $\mathcal{SSA}$ can be built from the weight enumerator of the $\mathcal{H}(\widetilde{C})$.

Unfortunately, it turns out that the closure $\widetilde{C}$ is a weakly self-dual code for every $q \geq 5$, considering both Euclidean and Hermitian duals, which are exactly the hard instances of $\mathcal{SSA}$ [29]. Moreover, for $\mathbb{F}_3$ and $\mathbb{F}_4$ equipped with the Euclidean and Hermitian inner product, respectively, the distribution of the dimension of $\mathcal{H}(\widetilde{C})$ follows the distribution of the dimension $\mathcal{H}(C)$, since the closure has the same dimension as C, and will be on average a small constant, [27], except in the cases where C is also a weakly self-dual code. Therefore, the LINEAR CODE EQUIVALENCE problem can be decided (and solved) in polyonomial time using $\mathcal{SSA}$ only in $\mathbb{F}_3$ and $\mathbb{F}_4$, as long as the hull of the given code is small (the worst-case being a weakly self-dual code). However, for $q \geq 5$ its complexity growth becomes exponential for all instances. Moreover, it was conjectured in [29], that for $q \geq 5$, CODE EQUIVALENCE is hard for almost all instances. This argument, was further supported by some impossibility results on the Tutte polynomial of a graph which corresponds to the weight enumerator of a code [34]. To conclude with, we would like to make clear that the hardness of the code equivalence arises from the absence of an easy computable invariant not the inexistence of an algorithm.

Table 1. Heuristic complexity for $\mathcal{SSA}$ and its extension over $\mathbb{F}_q$

Algorithm	Field (alphabet)	Random codes (average-case)	Weakly self-dual codes (worst-case)
$\mathcal{SSA}$	$\mathbb{F}_2$	$\mathcal{O}(n^3)$	$\mathcal{O}(2^k n^2 \log n)$
$\mathcal{SSA}$ extension	$\mathbb{F}_3$	$\mathcal{O}(n^3)$	$\mathcal{O}(3^k n^2 \log n)$
$\mathcal{SSA}$ extension	$\mathbb{F}_4$	$\mathcal{O}(n^3)$	$\mathcal{O}(2^{2k} n^2 \log n)$
$\mathcal{SSA}$ extension	$\mathbb{F}_q, q \geq 5$	$\mathcal{O}(q^k n^2 \log n)$	$\mathcal{O}(q^k n^2 \log n)$

4 Zero-Knowledge Protocols

A central concept in cryptography is zero-knowledge protocols. These protocols allow a prover to convince a verifier that it knows a secret without the verifier learning any information about the secret. In practice, this is used to allow one party to prove its identity to another by proving it has a particular secret. For a protocol to be zero-knowledge, no information can be revealed no matter what strategy a so-called cheating verifier, simply cheater, follows when interacting with the prover. Therefore, an important question is what happens to these protocols when the cheater is a quantum computer. Are there any zero-knowledge protocols sufficient to withstand such a powerful cheater in a post-quantum era?

In this section, we deal with protocols based on a particular type of alternative cryptography originating from error-correcting codes, called code-based cryptography. In this emerging field of cryptography the underlying hard problems which pose as its security assumptions, decoding in a random linear code and recovering the code structure, does not seem so far to be susceptible to attacks mounted by quantum computers [24]. In addition, as we shall mention in the following section there is a negative result regarding the connection between coding theory and the HIDDEN SUBGROUP problem, which is the starting point for designing efficient quantum algorithms.

The idea of using error-correcting codes for identification schemes is due to Harari [19], followed by Stern (first protocol) [31] and Girault [17]. Harari's protocol was broken and the security of Girault's one was severely weakened (we shall explain this shortly after) while the protocol of Stern was five-pass and unpractical. At Crypto'93, Stern proposed a new scheme [32], which is one of the main references in this area. Recently, there has been an upsurge on designing identification schemes mainly due to the work of several researchers [8,9,1], where their efforts concentrated on both reducing the communication cost and the probability of someone impersonating an honest prover.

4.1 Girault's Three-Pass Identification Scheme

Girault's identification scheme is a three-pass one with a cheating probability of $1/2$ (compared to the usual $2/3$ of Stern's protocol), and has the additional advantage that all computations are performed on the standard model instead of the random oracle model since there is no involvement of a hash function in the committments of the protocol. However, this advantage comes with a cost. At each round of the protocol a large number of bits has to be transmitted, which render the scheme unpractical. Its principle is as follows: Let H be an $(n-k) \times n$ matrix over the binary field $\mathbb{F}_2$ common to all users. Each prover $\mathcal{P}$ has an n-bit word e of small weight w randomly chosen by him and a public identifier $He = s$. Clearly, when H is a parity-check matrix of a linear code, computing e from H and s comes to finding a word of given small weight and given syndrome s, a well-known NP-hard problem. When $\mathcal{P}$ needs to authenticate to a verifier $\mathcal{V}$ as the owner of s, then $\mathcal{P}$ and $\mathcal{V}$ interact through the following scheme.

Step 1: $\mathcal{P}$ picks a random $n \times n$ permutation matrix P and a random $k \times k$ non-singular matrix S. P computes $H' = SHP$ and $s' = Ss$, and sends H' and s' to $\mathcal{V}$.

Step 2: $\mathcal{V}$ generates a random bit $c \in \{0, 1\}$ and sends it to $\mathcal{P}$.

Step 3a: If $c = 0$, $\mathcal{P}$ replies by delivering S, P to $\mathcal{V}$, who checks that $SHP = H'$ and $Ss = s'$.

Step 3b: If $c = 1$, $\mathcal{P}$ replies by delivering $e' = P^{-1}e$ to $\mathcal{V}$, who checks that the weight of e' is w and $H'e' = s'$.

The protocol is a multi-round one as it has to be repeated t times to reach a security level of $1 - (1/2)^t$ and was proved to be zero-knowledge on [17]. Its security is based on the hardness of two well-known problems in coding theory. The first one is the BINARY SYNDROME DECODING problem shown to be NP-complete in the worst case [5], but it is also widely believed that for the average case it still remains hard. The other assumption is related to the hardness of the PERMUTATION CODE EQUIVALENCE problem over the binary field since from the knowledge of H and H' someone must not be able to recover the scrambing matrix S and the permutation matrix P, as this would lead to information leakage about the secret key of $\mathcal{P}$. However, as we extensively discussed on section 3, $\mathcal{SSA}$ can recover the matrix P in (almost) polynomial time when the underlying code is chosen at random (see also the complexity figures in table 3), and then using elementary linear algebra the matrix S can also be found.

Still, the protocol can be used with weakly self-dual codes, the instances of PERMUTATION CODE EQUIVALENCE that the growth of $\mathcal{SSA}$ becomes exponential, however there is no significant advantage on decoding with self-dual codes and in addition this restrict too much the possibilities for the public key.

4.2 Improved Version of the Girault Protocol

We now consider, Girault's identification scheme in a q-ary setting. That is, the underlying finite field, will no longer be the binary field but the field $\mathbb{F}_q$

with q elements. For the security assumptions of the scheme we first have to consider syndrome decoding over $\mathbb{F}_q$. We define the decisional version of the q-ary SYNDROME DECODING problem, below,

Problem 3. Given an $m \times n$ matrix H over $\mathbb{F}_q$, a target vector $s \in \mathbb{F}_q^m$ and an integer $w > 0$ does there exist a vector $x \in \mathbb{F}_q^n$ of weight $\leq w$ such that $Hx = s$?

which was also proven to be NP-complete in [4]. There are two main families of algorithms for solving the latter problem: Information Set Decoding (ISD) and (Generalized) Birthday algorithm (GBA). ISD has the lowest complexity of the two, and in a recent work [25] the complexity of a generalization of Stern's algorithm from [33] is analyzed which permits the decoding of linear codes over arbitrary finite fields $\mathbb{F}_q$. For a general treatment of the topic we refer to [24], while for the security of the scheme it is sufficient to consider that all known decoding attacks have an exponential cost on the code length.

Moreover, in an attempt to repair the security of the scheme we consider the (semi)-linear code equivalence instead of the permutation code equivalence, depending on whether $\mathbb{F}_q$ is a prime field or not. As one of the purposes of this paper, is to state the implications of the hardness of the (SEMI)-LINEAR CODE EQUIVALENCE problem for designing cryptographic primitives, we choose the parameter q to be at least equal to 5, since we strongly believe that a random instance of the latter problem is hard for these cases (see also section 3).

The starting point of this improved version of Girault's scheme is the same as in the original one, with the exception that all operations now occur over $\mathbb{F}_q$, $q \geq 5$. Let H be an $(n-k) \times n$ matrix over $\mathbb{F}_q$ common to all users. Each prover $\mathcal{P}$ has an n-bit word e of small weight w randomly chosen by him and a public identifier $He = s$. As before, when a prover $\mathcal{P}$ needs to authenticate to a verifier $\mathcal{V}$ as the owner of s, then $\mathcal{P}$ and $\mathcal{V}$ interact through the following protocol.

Improved Version of Girault Identification Scheme

Key Generation: Random $[n, k]$ linear code with an $(n-k) \times n$ parity-check matrix H over $\mathbb{F}_q$
  - **Private key:** A word $e \in \mathbb{F}_q^n$ of small weight w
  - **Public key:** A public identifier $s \in \mathbb{F}_q^{n-k}$ such that $He = s$

Commitments:
  - $\mathcal{P}$ picks a random $n \times n$ monomial matrix M, a random $k \times k$ non-singular matrix S and a field automorphism α of $\mathbb{F}_q$.
  - $\mathcal{P}$ computes the commitments $s' = Ss$ and $H' = S\alpha(HM)$.
  - $\mathcal{P}$ sends s' and H' to $\mathcal{V}$.

Challenge: $\mathcal{V}$ chooses randomly $c \in \{0, 1\}$ and sends it to $\mathcal{P}$.

Response:
  - If $c = 0$ then $\mathcal{P}$ replies by delivering α, S, M to $\mathcal{V}$
  - If $c = 1$ then $\mathcal{P}$ replies by delivering $e' = \alpha^{-1}(M^{-1}e)$ to $\mathcal{V}$

Verification:
  - If $c = 0$ then $\mathcal{V}$ checks that $S\alpha(HM) = H'$ and $Ss = S'$.
  - If $c = 1$ then $\mathcal{V}$ checks that the weight of e' is w and $H'e' = s'$.

The scheme is again a three-pass one and has to be repeated t times to reach a security level of $1 - (1/2)^t$. The completeness, soundness and zero-knowledge of the scheme is a straight-forward verification of the proofs given by Girault in the original version [17], by replacing the permutation matrix P with the monomial matrix M and the field automorphism α (for non-prime fields) and therefore we avoid repeating them here to save space. Note that, the scheme is again usable in the standard model in contrast to the usual random oracle model.

We would like also to remark, that this q-ary version of Girault's scheme can be used again with the family of random linear codes for any field $\mathbb{F}_q$, $q \geq 5$. Moreover, we choose to commit the monomial matrix M instead of its (unique) factorization to a diagonal matrix D and a permutation matrix P (see also section 3) to reduce the (already) large cost of communication at each round (since we transmit matrices) as much as possible. A promising approach to circumvent this drawback could be to employ random structured codes as the public keys such as quasi-cyclic (QC) codes, similar to the work carried out in [1]. Although, there is no obvious advantage for an adversary mounting decoding attacks on QC codes, their rich structure may lead to structural attacks even when semi-linear code equivalence is considered (even though we are unaware of such kind of attacks) and a more careful analysis is required before proposing any specific parameters for the scheme.

5   A Note about Code Equivalence over $\mathbb{F}_q$ and Quantum Fourier Sampling

In [11], it was shown that permutation code equivalence over $\mathbb{F}_q$ has a direct reduction to a nonabelian HIDDEN SUBGROUP problem (HSP). It was further shown in the same paper that McEliece-type cryptosystems with certain conditions on the permutation automorphism groups of the underlying linear codes used as private keys, as is the case of rational Goppa codes, resist precisely the attacks to which the RSA and ElGamal cryptosystems are vulnerable, namely those based on generating and measuring coset states. This fact, eliminated the approach of strong Fourier sampling on which almost all known exponential speedups by quantum algorithms are based. In addition, these negative results have been extended in [12] for the case of Reed-Muller codes, which correspond to the particular case of the Sidelnikov cryptosystem.

There are two main questions arising from this framework: Whether there are any other families of codes suitable for cryptographic applications and what happens when we consider a more general notion of code equivalence over $\mathbb{F}_q$. We will investigate these matters, after briefly mentioning the conditions needed for the results of [11,12].

Recall from [11] that a linear code C is HSP-hard if strong quantum Fourier sampling, reveals negligible information about the permutation $\sigma \in \mathcal{S}_n$ of permutationally equivalent codes, i.e. $C' = \sigma(C)$. Moreover, the support of a permutation $\sigma \in \mathcal{S}_n$ is the number of points that are not fixed by σ, and the minimal degree of a subgroup $H \leq \mathcal{S}_n$ is the smallest support of any non-identity $\pi \in H$.

Theorem 2 (Theorem 1, [12]). *Let C be a q-ary $[n, k]$ linear code such that $q^{k^2} \leq n^{0.2n}$. If $|\mathrm{PAut}(C)| \leq e^{o(n)}$ and the minimal degree of $|\mathrm{PAut}(C)|$ is $\Omega(n)$ then C is HSP-hard.*

We now consider $\mathbb{F}_q$ to be a prime field (hence $\mathrm{Aut}(\mathbb{F}_q)$ is trivial) and the monomial group $\mathrm{MAut}(C)$ of a code $C \subseteq \mathbb{F}_q^n$ for the notion of linear code equivalence. Clearly, if the permutation part of $\mathrm{MAut}(C)$ satisfies the conditions of theorem 2, so does its closure $\widetilde{C}$ (see definition 5) which is a code of length $(q-1)n$ over the same field. Recall that two codes are linearly equivalent if and only if their closures are permutationally equivalent (c.f. theorem 1). In other words, the instances of codes that are HSP-hard for the PERMUTATION CODE EQUIVALENCE problem remain HSP-hard for the LINEAR CODE EQUIVALENCE. This remark, would further imply that someone could design a McEliece-type cryptosystem by considering a monomial transformation of the private key instead of just a permutation without having to worry about attacks originating from the quantum Fourier sampling, based on rational Goppa codes over $\mathbb{F}_q$ for instance. However, we should note that these results apply only to high-rate $[n, k]$ codes over $\mathbb{F}_q$ (as $q^{k^2} \leq n^{0.2n}$ must be satisfied).

6   Conclusion

In this paper, we presented an analysis of the hardness of the CODE EQUIVALENCE problem over $\mathbb{F}_q$ when the equivalence mapping is an isometry and not just a permutation of the code support. The hardness of the latter problem is of great importance when designing cryptographic primitives, such as public-key cryptosystems and identification schemes in the field of code-based cryptography. We stated the weaknesses of such an identification scheme (Girault's zero-knowledge protocol), and presented an improved version which relies on exactly these instances of the code equivalence that the problem is believed to be hard on average. Finally, we showed that some negative results regarding the possibility of attacking McEliece-type cryptosystems with quantum algorithms based on Fourier sampling apply also for other notions of code equivalence, besides the permutation equivalence, subject to certain conditions on the underlying family of codes used as private keys.

Acknowledgments. The work of the second author was carried out during the tenure of an ERCIM "Alain Bensoussan" Fellowship Programme. This Programme is supported by the Marie Curie Co-funding of Regional, National and International Programmes (COFUND) of the European Commission. In addition, the work of the second author has been supported by the Austrian Research Promotion Agency under grant 258376 and the Austrian COMET Program (FFG).

References

1. Aguilar, C., Gaborit, P., Schrek, J.: A new zero-knowledge code based identification scheme with reduced communication. In: 2011 IEEE Information Theory Workshop (ITW), pp. 648–652 (2011)
2. Assmus, E.F.J., Key, J.D.: Designs and their Codes. Cambridge Tracts in Mathematics, vol. 103. Cambridge University Press (1992), second printing with corrections (1993)
3. Babai, L., Codenotti, P., Grochow, J.A., Qiao, Y.: Code equivalence and group isomorphism. In: Proceedings of the Twenty-Second Annual ACM-SIAM Symposium on Discrete Algorithms, SODA 2011, pp. 1395–1408. SIAM (2011)
4. Barg, S.: Some new NP-complete coding problems. Probl. Peredachi Inf. 30, 23–28 (1994)
5. Berlekamp, E., McEliece, R., van Tilborg, H.: On the inherent intractability of certain coding problems (corresp.). IEEE Transactions on Information Theory 24, 384–386 (1978)
6. Betten, A., Braun, M., Fripertinger, H., Kerber, A., Kohnert, A., Wassermann, A.: Error-Correcting Linear Codes: Classification by Isometry and Applications. Algorithms and Computation in Mathematics, vol. 18. Springer, Heidelberg (2006)
7. Bouyukliev, I.: About the code equivalence. Ser. Coding Theory Cryptol. 3, 126–151 (2007)
8. Cayrel, P.L., Gaborit, P., Girault, M.: Identity-based identification and signature schemes using correcting codes. In: Augot, D., Sendrier, N., Tillich, J.P. (eds.) Workshop on Coding and Cryptography - WCC 2007, pp. 69–78. INRIA (2007)
9. Cayrel, P.-L., Véron, P., El Yousfi Alaoui, S.M.: A zero-knowledge identification scheme based on the q-ary syndrome decoding problem. In: Biryukov, A., Gong, G., Stinson, D.R. (eds.) SAC 2010. LNCS, vol. 6544, pp. 171–186. Springer, Heidelberg (2011)
10. Courtois, N.T., Finiasz, M., Sendrier, N.: How to achieve a McEliece-based digital signature scheme. In: Boyd, C. (ed.) ASIACRYPT 2001. LNCS, vol. 2248, pp. 157–174. Springer, Heidelberg (2001)
11. Dinh, H., Moore, C., Russell, A.: McEliece and niederreiter cryptosystems that resist quantum fourier sampling attacks. In: Rogaway, P. (ed.) CRYPTO 2011. LNCS, vol. 6841, pp. 761–779. Springer, Heidelberg (2011)
12. Dinh, H., Moore, C., Russell, A.: Quantum fourier sampling, code equivalence, and the quantum security of the mceliece and sidelnikov cryptosystems. Tech. rep. (2011), also available as arXiv:1111.4382v1
13. Feulner, T.: The automorphism groups of linear codes and canonical representatives of their semilinear isometry classes. Adv. Math. Commun. 3, 363–383 (2009)
14. Fripertinger, H.: Enumeration of linear codes by applying methods from algebraic combinatorics. Grazer Math. Ber. 328, 31–42 (1996)
15. Fripertinger, H.: Enumeration of the semilinear isometry classes of linear codes. Bayrether Mathematische Schriften 74, 100–122 (2005)
16. Fripertinger, H., Kerber, A.: Isometry classes of indecomposable linear codes. In: Giusti, M., Cohen, G., Mora, T. (eds.) AAECC 1995. LNCS, vol. 948, pp. 194–204. Springer, Heidelberg (1995)
17. Girault, M.: A (non-practical) three-pass identification protocol using coding theory. In: Seberry, J., Pieprzyk, J. (eds.) AUSCRYPT 1990. LNCS, vol. 453, pp. 265–272. Springer, Heidelberg (1990)

18. Grochow, J.A.: Matrix lie algebra isomorphism. Tech. Rep. TR11-168, Electronic Colloquium on Computational Complexity (2011), also available as arXiv:1112.2012, IEEE Conference on Computational Complexity (2012) (to appear)

19. Harari, S.: A new authentication algorithm. In: Wolfmann, J., Cohen, G. (eds.) Coding Theory 1988. LNCS, vol. 388, pp. 91–105. Springer, Heidelberg (1989)

20. Human, W.C.: Codes and groups. In: Pless, V., Human, W.C. (eds.) Handbook of Coding Theory, pp. 1345–1440. Elsevier, North-Holland (1998)

21. Kaski, P., Östergård, P.R.J.: Classification Algorithms for Codes and Designs. Algorithms and Computation in Mathematics, vol. 15. Springer, Heidelberg (2006)

22. MacWilliams, F.J.: Error-correcting codes for multiple-level transmission. Bell. Syst. Tech. J. 40, 281–308 (1961)

23. McEliece, R.J.: A public-key cryptosystem based on algebraic coding theory. Tech. Rep. DSN Progress Report 42-44, California Institute of Technology, Jet Propulsion Laboratory, Pasadena, CA (1978)

24. Overbeck, R., Sendrier, N.: Code-based cryptography. In: Bernstein, D.J., Buchmann, J., Dahmen, E. (eds.) Post-Quantum Cryptography, pp. 95–145. Springer (2009)

25. Peters, C.: Information-set decoding for linear codes over $\mathbb{F}_q$. In: Sendrier, N. (ed.) PQCrypto 2010. LNCS, vol. 6061, pp. 81–94. Springer, Heidelberg (2010)

26. Petrank, E., Roth, R.M.: Is code equivalence easy to decide? IEEE Trans. Inform. Theory 43, 1602–1604 (1997)

27. Sendrier, N.: On the dimension of the hull. SIAM J. Discrete Math. 10(2), 282–293 (1997)

28. Sendrier, N.: Finding the permutation between equivalent linear codes: The support splitting algorithm. IEEE Trans. Inform. Theory 26, 1193–1203 (2000)

29. Sendrier, N., Simos, D.E.: How easy is code equivalence over $\mathbb{F}_q$? In: WCC 2013: Proceedings of the 8th International Workshop on Coding and Cryptography (preprint 2012) (to appear, 2013), https://www.rocq.inria.fr/secret/PUBLICATIONS/codeq3.pdf

30. Skersys, G.: Calcul du groupe d'automorphisme des codes. Détermination de l' equivalence des codes. Thèse de doctorat, Université de Limoges (October 1999)

31. Stern, J.: An alternative to the fiat-shamir protocol. In: Quisquater, J.J., Vandewalle, J. (eds.) EUROCRYPT 1989. LNCS, vol. 434, pp. 173–180. Springer, Heidelberg (1990)

32. Stern, J.: A new identification scheme based on syndrome decoding. In: Stinson, D.R. (ed.) CRYPTO 1993. LNCS, vol. 773, pp. 13–21. Springer, Heidelberg (1994)

33. Stern, J.: A method for finding codewords of small weight. In: Wolfmann, J., Cohen, G. (eds.) Coding Theory 1988. LNCS, vol. 388, pp. 106–113. Springer, Heidelberg (1989)

34. Vertigan, D.: Bicycle dimension and special points of the Tutte polynomial. Journal of Combinatorial Theory, Series B 74, 378–396 (1998)

Timing Attacks against the Syndrome Inversion in Code-Based Cryptosystems*

Falko Strenzke

Cryptography and Computeralgebra, Department of Computer Science,
Technische Universität Darmstadt, Germany
fstrenzke@crypto-source.de

Abstract. In this work we present the first practical key-aimed timing attack against code-based cryptosystems. It arises from vulnerabilities that are present in the inversion of the error syndrome through the Extended Euclidean Algorithm that is part of the decryption operation of these schemes. Three types of timing vulnerabilities are combined to a successful attack. Each is used to gain information about the secret support, which is part of code-based decryption keys: The first allows recovery of the zero-element, the second is a refinement of a previously described vulnerability yielding linear equations, and the third enables to retrieve cubic equations.

Keywords: side channel attack, timing attack, post quantum cryptography, code-based cryptography.

1 Introduction

The McEliece PKC [1] and Niederreiter [2] Cryptosystems, built on error correcting codes, are considered immune to quantum computer attacks [3], and thus are of interest as candidates for future cryptosystems in high security applications. Accordingly, they have received growing interest from researchers in the past years and been analyzed with respect to efficiency on various platforms [4–8]. Furthermore, a growing number of works has investigated the side channel security of code-based cryptosystems [9–14].

Side channel security is a very important implementation aspect of any cryptographic algorithm. A side channel is given when a physical observable quantity that is measured during the operation of a cryptographic device, allows an attacker to gain information about a secret that is involved in the cryptographic operation. The usual observables used in this respect are the duration of the operation (timing attacks [15]), or the power consumption as a function over the time (power analysis attacks[16]).

So far, timing attacks against the decryption operation of the McEliece PKC targeting the plaintext have been developed [10, 12, 14]. In [11], a timing attack

---

* To the most part, this work was done in the author's private capacity, a part of the work was done at Cryptography and Computeralgebra, Department of Computer Science, Technische Universität Darmstadt, Germany

P. Gaborit (Ed.): PQCrypto 2013, LNCS 7932, pp. 217–230, 2013.

is proposed that targets the secret support that is part of the private key in code-based cryptosystems. From the time taken by the solving of the key equation the attacker learns linear equations about the support in this attack. But that work suffers from two major limitations: Neither is the information that is gained in itself sufficient for a practical attack, nor was the attack actually implemented.

This work extends on the analysis given in [11] in multiple ways: first of all, we find that a control flow ambiguity causing leakage in terms of the linear equations is manifest already in the syndrome inversion preceding the solving of the key equation in the decryption operation, and consequently the countermeasure proposed in that work is insufficient. We also show that there exists a timing side channel vulnerability in the syndrome inversion that allows the attacker to gain knowledge of the zero-element of the secret support. As an extension resp. generalization of the attack yielding linear equations, we derive a practical timing attack that lets the attacker gain cubic equations.

We then describe how to efficiently use these three vulnerabilities to build a practical attack that recovers the private key entirely. Lastly, we give results for practical executions of the timing attack on a personal computer.

2   Preliminaries

In this work, we give a brief description of the McEliece PKC, and stress those features of the decryption algorithm, that are necessary to understand the timing attack presented in this paper. A more detailed description and security considerations can be found e.g. in [17].

Goppa Codes. Goppa codes [18] are a class of linear error correcting codes. The McEliece PKC makes use of irreducible binary Goppa codes, so we will restrict ourselves to this subclass and to code lengths that are powers of two.

Definition 1. *Let the polynomial $g(Y) = \sum_{i=0}^{t} g_i Y^i \in \mathbb{F}_{2^m}[Y]$ be monic and irreducible over $\mathbb{F}_{2^m}[Y]$, and let m, t be positive integers. Then $g(Y)$ is called a* Goppa polynomial *(for an irreducible binary Goppa code).*

Then an irreducible binary Goppa code is defined as $\mathcal{C}(g(Y)) = \{c \in \mathbb{F}_2^n | S_c(Y)$ $:= \sum_{i=0}^{n-1} \frac{c_i}{Y - \alpha_i} = 0 \mod g(Y)\}$, where $n = 2^m$, $S_c(Y)$ is the syndrome of c, $\Gamma = (\alpha_i | i = 0, \ldots, n-1)$, the support of the code, where the α_i are pairwise distinct elements of $\mathbb{F}_{2^m}$, and c_i are the entries of the vector c.

The code defined in such way has length n, dimension $k \geq n - mt$ – however we restrict us to $k = n - mt$ in this work – and can correct up to t errors.

As for any linear error correcting code, for a Goppa code there exists a generator matrix $G \in \mathbb{F}_2^{k \times n}$ and a parity check matrix $H \in \mathbb{F}_2^{mt \times n}$ [19]. Given these matrices, a message $m \in \mathbb{F}_2^k$ can be encoded into a codeword c of the code by computing $c = mG$, and the syndrome $s \in \mathbb{F}_2^{mt}$ of a (potentially distorted) codeword can be computed as $s = cH^T$. Here, we do not give the formulas for the computation of these matrices as they are of no importance for the understanding of the attack developed in this work. The interested reader, however, is referred to [19].

Overview of the McEliece PKC. In this section we give a brief overview of the McEliece PKC. The McEliece *secret key* consists of the Goppa polynomial $g(Y)$ of degree t and the support $\Gamma = (\alpha_0, \alpha_1, \ldots, \alpha_{n-1})$, i.e. a permutation of $\mathbb{F}_{2^m}$, together they define the secret code $\mathcal{C}$. The *public key* is given by the public $n \times k$ generator matrix $G_p = SG$ over $\mathbb{F}_2$, where G is a generator matrix of the secret code $\mathcal{C}$ and S is a non-singular $k \times k$ matrix over $\mathbb{F}_2$, the purpose of which is to bring G_p into reduced row echelon form, i.e. $G_p = [\mathbb{I}|G_2]$, which results in a more compact public key [4].

Note that in the original definition of the McEliece PKC [1], the support is chosen to be in lexicographical ordering, but instead the public key is chosen as $G_p = SGP$, where P is a random permuation matrix. These two descriptions are completely equivalent: P corresponds to the permutation that has to be applied to a lexicographical ordered support Γ to produce the randomized secret support as it is defined above. Accordingly, the attack described in this work that attacks the secret support can alternatively be seen as an attack against the secret permutation.

The *encryption* operation allows messages $m \in \mathbb{F}_2^k$. A random vector $e \in \mathbb{F}_2^n$ with hamming weight wt $(e) = t$ has to be created. Then the ciphertext is computed as $z = mG_p + e$.

The Decryption is given in Algorithm 1. It makes use of the error correction algorithm, given by the Patterson Algorithm [20], shown in Algorithm 2. In Step 1 of this algorithm, the syndrome vector is computed by multiplying the ciphertext by the parity check matrix, and then turned into the syndrome polynomial $S(Y)$ by interpreting it as an $\mathbb{F}_{2^m}$ element and multiplying it with the vector of powers of Y. The Patterson Algorithm furthermore uses an algorithm for finding roots in polynomials over $\mathbb{F}_{2^m}$ (root_find()), and the Extended Euclidean Algorithm (EEA) for polynomials with a break condition based on the degree of the remainder, given in Algorithm 3. The root finding can e.g. be implemented as an exhaustive search on $\mathbb{F}_{2^m}$. Please note that all polynomials appearing in the algorithms have coefficients in $\mathbb{F}_{2^m}$.

The Niederreiter PKC [2] is a cryptosystem that is slightly different from the McEliece PKC, however there also an error vector is chosen during the encryption and decryption features the syndrome decoding. Since these features are, as we shall see, the preconditions for our attack, it is equally applicable to the Niederreiter PKC.

In the following, we turn to those details, that are relevant for the side channel issues we are going to address in Section 3. Please note that the error locator polynomial $\sigma(Y)$, which is determined in Step 4 of Algorithm 2, has the following form:

$$\sigma(Y) = \prod_{j \in \mathcal{E}} (Y - \alpha_j) = \sum_{i=0}^{t} \sigma_i Y^i. \tag{1}$$

where $\mathcal{E}$ is the set of those indexes i, for which $e_i = 1$, i.e. those elements of $\mathbb{F}_{2^m}$ that correspond to the error positions in the error vector. The determination of the error vector in Step 6 of Algorithm 2 makes use of this property. Accordingly, $\deg(\sigma(Y)) = $ wt (e) if wt $(e) \le t$ holds.

Algorithm 1. The McEliece Decryption Operation

Require: the McEliece ciphertext $z \in \mathbb{F}_2^n$
Ensure: the message $m \in \mathbb{F}_2^k$
1: $e \leftarrow \text{err\_corr}(z, g(Y))$
2: $m' \leftarrow z + e$
3: $m \leftarrow$ the first k bits of m'
4: return m

Algorithm 2 . The McEliece error correction with the Patterson Algorithm $(\text{err\_corr}(z, g(Y)))$

Require: the distorted code word $z \in \mathbb{F}_2^n$, the secret Goppa polynomial $g(Y)$ and
   secret support $\Gamma = (\alpha_0, \alpha_1, \ldots, \alpha_{n-1})$
Ensure: the error vector $e \in \mathbb{F}_2^n$
1: $S(Y) \leftarrow z H^\top \left(Y^{t-1}, \cdots, Y, 1 \right)^\top$
2: $\tau(Y) \leftarrow \sqrt{S^{-1}(Y) + Y} \bmod g(Y)$
3: $(a(Y), b(Y)) \leftarrow \text{EEA} \left(\tau(Y), g(Y), \lfloor \frac{t}{2} \rfloor \right)$
4: $\sigma(Y) \leftarrow a^2(Y) + Y b^2(Y)$
5: $\mathcal{E} = \{E_0, \ldots, E_{t-1}\} \leftarrow \text{rootfind}(\sigma(Y))$ // if α_i is a root, then $\mathcal{E}$ contains i
6: $e \leftarrow v \in \mathbb{F}_2^n$ with $v_i = 1$ if and only if $i \in \mathcal{E}$
7: **return** e

---

3 Analysis of Timing Side Channels in the Syndrome Inversion

We now explain three different vulnerabilities present in the syndrome decoding. To this end, we first explore certain properties of the syndrome inversion by the use of the EEA during the code-based decryption operation.

3.1 Properties of the Syndrome Inversion

The syndrome polynomial is defined as

$$S(Y) \equiv \sum_{i=1}^{w} \frac{1}{Y \oplus \epsilon_i} \equiv \frac{\Omega(Y)}{\sigma(Y)} \bmod g(Y) \tag{2}$$

Here, w is the Hamming weight of the error vector e and the $\{\epsilon_i | i \in \{1, \ldots w\}\}$ denote the support elements associated with the indexes of those bits in the error vector having value one in arbitrary ordering, i.e., for instance, if the bits found at the index j and k in the error vector have value one, then $\epsilon_1 = \alpha_j$, $\epsilon_2 = \alpha_k$ and so on. The identification of the error locator polynomial $\sigma(Y)$ in the denominator is simply a result of the form of the common denominator of all sum terms. In the McEliece PKC Decryption, during the error correction, Alg. 2, Step 2, $S^{-1}(Y)$ is computed by invoking Alg. 3 as $\text{EEA}(g(Y), S(Y), 0)$. But it is known that in case of $w \leq t/2$ instead it is possible to find $\sigma(Y)$ already at this

---

Algorithm 3. The Extended Euclidean Algorithm $(\text{EEA}(r_{-1}(Y), r_0(Y), d))$

Require: the polynomials $r_{-1}(Y)$ and $r_0(Y)$, with $\deg(r_0(Y)) < \deg(r_{-1}(Y))$
Ensure: two   polynomials $r_M(Y),$ $b_M(Y)$ satisfying $r_M(Y) = b_M(Y)r_0(Y) \bmod r_{-1}(Y)$ and $\deg(r_0(Y)) \leq \lfloor \deg(r_{-1})/2 \rfloor$
 1: $b_{-1} \leftarrow 0$
 2: $b_0 \leftarrow 1$
 3: $i \leftarrow 0$
 4: **while** $\deg(r_i(Y)) > d$ **do**
 5:   $i \leftarrow i+1$
 6:   $(q_i(Y), r_i(Y)) \leftarrow r_{i-2}(Y)/r_{i-1}(Y)$ // polynomial division with quotient q_i and remainder r_i
 7:   $b_i(Y) \leftarrow b_{i-2}(Y) + q_i(Y)b_{i-1}(Y)$
 8: **end while**
 9: $M \leftarrow i$
10: **return** $(r_M(Y), b_M(Y))$

---

stage by invoking Alg. 3 as $\text{EEA}(g(Y), S(Y), \lfloor t/2 \rfloor - 1)$, i.e. with $r_{-1}(Y) = g(Y)$ and $r_0(Y) = S(Y)$ and breaking once $\deg(r_i(Y)) \leq (t/2) - 1$. Then, it returns $\delta\sigma(Y) = b_M(Y)$ and furthermore $\delta\Omega(Y) = r_M(Y)$, where $\delta \in \mathbb{F}_{2^m}$ and M is the number of iterations performed by the EEA [21].

Given this form of the $S(Y)$, we can make a statement about the maximally possible number of iterations in the EEA used to compute $S^{-1}(Y) \equiv \sigma(Y)/\Omega(Y) \bmod g(Y)$. As already mentioned, the actual invocation of the syndrome inversion is $\text{EEA}(g(Y), S(Y), 0)$. But the above explained fact that we could stop at $\deg(r_i(Y)) \leq (t/2) - 1$ means that there is one iteration in the EEA where $r_i(Y) = \delta\Omega(Y)$ and $b_i(Y) = \delta\sigma(Y)$, in case of $w \leq (t/2) - 1$.

Theorem 1. *Assume a Goppa Code defined by $g(Y)$ and Γ. When Alg. 3 is invoked as $\text{EEA}(g(Y), S(Y), 1)$ with $S(Y) \equiv \frac{\Omega(Y)}{\sigma(Y)} \bmod g(Y)$, and the error vector e corresponding to $S(Y)$ satisfies $\text{wt}(e) \leq (\deg(g(Y))/2) - 1$, then for the number of iterations in Alg. 3 we find:*

$$M \leq M_{\max} = \deg(\Omega(Y)) + \deg(\sigma(Y))$$

Proof. Regard the iteration where $r_j(Y) = \delta\Omega(Y)$ and $b_j(Y) = \delta\sigma(Y)$. Since according to Alg. 3 the degree of $b_j(Y)$, starting from zero, increases at least by one in each iteration, we find $j \leq \deg(\sigma(Y))$. From here on, the degree of $r_j(Y) = \delta\Omega(Y)$ is decreased by at least one in each subsequent iteration down to $\deg(r_M(Y)) = 0$, i.e. $M - j \leq \deg(\Omega(Y))$, giving $M = M - j + j \leq \deg(\Omega(Y)) + \deg(\sigma(Y))$. $\qquad\blacksquare$

Because in the following we are only interested in the derivation of equations of the form $\sigma_i = 0$ for a specific value of i, we will ignore the constant δ from here on.

3.2   Linear Equations from $w = 4$ Error Vectors

We now investigate the effect of the above results for the case where ciphertexts created with error vectors of Hamming weight four are input to the decryption operation.

In the case of $w = 4$ the syndrome polynomial is of the form:

$$S(Y) \equiv \frac{\Omega(Y)}{\sigma(Y)} \equiv \sum_{i=1}^{4} \frac{1}{Y \oplus \epsilon_i} \equiv \frac{\sigma_3 Y^2 \oplus \sigma_1}{Y^4 \oplus \sigma_3 Y^3 \oplus \sigma_2 Y^2 \oplus \sigma_1 Y \oplus \sigma_0} \bmod g(Y), \quad (3)$$

where $\epsilon_i \in \mathbb{F}_{2^m}, i \in 1, \ldots, 4$ denote the four elements of the support associated with the error positions. Furthermore, in the right hand side of Eq. (3), which is found by bringing all four sum terms to their common denominator, we have

$$\sigma_3 = \epsilon_1 \oplus \epsilon_2 \oplus \epsilon_3 \oplus \epsilon_4.$$

With the aim of finding a timing vulnerability revealing certain coefficients of $\sigma(Y)$ and thus information about the secret support, we now analyze the connection between the number of iterations and their complexity on the one hand and the degree of $\Omega(Y)$ on the other. Regarding $\Omega(Y)$ for the case $w = 4$ we find that the coefficient to the highest power of Y is given by $\sigma_3 = \epsilon_1 \oplus \epsilon_2 \oplus \epsilon_3 \oplus \epsilon_4$. If $\sigma_3 = 0$, then the degree of $\Omega(Y)$ is zero, otherwise it is two. This means that in the case of $\sigma_3 = 0$ the maximal number of iterations in the inversion is four, in contrast to six in the general case. Table 1 gives an overview of the individual iterations in the syndrome inversion EEA when $w = 4$, where it is assumed that for each iteration $\deg(q_i(Y)) = 1$, i.e. the case where the maximal number of iterations $M_{\max}$ is executed. In the majority of the cases $M_{\max}$ iterations occur, i.e. six when $\deg(\Omega(Y)) = 2$ and four when $\deg(\Omega(Y)) = 0$. But with probability about $1/n$ in each iteration a larger degree of the quotient polynomial $q_i(Y)$ occurs, accordingly then $M < M_{\max}$. With the aim of assessing the reliability of the differences in running time allowing to identify the case $\deg(\Omega(Y)) = 0$, we examine whether $M < M_{\max}$ might lead to timings for $\deg(\Omega(Y)) = 2$ as low as for $\deg(\Omega(Y)) = 0$. We immediately find that the fifth iteration, which is only executed in the case $\sigma_3 = 0$, features a much more complex multiplication $q_5(Y) b_4(Y)$ than all the other iterations.

The control flow for the second EEA invocation, i.e. the solving of the key equation, for the case $w = 4$ has been analyzed in [11], there it is shown that in the case of $\sigma_3 = 0$ the number of iterations N is zero, whereas in the case $\sigma_3 \neq 0$ it is one. In that work, a countermeasure is proposed that removes the possibility to exploit the according timing differences in the second EEA invocation. However, due the fact that, as shown above, timing differences reveal $\sigma_3 = 0$ already in the syndrome inversion EEA, the countermeasure proposed in [11] is insufficient.

Experimental results confirm that taken together, the timing differences emerging in both EEA applications, i.e. the syndrome inversion and the key equation solving, actually allow for reliable distinction of $\deg(\Omega(Y))$ being zero or non-zero, and thus the attacker is able to learn linear equations of the form $\sigma_3 = \sum_{i=1}^{4} \epsilon_i = 0$. Remember that through the choice of the error vector during encryption, he chooses the indexes j_i with $i = 1, \ldots, 4$ of the support elements $\alpha_{j_i} = \epsilon_i$ according to the definition of the ϵ_i notation for the support elements.

Table 1. Overview of the iterations in the syndrome inversion EEA for Hamming weight four error vectors. If $\deg(\Omega(Y)) = 2$, $M_{\max}$, the maximal number of iterations is six, otherwise, if $\deg(\Omega(Y)) = 0$, we have $M_{\max} = 4$.

i	$\deg(q_i(Y))$	$\deg(b_i(Y))$	$\deg(r_i(Y))$
1	1	1	t-1
2	1	2	t-2
3	1	3	t-3
4	1	4	**2 (or 0)**
5	t - 5	t - 1	1
6	1	t	0

3.3   Cubic Equations from $w = 6$ Error Vectors

The vulnerability found for $w = 4$ error vectors can be generalized to any even value of w. For the attack that is subject of this work, we also employ the case $w = 6$. There, we find that the syndrome polynomial according to Eq. (2) is of the form

$$S(Y) \equiv \frac{\Omega(Y)}{\sigma(Y)} \equiv \frac{\sigma_5 Y^4 \oplus \sigma_3 Y^2 \oplus \sigma_1}{Y^6 \oplus \sigma_5 Y^5 \oplus \sigma_4 Y^4 \oplus \sigma_3 Y^3 \oplus \sigma_2 Y^2 \oplus \sigma_1 Y + \sigma_0} \mod g(Y),$$

(4)

where

$$\sigma_3 = \sum_{j=3}^{6} \sum_{k=1}^{j-1} \sum_{l=1}^{k-1} \epsilon_j \epsilon_k \epsilon_l,$$

(5)

$$\sigma_5 = \sum_{i=1}^{6} \epsilon_i.$$

(6)

As in case of $w = 4$, $\deg(\Omega(Y)) = 0$ implies zero iterations in the key equation EEA. Furthermore, it is again the most complex iteration of the syndrome inversion EEA that is skipped if $\deg(\Omega(Y)) = 0$. The difference to $w = 4$ is that here two coefficients of $\sigma(Y)$, i.e σ_3 and σ_5, have to be zero for this to happen.

Thus from detecting $\deg(\Omega(Y)) = 0$ the attacker can learn the equations $\sigma_3 = 0$ and $\sigma_5 = 0$. However, since from the vulnerability presented in Sec. 3.2 it is already possible for the attacker to learn linear equations about the secret support, the value of the "$w = 6$" vulnerability lies in the equation $\sigma_3 = 0$, which can be learned through a timing side channel analogously to the case "$w = 4$".

3.4   The Zero Element of the Support from $w = 1$ Error Vectors

For $w = 1$ the whole control flow in Patterson's Algorithm is very simple and unambiguous on a high level: $S(Y) \equiv \frac{1}{Y \oplus \epsilon_1} \mod g(Y)$, $S^{-1}(Y) = Y \oplus \epsilon_1$,

---

Algorithm 4. Polynomial Division poly_div$(n(Y), d(Y))$

Require: the polynomials $n(Y), d(Y)$ with $\deg(n(Y)) \geq \deg(d(Y))$
Ensure: two polynomials $s(Y), q(Y)$ with $q(Y)d(Y) + s(Y) = n(Y)$ and $\deg(s(Y)) < \deg(d(Y))$
  1: $s_{-1}(Y) \leftarrow n(Y)$
  2: $s_0(Y) \leftarrow d(Y)$
  3: $q_0(Y) \leftarrow 0$
  4: $i \leftarrow 0$
  5: **while** $\deg(s_i(Y)) \geq \deg(d(Y))$ **do**
  6:     $i \leftarrow i + 1$
  7:     $a_i \leftarrow s_{i-2,\deg(s_{i-2}(Y))} / s_{i-1,\deg(s_{i-1}(Y))}$
  8:     $f_i \leftarrow \deg(s_{i-2}(Y)) - \deg(s_{i-1}(Y))$
  9:     $q_i(Y) = q_{i-1} + a_i Y^{f_i}$
  10:     $s_i \leftarrow s_{i-2}(Y) - a_i s_{r-1}(Y) Y^{f_i}$
  11: **end while**
  12: **return** $(q_i(Y), s_i(Y))$

$\tau(Y) = \sqrt{\epsilon_1}$, $a(Y) = \tau(Y)$, $b(Y) = 1$, $\sigma(Y) = Y \oplus \epsilon_1$. The polynomial inversion is, according to Theorem 1, performed in exactly one iteration. But there is an ambiguous control flow within the polynomial division given in Alg. 4, that is executed within this EEA iteration: We find $q_1(Y) = Y$ because there is no alternative to $\deg(S(Y)) = t - 1$. In Alg. 4, $s_{i,j}$ denotes the coefficient to Y^j in $s_i(Y)$. If $\epsilon_1 = 0$, then the division has to stop at this point. Otherwise, a second iteration is performed giving $q_2(Y) = Y \oplus \epsilon_1$. Thus, if the timing difference resulting from the different number of iterations in the division is detectable, the index of z of the secret support element $\alpha_z = 0$ can be found.

4 Combining the "$w = 1$", "$w = 4$", and "$w = 6$" Vulnerabilities to a practical Attack

In this section we explain the construction of a practical attack based on the vulnerabilities shown in Sections 3.2, 3.3 and 3.4.

4.1 Description of the Attack Procedure

Step 1. By performing the respective queries on the decryption device with "$w = 4$" error vectors, a rank $n - m - 1$ linear equation system is build. The experimental results from [11] already showed that this is the maximal rank that can be achieved from the linear equations. Afterwards, the index of the zero element, α_z is determined through the "$w = 1$" vulnerability. In the majority of the cases, this information increases the rank of the equation system to $n - m$. In the rare cases when the rank remains at $n - m - 1$, the attack's on-line and off-line complexity is increased by a factor of n.

In the following, we assume that we have an equation system of rank $n - m$. This is the highest possible rank for a homogeneous linear equation system describing a permutation of $\mathbb{F}_{2^m}$, since there must be m linearly independent basis elements. Accordingly, by bringing the linear equation system into reduced row echelon form, we find the elements associated with the m rightmost columns must be a basis $\{\beta_i\}$:

$$
\begin{array}{ccccccc|ccc}
\alpha_0 & \alpha_1 & \dots & \alpha_i & \dots & \alpha_{n-m-3} & \alpha_{n-m-2} & \beta_0 & \dots & \beta_{m-1} \\
\hline
1 & 0 & \dots & 0 & \dots & 0 & 0 & X & \dots & X \\
 & & & \vdots & & & & & & \\
0 & 0 & \dots & 1 & \dots & 0 & 0 & X & \dots & X \\
 & & & \vdots & & & & & & \\
0 & 0 & \dots & 0 & \dots & 0 & 1 & X & \dots & X \\
\end{array}
$$

Step 2. At this point for each element α_i we know the corresponding B_i with $\alpha_i = \sum_{j \in B_j} \beta_i$, i.e. its representation in the chosen basis. If the values of all basis elements β_i were known, then the values of all α_i would be set as well and the support was recovered. Accordingly, the next step in the attack is to collect cubic equations according to Eq. (5) in a way that allows for efficient guessing resp. solving for the values of the β_i. To this end, the first set C_1 of "$w = 6$" equations is created by the employment of error vectors involving error positions corresponding to ϵ_i, $i = 1, \dots, 6$, where the following conditions hold:

1. $\epsilon_i \in \text{span}(\{\beta_{s_1}, \beta_{g_1}, \beta_{g_2}, \beta_{g_3}\})$. These are four arbitrarily chosen basis elements, where β_{s_1} denotes the one to be solved for in the resulting equation according to Eq. (5). The reason for this initial set of basis elements having cardinality four is that this is the lowest cardinality allowing to satisfy all the conditions in the following items.
2. $\sum_{i=1}^{6} \epsilon_i = 0$. This qualifies the error vector for the possibility of $\deg(\Omega(Y)) = N = 0$ according to Eq. (6) in the sense that $\sigma_5 = 0$ is already ensured. As a result, in contrast to the case of random $w = 6$ error vectors that have a probability for $\deg(\Omega(Y)) = N = 0$ in the domain of $1/n^2$, for these candidates this probability is about $1/n$.
3. Exactly two of the ϵ_i contain β_{s_1}. The reason for this constraint is to keep the process of solving, the details of which we shall see shortly, as simple as possible. Specifically, the twofold occurrence of β_{s_i} leads to a quadratic equation for β_{s_i}.

Candidate error vectors e that meet these conditions are used to build ciphertexts which are input to the decryption device; and from the timing of the decryption, it is inferred whether actually $\deg(\Omega(Y)) = N = 0$ occurred. The number of such equations to be collected for one β_{s_i} is given by c_i, which is a parameter for the attack.

After c_1 equations are found for β_{s_1}, the second set of equations is build in the same way as the first, the only differences being that the basis element to be solved for now is $\beta_{s_2} \notin \{\beta_{s_1}, \beta_{g_1}, \beta_{g_2}, \beta_{g_3}\}$, and first condition becomes $\epsilon_i \in \text{span}(\{\beta_{s_1}, \beta_{g_1}, \beta_{g_2}, \beta_{g_3}, \beta_{s_2}\})$. In this manner successively sets of cubic equations for $m - 3$ different β_{s_i} are collected until the equations in the last set involve all β_i.

Step 3. In this step the solving resp. guessing is performed. Let those two ϵ_i that contain β_{s_i} according to the third condition in Step 4 always be ϵ_1 and ϵ_2. From the conditions given in Step 4, and Eq. (5) we have for each β_{s_i} a quadratic equation

$$a\beta_{s_i}^2 + b\beta_{s_i} + c = 0, \tag{7}$$

with $a = \sum_{j=3}^{6} \epsilon_i$, $b = (\epsilon_1 + \epsilon_2)a$, $c = (\epsilon_1 - \beta_{s_i})(\epsilon_2 - \beta_{s_i})a + (\epsilon_1 + \epsilon_2) \sum_{j=4}^{6} \sum_{k=3}^{j-1} \epsilon_j \epsilon_k$ $+ \sum_{j=5}^{6} \sum_{k=4}^{j-1} \sum_{l=3}^{k-1} \epsilon_j \epsilon_k \epsilon_l$ (for clarity, in these formulas we provide "+" and "−" even though both amount to "$\oplus$"). Such a quadratic equation has two solutions for β_{s_i}.

The solving is performed as follows: enumerate the initial guesses, i.e. all the possible combinations of the values for $\beta_{g_1}, \beta_{g_2}, \beta_{g_3}$. Here, and for the subsequent guesses, since we are looking for linearly independent $\mathbb{F}_{2^m}$ elements, it holds that

$$\beta_{g_i} \notin \mathrm{span}(\{\beta_{g_1}, \ldots, \beta_{g_{i-1}}\}), \tag{8}$$

where we imply the convention $\beta_{s_i} = \beta_{g_{i+3}}$.

For each such combination of values for $\beta_{g_1}, \beta_{g_2}, \beta_{g_3}$ the roots of each equation in C_1 are potential candidates for the value of β_{s_1}. However, additionally to the restriction from Eq. (8), those roots that are found only for a subset of C_1 are discarded. This is wherein the value of a choice $c_i = |C_i| > 1$ lies. The larger the c_i are chosen, the higher is the on-line effort of the attack (more cubic equations have to be collected), but the off-line effort is reduced as the number possible solutions for each β_{s_i} is decreased.

The remaining roots are iterated over to find the possible solutions for β_{s_2} by solving the equations in C_2, which in turn are used to compute the possible values of β_{s_3}, etc. Whenever in such a chain of guesses a solution for all β_{s_i} is found, a guess for the whole support $\Gamma = (\alpha_i | \alpha_i = \sum_{j \in B_i} \beta_j)$ is implied, which has to be checked by a means of key recovery, as described in [13].

4.2   Experimental Results

We conducted the attack with the following measurement setup on an Intel Core 2 Duo x86 platform: from the attack program, the decryption function was called with the attack ciphertexts as input, and the decryption time was measured with the CPU's cycle counter.

Because the cycle counts measured for a deterministic operation of the duration of a code-based decryption vary considerably on such CPUs, a specific strategy has to be used to identify positives, by which we refer to $\epsilon_1 = 0$ for $w = 1$ and $\deg(\Omega(Y)) = 0$ for $w = 4$ and $w = 6$, i.e. those cases that yield an equation for the attacker. Specifically, an approximate model for cycle counts on modern x86 CPUs like the Core2 Duo is a hypothetical constant cycle count associated with the operation which is increased by a random delay on every execution of an operation. Because in all three different attack types the positives, from the algorithmic point of view, are executed faster than the negatives, the following classification strategy can be used: Prior to the attack a training phase is carried out where the minimal cycle counts for positives are determined as well as the minimal cycle counts for negatives (using a different secret key than during the attack). Then the border below which an operation is classified as a positive during the attack is set as the mean of these two values. We refer to the distance between the minimal cycle counts for positives and the minimal cycle counts for negatives as the cycles gap. Clearly, a larger such gap increases the probability for finding positives. Furthermore, the above approximate model for the cycle counts on the employed CPU is lacking other effects that could be observed in our experiments: during the execution of the attack the previously determined maximal and minimal cycle counts for the two classes of operations seem to be subject to an "upwards drift", i.e. they tend to successively increase over time but sometimes also drop again approximately to the initial levels after some time.

Table 2 summarizes the results for single attack runs with different code parameters. The rows labeled "cycles gap ..." indicate the above discussed gaps. We found that gaps of a couple of hundreds cycles that are characteristic for the $w = 1$ vulnerability tend to cause problems in the detection of positives, i.e. in some runs due to the mentioned drift of the cycle counts the zero support element could not be determined, while the considerably larger gaps for the $w = 4$ and $w = 6$ vulnerabilities allow for reliable detection of positives.

The rows labeled "number of queries ..." show the number of decryption operations that had to be executed with ciphertexts created with error vectors of the respective weight in the course of a single run of the attack.

"number of final verifications" is the number of the guesses for the complete support that are output by the attack. We did not implement an actual verification, but simply compare the guess for the support with the correct support Γ. As already mentioned, in [13] the procedure that had to be used in a real life attack is described. It involves only some linear algebra operations on the public key and the invocation of an EEA and would not perceptibly increase the time for solving, given the small numbers of such final verifications occurring in the attacks.

The time for the solving step is given in the last row. From the theory, one expects an increase of the solving time by a factor of about eight for each increase of m by one. The reason is that the number of initial guesses, i.e. the number of combinations of values that can be chosen for β_{g_1}, β_{g_2} and β_{g_3} is roughly n^3, and all $\mathbb{F}_{2^m}$ operations, including the solving of the quadratic equations [22] Eq. (7), are done with the help of lookup tables, and thus execute in constant time.

The number of equations gathered per β_{s_i} were chosen as $C_1 = 1$, $C_2 = 2$, $C_3 = 4$, i.e. chosen as the double of the previous count, up to a maximal value of 16, i.e. $C_i = 16$ for $i \geq 5$.

As previously mentioned, in the rare cases where the knowledge about the zero-element of the support does not increase the rank of the equation system, Steps 2 and 3 would have to be repeated about up to n times, for these cases stronger hardware would be needed to keep the solving time in reasonable margins.

Table 2. Experimental results for single runs of the attack. Refer to the text for explanations

	$m = 9, t = 33$	$m = 10, t = 40$
cycles gap $w = 1$	≈ 400	≈ 600
cycles gap $w = 4$	$\approx 13,000$	$\approx 19,000$
cycles gap $w = 6$	$\approx 17,000$	$\approx 23,000$
number of queries for $w = 1$ (Step 1)	3,575,494	11,782,695
number of queries for $w = 4$ (Step 1)	1,517,253	2,869,424
number of queries for $w = 6$ (Step 2)	374,927	1,837,125
number of final verifications (Step 3)	$\approx 8,000$	$\approx 2,000$
running time for solving on 1 GHz x86 CPU (Step 3)	3h	28h

5   Conclusion

The results of this work show that timing attacks based on control flow vulnerabilities in the syndrome inversion and the key equation EEA are a threat to the confidentiality of the secret key. In the chosen measurement setup, the attack has been proved to be practical. Apart from the recovery of the zero-element of the support, the cycles gaps between the controls flows that have to be distinguished are rather large, and thus remote timing attacks seem feasible too. If the zero-element remains unknown, the on-line and off-line attack complexity can still be managed with appropriate hardware.

The question of countermeasures against this attack has not been explicitly addressed in this work, but two possibilities seem to suggest themselves: the first would be similar to the countermeasures given in [12], where "premature" abortion of the key equation solving EEA is prevented by enforcing the "missing" iterations. This however is a delicate undertaking, as even the smallest timing differences have to be prohibited and thus the complexity of the individual iterations must be accounted for (consider for instance the "$w = 1$ attacks" from Section 3.4).

The second option would be stronger in this respect: there, we alter the cryptosystem's parameter specification: during the encryption, only $t - 1$ errors are added, and prior to the standard decryption operation, another "bit flip error" is applied, the position of which should be the same for repeated decryptions of a certain ciphertext but otherwise appear as random, and thus should be pseudo-randomly derived from the ciphertext and a constant secret value (for instance a hash value of the secret key). This approach would guarantee a pervasive alteration of the decryption operation, however it demands an increase of security parameters to compensate for the lower error weight used during encryption.

References

1. McEliece, R.J.: A public key cryptosystem based on algebraic coding theory. DSN Progress Report 42-44, 114–116 (1978)
2. Niederreiter, H.: Knapsack-type cryptosystems and algebraic coding theory. Problems Control Inform. Theory 15(2), 159–166 (1986)

3. Bernstein, D.J., Buchmann, J., Dahmen, E.: Post Quantum Cryptography. Springer Publishing Company, Incorporated (2008)

4. Biswas, B., Sendrier, N.: McEliece Cryptosystem Implementation: Theory and Practice. In: Buchmann, J., Ding, J. (eds.) PQCrypto 2008. LNCS, vol. 5299, pp. 47–62. Springer, Heidelberg (2008)

5. Heyse, S.: Low-Reiter: Niederreiter Encryption Scheme for Embedded Microcontrollers. In: Sendrier, N. (ed.) PQCrypto 2010. LNCS, vol. 6061, pp. 165–181. Springer, Heidelberg (2010)

6. Eisenbarth, T., Güneysu, T., Heyse, S., Paar, C.: MicroEliece: McEliece for Embedded Devices. In: Clavier, C., Gaj, K. (eds.) CHES 2009. LNCS, vol. 5747, pp. 49–64. Springer, Heidelberg (2009)

7. Shoufan, A., Wink, T., Molter, G., Huss, S., Strenzke, F.: A Novel Processor Architecture for McEliece Cryptosystem and FPGA Platforms. In: Proceedings of the 2009 20th IEEE International Conference on Application-Specific Systems, Architectures and Processors, ASAP 2009, pp. 98–105. IEEE Computer Society, Washington, DC (2009)

8. Strenzke, F.: A Smart Card Implementation of the McEliece PKC. In: Samarati, P., Tunstall, M., Posegga, J., Markantonakis, K., Sauveron, D. (eds.) WISTP 2010. LNCS, vol. 6033, pp. 47–59. Springer, Heidelberg (2010)

9. Molter, H.G., Stöttinger, M., Shoufan, A., Strenzke, F.: A Simple Power Analysis Attack on a McEliece Cryptoprocessor. Journal of Cryptographic Engineering (2011)

10. Strenzke, F., Tews, E., Molter, H., Overbeck, R., Shoufan, A.: Side Channels in the McEliece PKC. In: Buchmann, J., Ding, J. (eds.) PQCrypto 2008. LNCS, vol. 5299, pp. 216–229. Springer, Heidelberg (2008)

11. Strenzke, F.: A Timing Attack against the Secret Permutation in the McEliece PKC. In: Sendrier, N. (ed.) PQCrypto 2010. LNCS, vol. 6061, pp. 95–107. Springer, Heidelberg (2010)

12. Shoufan, A., Strenzke, F., Molter, H., Stöttinger, M.: A Timing Attack against Patterson Algorithm in the McEliece PKC. In: Lee, D., Hong, S. (eds.) ICISC 2009. LNCS, vol. 5984, pp. 161–175. Springer, Heidelberg (2010)

13. Heyse, S., Moradi, A., Paar, C.: Practical power analysis attacks on software implementations of mceliece. In: Sendrier, N. (ed.) PQCrypto 2010. LNCS, vol. 6061, pp. 108–125. Springer, Heidelberg (2010)

14. Avanzi, R., Hoerder, S., Page, D., Tunstall, M.: Side-channel attacks on the mceliece and niederreiter public-key cryptosystems. J. Cryptographic Engineering 1(4), 271–281 (2011)

15. Kocher, P.C.: Timing Attacks on Implementations of Diffie-Hellman, RSA, DSS, and Other Systems. In: Koblitz, N. (ed.) CRYPTO 1996. LNCS, vol. 1109, pp. 104–113. Springer, Heidelberg (1996)

16. Kocher, P.C., Jaffe, J., Jun, B.: Differential Power Analysis. In: Wiener, M. (ed.) CRYPTO 1999. LNCS, vol. 1666, pp. 388–397. Springer, Heidelberg (1999)

17. Engelbert, D., Overbeck, R., Schmidt, A.: A Summary of McEliece-Type Cryptosystems and their Security. Journal of Mathematical Cryptology 1(2), 151–199 (2006)

18. Goppa, V.D.: A new class of linear correcting codes. Problems of Information Transmission 6, 207–212 (1970)

19. MacWilliams, F.J., Sloane, N.J.A.: The theory of error correcting codes. North Holland (1997)
20. Patterson, N.: Algebraic decoding of Goppa codes. IEEE Trans. Info. Theory 21, 203–207 (1975)
21. Sugiyama, Y., Kasahara, M., Hirasawa, S., Namekawa, T.: A method for solving key equation for decoding goppa codes. Information and Control 27(1), 87–99 (1975)
22. Biswas, B., Herbert, V.: Efficient Root Finding of Polynomials over Fields of Characteristic 2. In: WEWoRK (2009), hal.inria.fr/hal-00626997/PDF/tbz.pdf

Simple Matrix Scheme for Encryption

Chengdong Tao[1], Adama Diene[2], Shaohua Tang[3], and Jintai Ding[4],*

[1] South China University of Technology, China
chengdongtao2010@gmail.com
[2] Department of Math. Sciences, UAE University - Al-Ain, United Arab Emirates
adiene@uaeu.ac.ae
[3] South China University of Technology, China
csshtang@gmail.com
[4] University of Cincinnati, Ohio, USA and ChongQing University,China
jintai.ding@gmail.com

Abstract. There are several attempts to build asymmetric pubic key encryption schemes based on multivariate polynomials of degree two over a finite field. However, most of them are insecure. The common defect in many of them comes from the fact that certain quadratic forms associated with their central maps have low rank, which makes them vulnerable to the MinRank attack. We propose a new simple and efficient multivariate pubic key encryption scheme based on matrix multiplication, which does not have such a low rank property. The new scheme will be called Simple Matrix Scheme or ABC in short. We also propose some parameters for practical and secure implementation.

Keywords: Multivariate Public Key Cryptosystem, Simple Matrix Scheme, MinRank Attack.

1 Introduction

Public key cryptography plays an important role in secure communication. The most widely used nowadays are the number theoretical based cryptosystems such as RSA, DSA, and ECC. However, due to Shor's Algorithm, such cryptosystems would become insecure if a large Quantum computer is built. Recent progress made in this area makes this threat realer than ever before. Moreover, the computing capacity of these Number Theoretic based systems is proved to be limited. These are some reasons which motivate researchers to develop a new family of cryptosystems that can resist quantum computers attacks and that are more efficient in terms of computation. Researchers usually use Post Quantum Cryptography (PQC) to denote this new family.

Multivariate public key cryptosystems (MPKC) belong to the PQC family. If well designed, they can be a good candidate for PQC. The public key of an MPKC is a system of multivariate polynomials, usually quadratic, over a finite field. The security of MPKCs is based on the knowledge that solving a set of

---

* Corresponding author.

P. Gaborit (Ed.): PQCrypto 2013, LNCS 7932, pp. 231–242, 2013.

multivariate polynomial equations over a finite field, in general, is proven to be an NP-hard problem [9]. In fact quantum computers do not appear to have an advantage when dealing with NP-hard problems. However, this does not guarantee that these cryptosystems are secure. The first such practical system was proposed in 1988 by Matsumoto and Imai with their scheme called C* or MI. Nonetheless, Jacques Patarin proved it insecure using linearization equations attack a few years later [18].

In [5], the authors showed that the rank of the quadratic form associated to the central map of C* is only two and therefore the private key could be also recovered with the help of the MinRank Attack.

In [19] Patarin extended the C* scheme by using a new central map to construct a new encryption scheme called Hidden Field Equations (HFE). But Kipnis and Shamir found a way to recover the private keys using the MinRank Attack [13]. Furthermore, it is showed in [8] that inverting HFE is quasi-polynomial if the size of the field and the degree of the HFE polynomials are fixed.

In [15], T.T. Moh proposed a multivariate asymmetric encryption scheme called TTM.

But again, it was broken by exploiting the fact that some quadratic form associated to the central map is of low rank [3].

In the last two decades, many other MPKCs have been proposed for encryption but almost all of them are proven to be insecure and many of them share a common defect; that is some quadratic forms associated to their central maps have low rank and therefore are vulnerable to the MinRank Attack. In consequence, for a MPKC to be secure, it is necessary that all quadratic forms associated with the central map have a rank high enough.

This paper will propose a new multivariate public key scheme for encryption having the property that the quadratic forms associated to the central map do not have a low rank but a rank related to a certain parameter n. The scheme is constructed using some simple matrix multiplications and it will be called Simple Matrix encryption scheme or ABC in short.

This paper is organized as follows. In Section 2 we give an illustration of the MinRank attack using HFE. In Section 3, we describe the construction of the ABC scheme. The security analysis is presented in Section 4. Section 5 shows a practical implementation of the ABC scheme while Section 6 discusses the efficiency and Section 7 concludes the paper.

2   MinRank Attack

The MinRank attack is a cryptanalysis tool that can be used to recover the secret key of MPKCs whose quadratic form associated to the central map is of low rank. In this section, we give an illustration by describing the MinRank attack on the HFE scheme. The attack was first performed by Kipnis and Shamir [13] who showed that the security of HFE can be reduced to a MinRank problem.

2.1   The HFE Scheme

The HFE cryptosystem was proposed by Jacques Patarin in [19]. It can be described as follow. Let $q = p^e$, where p is a prime and $e \geq 1$. Let K be an extension of the finite field $k = \mathbb{F}_q$ of degree n. Clearly, $K \cong k^n$.

Let $\phi : K \to k^n$ be the k-linear isomorphism map between the finite field K and the n-dimensional vector space k^n. The central map of HFE is a univariate polynomial $P(x)$ of the following form

$$P(x) = \sum_{i=0}^{r-1}\sum_{j=0}^{r-1} p_{ij} x^{q^i + q^j} \in K[x],$$

where $p_{ij} \in K$ and r is a small constant chosen in a way such that $P(x)$ can efficiently inverted. The public key is given to be

$$\bar{F} = T \circ \phi \circ P \circ \phi^{-1} \circ S,$$

where $T : k^n \longrightarrow k^n$ and $S : k^n \longrightarrow k^n$ are two invertible linear transformations and the private key consist of T, P and S.

2.2   MinRank Attack on HFE

In [14], Kipnis and Shamir showed that the public key $\bar{F}$ and the transformations S, T, T^{-1} can be viewed as maps G^*, S^*, T^*, T^{*-1} over K. More precisely,

$$S^*(x) = \sum_{i=0}^{n-1} s_i x^{q^i}, \qquad T^{*-1}(x) = \sum_{i=0}^{n-1} t_i x^{q^i}.$$

and $G^*(x) = T^*(P(S^*(x)))$. We can express $G^*(x)$ in the form:

$$G^*(x) = \sum_{i=0}^{n-1}\sum_{j=0}^{n-1} g_{ij} x^{q^i + q^j} = \underline{x} G \underline{x}^t,$$

where $\underline{x} = (x, x^q, \ldots, x^{q^{n-1}})$ is a vector over K, $\underline{x}^t$ is the transposition of $\underline{x}$ and $G = [g_{ij}]$ is a matrix over K. The identity $T^{*-1}(G^*(x)) = P(S^*(x))$ implies that

$$G' = \sum_{i=0}^{n-1} t_k G^{*k} = WPW^t,$$

where $P = [p_{ij}]$ over K, G^{*k} and W are two matrices over K whose repective (i, j) entries are $g_{i-k,j-k}^{q^k}$ and $s_{i-j}^{q^i}$, with $i - k, j - k$ and $i - j$ computed modulo n. Since the rank of WPW^t is not more than r, recovering $t_0, t_1, \ldots, t_{n-1}$ can be reduced to solving a MinRank problem, that is, to find $t_0, t_1, \ldots, t_{n-1}$ such that

$$Rank(\sum_{i=0}^{n-1} t_k G^{*k}) \leq r.$$

Methods to solve the MinRank problem for small r can be found in [11]. Once the values $t_0, t_1, \ldots, t_{n-1}$ are found, T and S will be then easily computed. Therefore, the key point to attack HFE is to solve the MinRank problem.

The Kipnis-Shamir attack was improved by Courtois using a different method to solve the MinRank problem [3]. However, Ding et al. showed that the original Kipnis-Shamir attack and the improvement of Courtois are not valid in [4]. Later, Faugère et al. proposed a more comprehensive improvement of the Kipnis-Shamir attack against HFE [2].

3 Construction of ABC Cryptosystem

Let $n, m, s \in \mathbb{Z}$ be integers satisying $n = s^2$ and $m = 2n$. For a given integer s, let k^s denote the set of all s-tuples of elements of k. We denote the plaintext by $(x_1, x_2, \ldots, x_n) \in k^n$ and the ciphertext by $(y_1, y_2, \ldots, y_m) \in k^m$. The polynomial ring with n variables in k will be denoted by $k[x_1, \ldots, x_n]$. Let $\mathcal{L}_1 : k^n \to k^n$ and $\mathcal{L}_2 : k^m \to k^m$ be two linear transformations, i.e.

$$\mathcal{L}_1(x) = L_1 x \quad \text{and} \quad \mathcal{L}_2(y) = L_2 y,$$

where L_1 and L_2 are respectively an $n \times n$ matrix and an $m \times m$ matrix with entries in k, $x = (x_1, x_2, \ldots, x_n)^t$, $y = (y_1, y_2, \ldots, y_m)^t$, and t denote the matrix transposition.

The Central map Let

$$A = \begin{pmatrix} x_1 & x_2 & \cdots & x_s \\ x_{s+1} & x_{s+2} & \cdots & x_{2s} \\ \vdots & \vdots & \ddots & \vdots \\ x_{(s-1)s+1} & x_{(s-1)s+2} & \cdots & x_{s^2} \end{pmatrix}; \quad B = \begin{pmatrix} b_1 & b_2 & \cdots & b_s \\ b_{s+1} & b_{s+2} & \cdots & b_{2s} \\ \vdots & \vdots & \ddots & \vdots \\ b_{(s-1)s+1} & b_{(s-1)s+2} & \cdots & b_{s^2} \end{pmatrix};$$

and $C = \begin{pmatrix} c_1 & c_2 & \cdots & c_s \\ c_{s+1} & c_{s+2} & \cdots & c_{2s} \\ \vdots & \vdots & \ddots & \vdots \\ c_{(s-1)s+1} & c_{(s-1)s+2} & \cdots & c_{s^2} \end{pmatrix}$ be three $s \times s$ matrices, where $x_i \in$

k, b_i and c_i are randomly chosen as linear combination of elements from the set $\{x_1, \ldots, x_n\}$, where $i = 1, 2, \ldots, n$. Define $E_1 = AB$, $E_2 = AC$ and let $f_{(i-1)s+j}$ and $f_{s^2+(i-1)s+j} \in k[x_1, \ldots, x_n]$ be respectively the (i, j) element of E_1 and E_2 ($i, j = 1, 2, \ldots, s$). Then we obtain with this notation m polynomials $f_1, f_2, \ldots, f_m$, and we define the central map to be

$$\mathcal{F}(x_1, \ldots, x_n) = (f_1(x_1, x_2, \ldots, x_n), \ldots, f_m(x_1, x_2, \ldots, x_n)).$$

We note that for any $1 \le i \le m$, the rank of the quadratic form f_i which is associated with the central map $\mathcal{F}$ is close to or equal to $2s$. Define

$$\bar{\mathcal{F}} = \mathcal{L}_2 \circ \mathcal{F} \circ \mathcal{L}_1 = (\bar{f}_1, \bar{f}_2, \ldots, \bar{f}_m),$$

where $\mathcal{L}_1 : k^n \to k^n$ and $\mathcal{L}_2 : k^m \to k^m$ are as above, $\bar{f}_i \in k[x_1, \ldots, x_n]$ are m multivariate polynomials of degree two. The secret key and the public key are given by:

Secret Key   The secret key is made of the following two parts:

1) The invertible linear transformations $\mathcal{L}_1, \mathcal{L}_2$.
2) The coefficients of x_i of the elements in matrices B, C.

Public Key   The public key is made of the following two parts:

1) The field k, including the additive and multiplicative structure;
2) The maps $\bar{\mathcal{F}}$ or equivalently, its m total degree two components

$$\bar{f}_1(x_1, x_2, \ldots, x_n), \ldots, \bar{f}_m(x_1, x_2, \ldots, x_n) \in k[x_1, \ldots, x_n].$$

Encryption
Given a message $x_1, x_2, \ldots, x_n$, the corresponding ciphertext is

$$(y_1, y_2, \ldots, y_m) = \bar{\mathcal{F}}(x_1, x_2, \ldots, x_n).$$

Decryption
To decrypt the ciphertext $(y_1, y_2, \ldots, y_m)$, one need to perform the following steps:

1 Compute $(\bar{y}_1, \bar{y}_2, \ldots, \bar{y}_m) = \mathcal{L}_2^{-1}(y_1, y_2, \ldots, y_m)$.
2 Put

$$E_1 = \begin{pmatrix} \bar{y}_1 & \bar{y}_2 & \cdots & \bar{y}_s \\ \bar{y}_{s+1} & \bar{y}_{s+2} & \cdots & \bar{y}_{2s} \\ \vdots & \vdots & \ddots & \vdots \\ \bar{y}_{(s-1)s+1} & \bar{y}_{(s-1)s+2} & \cdots & \bar{y}_{s^2} \end{pmatrix};$$

$$E_2 = \begin{pmatrix} \bar{y}_{s^2+1} & \bar{y}_{s^2+2} & \cdots & \bar{y}_{s^2+s} \\ \bar{y}_{s^2+s+1} & \bar{y}_{s^2+s+2} & \cdots & \bar{y}_{s^2+2s} \\ \vdots & \vdots & \ddots & \vdots \\ \bar{y}_{s^2+(s-1)s+1} & \bar{y}_{s^2+(s-1)s+2} & \cdots & \bar{y}_{2s^2} \end{pmatrix}.$$

Since $E_1 = AB, E_2 = AC$, we consider the following cases:

(i) If E_1 is invertible, then $BE_1^{-1}E_2 = C$. We have n linear equations with n unknowns $x_1, \ldots, x_n$.

(ii) If E_2 is invertible, but E_1 is not invertible, then $CE_2^{-1}E_1 = B$. We also have n linear equations with n unknowns $x_1, \ldots, x_n$.

(iii) If both E_1 and E_2 are not invertible but A is invertible, then $A^{-1}E_1 = B$, $A^{-1}E_2 = C$. We interpret the elements of A^{-1} as the new variables W_i and we end up with $m = 2n$ linear equations in m unknowns. Then we eliminate the new variables to derive n linear equations in the x_i.

(iv) If A is a singular matrix and the rank of A is $n - r$, then there exits a nonsingular matrix W such that $WA = \begin{pmatrix} I & 0 \\ 0 & 0 \end{pmatrix}$, where I is a $(n - r) \times (n - r)$ identity matrix, 0 is a zero matrix. Let $W = \begin{pmatrix} W_1 & W_2 \\ W_3 & W_4 \end{pmatrix}$, $B = \begin{pmatrix} B_1 & B_2 \\ B_3 & B_4 \end{pmatrix}$, $C = \begin{pmatrix} C_1 & C_2 \\ C_3 & C_4 \end{pmatrix}$, $E_1 = \begin{pmatrix} E_{11} & E_{12} \\ E_{13} & E_{14} \end{pmatrix}$, $E_2 = \begin{pmatrix} E_{21} & E_{22} \\ E_{23} & E_{24} \end{pmatrix}$, where $W_1, B_1, C_1, E_{11}, E_{21}$ are a $(n-r) \times (n-r)$ matrices. Since $WE_1 =$

$WAB, WE_2 = WAC$, that is $W_1E_{11} + W_2E_{13} = B_1$, $W_1E_{12} + W_2E_{14} = B_2$, $W_1E_{21} + W_2E_{23} = C_1$, $W_1E_{22} + W_2E_{24} = C_2$.

We interpret the elements of W_1, W_2 as the new variables and we end up with $2s(s - r)$ linear equations in $s(s - r) + n$ unknowns. Then we eliminate the $s(s - r)$ elements of W_1, W_2 in these equations. If these $2s(s-r)$ linear equations are independent, we gain $n - sr$ linear equations with the variables $x_1, x_2, ..., x_n$.

The dimension of the solution space of the linear equations with the variables $x_1, x_2, ..., x_n$ is in general very small. Solving this system by Gaussian elimination enables us to eliminate most of the unknowns, say Z of them. Then we write these Z variables as linear combinations of the remaining unknown variables and then substitute them into the central equations. We then obtain a new system of equations of degree two in the remaining $n - Z$ unknowns which can be easily solved since the number of variables of this new system of equations is very small. Sometimes we may have more than one solution, but the probability is very small.

3 Compute the plaintext $(x_1, x_2, \ldots, x_n) = \mathcal{L}_1^{-1}(\tilde{x}_1, \tilde{x}_2, \ldots, \tilde{x}_n)$.

Our experiments show that even if A is a singular matrix, decryption remains successful as long as the rank of A is no less than $s - 2$. When the rank of A is less than $s - 2$, decryption may fail. Let $r > 0$ be the rank of A, then the number of $s \times s$ matrix of rank r over k is $\dfrac{q^{r(r-1)/2} \prod\limits_{i=s-r+1}^{s} (q^i - 1)^2}{\prod\limits_{i=1}^{r} (q^i - 1)}$, thus for any $s \times s$

matrix A, the probability of A of rank r is $\dfrac{q^{r(r-1)/2} \prod\limits_{i=s-r+1}^{s} (q^i - 1)^2}{q^{s^2} \prod\limits_{i=1}^{r} (q^i - 1)}$. Therefore,

the probability of A of rank less than r is $1 - \sum\limits_{j=r}^{s} \dfrac{q^{j(j-1)/2} \prod\limits_{i=s-j+1}^{s} (q^i - 1)^2}{q^{s^2} \prod\limits_{i=1}^{j} (q^i - 1)}$. For

example, let $q = 2^8, s = 8$, then the probability of A of rank less than 6 is about 2.125919×10^{-22}, thus, in this case, the probability of decryption failure is about 2.125919×10^{-22}. This means that we can adjust the parameters to make sure that decryption will not be a problem.

4   Security Analysis

In this section, we will study the security of the ABC scheme in order to able us to choose the appropriate parameters for a secure encryption.

4.1   High Order Linearization Equation Attack

Linearization equation attack was first discussed in [18] to attack MI [16]. Later, high order linerlization equation attack was proposed to attack MFE cryptosystem [6]. We use this method to attack our scheme. Since $BE_1^{-1}E_2 = C$ (the case where $CE_2^{-1}E_1 = B$ is similar), there exists polynomial g_1, with $deg(g_1) \leq s$, such that $Bg_1(E_1)E_2 = Cdet(E_1)$. Therefore, the plaintext and the ciphertext satisfy the equation:

$$\sum_{i_0=1}^{n} \sum_{i_1,\ldots,i_s=1}^{m} \mu_{i_0,i_1,\ldots,i_s} x_{i_0} y_{i_1} \cdots y_{i_s} +$$

$$+ \sum_{i_0=1}^{n} \sum_{i_1,\cdots,i_{s-1}=1}^{m} \nu_{i_0,i_1,\ldots,i_{s-1}} x_{i_0} y_{i_1} \cdots y_{i_{s-1}} + \cdots +$$

$$+ \sum_{i_0=1}^{n} \gamma_{i_0} x_{i_0} + \sum_{i_1=1}^{m} \xi_{i_1} y_{i_1} + \theta = 0,$$

which means that we derive linearization equations with order $n+1$. The coefficients $\mu_{i_0,i_1,\ldots,i_s}, \nu_{i_0,i_1,\cdots,i_{s-1}}, \ldots, \gamma_{i_0}, \xi_{i_1}, \theta$ are variables taking value in k. The number of variables is

$$n \sum_{j=0}^{s} \binom{m}{j} + m + 1 = n\binom{m+s}{s} + m + 1.$$

Using the public key we can generate many plaintext-ciphertxet pairs. By substituting these plaintext-ciphertxet pairs into the equations, we have $n\binom{m+s}{s}+m+1$ linear equations with $n\binom{m+s}{s} + m + 1$ variables. However, the computation complexity of solving this linearization equation is $\left(n\binom{m+s}{s} + m + 1\right)^{\omega}$, where $\omega = 3$ in the usual Gaussian elimination algorithm and $\omega = 2.3766$ in improved algorithm which is impractical for a bit size greater than or equal to 64. Note here that the computation complexity is even high in the case where E_1 and E_2 are not invertible.

4.2   Rank Attack

There are two different methods of using the rank attack. The first one is called MinRank attack or Low Rank attack and an illutration was discussed in section 2. The other one is called the High Rank Attack. We will look at these two attacks against the ABC scheme. For the MinRank attack, let us assume without lost of generality that the public key polynomials and the secret polynomials are homogeneous quadratic polynomials. Let $\mathcal{L}_1, \mathcal{L}_2$ be two invertible linear transformations. Let $\bar{Q}_1, \bar{Q}_2, \ldots, \bar{Q}_m$ be the symmetric matrices associated with the public key quadratic polynomials and $Q_1, Q_2, \ldots, Q_m$ be the symmetric matrices associate with the secret key quadratic polynomials. Clearly, the rank of Q_i is bounded by $2s$. With the MinRank attack, one tries to find $(t_1, t_2, \ldots, t_m) \in k^m$ such that the rank of the linear combinations $\sum_{i=1}^{m} t_i \bar{Q}_i$ is no more than $2s$. In order to find such a linear combination, one can choose any vector $v \in k^n$ and try to solve the equations $(\sum_{i=1}^{m} t_i \bar{Q}_i)v = 0$ with the unknowns $t_1, \ldots, t_m$. After finding at least one linear combination of this form, attacker can recover $\mathcal{L}_2$. The attacker can recover $\mathcal{L}_1$ and $Q_1, \ldots, Q_m$ when $\mathcal{L}_2$ is known. More detail about the MinRank attack can be found in [3,10]. The complexity of this attack against the ABC scheme is $O(q^{\lceil \frac{m}{n} \rceil 2s} m^3)$.

For the High Rank Attack, we form an arbitrary linear combinations $Q = \sum_{i=1}^{m} \alpha_i \bar{Q}_i$, then we find $V = Ker(Q)$. If Q have a nontrivial kernel, set $\sum_{i=1}^{m} \lambda_i \bar{Q}_i V = 0$ and check if the solution set $\hat{V}$ of λ_i has a dimension $n - 2s$. This attack uses about $O(n^6 q^{2s})$ field multiplications. Moreover, note that for every vector v of dimension n, there exists a linear combination of the 2^n secret polynomials that yields zero with probability roughly $1 - \frac{1}{q^n}$. So we are faced with a lot of parasitic solutions, which have to be ruled out at the end. Also as it was mentioned earlier the rank of the Q_i is associated with $2\sqrt{n}$ which means that the complexity of the rank attack may not be polynomial time in the number of variables. These facts prove that the Rank attack is really inefficient against the ABC scheme.

4.3 Algebraic Attack

Let $\bar{f}_1(x_1, \ldots, x_n), \ldots, \bar{f}_m(x_1, \ldots, x_n) \in k[x_1, \ldots, x_n]$ be the public key polynomials. Let $y_1, y_2, \ldots, y_m$ be the ciphertext. We try to solve the system of equations

$$\begin{cases} \bar{f}_1(x_1, x_2, \ldots, x_n) = y_1; \\ \bar{f}_2(x_1, x_2, \ldots, x_n) = y_2; \\ \cdots\cdots\cdots\cdots\cdots \\ \bar{f}_m(x_1, x_2, \ldots, x_n) = y_m, \end{cases}$$

directly by Gröbner bases or XL method and its variations Mutant XL algorithm[25][26][27].

We carried out a number of experiments with MAGMA [1], which contains an efficient implementation of F4 algorithm [9] for computing Gröbner bases. Table 1 shows the results of our experiments to attack an instance of ABC scheme in a finite field k of 3 elements.

Table 1. Result of experiments with direct attack using MAGMA(2.12-16) on a 1.80GHz Intel(R) Atom(TM) CPU

n	9	16	25
time(s)	0.016	3.494	17588.380
memory(MB)	3.4	8.1	1111.7
degree of regularity	4	5	6

As the table 1 shows, the time and memory complexity increase as n grows. Also the degree of regularity increases as n grows which indicated that complexity is exponential.

4.4   Special Attacks

In terms of the design, one may think that maybe we can choose B and C such that their entries are randomly selected sparse linear functions or even monomials, which will allow us to have smaller secret key. However in the case of using only monomials, there is a possible new risk, namely there is a possibility that the central map polynomials are so sparse that they may have hidden UOV structures, that is there are no quadratic terms of a set of variables in the central map polynomials. One may then use UOV Reconciliation attack to find such structure [23][24]. It is not a good ideal to use monomials for B and C, such a distinguished feature is in general not desired. But in the case of general B and C such a feature does not exist. It is an open interesting problem to find out what really happens in the case of sparse B and C.

On the other hand, one may say that how about making A also more general, namely entries are selected as random linear functions. It is clear this is not needed since a linear transformation will easily remove such a feature. Using a matrix A of variables and L_1 is equivalent to using a matrix A of linear functions, without any transformation L_1. In the case of A also more general, one may consider certain tensor related attack, but we cannot see yet any effective way to do so.

5   A Practical Implementation for Encryption

For a practical implementation, we let k be the finite field of $q = 2^8$ elements and $n = 64$. In this case, the plaintext consist of the message $(x_1, \ldots, x_{64}) \in k^{64}$. The public map is $\bar{F} : k^{64} \to k^{128}$ and the central map is $F : k^{64} \to k^{128}$.

The public key consists of 128 quadratic polynomials with 64 variables. The number of coefficients for the public key polynomials is

$$128 \times 66 \times 65/2 \in \{274560, \text{or about } 280KB \text{ of storage}\}.$$

The private key consists of the coefficients of the x_i of the entries of the matrices B and C. and the two linear transformations $\mathcal{L}_1, \mathcal{L}_2$. The total size is about $30KB$.

The size of a document is $8n = 8 \times 64 = 512bits$ and the total size of the ciphertext is $1024bits$.

Based on the preceding discussion in section 4, security level for this implementation is lager than 2^{86}. Using odd characteristic field may be good to resist algebraic attack, but it requires more storage.

6   Efficiency of ABC Scheme

In this section, we will compare the efficiency of decryption in ABC scheme with HFE challenge 1 by Patarin [19]. This HFE was broken using algebraic attack [13]. In this HFE scheme, J.Patarin chose the parameters as follow: $q = 2, n = 80$, the degree of central map is 96. Let $P(x)$ be the central map of HFE, the main computation of decryption is to solve the equation $P(x) = y$ over the finite field of 2^{80} elements. In [20], J.Patarin estimated that the complexity of solving this equation is about $O(d^2n^3)$ or $O(dn^3+d^3n^2)$–depending on the chosen algorithms, where d is the degree of $P(x)$. Thus the decryption process needs about 6.4×10^9 times field multiplication over the finite field of 2^{80} elements.

For the proposed parameters of the ABC scheme above, $q = 2^8, n = 64$ and $m = 128$, the steps of decryption were presented in section 3. The computation of step 1) and step3) of decryption are very fast. The main computation of decryption is step 2), solving a set of linear equations. Therefore, we only need about $128^3 = 2^{21} \approx 2.1 \times 10^6$ times field multiplications over the finite field of 2^8 elements for decryption. It is much faster than HFE scheme.

7   Conclusion

In this paper, we propose a new multivariate algorithm for encryption called ABC. A highlight of ABC scheme is that all the quadratic forms associated with the central map are not of low rank but related to some variable integer n. Therefore, it is immune to the MinRank Attack. Another highlight of ABC scheme is that the computation of decryption is very fast, because the main computation is to solve certain linear equations. However we still cannot show that ABC is provably secure.

Acknowledgment. The authors are grateful to the referees for constructive suggestion on special attack and the design which helped to improve the paper. The work was partially supported by **Charles Phelps Taft Foundation** and the NSF of China under the grant #60973131 and the National Natural Science Foundation of China under grant No. U1135004.

References

1. Bosma, W., Cannon, J.J., Playoust, C.: The Magma algebra system I: the user language. J. Symb. Comput. 24(3-4), 235–265 (1997)
2. Bettale, L., Faugère, J.-C., Perret, L.: Cryptanalysis of multivariate and odd-characteristic HFE variants. In: Catalano, D., Fazio, N., Gennaro, R., Nicolosi, A. (eds.) PKC 2011. LNCS, vol. 6571, pp. 441–458. Springer, Heidelberg (2011)
3. Goubin, L., Courtois, N.T.: Cryptanalysis of the TTM cryptosystem. In: Okamoto, T. (ed.) ASIACRYPT 2000. LNCS, vol. 1976, pp. 44–57. Springer, Heidelberg (2000)
4. Ding, J., Schmidt, D., Werner, F.: Algebraic attack on HFE revisited. In: Wu, T.-C., Lei, C.-L., Rijmen, V., Lee, D.-T. (eds.) ISC 2008. LNCS, vol. 5222, pp. 215–227. Springer, Heidelberg (2008)
5. Ding, J., Gower, J., Schmidt, D.: Multivariate Public Key Cryptography. Advances in Information Security series. Springer, Heidelberg (2006)

6. Ding, J., Yang, B.-Y., Chen, C.-H.O., Chen, M.-S., Cheng, C.-M.: New Differential-Algebraic Attacks and Reparametrization of Rainbow. In: Bellovin, S.M., Gennaro, R., Keromytis, A.D., Yung, M. (eds.) ACNS 2008. LNCS, vol. 5037, pp. 242–257. Springer, Heidelberg (2008)

7. Ding, J., Hu, L., Nie, X., Li, J., Wagner, J.: High Order Linearization Equation (HOLE) Attack on Multivariate Public Key Cryptosystems. In: Okamoto, T., Wang, X. (eds.) PKC 2007. LNCS, vol. 4450, pp. 233–248. Springer, Heidelberg (2007)

8. Ding, J., Hodges, T.J.: Inverting HFE systems is quasi-polynomial for all fields. In: Rogaway, P. (ed.) CRYPTO 2011. LNCS, vol. 6841, pp. 724–742. Springer, Heidelberg (2011)

9. Faugère, J.C.: A new efficient algorithm for computing Gröbner bases (F4). J. Pure Appl. Algebra 139, 61–88 (1999)

10. Faugère, J.-C., Levy-dit-Vehel, F., Perret, L.: Cryptanalysis of minRank. In: Wagner, D. (ed.) CRYPTO 2008. LNCS, vol. 5157, pp. 280–296. Springer, Heidelberg (2008)

11. Kipnis, A., Shamir, A.: Cryptanalysis of the Oil & Vinegar Signature Scheme. In: Krawczyk, H. (ed.) CRYPTO 1998. LNCS, vol. 1462, pp. 257–267. Springer, Heidelberg (1998)

12. Kipnis, A., Patarin, J., Goubin, L.: Unbalanced Oil and Vinegar Signature Schemes. In: Stern, J. (ed.) EUROCRYPT 1999. LNCS, vol. 1592, pp. 206–222. Springer, Heidelberg (1999)

13. Kipnis, A., Shamir, A.: Cryptanalysis of the HFE public key cryptosystem by relinearization. In: Wiener, M. (ed.) CRYPTO 1999. LNCS, vol. 1666, pp. 19–30. Springer, Heidelberg (1999)

14. Lidl, R., Niederreiter, H.: Finite Fields. Encyclopedia of Mathematics and its applications, vol. 20. Cambridge University Press

15. Moh, T.T.: A fast public key system with signature and master key functions. In: Proceedings of CrypTEC 1999, International Workshop on Cryptographic Techniques and E-Commerce, pp. 63–69. Hong-Kong City University Press (July 1999), http://www.usdsi.com/cryptec.ps

16. Matsumoto, T., Imai, H.: Public quadratic polynomial-tuples for efficient signature-verification and message-encryption. In: Günther, C.G. (ed.) EUROCRYPT 1988. LNCS, vol. 330, pp. 419–453. Springer, Heidelberg (1988)

17. Patarin, J.: The Oil and Vinegar Signature Scheme. Presented at the Dagstuhl Workshop on Cryptography (September 1997) (transparencies)

18. Patarin, J.: Cryptoanalysis of the Matsumoto and Imai public key scheme of Eurocrypt'88. In: Coppersmith, D. (ed.) CRYPTO 1995. LNCS, vol. 963, pp. 248–261. Springer, Heidelberg (1995)

19. Patarin, J.: Hidden fields equations (HFE) and isomorphisms of polynomials (IP): Two new families of asymmetric algorithms. In: Maurer, U.M. (ed.) EUROCRYPT 1996. LNCS, vol. 1070, pp. 33–48. Springer, Heidelberg (1996)

20. Rivest, R., Shamir, A., Adleman, L.M.: A method for obtaining digital signatures and public-key cryptosystems. Communications of the ACM 21(2), 120–126

21. Shor, P.: Polynomial-time algorithms for prime factorization and discrete logarithms on a quantum computer. SIAM Journal on Computing 26(5), 1484–1509 (1997)

22. Wang, L.-C., Yang, B.-Y., Hu, Y.-H., Lai, F.: A "Medium-Field" Multivariate Public-Key Encryption Scheme. In: Pointcheval, D. (ed.) CT-RSA 2006. LNCS, vol. 3860, pp. 132–149. Springer, Heidelberg (2006)

23. Ding, J., Yang, B.-Y., Chen, C.-H.O., Chen, M.-S., Cheng, C.-M.: New differential-algebraic attacks and reparametrization of rainbow. In: Bellovin, S.M., Gennaro, R., Keromytis, A.D., Yung, M., et al. (eds.) ACNS 2008. LNCS, vol. 5037, pp. 242–257. Springer, Heidelberg (2008)
24. Thomae, E.: A Generalization of the Rainbow Band Separation Attack and its Applications to Multivariate Schemes. IACR Cryptology ePrint Archive (2012)
25. Buchmann, J.A., Ding, J., Mohamed, M.S.E., et al.: MutantXL: Solving multivariate polynomial equations for cryptanalysis. Symmetric Cryptography, 09031 (2009)
26. Mohamed, M.S.E., Mohamed, W.S.A.E., Ding, J., Buchmann, J.: *MXL2*: Solving polynomial equations over GF(2) using an improved mutant strategy. In: Buchmann, J., Ding, J., et al. (eds.) PQCrypto 2008. LNCS, vol. 5299, pp. 203–215. Springer, Heidelberg (2008)
27. Mohamed, M.S.E., Cabarcas, D., Ding, J., Buchmann, J., Bulygin, S.: MXL3: An efficient algorithm for computing gröbner bases of zero-dimensional ideals. In: Lee, D., Hong, S., et al. (eds.) ICISC 2009. LNCS, vol. 5984, pp. 87–100. Springer, Heidelberg (2010)

Multivariate Signature Scheme
Using Quadratic Forms

Takanori Yasuda[1], Tsuyoshi Takagi[2], and Kouichi Sakurai[1,3]

[1] Institute of Systems, Information Technologies and Nanotechnologies
[2] Institute of Mathematics for Industry, Kyushu University
[3] Department of Informatics, Kyushu University

Abstract. Multivariate Public Key Cryptosystems (MPKC) are candidates for post-quantum cryptography. MPKC has an advantage in that its encryption and decryption are relatively efficient. In this paper, we propose a multivariate signature scheme using quadratic forms. For a finite dimensional vector space V, it is known that there are exactly two equivalence classes of non-degenerate quadratic forms over V. We utilize the method to transform any non-degenerate quadratic form into the normal form of either of the two equivalence classes in order to construct a new signature scheme in MPKC. The signature generation of our scheme is between eight and nine times more efficient more than the multivariate signature scheme Rainbow at the level of 88-bit security. We show that the public keys of our scheme can not be represented by the public keys of other MPKC signature schemes and this means our scheme is immune to many attacks that depend on the form of the central map used by these schemes.

Keywords: Multivariate Public Key Cryptosystem, Digital signature, Rainbow, Post-quantum cryptography.

1 Introduction

Multivariate Public Key Cryptosystems (MPKC) [9] can be potentially applied to post-quantum cryptography. MPKC can be used for encryption and digital signature, and its encryption and decryption processes (and signature generation and verification) are relatively efficient in comparison with RSA and elliptic curve cryptography[8]. The security of MPKC depends on the difficulty of solving a system of multivariate polynomials that form its secret key and public key, and the security of MPKC depends on the difficulty in solving a system of multivariate polynomials. At present, the most efficient way to solve a system of multivariate polynomials is to compute the Gröbner basis. The attacks against this method are called direct attacks, and they are applicable against any MPKC scheme. For UOV[18] and Rainbow[10] signature scheme, the direct attacks determine their security level.

In this paper, we propose a new signature scheme. It is known that there are two isometry classes of non-degenerate quadratic forms on a vector space with

P. Gaborit (Ed.): PQCrypto 2013, LNCS 7932, pp. 243–258, 2013.

a prescribed dimension[32]. We use a computational method whereby any non-degenerate quadratic form is transformed into either of the canonical forms of two classes of the signature generation of our scheme. We estimate the efficiency of the signature generation in terms of the number of multiplications of base field. The signature generation scheme consists of two affine transformations and the inverse computation of the central map. The inverse computation of the central map of our scheme is cheaper than the two affine transformations. We compare of the efficiency of our signature generation with that of Rainbow. We choose the parameters of both schemes under the assumption that the security level of these schemes against direct attacks are same, As a result, we find that the signature generation of our scheme is between eight and nine times more efficient than Rainbow at the level of 88-bit security.

A lot of MPKC signature schemes have been proposed: however, not much is known about relations between the different schemes (*ref.* [20]). For example, it is still an open problem as to whether the public key of the Matsumoto-Imai scheme can be expressed as a public key of Rainbow. Our scheme uses two systems of multivariate polynomials. These systems have a property whereby the regions of their values are exclusive. In particular, the two system are not surjective. On the other hand, schemes that have already proposed (e.g. UOV and Rainbow) use only one system of multivariate polynomials. Moreover, their system is surjective. Accordingly, the public key of our scheme is not able to be expressed in terms of the public keys previously proposed schemes. As far as we know, this is the first report of a public key of a scheme that can not be expressed by using the public keys of other schemes.

We can explain the importance of the public key of a scheme not being expressed by the keys of other schemes by using an example. For Rainbow, UOV attack, UOV-Reconciliation attack, Rainbow-band-separation attack, etc., have been proposed against Rainbow. These attacks all transform the public key into a central map of Rainbow. If the public key can be transformed into the central map, the signature can be forged using the same method as the signature generation of Rainbow. However, it can be proved that these attacks can not be applied to our scheme. The public key of our scheme can not be transformed into the central maps of Rainbow. Of course, attacks that are independent of the signature scheme like as direct attacks can be launched against our scheme. In addition, there is a possibility that there is an attack which works well against our scheme. In fact, MinRank attack and a method for solving Isomorphisms of Polynomials can be applied to our scheme. The analysis of these attacks will be tackled more elaborately in the future.

2   Construction of Signature Scheme in MPKC

A lot of MPKC signature schemes e.g., UOV[18] and Rainbow[10], have been proposed.

The security of MPKC is based on the difficulty of solving the *MQ problem*. An MQ problem is to find a solution of the following system of quadratic polynomials with n variables and m polynomials.

$$\begin{cases} a_{11}^{(1)} x_1^2 + a_{12}^{(1)} x_1 x_2 + \ldots + c^{(1)} = 0, \\ a_{11}^{(2)} x_1^2 + a_{12}^{(2)} x_1 x_2 + \ldots + c^{(2)} = 0, \\ \quad \vdots \\ a_{11}^{(m)} x_1^2 + a_{12}^{(m)} x_1 x_2 + \ldots + c^{(m)} = 0. \end{cases}$$

Here, the coefficients $a_{ij}^{(k)}$ belong to a finite field K. For large m and n and if the coefficients are chosen randomly, the MQ problem is considered to be NP-hard[13].

MPKC aims to design secure encryption and signature schemes by using a system of quadratic multivariate polynomials as the public key. To design a MPKC scheme, we start by constructing a secret key from a system of multivariate polynomials which is easy to solve, and next, we transform the secret key into public key by using affine transformations. Note that not every system of multivariate polynomials can be used in MPKC. More concretely, the secret and public keys are constructed as follows.

Secret Key: A system g of multivariate polynomials with n variables and m polynomials satisfying the following condition, two affine transformations $L : K^m \to K^m$, $R : K^n \to K^n$.

    Condition: For any $\mathbf{c} \in K^m$, we can efficiently compute $\mathbf{x} \in K^n$ such that $g(\mathbf{x}) = \mathbf{c}$.

Public Key: A system of multivariate polynomials defined by $f = L \circ g \circ R$.

g appearing in the secret key is regarded as a map $K^n \to K^m$. This map is called the *central map* of this scheme. The signature generation and verification are as follows:

Signature Generation: Let $\mathbf{M} \in K^m$ be a message. Compute $\mathbf{A} = L^{-1}(\mathbf{M})$, $\mathbf{B} = g^{-1}(\mathbf{B})$ and $\mathbf{C} = R^{-1}(\mathbf{B})$ in this order. $\mathbf{C}$ is a signature

Verification: If $F(\mathbf{C}) = \mathbf{M}$, the signature is accepted: it is rejected otherwise.

It is natural to choose a surjective map as a central map of a signature scheme because for any possible message, the corresponding signature must be generated. In fact, all of the previously proposed signature schemes use surjective map.

3 Quadratic Forms

In this section, we summarize the fundamental facts about quadratic forms that are necessary for our scheme. The details of quadratic forms and their properties is refered to [32].

3.1   Definition and Facts

Let K be a finite field with odd order q. Let V be an r-dimensional vector space over K.

Definition 1. *A map $q : V \to K$ is said to be a quadratic form if the following is satisfied:*

1. $q(ax) = a^2 q(x)$ $(a \in K,\ x \in V)$,
2. $V \times V \ni (x, y) \mapsto q(x + y) - q(x) - q(y)$ *is bilinear.*

For a quadratic form q on V, there is an $r \times r$-matrix $A = (a_{ij})_{ij}$ such that

$$q(x + y) - q(x) - q(y) = \mathbf{x} A \mathbf{y}^T \ (x, y \in V),$$

where $\mathbf{x}, \mathbf{y}$ are row vectors in K^r corresponding x, y, respectively. The matrix A is called a matrix expression of the quadratic form q. Note that a matrix expression of q is not unique. In fact, $\frac{1}{2}(A + A^T)$ is also a matrix expression of q. Since this matrix is symmetric, we will take a symmetric matrix as a matrix expression of q. Conversely, for a $r \times r$-symmetric matrix A,

$$q(x) = \frac{1}{2} \mathbf{x} A \mathbf{x}^T$$

is a quadratic form on V.

It looks like a quadratic form corresponds to a one-to-one and onto symmetric matrix. However, this is not true because the choice of matrix expression of a quadratic form is not determined under the condition that the matrix must be symmetric. In fact, the choice depend on the choice of the basis of V over K. Accordingly, we define the concept of equivalence of quadratic forms as follows.

Definition 2. *Let q_1 and q_2 be quadratic forms on V. Let A_1 and A_2 be matrix expressions of q_1 and q_2, respectively. We say that q_1 and q_2 are isometric if there is an $r \times r$-regular matrix C such that*

$$A_1 = C A_2 C^T.$$

The relation of isometricity is independent of the choice of the basis of V, and hence, it is an equivalence relation.

Definition 3. *Let q be a quadratic form on V that can be expressed as a matrix A. We say that q is non-degenerate if $\det A \neq 0$.*

This definition is independent of the choice of the basis of V.

Lemma 1. *Any quadratic form q with more than 1 dimension, can represents any element of K.*

The following classification theorem is for quadratic forms over finite fields.

Theorem 1 (Classification theorem). *Let V be a vector space of dimension r over K. Let q be a non-degenerate quadratic form on V. Set a non-square δ in K. Then q is isometric to either of the following two quadratic forms:*

$$q_1(x) = \mathbf{x}A_1\mathbf{x}^T, \quad q_\delta(x) = \mathbf{x}A_\delta\mathbf{x}^T,$$

$$A_1 = \mathrm{I}_r(: \textit{identity matrix}), \quad A_\delta = \begin{pmatrix} 1 & & & \\ & \ddots & & \\ & & 1 & \\ \hline & & & \delta \end{pmatrix}.$$

Here, q_1 and q_δ are not isomorphic.

3.2   Idea Behind Our Scheme

Let us explain how Theorem 1 is applied to our scheme. The following is a corollary of the theorem.

Corollary 1. *Let A be an $r \times r$-symmetric matrix with $\det A \neq 0$. Then, there is an η in $\{1, \delta\}$ and an $r \times r$-regular matrix C such that*

$$CAC^T = A_\eta.$$

(Note that C is not uniquely determined.) If η' is another η in $\{1, \delta\}$ there is no $r \times r$-symmetric matrix C' such that

$$C'AC'^T = A_{\eta'}.$$

Next, let us consider the case that a quadratic form is degenerate. In this case, there is a regular matrix B such that

$$BAB^T = \begin{pmatrix} * & \cdots & * & 0 \\ \vdots & \ddots & \vdots & \vdots \\ * & \cdots & * & \vdots \\ \hline 0 & \cdots & \cdots & 0 \end{pmatrix}.$$

Lemma 1 and induction on r enable us to prove the following.

Proposition 1. *Let A be an $r \times r$-symmetric matrix with $\det A = 0$. Then for each $\eta = 1, \delta$, there is an $r \times r$-matrix C,*

$$CAC^T = A_\eta.$$

Theorem 1 and Corollary 1 can be rewritten in terms of a system of multivariate polynomials. The input and output of the system are expressed as matrices. In fact, for an $r \times r$-matrix variable X, the system is described as

$$f_1(X) = XA_1X^T, \quad f_\delta(X) = XA_\delta X^T$$

The scalar variables of f_1 and f_δ consists of the components of the matrix variable X. Let x_{ij} $(1 \leq i,j \leq r)$ be r^2 variables, and $X = (x_{i,j})$. f_1 and f_δ can be regarded as polynomials with respect to r^2 variables x_{ij}. Since the output matrices of f_1 and f_δ are symmetric, both f_1 and f_δ can be regarded as consisting of $r(r+1)/2$ polynomials.

Theorem 1 and Corollary 1 can be rewritten using f_1 and f_2.

Proposition 2. *For any $w \in K^{r(r+1)/2}$, either of the two systems of multivariate polynomials,*

$$f_1(X) = w, \quad f_\delta(X) = w \tag{1}$$

has a solution. (The solution is not uniquely determined.) Moreover, if the symmetric matrix w is regular, then only one of the above equations has a solution.

This proposition guarantees that when f_1, and f_δ are used in a signature system, there exists its signature for any message. The method of computing (1) will be explained later.

Example. Let us consider the case of $r = 2$ as an example. For a $\alpha \in K^\times$, we define a quadratic multivariate polynomial $f_\alpha : K^4 \to K^3$ by

$$f_\alpha(x_{11}, x_{12}, x_{21}, x_{22})$$
$$= (x_{11}^2 + x_{12}^2\alpha, \; x_{11}x_{21} + x_{12}x_{22}\alpha, \; x_{21}^2 + x_{22}^2\alpha)$$

This map is not surjective, and from Theorem 1 and Proposition 1, we have

1. For $\mathbf{w} = (w_1, w_2, w_3) \in K^3$, if $w_1 w_3 - w_2^2 \in \alpha \cdot \{b^2 \mid b \in K\}$, $f_\alpha(X) = \mathbf{w}$ has a solution.
2. Otherwise, $f_\alpha(X) = \mathbf{w}$ has no solution.

In particular, if $\delta \in K^\times$ is a non-square and $\alpha = 1, \delta$, f_1 and f_δ are not surjective. However, the union of the regions of values of the two maps coincides with whole K^3 (Prop. 2), that is, for any $\mathbf{w} = (w_1, w_2, w_3) \in K^3$, either of the following has a solution.

$$\begin{cases} f_1(X) = \mathbf{w} \\ f_\delta(X) = \mathbf{w}. \end{cases}$$

3.3 Computing the Inverse of f_1, f_δ

We will explain how to compute X by assuming that $f_1(X) = \mathbf{w}$ has a solution $X \in K^{r^2}$ for $\mathbf{w} \in K^{r(r+1)/2}$. Since a solution of $f_\delta(X) = \mathbf{w}$ can be computed in the similar fashion, we will explain only the case in which $f_1(X) = \mathbf{w}$.

Case of $r = 2$. The case of $r = 1$ is easy. ($X = \sqrt{\mathbf{w}}$ is a solution.) Thus, we will start with the case of $r = 2$. In this case, any $\mathbf{w}(\neq 0) \in K^3$ can be expressed by one of the following three 2×2-symmetric matrices:

$$\begin{pmatrix} 0 & b \\ b & 0 \end{pmatrix}, \quad \begin{pmatrix} a & b \\ b & c \end{pmatrix} \ (a \neq 0), \quad \begin{pmatrix} a & b \\ b & c \end{pmatrix} \ (c \neq 0).$$

We can diagonalize these matrices using the following operation (at most twice):

1. $\begin{pmatrix} 1/2 & 1 \\ -1/2 & 1 \end{pmatrix} \begin{pmatrix} 0 & 1 \\ 1 & 0 \end{pmatrix} \begin{pmatrix} 1/2 & 1 \\ -1/2 & 1 \end{pmatrix}^T = \begin{pmatrix} 1 & 0 \\ 0 & -1 \end{pmatrix}$,

2. $\begin{pmatrix} 1 & 0 \\ a^{-1}b & -1 \end{pmatrix} \begin{pmatrix} a & b \\ b & c \end{pmatrix} \begin{pmatrix} 1 & 0 \\ a^{-1}b & -1 \end{pmatrix}^T = \begin{pmatrix} a & 0 \\ 0 & c - a^{-1}b^2 \end{pmatrix}$
   $(a \neq 0)$,

3. $\begin{pmatrix} 0 & 1 \\ 1 & 0 \end{pmatrix} \begin{pmatrix} a & b \\ b & c \end{pmatrix} \begin{pmatrix} 0 & 1 \\ 1 & 0 \end{pmatrix}^T = \begin{pmatrix} c & b \\ b & a \end{pmatrix} \ (c \neq 0)$.

Let

$$\begin{pmatrix} a' & 0 \\ 0 & c' \end{pmatrix} \ (a' \neq 0) \tag{2}$$

be the diagonalized matrix. In addition, we assume that $c' \neq 0$. Accordingly, there is a $d \in K$ such that $c'd^2 = a'^{-1}$. (In the case of f_δ, $c'd^2 = a'^{-1}\delta$.) d can be computed efficiently by precomputing of the square roots of the elements of K. We have

$$\begin{pmatrix} 1 & 0 \\ 0 & d \end{pmatrix} \begin{pmatrix} a' & 0 \\ 0 & c' \end{pmatrix} \begin{pmatrix} 1 & 0 \\ 0 & d \end{pmatrix}^T = \begin{pmatrix} a' & \\ & a'^{-1} \end{pmatrix}.$$

Since for any a', we can precompute a matrix $C_{a'}$ such that

$$C_{a'} \begin{pmatrix} a' & 0 \\ 0 & a'^{-1} \end{pmatrix} C_{a'}^T = \begin{pmatrix} 1 & \\ & 1 \end{pmatrix}, \tag{3}$$

(2) can be transformed into A_1. In the case that $c' = 0$, from (3), we have

$$\left(\begin{pmatrix} 1 & 0 \\ 0 & 0 \end{pmatrix} C_{a'}^{-1} \right) A_1 \left(\begin{pmatrix} 1 & 0 \\ 0 & 0 \end{pmatrix} C_{a'}^{-1} \right)^T = \begin{pmatrix} a' & 0 \\ 0 & 0 \end{pmatrix}.$$

In either case, we can make an equation of the form,

$$(D_l \cdots D_1) A_1 (D_l \cdots D_1)^T = \mathbf{w}$$

and, $\mathbf{x} = D_l \cdots D_1$ is a solution. Here, $D_1, \ldots, D_l$ are the transformation matrices described above.

Case of $r \geq 3$. Let $r \geq 3$. $\mathbf{w}(\neq 0) \in K^{r(r+1)/2}$ can be expressed as an $r \times r$-symmetric matrix

$$\begin{pmatrix} a * \cdots * b * \cdots \\ * & & * \\ \vdots & \ddots & \vdots \\ * & & * \\ b * \cdots * c \\ * \\ \vdots & & \ddots \end{pmatrix}. \tag{4}$$

We can apply the above operations in the case of two dimensions to the 2×2-matrix composed by rows and columns including a, b and c:

$$\begin{pmatrix} a & b \\ b & c \end{pmatrix},$$

By iterating these operations, the matrix (4) can be transformed into

$$\begin{pmatrix} a & 0 \cdots 0 \\ \hline 0 & * \cdots * \\ \vdots & \vdots \ddots \vdots \\ 0 & * \cdots * \end{pmatrix}.$$

We can apply the same operation to the $(r-1) \times (r-1)$-miner matrix. Induction shows that the matrix (4) can be diagonalized. The diagonal matrix can be transformed into A_1 in the similar fashion as in the case of two dimensions. Consequently, we obtain a solution X of $f_1(X) = \mathbf{w}$. This solution is not unique.

4   Our Scheme

Let $n = r^2$, $m = r(r+1)/2$. The key generation, signature generation and verification of our scheme are as follows:

- **Key generation**

    Secret key. The secret key consists of a non-square $\delta \in K$, a $r \times r$-regular matrix B, and two randomly chosen affine transformations $L : K^m \to K^m$ and $R : K^n \to K^n$.

    Public key. The public key consists of the composite maps $F_1 = L \circ f_1 \circ R$, $F_\delta = L \circ f_{\delta,B} \circ R : K^n \to K^m$, where $f_{\delta,B}(X) = XBA_\delta B^T X^T$.

- **Signature Generation.** Let $\mathbf{M} \in K^m$ be a message. To generate a signature $\mathbf{S}$ from $\mathbf{M}$, first compute $\mathbf{M}' = L^{-1}(\mathbf{M})$, After that, compute an $r \times r$-matrix $\mathbf{S}'$ such that

$$\mathbf{M}' = \mathbf{S}' A_1 \mathbf{S}'^T \text{ or } \mathbf{M}' = \mathbf{S}' A_\delta \mathbf{S}'^T$$

In the former case, compute $\mathbf{S} = R^{-1}(\mathbf{S}')$. In the latter case, compute $\mathbf{S} = R^{-1}(\mathbf{S}'B^{-1})$. $\mathbf{S}'$ is computed using the improved algorithm described above. $L^{-1}(\mathbf{M})$ and $R^{-1}(\mathbf{S}')$ can be easily computed since L and R are affine transformations.

- **Verification** If $F_1(\mathbf{S}) = \mathbf{M}$ or $F_\delta(\mathbf{S}) = \mathbf{M}$, the signature is accepted. Otherwise, it is rejected.

5   The Security of Our Scheme

5.1   Application of Attacks against Rainbow

Here, we show whether the following famous attacks against Rainbow can be launched to our scheme.

- direct attacks ([2,35])
- MinRank attack ([15,34,4])
- HighRank attack ([15,11,25])
- UOV attack ([19,18])
- Rainbow-Band-Separation(RBS) attack ([11,24])
- UOV-Reconciliation(UOV-R) attack ([11,24])

Direct Attacks. For a message $\mathbf{M}$, if $\mathbf{C}'$ is found such that $F(\mathbf{C}') = \mathbf{M}$, then one can forge a signature for $\mathbf{M}$. (F is the public key.) Since this equation involves multivariate polynomial equations, it can be solved by computing gröbner basis[2,35]. Direct attacks depend on the message. Since they are attacks against the MQ problem on which the security of MPKC is based, these attacks are applicable to any MPKC scheme, including ours.

Other Attacks against Rainbow. Beside the direct ones, there are 5 attacks that transform the public key into a central map of Rainbow. These attacks are not effective against our scheme. For these attacks to be applicable to our scheme, the public key of our scheme must be able to transformed into the central map of Rainbow. Since the central map of Rainbow is surjective, the public key of our scheme must also be surjective. However, the public key of our scheme is not surjective, so this is a contradiction that proves such attacks can not be launched.

5.2   Attacks Applicable to Only Our Scheme

As we explained before, the attacks in which the public map is reduced to a central map surjective can not be applied to our scheme. However, there are attacks of other type. Here, we explain two attacks applicable to our scheme.

MinRank Attack. This attack is different from the MinRank attack against Rainbow explained above. The quadratic polynomials composing the central maps of our scheme correspond to matrices of rank r. This property can be applied to an attack against our scheme. We denote by $\mathcal{A}$ the space spanned by the square matrices associated to quadratic parts of components of the public key. Most of elements have rank r^2 by randomness of the affine transformations in the secret key. However, the matrices associated to components of the central map have rank r, and r is the minimal rank among elements in $\mathcal{A}$. On the other hand, a matrix of minimal rank in $\mathcal{A}$ can be found by solving MinRank problem[21]. Therefore, this method reveals the secret key of our scheme. The complexity of solving MinRank problem is estimated by [4]

$$q^r \cdot m(n^2/2 - m^2/6) \; \mathbf{m},$$

where m, n are the number of equations and variables, respectively, and $\mathbf{m}$ means the multiplication in K.

Isomorphism of Polynomials Problem. Isomorphism of Polynomials(IP) Problem[28] is related to an attack against our scheme. The IP problem is the following: given multivariate polynomial maps F and G, find affine transformations A and B such that $F = A \circ G \circ B$ (if they exist). There are several papers which treat IP problem and its variants[6,7,27]. Since our scheme uses special central maps, f_1 and $f_{\delta,B}$, the problem for finding the sceret key of our scheme can be replaced by the IP problem in the case that F and G are the public and secret key, respectively. Patarin estimated the complexity of solving general IP problem by $\mathcal{O}(q^{3n/2})$ for any system of n quadratic equations with n variables, (and by $\mathcal{O}(q^{n/2})$ for a system of Matsumoto-Imai scheme)[28].

6 Efficiency of Signature Generation

In this section, we estimate the efficiency of signature generation of our scheme in terms of the number of multiplications of the base field. After that, we compare it with the efficiency of signature generation of Rainbow.

6.1 Efficiency of Signature Generation of Our Scheme

In our scheme, the number of variables is $n = r^2$ and the number of polynomials is $m = r(r+1)/2$. The signature generation is computed in the following order: For $\mathbf{M} \in K^m$, (i) $\mathbf{A} = L^{-1}(\mathbf{M})$ is computed for an affine isomorphism $L : K^m \to K^m$ in the secret key, (ii) a solution $\mathbf{B}$ of $f_1(\mathbf{X}) = \mathbf{A}$ or $f_\delta(\mathbf{X}) = \mathbf{A}$ is computed by regarding $\mathbf{A}$ as $r \times r$-symmetric matrix, . (iii) if $f_\delta^{-1}(\mathbf{A})$ has a solution, $\mathbf{B}$ is replaced by $\mathbf{B}B^{-1}$. (iv) $\mathbf{C} = R^{-1}(\mathbf{B})$ is computed for an affine isomorphism $R : K^n \to K^n$ in the secret key.

The number of multiplications of the base field for $\mathbf{A} = L^{-1}(\mathbf{M})$ and $\mathbf{C} = R^{-1}(\mathbf{B})$ is

$$L^{-1} : \; \left(\frac{r(r+1)}{2}\right)^2, \quad R^{-1} : \; r^4. \tag{5}$$

The $r \times r$-symmetric matrix $\mathbf{A}$ in the computation of (ii) is diagonalized. The following computation of transformation is dominant.

$$
\begin{pmatrix}
0 \cdots & 0 & & \cdots & 0 \\
\vdots \ddots & & \ddots & & \vdots \\
0 & a & b & & \\
& \ddots & & \ddots & \\
\vdots & b & c & & \\
0 \cdots & & \ddots & & \ddots
\end{pmatrix}
\rightarrow
\begin{pmatrix}
0 \cdots & 0 & & \cdots & 0 \\
\vdots \ddots & & \ddots & & \vdots \\
0 & a & & 0 & \\
& \ddots & & \ddots & \\
\vdots & 0 & c - b^2/a & & \\
0 \cdots & & \ddots & & \ddots
\end{pmatrix}
$$

$(a \neq 0)$. This computation is executed using the transformation **2** in § 3.3. The number of a's is equal to r because a is on the diagonal, The number of b's is equal to the number of positions in the upper triangle of the matrix. The transformation **2** in § 3.3 needs $3r$ multiplications of the base field. Therefore, the total number of multiplications is equal to $3r^2(r-1)/2$.

We want to use the same matrix transformation for $a = 0$, too. To do this, we can use the transformation **1,3** in § 3.3, in order to transform the matrix such that a non-zero element appears at the position of a.

The transformation **3** needs no multiplications of K. The transformation **1** needs $6r$ multiplications. Since the number of a's is equal to r, the total number of multiplications is equal to $6r^2$.

The above computation complete the diagonalization. After that, the diagonal components have to be transformed into 1. For each diagonal component, the computation to transform them into 1 needs $12r$ multiplications. The number of components that have to be transformed into 1 is equal to $r - 1$. The total number of multiplications is at most $12r(r - 1)$. Therefore the total number of multiplications in the computation of (ii) is at most

$$
\frac{3r^2(r-1)}{2} + 6r^2 + 12r(r-1) = \frac{3r^3 - 33r^2 - 24r}{2}. \tag{6}
$$

r^2 multiplication is needed in the computation of $\mathbf{B}B^{-1}$ in (iii). Consequently, we have the following theorem.

Theorem 2. *In the signature generation of our scheme, the number of multiplication of K needs at most*

$$
\frac{5}{4}r^4 + 2r^3 - \frac{61}{4}r^2 - 12r.
$$

The term contributing to (ii) is cubic orders of r. On the other hand, the terms contributing to (i) and (iv) are quartic orders of r. Therefore, the inverse of the central map is computed using fewer multiplications than those of affine transformations.

6.2   Comparison of Efficiencies

Here, we compare the efficiency of the signature generation of our scheme and with that of Rainbow. We will not describe the scheme of Rainbow in any detail, save for the notation that is necessary to describe the number of multiplications of K.

Let n, t be natural numbers (t is called the layer number.) Let $v_1, \ldots, v_{t+1}$ be natural numbers satisfying

$$0 < v_1 < v_2 < \cdots < v_t < v_{t+1} = n.$$

For $i = 1, \ldots, t$, we write $o_i = v_{i+1} - v_i$ and $m = n - v_1$. This defines Rainbow with a parameter $(v_1, o_1, \ldots, o_t)[10]$,

$$\text{Rainbow}(K; v_1, o_1, \ldots, o_t).$$

This is a signature scheme using a system of multivariate polynomials with n variables and m polynomials.

The signature of $\text{Rainbow}(K; v_1, o_1, \ldots, o_t)$ is generated as follows: For a message $\mathbf{M} \in K^m$, (R1) compute $\mathbf{A} = L^{-1}(\mathbf{M})$, (R2) $\mathbf{B} = G^{-1}(\mathbf{A})$, and (R3) $\mathbf{C} = R^{-1}(\mathbf{B})$. Here, G is the central map of Rainbow, and L and R are affine transformations. Therefore, the number of multiplications of $\mathbf{A} = L^{-1}(\mathbf{M})$ and $\mathbf{C} = R^{-1}(\mathbf{B})$ are the same as those of the affine transformations of our scheme. The number of multiplications in the computation of (R2) is estimated as follows:

$$\sum_{h=1}^{t} \left(\frac{o_h v_h^2}{2} + \frac{o_h^3}{3} + (v_h + 1)o_h^2 + \frac{3o_h v_h}{2} + \frac{v_h(v_h + 1)}{2} - \frac{o_h}{3} \right).$$

The $o_h v_h^2/2$ part is the cost for setting the linear equations to compute the inverse of the central map of Rainbow. The $o_h^3/3$ part is the cost for solving the linear equations with the $o_h \times o_h$-matrix. Using these results, we can compare the efficiency of signature generation. A signature generated with two layers and v_1, o_1, o_2 whose values are almost the same is often used in Rainbow[9,24]. Therefore, we can choose $v_1 = o_1 = o_2$ for simplicity. Moreover, we can determine other parameters based on the security against direct attacks. For our scheme to be more secure than Rainbow against direct attacks, it is sufficient that the number of polynomials of our scheme equals the number of polynomials of Rainbow, and the number of variables of our scheme is in fact greater than or equal to the number of variables of Rainbow. We can choose the parameter r of our scheme, and the parameters of Rainbow as $v_1 = o_1 = o_2 = r(r + 1)/4$. The number of polynomials of these schemes are the same. Here, we have assumed that $r(r + 1)/4$ is an integer. The number of variables of our scheme is equal to r^2, and that of Rainbow is equal to $3r(r + 1)/4$. If $r \geq 3$ then the condition placed upon the number of variables is satisfied. We call this Rainbow, Rainbow A, and we compare it with our scheme.

The number of multiplications of K in the computation of (R2) is estimated as follows.

$$\frac{37}{348}V^3 + \frac{1}{2}V^2 + \frac{5}{24}V \quad (V = r(r+1)).$$

We needs the following multiplications in (R1) and (R3).

$$\left(\frac{3}{4}V\right)^2 + \left(\frac{1}{2}V\right)^2 = \frac{17}{16}V^2 \quad (V = r(r+1)),$$

respectively. The total number of multiplications in the signature generation of Rainbow A is equal to

$$\frac{37}{384}r^3(r+1)^3 + \frac{25}{16}r^2(r+1)^2 + \frac{5}{24}r(r+1).$$

Comparing this with Theorem 2, we can see that the orders of r of these schemes are different. For our scheme, the maximal order is 4: on the other hand, it is 6 for Rainbow A. To compute the inverse of the central map of Rainbow A, we need to solve linear equations of size $O(r^2)$; the cost is $O(r^6)$. On the other hand, to compute the inverse of the central map of our scheme, we need to diagonalize a matrix of size $O(r)$; the cost is $O(r^3)$. In addition, the cost of computing of affine transformations is $O(r^4)$. That is, the affine transformations are more computationally expensive than that the inverse of the central map. As r grows, the signature generation of our scheme becomes more efficient than that of Rainbow A. The tables below compares the efficiencies of these signature generations as well as the secret key and public key lengths for $r = 8$ and 11. The efficiency of the signature generation is estimated in terms of the number of multiplications of K. The secret key and public key lengths are represented by the number of elements of K. Tables 1 and 2 indicate that the signature generation of our scheme is eight to nine times more efficient than that of Rainbow A. The secret key of our scheme is shorter than that of Rainbow A; however, the public key of our schemed is longer.

From [24], Rainbow A with $K = GF(31)$ and $r = 8$ has the same security level against direct attacks as a symmetric key with 88-bits. In the case of $K = GF(31)$ and $r = 11$, Rainbow A with $r = 6$ is more secure against direct attacks than a symmetric key with 140-bits. In this case, our scheme will have higher security level against direct attacks than 88-bits if $r = 8$, and 140-bits if $r = 11$. Note that in our scheme, the order of K may have to be lager than 31 to be secure against MinRank attack. Concretely, K needs an order more than 2048 for $r = 8$, and with more than 6781 for $r = 11$.

Table 1. Efficiencies of Our Scheme and Rainbow ($r = 8$)

	Our scheme	Rainbow A
m	36	36
n	64	54
Efficiency of the signature generation	5072	44079
Secret key length	5532	38520
Public key length	154440	55400

Table 2. Efficiencies of Our Scheme and Rainbow ($r = 11$)

	Our scheme	Rainbow A
m	66	66
n	121	99
Efficiency of the signature generation	18733	248864
Secret key length	19305	219120
Public key length	990396	333300

7   Concluding Remarks

We proposed a new construction of Rainbow using quadratic forms. Our scheme uses a non-surjective multivariate map as a central map. Since previously proposed signature schemes in MPKC use surjective multivariate maps, the public key of our scheme can not be described by the public keys of other signature schemes in MPKC. The signature generation of our scheme is eight to nine times more efficient than that of Rainbow at the level of 88-bit security.

In the future, we will analyze the security of our scheme more elaborately.

Acknowledgements. This work was partially supported by the Japan Science and Technology Agency (JST) Strategic Japanese-Indian Cooperative Programme for Multidisciplinary Research Fields, which aims to combine Information and Communications Technology with Other Fields. The first author is supported by Grant-in-Aid for Young Scientists (B), Grant number 24740078. The authors would like to thank Prof. Tsutomu Matsumoto for helpful comments on Graph Isomorphism. Dr. Yasufumi Hashimoto read carefully the preliminary version of this paper, and pointed out that MinRank attack can be applied to our scheme. The authors would like to thank him.

References

1. Anshel, I., Anshel, M., Goldfeld, D.: An Algebraic Method for Public-Key Cryptography. Math. Res. Lett. 6(3-4), 287–291 (1999)
2. Bernstein, D.J., Buchmann, J., Dahmen, E.: Post Quantum Cryptography. Springer, Heidelberg (2009)

3. Berger, T.P., Cayrel, P.-L., Gaborit, P., Otmani, A.: Reducing Key Length of the McEliece Cryptosystem. In: Preneel, B. (ed.) AFRICACRYPT 2009. LNCS, vol. 5580, pp. 77–97. Springer, Heidelberg (2009)

4. Billet, O., Gilbert, H.: Cryptanalysis of Rainbow. In: De Prisco, R., Yung, M. (eds.) SCN 2006. LNCS, vol. 4116, pp. 336–347. Springer, Heidelberg (2006)

5. Boneh, D., Durfee, G.: Cryptanalysis of RSA with Private Key d Less Than $N^{0.292}$. IEEE Trans. Inform. Theory 46(4), 1339–1349 (2000)

6. Bouillaguet, C., Fouque, P.-A., Véber, A.: Graph-Theoretic Algorithms for the Isomorphism of Polynomials Problem. IACR Cryptology ePrint Archive Report 2012/607

7. Bouillaguet, C., Faugère, J.-C., Fouque, P.-A., Perret, L.: Practical Cryptanalysis of the Identification Scheme Based on the Isomorphism of Polynomial with One Secret Problem. In: Catalano, D., Fazio, N., Gennaro, R., Nicolosi, A. (eds.) PKC 2011. LNCS, vol. 6571, pp. 473–493. Springer, Heidelberg (2011)

8. Chen, A.I.-T., Chen, M.-S., Chen, T.-R., Cheng, C.-M., Ding, J., Kuo, E.L.-H., Lee, F.Y.-S., Yang, B.-Y.: SSE Implementation of Multivariate PKCs on Modern x86 CPUs. In: Clavier, C., Gaj, K. (eds.) CHES 2009. LNCS, vol. 5747, pp. 33–48. Springer, Heidelberg (2009)

9. Ding, J., Gower, J.E., Schmidt, D.S.: Multivariate Public Key Cryptosystems. Advances in Information Security, vol. 25. Springer (2006)

10. Ding, J., Schmidt, D.: Rainbow, a New Multivariable Polynomial Signature Scheme. In: Ioannidis, J., Keromytis, A.D., Yung, M. (eds.) ACNS 2005. LNCS, vol. 3531, pp. 164–175. Springer, Heidelberg (2005)

11. Ding, J., Yang, B.-Y., Chen, C.-H.O., Chen, M.-S., Cheng, C.-M.: New Differential-Algebraic Attacks and Reparametrization of Rainbow. In: Bellovin, S.M., Gennaro, R., Keromytis, A.D., Yung, M. (eds.) ACNS 2008. LNCS, vol. 5037, pp. 242–257. Springer, Heidelberg (2008)

12. Farb, B., Dennis, K.: Noncommutative Algebra. In: Graduate Texts in Mathematics. Springer (1993); ACNS 2008

13. Garey, M.R., Johnson, D.S.: Computers and Intractability: A Guide to the Theory of NP-Completeness. W.H. Freeman & Co., Ltd. (1979)

14. Galbraith, S.D., Ruprai, R.S.: Using Equivalence Classes to Accelerate Solving the Discrete Logarithm Problem in a Short Interval. In: Nguyen, P.Q., Pointcheval, D. (eds.) PKC 2010. LNCS, vol. 6056, pp. 368–383. Springer, Heidelberg (2010)

15. Goubin, L., Courtois, N.T.: Cryptanalysis of the TTM Cryptosystem. In: Okamoto, T. (ed.) ASIACRYPT 2000. LNCS, vol. 1976, pp. 44–57. Springer, Heidelberg (2000)

16. Hashimoto, Y., Sakurai, K.: On Construction of Signature Schemes based on Birational Permutations over Noncommutative Rings. In: Proceedings of the 1st International Conference on Symbolic Computation and Cryptography (SCC 2008), pp. 218–227 (2008)

17. Ko, K.H., Lee, S.-J., Cheon, J.H., Han, J.W., Kang, J.-S., Park, C.-S.: New Public-Key Cryptosystem Using Braid Groups. In: Bellare, M. (ed.) CRYPTO 2000. LNCS, vol. 1880, pp. 166–183. Springer, Heidelberg (2000)

18. Kipnis, A., Patarin, J., Goubin, L.: Unbalanced Oil and Vinegar Signature Schemes. In: Stern, J. (ed.) EUROCRYPT 1999. LNCS, vol. 1592, pp. 206–222. Springer, Heidelberg (1999)

19. Kipnis, A., Shamir, A.: Cryptanalysis of the Oil & Vinegar Signature Scheme. In: Krawczyk, H. (ed.) CRYPTO 1998. LNCS, vol. 1462, pp. 257–266. Springer, Heidelberg (1998)

20. Lin, D., Faugère, J.-C., Perret, L., Wang, T.: On Enumeration of Polynomial Equivalence Classes and Their Application to MPKC. Cryptology ePrint Archive: Report 2011/055

21. Faugère, J.-C., Levy-dit-Vehel, F., Perret, L.: Cryptanalysis of MinRank. In: Wagner, D. (ed.) CRYPTO 2008. LNCS, vol. 5157, pp. 280–296. Springer, Heidelberg (2008)

22. van Oorschot, P.C., Wiener, M.J.: Parallel Collision Search with Cryptanalytic Applications. Journal of Cryptology 12, 1–28 (1999)

23. Petzoldt, A., Bulygin, S., Buchmann, J.: A Multivariate Signature Scheme with a Partially Cyclic Public Key. In: Proceedings of the Second International Conference on Symbolic Computation and Cryptography (SCC 2010), pp. 229–235 (2010)

24. Petzoldt, A., Bulygin, S., Buchmann, J.: Selecting Parameters for the Rainbow Signature Scheme. In: Sendrier, N. (ed.) PQCrypto 2010. LNCS, vol. 6061, pp. 218–240. Springer, Heidelberg (2010)

25. Petzoldt, A., Bulygin, S., Buchmann, J.: CyclicRainbow – A Multivariate Signature Scheme with a Partially Cyclic Public Key. In: Gong, G., Gupta, K.C. (eds.) INDOCRYPT 2010. LNCS, vol. 6498, pp. 33–48. Springer, Heidelberg (2010)

26. Petzoldt, A., Bulygin, S., Buchmann, J.: Linear Recurring Sequences for the UOV Key Generation. In: Catalano, D., Fazio, N., Gennaro, R., Nicolosi, A. (eds.) PKC 2011. LNCS, vol. 6571, pp. 335–350. Springer, Heidelberg (2011)

27. Faugère, J.-C., Perret, L.: Polynomial Equivalence Problems: Algorithmic and Theoretical Aspects. In: Vaudenay, S. (ed.) EUROCRYPT 2006. LNCS, vol. 4004, pp. 30–47. Springer, Heidelberg (2006)

28. Patarin, J., Goubin, L., Courtois, N.T.: Improved Algorithms for Isomorphisms of Polynomials. In: Nyberg, K. (ed.) EUROCRYPT 1998. LNCS, vol. 1403, pp. 184–200. Springer, Heidelberg (1998)

29. Petzoldt, A., Thomae, E., Bulygin, S., Wolf, C.: Small Public Keys and Fast Verification for Multivariate Quadratic Public Key Systems. In: Preneel, B., Takagi, T. (eds.) CHES 2011. LNCS, vol. 6917, pp. 475–490. Springer, Heidelberg (2011)

30. Pollard, J.M.: Monte Carlo Methods for Index Computation mod p. Mathmatics of Computation 143(32), 918–924 (1978)

31. Rai, T.S.: Infinite Gröbner Bases and Noncommutative Polly Cracker Cryptosystems. PhD Thesis, Virginia Polytechnique Institute and State Univ. (2004)

32. Scharlau, W.: Quadratic and Hermitian Forms. Springer (1987)

33. Wiener, M.J.: Cryptanalysis of Short RSA Secret Exponents. IEEE Trans. Inform. Theory 36(3), 553–558 (1990)

34. Yang, B.-Y., Chen, J.-M.: Building Secure Tame-like Multivariate Public-Key Cryptosystems: The New TTS. In: Boyd, C., González Nieto, J.M. (eds.) ACISP 2005. LNCS, vol. 3574, pp. 518–531. Springer, Heidelberg (2005)

35. Yang, B.-Y., Chen, J.-M.: All in the XL Family: Theory and Practice. In: Park, C.-s., Chee, S. (eds.) ICISC 2004. LNCS, vol. 3506, pp. 67–86. Springer, Heidelberg (2005)

36. Yang, B.-Y., Chen, J.-M.: All in the XL Family: Theory and Practice. In: Park, C.-s., Chee, S. (eds.) ICISC 2004. LNCS, vol. 3506, pp. 67–86. Springer, Heidelberg (2005)

37. Yasuda, T., Sakurai, K.: A security analysis of uniformly-layered Rainbow — Revisiting Sato-Araki's Non-commutative Approach to Ong-Schnorr-Shamir Signature Towards PostQuantum Paradigm. In: Yang, B.-Y. (ed.) PQCrypto 2011. LNCS, vol. 7071, pp. 275–294. Springer, Heidelberg (2011)

Author Index